AF900155

N-Heterocyclic Carbenes in Organocatalysis

N-Heterocyclic Carbenes in Organocatalysis

With a Foreword by Ronald Breslow

Edited by
Akkattu T. Biju

WILEY-VCH

Editor
Prof. Akkattu T. Biju
Department of Organic Chemistry
Indian Institute of Science (IISc)
Bangalore-560012
India

Cover © fotohunter / iStock.com

All books published by **Wiley-VCH** are carefully produced. Nevertheless, authors, editors, and publisher do not warrant the information contained in these books, including this book, to be free of errors. Readers are advised to keep in mind that statements, data, illustrations, procedural details or other items may inadvertently be inaccurate.

Library of Congress Card No.:
applied for

British Library Cataloguing-in-Publication Data
A catalogue record for this book is available from the British Library.

Bibliographic information published by the Deutsche Nationalbibliothek
The Deutsche Nationalbibliothek lists this publication in the Deutsche Nationalbibliografie; detailed bibliographic data are available on the Internet at <http://dnb.d-nb.de>.

© 2019 Wiley-VCH Verlag GmbH & Co. KGaA, Boschstr. 12, 69469 Weinheim, Germany

All rights reserved (including those of translation into other languages). No part of this book may be reproduced in any form – by photoprinting, microfilm, or any other means – nor transmitted or translated into a machine language without written permission from the publishers. Registered names, trademarks, etc. used in this book, even when not specifically marked as such, are not to be considered unprotected by law.

Print ISBN: 978-3-527-34310-2
ePDF ISBN: 978-3-527-80905-9
ePub ISBN: 978-3-527-80907-3
oBook ISBN: 978-3-527-80904-2

Typesetting SPi Global, Chennai, India
Printing and Binding C.O.S. Printers Pte Ltd, Singapore
Printed on acid-free paper

10 9 8 7 6 5 4 3 2 1

Contents

Preface

Ever since the seminal independent reports by the Bertrand and Arduengo groups on the isolation of free carbenes, N-heterocyclic carbene (NHC)-based organocatalysis has emerged as one of the efficient and sophisticated synthetic tool for the rapid construction of medicinally and biologically important molecules from simple starting materials. Employing chiral carbenes can result in the enantioselective synthesis of target compounds. Primarily, NHCs are used for the umpolung (*reversal of normal mode of reactivity*) of aldehydes, leading to the generation of acyl anion equivalents (*Breslow intermediates*). This nucleophilic intermediates can react with various electrophiles, such as aldehydes, ketones, imines, activated, polarized C=C double bonds, and even unactivated C=C double bonds. The benzoin condensation and Stetter reaction are the two well-known transformations, which employ acyl anions as key intermediates. Historically, the advancement of umpolung catalysis dates back to the seminal discovery of the cyanide-catalyzed benzoin reaction by Wöhler and Liebig in 1832. The origin of NHC reactivity in this field can be attributed to the demonstration that thiazolium salts can catalyze the benzoin reaction by Ukai in 1943, which was followed by the rationalization of the mechanism proceeding via the acyl anions (*via umpolung*) by Breslow in 1958.

Generally, imidazolium, thiazolium, or triazolium salt-derived NHCs have been used successfully for this umpolung, and during recent years, there has been an increased interest in NHC-catalyzed transformations, and many new reactions have been developed. The phenomenal success of NHCs in organocatalysis can be attributed primarily to their electronic properties leading to different modes of action in catalysis. The different modes of NHC reactivity include the umpolung of aldehydes, generation of homoenolate equivalents from enals, NHC-induced generation of activated carboxylates, and α,β-unsaturated acylazolium intermediates. Moreover, NHCs are versatile catalysts for (dynamic) kinetic resolutions, desymmetrization strategies, inducing polymerization reactions, and have found several applications in the synthesis of natural products and biologically active compounds. In addition, NHCs are useful in domino reactions and catalyzing acyl–anion-free transformations. The lack of a book in this area has prohibited chemists from appreciating the recent developments in the concept of NHC catalysis. The focus of the present book is on recent developments in the NHC-catalyzed carbon–carbon and carbon–heteroatom bond-forming reactions for the synthesis of various heterocycles, carbocycles, and acyclic molecules. Examples from the literature, which have demonstrated the synthetic utility of the process as well as original contributions, which have spearheaded research in the area, are included in the book.

As the field of NHC organocatalysis has grown considerably and continues growing further, with the strong support of Wiley-VCH, we decided to bring out a new book under the title "N-Heterocyclic Carbenes in Organocatalysis" to highlight the developments occurred in this rapidly evolving area. Twelve chapters highlighting the different modes of reactivity of NHCs in catalysis are presented in this book. NHCs are also widely used as ligands for transition metals and are used in a variety of metal-catalyzed transformations. As there are several reviews/books available in this topic, the application of NHCs in metal catalysis are not covered in this book.

This book is a collective effort of a number of organic chemists working in the area of NHC catalysis. Most importantly, I would like to thank all the authors, who have contributed immensely to this project, for their valuable time, efforts, as well as the expertise to make this book a source of inspiration for beginners as well as advanced chemists practicing synthetic organic chemistry. It is anticipated that the simplicity of the NHC-catalyzed methods will inspire a broad range of organic chemists to explore new possibilities and imaginative applications of this concept and thereby solving some of the remaining challenges in this area. I am also thankful to Drs Elke Maase (*Executive Commissioning Editor*) and Lesley Jebaraj (*Project Editor*) at Wiley-VCH for their wholehearted support and useful advices in preparing/developing this book.

12 August 2018

Akkattu T. Biju
Associate Professor
Department of Organic Chemistry
Indian Institute of Science
Bangalore 560012
India

Discovery of Catalysis by Nucleophilic Carbenes

Thiamine pyrophosphate is a catalyst in biochemistry for processes that perform reactions similar to those catalyzed by cyanide ion in ordinary chemistry (thiamine pyrophosphate is in sense a nontoxic analog of cyanide ion). Possible enzyme reaction mechanisms suggested by Breslow included reactions catalyzed by the amino group of thiamine, but those mechanisms were excluded by the work of Westheimer and of Breslow. Also, Breslow believed that Nature does not make unique difficult compounds for no special biochemical need, so he decided to examine thiazolium salts in D_2O to see if the C-2 proton was easily removed reversibly. Thiamine is a thiazolium salt. He used a homemade 30 MHz NMR machine built by his Columbia colleague Benjamin Dailey and saw that the C-2 proton of the thiazolium ion ring did indeed exchange with D_2O under mild basic conditions, via the C-2 carbanion. This explained why and how Nature used a thiazolium cation in biochemical reactions catalyzed by thiamine pyrophosphate.

Breslow published this result and then pointed out its significance for chemical or biochemical catalysis by thiamine derivatives and other related cyclic cations, such as imidazolium cations. He postulated that there was stabilization of the C-2 carbanion structure by resonance with a neutral carbene form, a unique process. The carbene resonance form was nucleophilic because of electron donation by nitrogen groups, in contrast with the normal electrophilic character of simple carbenes. Since that time, such nucleophilic carbenes (one of the two resonance forms) have been exploited by chemists worldwide as a new class of catalysts with many novel forms and catalytic or other chemical reactions. The investigation of why Nature uses such species to catalyze biochemical reactions has taught us remarkable new chemistry as well.

N-heterocyclic carbenes (NHCs) have emerged as versatile organocatalysts for various synthetic transformations, and the primary mode of NHC reactivity is the umpolung of aldehydes. The benzoin reaction and the Stetter reaction are the common transformations catalyzed by NHCs. This umpolung concept was extended in 2004 to α,β-unsaturated aldehydes for the generation of homoenolate equivalents. The field of NHC catalysis is expanding rapidly, and there is no book available highlighting the application of NHCs in organocatalysis. Against this background, Akkattu T. Biju has assembled an outstanding collection of work of talented authors in the area of NHC catalysis under the title "*N-heterocyclic Carbenes in organocatalysis.*" The recent developments in the NHC-catalyzed transformations for the synthesis of various heterocycles, carbocycles, and acyclic molecules form the subject matter of this book.

The first chapter, an authoritative contribution by F. Glorius (Westfälische Wilhelms-Universität Münster) and M.N. Hopkinson (Freie Universität Berlin), describes an overview on NHCs. The structure of NHCs, methods of preparation, the electronic, and steric properties are clearly demonstrated in this chapter. A detailed account of NHC-catalyzed benzoin reaction has been provided in the second chapter by M. Gravel from University of Saskatchewan. Description on cross-benzoin reaction, aza-benzoin reaction, as well as the enantioselective benzoin reaction has been highlighted in this chapter. NHC-catalyzed umpolung of aldehydes followed by the interception of the acyl anions with Michael acceptors, the Stetter reaction is explained in detail in the third chapter by A. T. Biju from IISc Bangalore. The inter/intramolecular versions of Stetter reaction and the NHC-catalyzed hydroacylation reactions are well documented in this chapter. One of the important mode of reactivity of NHC is the generation of homoenolates (conjugate umpolung) from α,β-unsaturated aldehydes. V. Nair from NIIST, Trivandrum, has provided an update on NHC-catalyzed reactions proceeding via the homoenolates in the fourth chapter. The interception of homoenolates with various carbonyls, imines, as well as activated carbon–carbon bonds are summarized in this chapter. Recent advances in the NHC-catalyzed domino/cascade processes form the subject matter of the fifth chapter by D. Enders from RWTH Aachen University and P. Chauhan IIT – Jammu, India. The majority of the cascade reactions are initiated by the addition of NHCs to aldehydes generating the Breslow intermediates/(homo)enolates/α,β-unsaturated acylazolium intermediates. A detailed account on NHC-catalyzed generation of α,β-unsaturated acylazolium intermediates followed by their application in heterocycles/carbocycles synthesis has been outlined in the sixth chapter contributed by D. Lupton from Monash University.

The development of new activation modes in NHC catalysis is important as this provides chemists a platform for developing new reactions. The central theme of the seventh chapter by Y. R. Chi (Nanyang Technological University) is to provide an overview on the new activation modes using NHC catalysis. Recent advances in NHC catalysis proceeding via the radical intermediates are also summarized in this chapter. In the eighth chapter of this book by S. Ye from Chinese Academy of Sciences, Beijing, a summary of various transformations proceeding via NHC-bound enolates and dienolates has been provided. In addition to the chemistry of NHC-acyl anions and related species, NHCs are also used as Brønsted base catalysts in organic chemistry. A detailed account on the role of NHCs as Brønsted base catalysts in various organic reactions are provided in the ninth chapter by Y. Huang from Peking University. In the 10th chapter by Y. Zhao from the National University of Singapore, NHC-catalyzed transformations proceeding via the (dynamic) kinetic resolution and desymmetrization pathways are explained. The application of NHCs in metal-free polymerization reactions are accounted in the 11th chapter by D. Taton from University of Bordeaux. The synthetic potential of NHCs in a natural product, and complex target synthesis has been demonstrated by K. A. Scheidt (Northwestern University) in the last chapter of this book.

It is anticipated that this book will serve as a tool for synthetic organic chemists to appreciate the potential of NHC catalysts in various bond-forming reactions. Moreover, it is believed that this book will allow chemists to solve the many more challenges remaining in the area of NHC catalysis and discover new/unprecedented reactions in this process.

27 April 2017

Ronald Breslow
Department of Chemistry
Columbia University
New York, NY

About the Editor

Akkattu T. Biju received his M.Sc. from the Sacred Heart College Thevara (affiliated to MG University, Kerala, India) and Ph.D. under the guidance of Dr. Vijay Nair at the CSIR-NIIST (formerly RRL), Trivandrum, India. Subsequently, he has been a postdoctoral fellow with Prof. Tien-Yau Luh at the National Taiwan University, Taipei, and an Alexander von Humboldt fellow with Prof. Frank Glorius at the Westfälische Wilhelms-Universität Münster, Germany. In June 2011, he began his independent research career at the CSIR-National Chemical Laboratory, Pune. From June 2017 onward, he has been an Associate Professor at the Department of Organic Chemistry, Indian Institute of Science, Bangalore. His research focuses on the development of transition-metal-free carbon–carbon and carbon–heteroatom bond-forming reactions using aryne chemistry and N-heterocyclic carbene (NHC) organocatalysis, and their application in organic synthesis. He is the recipient of AVRA Young Scientist Award (2016), CRSI Young Scientist Award (2015), NCL-Research Foundation Scientist of the Year Award (2014), ISCB Young Scientist Award (2014), Thieme Chemistry Journals Award (2014), OPPI Young Scientist Award (2012), Alexander von Humboldt Fellowship (2009), and is a member of the National Academy of Sciences, India (NASI), Allahabad (2012).

1

An Overview of NHCs

Matthew N. Hopkinson[1] *and Frank Glorius*[2]

[1] *Freie Universität Berlin, Institut für Chemie und Biochemie, Takustr. 3, 14195 Berlin, Germany*
[2] *Westfälische Wilhelms-Universität Münster, Organisch-Chemisches Institut, Corrensstr. 40, 48149 Münster, Germany*

Since the unambiguous isolation of the first free N-heterocyclic carbene (NHC) IAd in 1991 by Arduengo et al. [1], these compounds have received a huge amount of attention from across the chemical community [2]. As stable examples of coordinatively unsaturated, electronically deficient carbene compounds, much of the earlier interest resulted from their status as academic curiosities. However, as studies were conducted on their properties and reactivity, the full potential of NHCs in many different areas of chemistry was revealed. As strong σ-donors to metal centers, NHCs are nowadays widely used as ancillary ligands in organometallic chemistry including in industrially important catalytic transformations [3], rivaling phosphines, and cyclopentadienyls in this role. Many NHC–metal adducts are also attracting attention in materials science and as potential metallopharmaceuticals [4]. The strong binding properties and stabilizing features of NHCs have also led to many applications not involving metals. For example, boron–NHC adducts have been widely studied in a number of different contexts [5], whereas some classes of π-accepting NHCs have been shown to activate small molecules such as ammonia [6]. It is the reactivity of NHCs as organocatalysts, however, that forms the basis of this book [7]. First observed by Ukai et al. in a thiazolium salt-mediated benzoin condensation in 1943 [8], NHCs have proved efficient catalysts for umpolung reactions of aldehyde substrates, reacting via so-called Breslow intermediates, which can be considered acyl anion equivalents [9]. Alongside these processes, alternative transformations involving a wide range of different reactivity modes with both aldehydes and other substrate classes are possible with the application of chiral NHCs often allowing for high levels of enantioselectivity [10]. The scope and diversity of NHC organocatalysis continue to expand at a rapid pace and detailed summaries highlighting that the major reaction classes and current research trends are to be found in the subsequent chapters of this book. In this introductory chapter, we instead provide a general overview of NHCs, highlighting their common structural features and properties and briefly summarizing some of their uses outside of organocatalysis. The major synthetic routes to commonly

N-Heterocyclic Carbenes in Organocatalysis, First Edition. Edited by Akkattu T. Biju.
© 2019 Wiley-VCH Verlag GmbH & Co. KGaA. Published 2019 by Wiley-VCH Verlag GmbH & Co. KGaA.

employed NHC organocatalysts are discussed while a particular focus is given to understanding and comparing the various electronic and steric properties of different classes of NHC.

1.1 General Structure of NHCs

1.1.1 Classes of NHCs and Related Stable Carbenes

The definition of what constitutes an NHC is often subject to different interpretations, and many classes of carbene compounds have been labeled NHCs in the literature. For the sake of clarity, we consider an NHC as any compound featuring a carbene center as part of a heterocyclic ring containing at least one nitrogen atom. Many different carbenes satisfy these criteria, and a selection of some of the most important classes of NHC is shown in Figure 1.1. The first reported compound IAd (**1a**), isolated by Arduengo et al. [1], is an example of an imidazolylidene NHC (**1**). The carbene center in this species is situated between the two nitrogen atoms in an aromatic imidazole heterocycle. Imidazolylidenes of this type were the focus of many of the early studies on NHCs, and they continue to find applications across many areas of chemistry. Derivatives featuring aromatic nitrogen substituents such as mesityl (IMes, **1b**) and 2,6-diisopropyl (IPr or IDipp, **1c**) are particularly widely used as ligands for transition metals and feature in catalysts for cross-coupling reactions and other important transformations. Related to imidazolylidenes but often displaying different reactivity are their saturated imidazolinylidene analogs **2**. These species also feature two nitrogen atoms adjacent to the carbene center in a five-membered heterocycle, yet are not aromatic. The first free imidazolinylidene NHC SIMes (**2a**) was prepared by Arduengo et al. [11], and this compound, which again features *N*-mesityl substituents, has since found a widespread use as a ligand for ruthenium in Grubbs' second-generation olefin metathesis catalysts [3f, u]. Benzimidazolylidenes **3** possess a benzene ring fused onto an imidazolylidene NHC while derivatives with larger ring sizes such as six-membered tetrahydropyrimidinylidenes **4** or various ring-sized *N*,*N*′-diamidocarbenes (DACs, **5**) have been prepared. These latter compounds, pioneered by Bielawski and coworkers, have been shown to activate small molecules such as ammonia and undergo insertion into alkenes in a similar manner to classical triplet carbenes [12].

There is also no requirement for two nitrogen atoms in the heterocycle, and various NHC classes featuring an alternative heteroatom such as oxygen (oxazolylidenes, **6**) or sulfur (thiazolylidenes, **7**) in place of one nitrogen are accessible. Thiazolylidenes (**7**) have been particularly widely used as organocatalysts with the original report by Ukai et al. in 1943 making use of such a species, although the involvement of a free carbene as a catalyst was at that time not acknowledged [8]. NHCs **8**, which feature one nitrogen as the only heteroatom in a pyrrolidinylidene ring, were introduced by Bertrand and coworkers [13] and are commonly referred to as cyclic (amino)(alkyl)carbenes (CAACs) or cyclic (amino)(aryl)carbene (CAArC) depending on the nature of the carbon substituent adjacent to the carbene center [14]. These compounds are more π-accepting than other classes

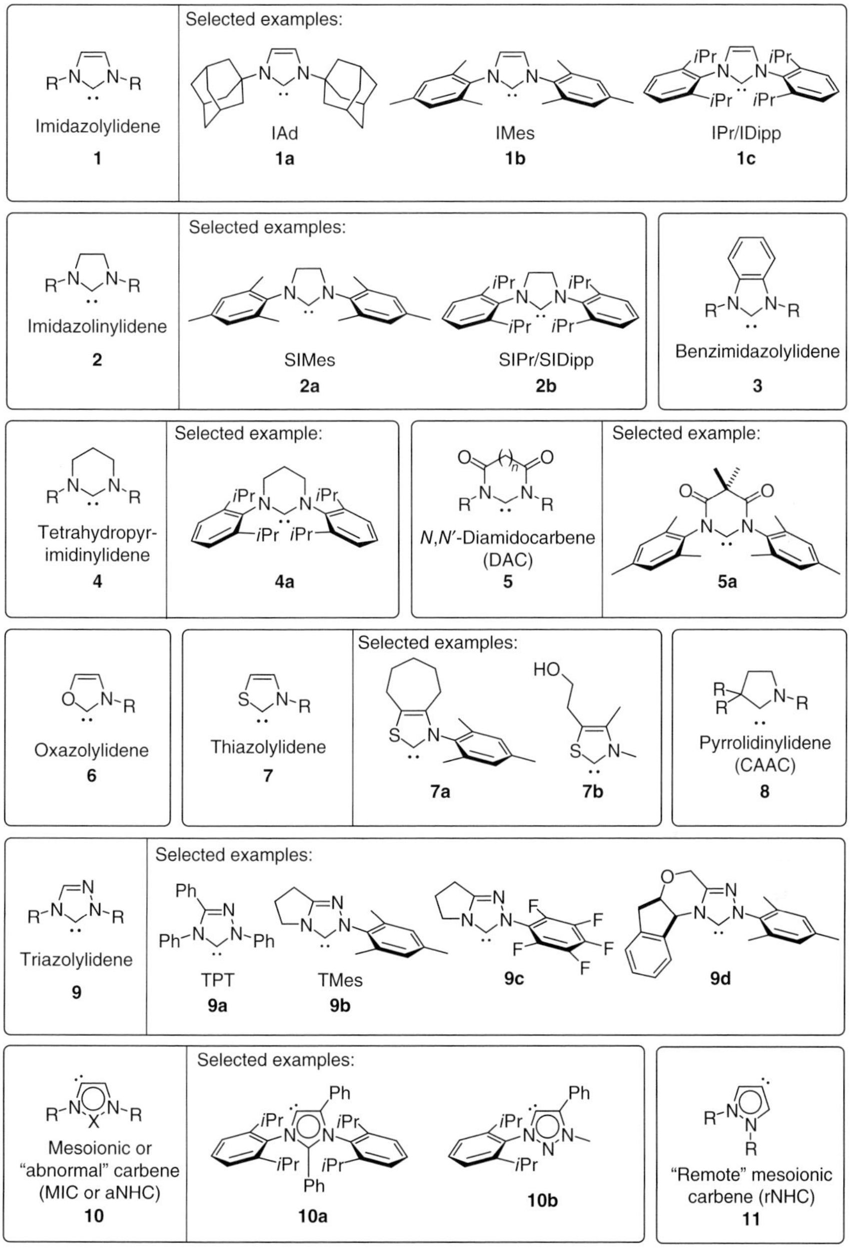

Figure 1.1 Some important classes of NHC with selected examples.

of NHCs and have proved useful for stabilizing sensitive p-block species and even organic radicals [15]. There are also classes of NHCs that feature more than two nitrogen atoms in the heterocyclic ring. Triazolylidenes **9** have found a particularly widespread use as organocatalysts and can be considered the NHCs of choice for many transformations. Examples of widely employed triazolylidenes include 1,3,4-triphenyl-4,5-dihydro-1*H*-1,2,4-triazol-5-ylidene (TPT) (**9a**) [16], which is also often referred to as the Enders carbene and *N*-pentafluorophenyl (**9b**) or *N*-mesityl-substituted bicyclic systems (TMes, **9c**). A further attractive feature of triazolylidene organocatalysts is the relative abundance of chiral derivatives such as compound **9d**, which allows enantioselective reactions to be realized. In addition to considering the range of different NHC ring sizes and substitution patterns, distinct NHC structures can be accessed by generating the carbene center at different positions. Although the carbene carbon is normally situated between the two nitrogen atoms in imidazolylidenes **1**, it is also possible to generate the carbene at the 4-position. In this case, a neutral, non-zwitterionic resonance structure cannot be drawn, and the species is referred as a mesoionic or "abnormal" carbene (aNHC) **10** [17]. When the carbene center is not situated adjacent to a nitrogen atom, the species is called a "remote" NHC (rNHC) **11** [18].

Although the above classes of compounds satisfy all the criteria for NHCs listed above, it is important to remember that many related non-NHC compounds have also been reported that possess similar structural features (Figure 1.2) [19]. For example, acyclic derivatives often known as acyclic diamino carbenes (ADCs) (**12**) that do not feature a nitrogen heterocycle or species where the nitrogen has been replaced by a different heteroatom such as phosphorus have been synthesized [19, 20]. A free acyclic carbene species **13** stabilized by adjacent phosphorus and silicon substituents was in fact reported by Bertrand and coworkers in 1988, three years before the isolation of a free NHC [21]. Cyclopropenylidene compounds (**14**) featuring exocyclic nitrogens were also more recently synthesized by Bertrand and coworkers [22], while extended "bent allene" species such as **15** have themselves been used as ligands [23]. These compounds can be considered as acyclic carbenes stabilized by NHCs.

1.1.2 Structural Features Common to All NHCs

As shown in Figure 1.3 for the representative imidazolylidene IMes (**1b**), there are several structural features that are common to all classes of NHCs. These features each play a role in stabilizing the carbene moiety and variations in their structure or substitution pattern result in different properties of the overall NHC. Apart from remote NHCs, the carbene carbon is situated adjacent to at least one nitrogen atom incorporated into the ring structure. The third group on the nitrogen atom(s) is referred to as the *N*-substituent(s) and can be aliphatic

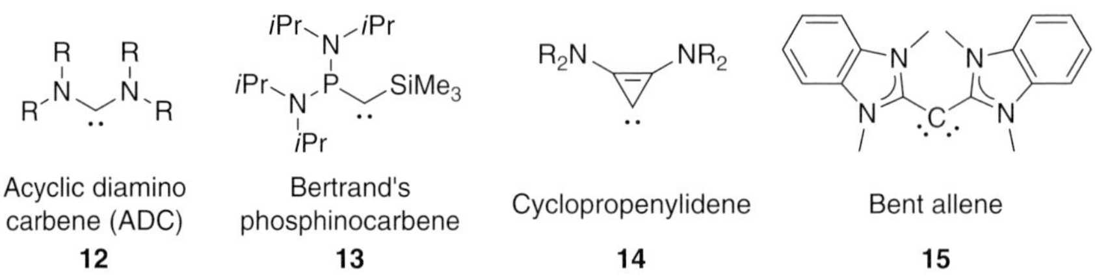

Figure 1.2 Selected related classes of stable carbene.

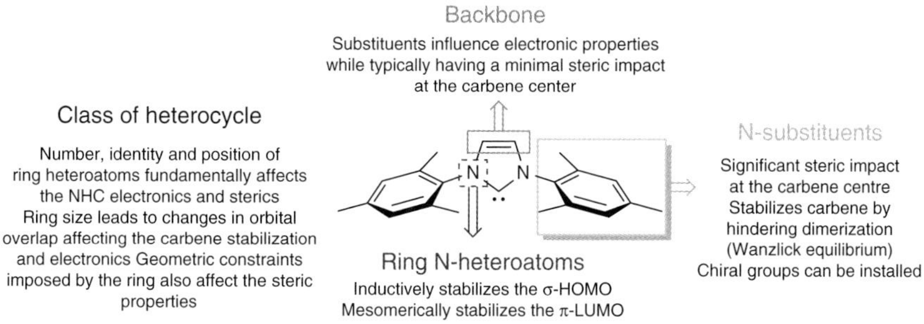

Figure 1.3 General structural features of NHCs exemplified for IMes (**1b**).

or aromatic in nature. The remaining positions on the ring structure (i.e. the 4- and 5-positions in imidazolylidenes) are typically collectively called the NHC backbone. The class of nitrogen heterocycle is fundamental in determining the properties of NHC with each ring having inherently different substitution patterns accessible to it depending on the ring size and identity and position of the heteroatoms. Thiazolylidene NHCs (**7**), which are commonly used as organocatalysts, for example, have a sterically unhindered site adjacent to the carbene carbon by virtue of the two-coordinate sulfur ring heteroatom, whereas CAACs (**8**) have a tetrahedral, sp^3-hybridized carbon at this site.

The structural features of NHCs also allow for relatively independent variation of the overall electronic and steric properties. The substituents on the ring backbone, for example, have a comparatively small effect on the steric environment at the carbene center but more strongly influence the overall electronic character. The nitrogen substituents or other groups situated directly adjacent to the carbene on the other hand are crucial in determining the steric properties. In this respect, NHCs differ greatly from phosphines, for which changing the substituents bound to the central phosphorus atom is both often nontrivial and inherently affects both the steric and electronic properties of the compound. The effects of structural changes on the electronic and steric properties of the NHC are discussed in more detail in Sections 1.4 and 1.5.

1.1.3 Stabilization of the Carbene Center

The remarkable stability of the carbene center in NHCs results from a combination of kinetic and thermodynamic factors. As carbon atoms with an incomplete electron octet, classical carbene species formed as transient intermediates are typically unstable toward dimerization to the corresponding alkenes. Steric clashing between the bulky aryl or alkyl groups commonly found on nitrogen or other substituents situated adjacent to the carbene carbon in NHCs, however, can kinetically disfavor this dimerization process, which for NHCs is referred to as the Wanzlick equilibrium. Electronic factors resulting from the NHC structure also play a major part in stabilizing the free carbene structure. Unlike classical carbenes, which most often have triplet ground states, NHCs are singlet carbenes with a lone pair of electrons localized in a sp^2-hybridized orbital located in the plane of the ring (HOMO, highest occupied molecular orbital) and an empty

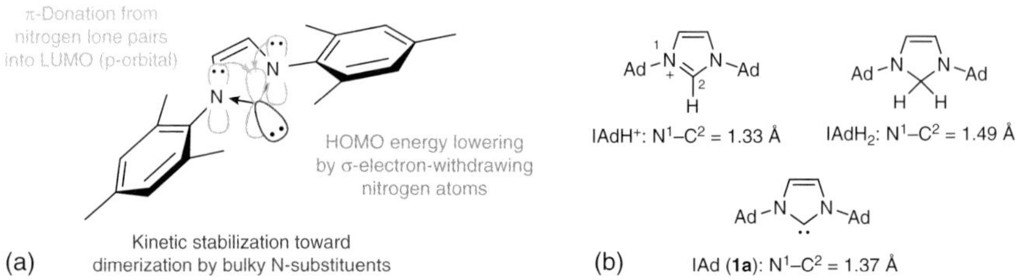

Figure 1.4 (a) Stabilization of the carbene by adjacent ring nitrogens. (b) Comparison of the N^1—C^2 bond lengths in IAd derivatives.

p-orbital situated perpendicular to it (LUMO, lowest unoccupied molecular orbital, Figure 1.4a) [2c]. As demonstrated in studies by Goddard and coworker [24], singlet carbenes are less prone to dimerization, and inductive lowering of the HOMO energy by the adjacent electronegative nitrogen heteroatom(s) in NHCs leads to large singlet-triplet energy gaps. Mesomeric donation of electron density through overlap of the occupied nitrogen p-orbitals and the empty LUMO also leads to electronic stabilization of the singlet ground-state structure. Furthermore, the structural confinements imposed by the cyclic nature of NHCs prevents the carbene carbon from adopting the linear geometry favored for the p_x,p_y degeneracy in triplet carbenes, further increasing the singlet–triplet energy gap. As shown in Figure 1.4b for IAd (**1a**), the singlet ground-state bonding situation with mesomeric electron density donation by the ring nitrogens can be inferred by comparing the different N^1—C^2 bond lengths in the free carbene, the corresponding cationic azolium salt ($IAdH^+$) and the C^2-saturated analog ($IAdH_2$). In the azolium species $IAdH^+$, the N^1—C^2 bond is a formal double bond with a bond length of 1.33 Å [1], whereas the corresponding single N^1—C^2 σ-bond in $IAdH_2$ is expectedly longer with a value of 1.49 Å [25]. The N^1—C^2 bond length of 1.37 Å obtained from the X-ray crystal structure of the free NHC IAd (**1a**) lies in between these two extremes and is consistent with a partial double-bond character [1].

The relative stability of the free carbene form of different NHCs and variations in their reactivity can often be explained by analyzing the degree to which each of the stabilizing features described above apply. Carbenes such as CAACs (**8**), which contain just one nitrogen atom adjacent to the carbene carbon, for example, accordingly have a lower degree of mesomeric donation into the LUMO and are more accepting of π-electron density from metal or other groups bound to the carbene center in complexes or adducts [14a]. The amide functionality in DACs (**5**) also leads to less π-electron donation into the LUMO, resulting in smaller singlet–triplet gaps and more triplet carbene-like reactivity [12]. The number and identity of the heteroatoms as well as the ring size and the nitrogen and backbone substitution all play a role in influencing the carbene stability and reactivity. In addition to these universal considerations, compounds with unsaturated backbones such as imidazolylidenes (**1**), thiazolylidenes (**7**), and triazolylidenes (**9**) benefit from increased stability because of their partial aromatic character. As revealed by theoretical studies by the groups of Apeloig Schwarz and coworkers [26] and Frenking and coworker [27] on model imidazolylidenes, this effect explains the increased persistence of carbenes such as IMes (**1b**) relative to their saturated imidazolinylidene analogs (e.g. SIMes, **2a**). Overall, the combination of these electronic effects is even sufficient to stabilize the free carbene 1,3-dimethylimidazol-2-ylidene (IDM, **1d**), which is persistent in solution despite its lack of bulky *N*-substituents [28].

1.2 NHCs as σ-Donating Ligands

The most important feature of NHCs arising from this singlet ground-state electronic structure is their remarkable σ-donation abilities. It is this property that has led to the widespread interest in NHCs and has elevated them from mere exotic curiosities to workhorses across many areas of modern chemistry. In particular, NHCs now rank among the most widely employed spectator ligands for transition metals usurping phosphines and cyclopentadienyls for many applications [3]. In fact, the first examples of metal–imidazolylidene complexes were reported over 20 years before the first successful isolation of a free NHC. In 1968, independent reports by the groups of Wanzlick and Schönherr [29] and Öfele [30] introduced Hg^{II} and Cr^{III} complexes, respectively, whereas extensive studies were also conducted throughout the 1970s and 1980s, notably by Lappert and coworkers [31]. Nowadays, NHC-containing complexes are known for all transition metals, while extensive coordination chemistry has also been developed for the f-block [3s] and group 1 and 2 elements [3c].

A selection of NHC–metal complexes that have found applications in catalysis is shown in Figure 1.5. Among the most important NHC-containing species are the series of second-generation ruthenium metathesis catalysts such as the Hoveyda–Grubbs II complex (**16**) [3f, u]. In comparison to the first-generation Grubbs catalysts, which feature a phosphine ligand in place of the imidazolinylidene SIMes (**2a**), the second-generation catalysts are typically active at much lower loadings and are more thermally stable by virtue of the increased stability offered by the strongly binding NHC. Furthermore, the electron-donating carbene results in improved catalytic activity because of an increased rate of formation of the coordinatively unsaturated catalytically active intermediate [32]. NHCs have also proved highly beneficial as ligands for cross-coupling catalysts based on palladium or other transition metals [3i–l, o]. Imidazolylidene-bearing palladium species such as the Pd–PEPPSI–NHC series

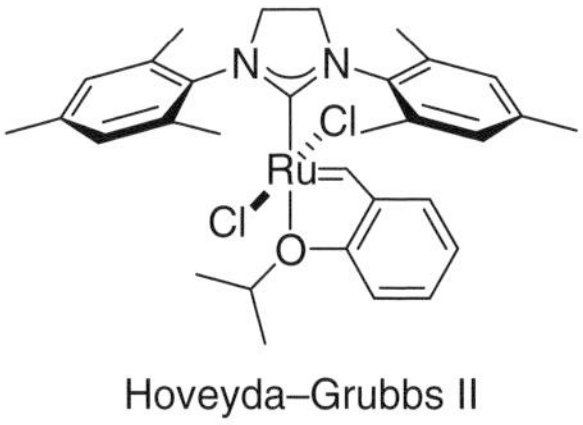

Hoveyda–Grubbs II olefin metathesis catalyst (**16**)

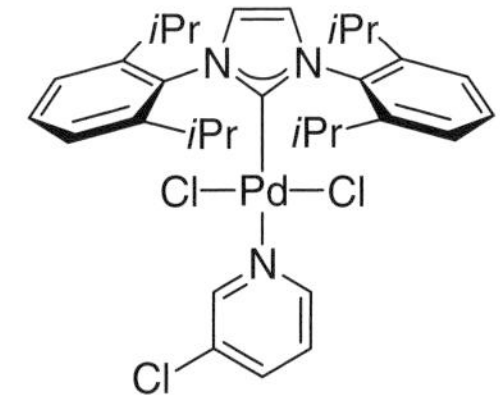

Organ's Pd–PEPPSI–IPr cross-coupling catalyst (**17**)

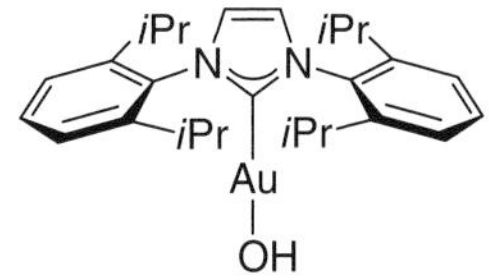

Nolan's gold(I) hydroxide catalyst (**18**)

Figure 1.5 Selected examples of important NHC-containing transition metal catalysts.

of precatalysts (e.g. Pd–PEPPSI–IPr, **17**) developed by Organ and coworkers are nowadays widely applied in organic synthesis [3]. In these complexes, the strongly binding NHC typically leads to improved stabilization of low-valent intermediates, reducing catalyst decomposition. The electron richness of NHCs can also facilitate challenging oxidative addition steps, while their steric bulkiness can aid reductive elimination. NHC-bearing complexes of late transition metals have also found widespread use as catalysts for other organic transformations with gold(I) species such as **18** developed by Nolan and coworkers, for example, mediating nucleophilic addition reactions to carbon–carbon multiple bonds [3g, 33]. Applications of NHC–metal complexes, however, are not limited to homogeneous catalysis. As pioneered by Youngs and coworkers, various complexes have shown promise as metallopharmaceuticals, with imidazolylidene adducts of gold(I) being investigated as potential antitumor agents [4]. NHC-containing complexes have also found many uses in materials sciences, being employed in such diverse roles as self-healing polymers, liquid crystals, metal–organic frameworks (MOFs) [34], and photoactive materials [35]. Recent trends have also seen the strong σ-donating properties of NHCs being exploited for binding to metal surfaces or nanoparticles [36]. Immobilization of the ligands on solid supports has allowed for catalysis with NHC–metal complexes to be conducted in a heterogeneous manner [37], while water-soluble derivatives have also been developed [38].

In addition to binding to metal centers, NHCs have been widely employed as ligands for non- or semimetallic species (Figure 1.6). In this context, NHCs have often allowed for the stabilization and isolation of hitherto unprecedented forms of these elements. For example, Robinson and coworkers used the imidazolylidene IPr as a stabilizing moiety for an otherwise inaccessible P_2 fragment, in which phosphorus resembles the form of nitrogen in N_2 [39]. A similar species was also prepared by the same group with silicon [40], while more recent work has led to the isolation of silicon oxide fragments (e.g. **19**) analogous to silica-based materials used in advanced electronic devices [41]. CAACs have proved particularly versatile for stabilizing novel non- and semimetallic species [14a]. As demonstrated by Bertrand and coworkers, the π-accepting character of these ligands has allowed for the stabilization of open-shell species including the carbon-based radical **20** [15, 42]. This compound, which was generated upon single-electron reduction of the corresponding acyl azolium salt, was found to be stable over several months and can be considered a "bottleable" radical. NHC–boron adducts have also received significant research attention [5]. In an early work by Fensterbank , Lacôte, Malacria, Curran, and coworkers, the triazolylidene–borane species **21** was used as a hydrogen atom donor in place of highly toxic tin hydride derivatives in radical deoxygenation reactions of alcohols [43]. The NHC was shown to lead to a significant weakening of the boron–hydrogen bond relative to uncoordinated or amine–borane or phosphorus–borane adducts because of the stabilization of boryl radicals formed upon hydrogen atom transfer through spin density delocalization into the NHC π-system. Studies conducted by Bielawski and coworkers have focused on DACs as activators of small molecules [12]. The free carbene **5a** reacts with aldehydes in a formal [2+1] cycloaddition process affording an epoxide species [44].

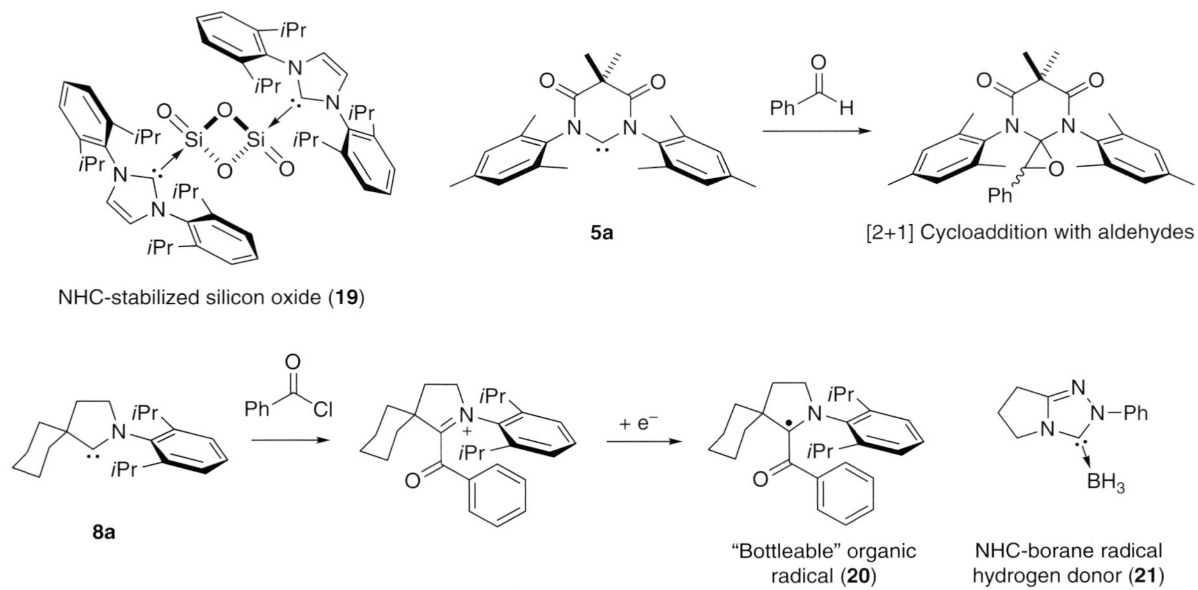

Figure 1.6 Selected examples of adducts between NHCs and non- or semimetallic species.

This reactivity stands in stark contrast to that observed with imidazolylidene, triazolylidene, and other NHC classes commonly employed as organocatalysts, which mediate umpolung reactions of aldehydes via the Breslow intermediate.

1.2.1 The Nature of Bonding in NHC Adducts

The full nature of the bonding interaction between carbene centers and metals or nonmetals has been widely studied over the last 20 years [45]. The dominant feature of this bonding is the remarkably strong σ-donation from the carbene lone pair into a vacant σ-orbital at the metal or nonmetal. In graphical representations of NHC–metal complexes such as those shown in Figure 1.5, a single rather than a double bond is typically drawn and, in the absence of steric constraints or chelation, NHC–metal bonds can usually rotate. In this regard, NHCs behave differently from traditional Fischer or Schrock carbenes, which are more accurately considered as forming formal double bonds with metals. For the most part, π-orbital overlap in NHC–metal complexes is thought of as being restricted to within the NHC ring and a curved line is sometimes drawn between the ring heteroatoms to reflect this. In recent years, however, the contribution of π-electron donation to the ligand–metal bonding interaction itself has been more widely acknowledged. Specifically, filled orbitals of π-symmetry on the metal can donate electron density into the empty p-orbital LUMO of the carbene. Donation of electron density from the carbene π-system into vacant metal π-orbitals can also occur. The importance of each factor is dependent on the metal, the identity and geometry of other ligands at the complex, and, importantly, on the type of carbene itself. Recent research has provided new spectroscopic methods for measuring and quantifying the π-accepting ability of different NHCs, and these will be discussed in more detail in Section 1.4.3 [46].

1.2.2 Comparing NHC and Phosphine Ligands

In comparison to phosphines, NHCs are generally more electron rich and, upon binding to metals, afford complexes that are more thermodynamically stable. Indeed, metal–ligand bond distances measured in crystal structures are typically shorter in NHC–metal carbene complexes than in the corresponding phosphine-ligated species, whereas bond dissociation energies are often higher. This can lead to improved resistance to chemical or thermal decomposition and goes some way to explaining the superiority of NHC–metal complexes in many applications, notably in catalysis. In a systematic study in 2003, Nolan and coworkers compared bond disruption enthalpies (BDEs) for the coordination of a series of imidazolylidene and imidazolinylidene ligands and the phosphine PCy_3 (Cy = cyclohexyl) to a model ruthenium(II) complex (Figure 1.7) [47]. Even in comparison to such a highly Lewis-basic phosphine, for nearly all the NHCs tested, the metal–ligand binding strength was found to be greater with the carbene. The only exception to this general trend was observed with the extremely sterically hindered imidazolylidene IAd. In this case, clashing with the other ligands bound to the metal likely hinders close approach of the NHC and leads to a weaker metal–ligand bond.

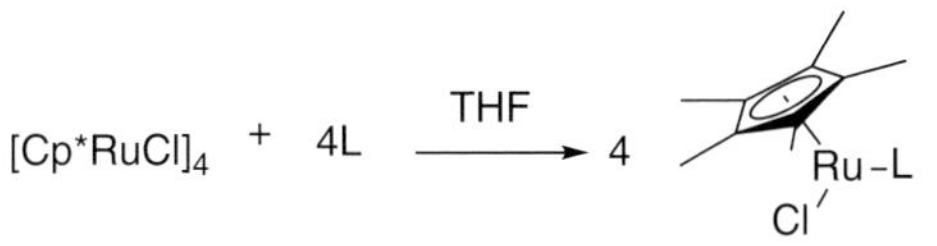

L	Relative BDE (kcal mol^{-1})
L = IMes (**1b**)	15.6
L = IPr (**1c**)	11.1
L = SIMes (**2a**)	16.8
L = SIPr (**2b**)	12.1
L = PCy_3	10.5

Figure 1.7 Comparison of the binding strength of different NHCs and PCy_3 in [Cp*Ru(L)Cl] complexes (L = ligand) by Nolan and coworkers. Source: Hillier et al. 2003 [47]. Reproduced with permission of ACS.

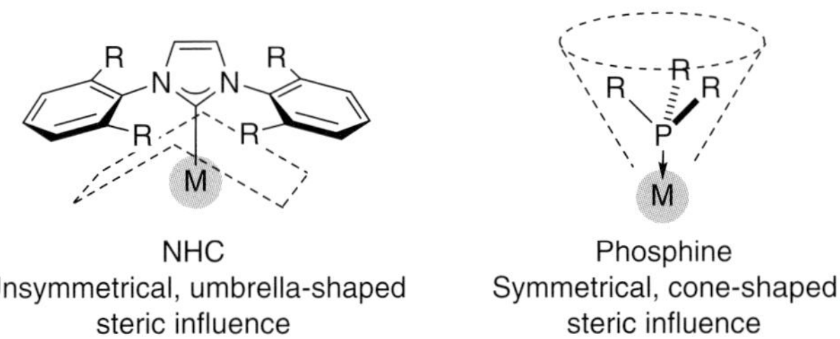

Figure 1.8 Comparing the steric properties of NHCs and phosphines.

A major difference between NHCs and phosphines concerns their steric properties. Although phosphines have a cone-shaped spatial arrangement with three substituents bound to an sp^3-hybridized phosphorus atom, the steric footprint of NHCs has been described as "fan" or, somewhat more accurately, "umbrella" shaped (Figure 1.8). This results in a greater steric influence of the ligand at the metal center as bulky substituents adjacent to the carbene in NHCs point toward the metal rather than away from it. Indeed, the increased steric bulk of the ligand can lead to greater noncovalent interactions with other species in the complex, increasing metal–ligand binding [48]. Furthermore, although the steric requirement of phosphines is essentially isotropic, the planar heterocyclic nature of NHCs means that the spatial footprint of NHCs varies considerably upon rotation of the metal–carbene bond. As a result, NHCs may rotate to reduce steric crowding and accommodate other bulky ligands in a complex. The anisotropy of NHCs also raises challenges in designing enantioselective catalytic processes as strategies must be adopted that restrict the effect of rotation and fix the steric environment at the catalytically active metal center [3t].

1.3 Synthesis of NHCs

1.3.1 Generation of the Free Carbene

Accessing NHCs for use as ligands or organocatalysts requires the synthesis of a suitable precursor. Most commonly, the corresponding cationic azolium salt is prepared with the free carbene being obtained upon simple deprotonation usually under mild conditions with various organic or inorganic bases. Separating the desired NHC from the protonated base by-product, however, is often

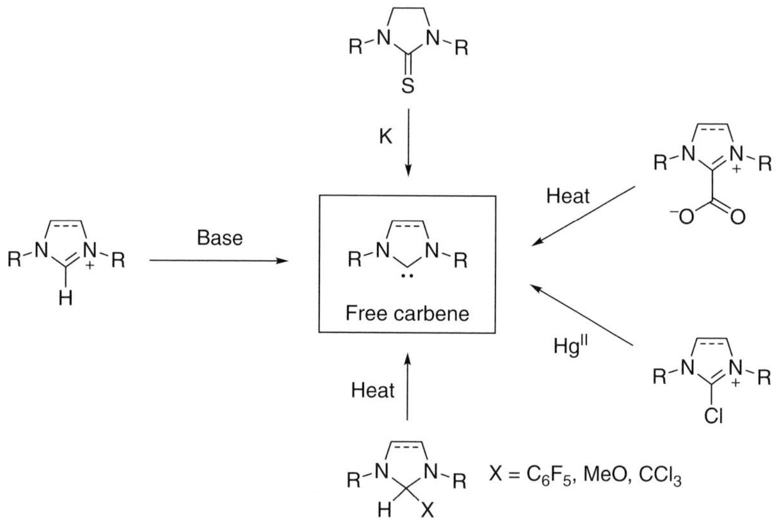

Figure 1.9 Some commonly employed methods of generating free NHCs.

not trivial, and for most applications, including in the vast majority of NHC organocatalytic reactions, preisolation of the free carbene is not performed. Instead, the NHC is generated *in situ* with the usually stable, crystalline azolium salt precursors being employed directly in the presence of a suitable base. Some alternative strategies to generate free NHCs from various precursors are shown in Figure 1.9. These methods include the desulfurization of thioureas using molten potassium metal [49]. Although requiring rather harsh conditions, the potassium sulfide by-product of this process is insoluble in the tetrahydrofuran (THF) reaction medium, facilitating isolation of the free carbene. Alternatively, a vacuum pyrolysis approach can be conducted, whereby extrusion of a small volatile molecule such as chloroform or pentafluorobenzene from a suitable precursor provides a clean route to the free carbene [50]. This method was employed by Enders et al. to generate the free triazolylidene TPT (**9a**) upon extrusion of a molecule of methanol from the corresponding adduct [16]. Similarly, heating of zwitterionic azolium carboxylate species can afford free NHCs upon loss of CO_2 [51]. 2-Chloro-substituted azolium salts have also been employed as NHC precursors, delivering the free carbene upon reduction with bis(trimethylsilyl)mercury [52]. These precursors can also be used to prepare NHC–metal complexes directly via oxidative addition into the C—Cl bond with a low-valent metal salt [53]. Alternative methods to prepare NHC–metal complexes that do not require generation of the free carbene include ligand transfer from preformed silver(I) or copper(I) complexes or *de novo* templated construction of the NHC from a precursor already containing the metal.

1.3.2 Synthetic Routes Toward Azolium Salt NHC Precursors

For the vast majority of applications including in NHC organocatalysis, the desired NHC is generated *in situ* upon deprotonation of the corresponding cationic azolium salt. As heterocyclic organic molecules, synthetic routes toward these species have been widely studied in a variety of contexts over many decades. Nowadays, there exist many different approaches to synthesizing azolium salts for all the core ring structures of NHCs and, for the most part, modular routes are available where simple variation of the starting materials provides facile access to a large range of diversely substituted derivatives. As a result, libraries of subtly different NHCs can be designed, synthesized, and tested in a fast and systematic manner, allowing for intricate fine-tuning of the NHCs' properties for any given application. Moreover, enantioenriched NHCs useful as chiral ligands or organocatalysts can be readily obtained by simply incorporating starting materials from the chiral pool into the general synthetic approach.

One attractive approach to synthesize azolium salt precursors involves quaternization of the nitrogen heteroatom of the corresponding neutral compound. Nitrogen heterocycles of this type rank among the most studied classes of compounds in organic chemistry, and, as a result, a huge number of diversely substituted derivatives are readily available. This method is particularly useful for preparing *N*-alkyl substituted NHC precursors with simple nucleophilic substitution in the presence of a suitable alkyl electrophile (e.g. alkyl halide or pseudo-halide) generally resulting in clean formation of the desired azolium salt. Synthetic strategies to directly access cationic azolium salts invariably involve a cyclization process as a key step. As outlined in an excellent review by Lavigne, Bellemin-Laponnaz, César, and coworkers [54], these synthetic approaches can be classified according to whether the cyclization involves installation of the "precarbenic" carbon, a ring nitrogen or the NHC backbone. Arguably, the most prevalent of these strategies involves the incorporation of the precarbenic carbon. Most commonly, a tethered *N*,*N*′-disubstituted diimine and diamine substrate is reacted with a C1 synthon such as triethyl orthoformate ($HC(OEt)_3$). This process constitutes the last step of many synthetic routes to NHCs and is often high yielding and operationally simple. For symmetrical imidazolylidenes, a general two-step strategy can be used with an initial condensation of 2 equiv of the desired aniline starting material with glyoxal (**22**) affording the diimine intermediate suitable for cyclization with $HC(OEt)_3$ (Figure 1.10) [55]. Reduction of the diimine species (**23**) with sodium borohydride or lithium aluminum

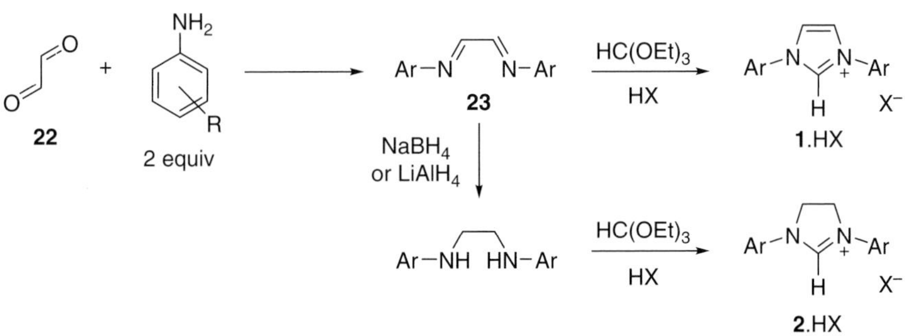

Figure 1.10 Synthesis of imidazolium and imidazolinium salt precursors from glyoxal (**22**).

hydride before cyclization leads instead to imidazolinium salt precursors for imidazolinylidene NHCs.

Several reagents have been employed in place of $HC(OEt)_3$ including paraformaldehyde, chloromethyl ethers or pivalates, and diiodomethane, while alternative routes to prepare the key diimine intermediates, including unsymmetrical derivatives, may be applied [54]. A different strategy to construct imidazolium salts makes use of a 3 + 2 annulation process, whereby the backbone fragment is coupled onto a formamidine starting material **24** (Figure 1.11) [56]. This approach can be used to prepare unsymmetrical *N*-aryl imidazolylidene NHCs as formamidines bearing different aromatic substituents can be prepared upon sequential condensation of anilines with triethyl orthoformate under first acidic and then acid-free conditions [57].

Some synthetic strategies commonly employed to access precursors for the other two classes of NHCs most commonly employed as organocatalysts, thiazolylidenes, and triazolylidenes are shown in Figures 1.12 and 1.13. Thioformamide substrates **25** can be condensed with α-chloroketones **26** to afford thiazolium salts featuring different *N*-substituents [58]. An alternative approach starting from readily prepared cyclic dithiocarbamates **27** has been employed to synthesize a variety of bicyclic derivatives using hydrogen peroxide and acetic acid [59]. Comparatively fewer synthetic routes have been reported toward triazolium salt precursors for triazolylidene NHCs, and, as for imidazolium salts, a final cyclization step using triethyl orthoformate is generally employed. The bicyclic derivatives **9b** and **9c** featuring an *N*-mesityl or *N*-pentafluorophenyl substituent, which are commonly employed as organocatalysts, can be prepared in a three-step sequence from pyrrolidinone **28** [60]. Initial methylation at the oxygen with trimethyloxonium tetrafluoroborate is followed by addition–elimination of the appropriate arylhydrazine to afford an intermediate susceptible to cyclization. This strategy is also convenient for preparing chiral triazolylidenes, which have proven to be the most successful

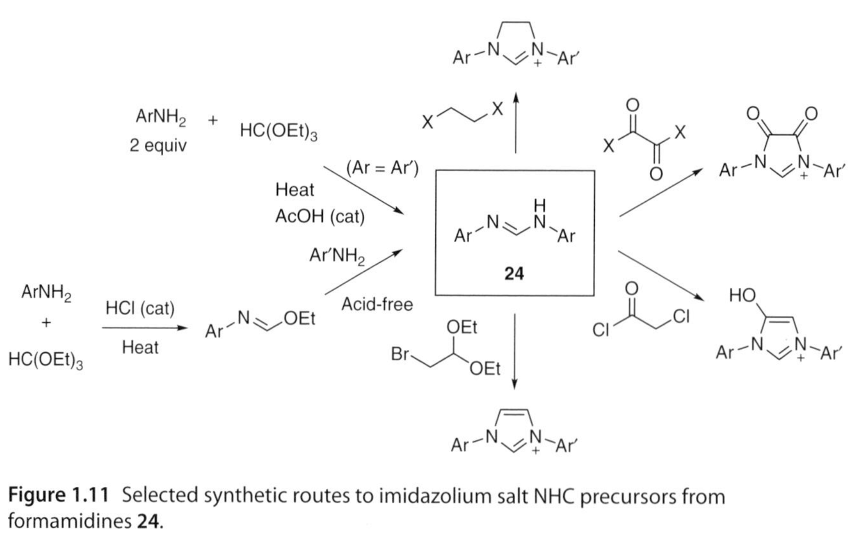

Figure 1.11 Selected synthetic routes to imidazolium salt NHC precursors from formamidines **24**.

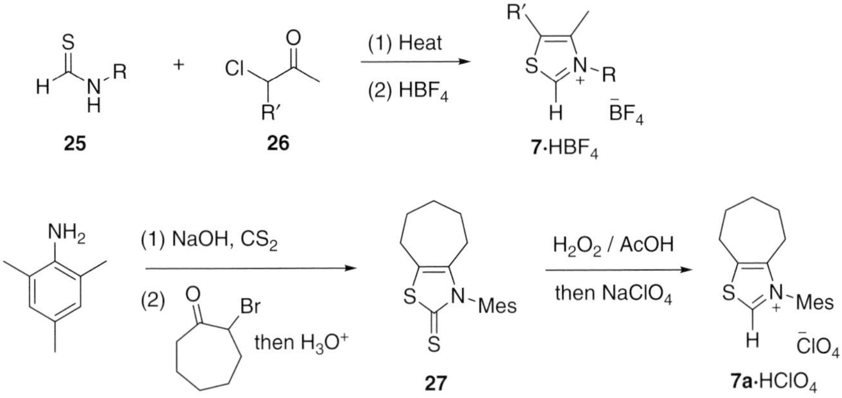

Figure 1.12 Selected synthetic routes to thiazolium salt NHC precursors.

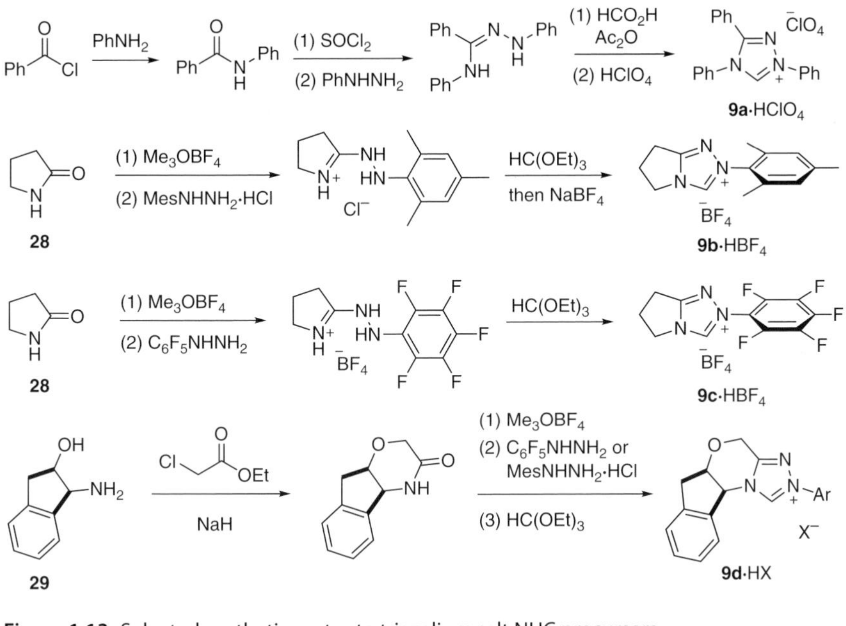

Figure 1.13 Selected synthetic routes to triazolium salt NHC precursors.

class of organocatalysts for many enantioselective reactions. In particular, chiral bicyclic derivatives featuring substituted pyrrolidine or morpholine rings fused to the triazolylidene are widely used, whereas the series of aminoindane-based catalysts such as **9d** has proved highly efficient for many transformations. For these latter catalysts, chiral 1-amino-2-indanol starting materials **29** are first reacted with an α-chloroester to afford a morpholinone species that can then undergo the same sequence of methylation, coupling with an arylhydrazine and cyclization as for the achiral compounds **9b** and **9c** [60]. An alternative synthetic strategy was used by Enders, Teles, and coworkers to prepare the triazolium salt precursor for the commonly used tris(aryl)-substituted NHC TPT (**9a**) [61].

A five-step sequence was employed starting from benzoyl chloride, wherein an initial amide formation followed by activation with thionyl chloride and hydrazine coupling yielded the key diamine intermediate, which in this case was reacted with formic acid as the C1 synthon. Anion exchange then afforded the NHC precursor as the perchlorate salt.

1.4 Quantifying the Electronic Properties of NHCs

The scope of available NHC scaffolds has grown rapidly since the early 1990s, and there now exists a large number of diverse derivatives with different steric and electronic properties. In order to facilitate the selection of a suitable NHC for any given application and to more effectively compare NHCs with other classes of compounds such as phosphines, significant efforts have been devoted to quantifying the properties of NHCs using a variety of different parameters. In this section, the various approaches used to quantify the electronic properties of NHCs are discussed [62], whereas an overview of the methods for quantifying their steric properties is provided in Section 1.5 [63]. In each case, the method of quantification will be discussed followed by a comparison of the data obtained for different NHC classes and nitrogen and backbone substituents.

1.4.1 pK_a Measurements of Azolium Salts

As free NHCs are most commonly obtained upon deprotonation of an azolium salt, knowing the pK_a values of these species is useful for selecting an appropriate base. The variation in the pK_a as a function of the substitution pattern and class of azolium salt also provides a window into the electronic properties of the corresponding NHC. A selection of pK_a values measured by NMR spectroscopy in D_2O for azolium precursors to various NHC classes are shown on a scale in Figure 1.14 [64]. As can be clearly observed from comparing the values obtained for *N*-methyl-substituted derivatives, the class of NHC has a significant effect on the acidity at the precarbenic carbon. Although the pK_a for the benzimidazolium salt **3a**·H^+ (21.6) is only marginally lower than the value obtained for the imidazolium salt of IDM (**1d**·H^+, 22.0), the corresponding thiazolium salt **7c**·H^+ is significantly more acidic (19.5). Moreover, the oxazolium salt **6a**·H^+, which features a more electronegative oxygen atom, has an even lower pK_a value of 16.9. The inclusion of a third electronegative nitrogen atom into the ring structure in triazolium salts also leads to a lowering of the pK_a value. Thus, the azolium salt of TPT (**9a**·H^+) has a value of 16.8 (measured in a 33% v/v CD_3CN/D_2O mixture), while other precursors to common triazolylidene organocatalysts have similar pK_a's (Figure 1.15). The increased acidity of thiazolium and triazolium salts relative to imidazolium derivatives means that organocatalytic reactions using these NHCs can be conducted with milder bases. Conversely, increasing the heterocycle ring size or reducing the degree of unsaturation results in an increase in the pK_a value. Switching from the imidazolium salt of IMes to the imidazolinium salt of SIMes leads to a small decrease in acidity of the

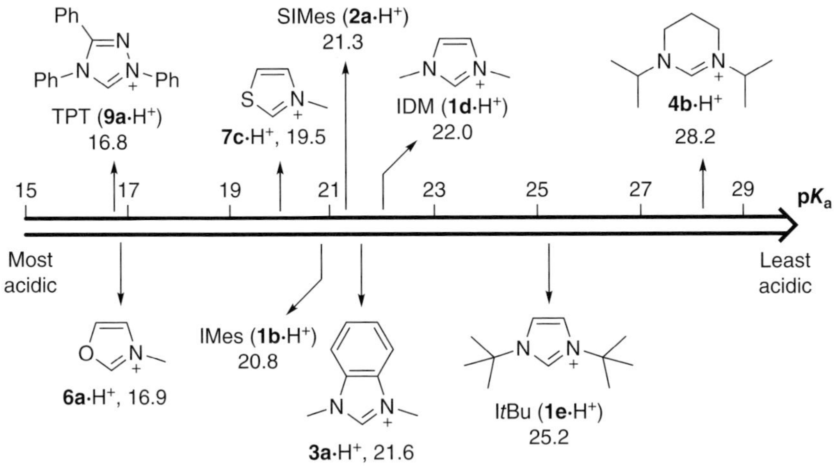

Figure 1.14 Scale showing pK_a values measured in D_2O for representative examples of different classes of NHC.

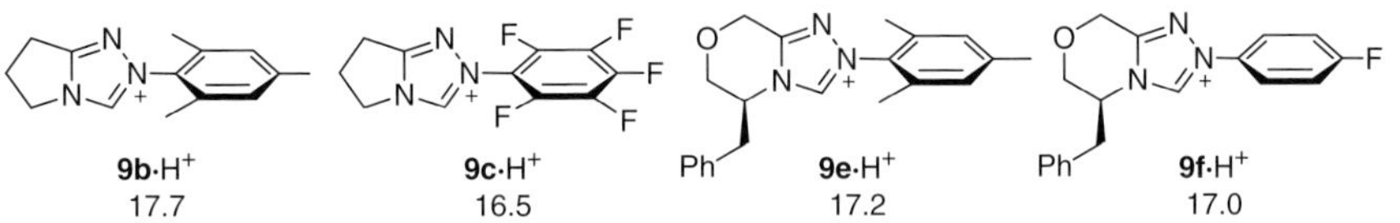

Figure 1.15 Literature pK_a values for azolium salts of some triazolylidene organocatalysts with different *N*-aryl groups (measured by NMR in D_2O).

precarbenic carbon equivalent to an increase of 0.5 pK_a units (**1b**·H^+ = 20.8, **2a**·H^+ = 21.3), whereas the six-membered azolium species **4b** is substantially harder to deprotonate with a measured pK_a value of 28.2.

Notable, albeit somewhat less dramatic, changes in pK_a values can be observed for different substitution patterns within the same class of NHC. The N-substituents, which are situated closest to the carbene carbon, predictably have the greatest effect. Changing the *N*-aryl groups in the series of bicyclic pyrrolidine-based triazolylidenes from pentafluorophenyl (**9c**) to mesityl (**9b**) leads to an increase in pK_a of the corresponding azolium salts of 1.2 units reflecting the greater electron-donating ability of the mesityl group [64c]. Exceptionally bulky *N*-substituents such as the *N*-*tert*-butyl groups in the azolium salt of I*t*Bu (**1e**) lead to an increase in the pK_a values relative to analogs bearing sterically less demanding groups (**1e**·H^+ = 25.2 cf. **1d**·H^+ = 22.0) [64a, b].

1.4.2 Tolman Electronic Parameter (TEP)

The most commonly measured parameter for evaluating the electronic properties of NHCs is the Tolman electronic parameter (TEP). Originally developed by Tolman for phosphines [65], this metric specifically investigates the electron-donating ability of ligands in transition metal carbonyl complexes. According to Tolman's rationale, an electron-donating phosphine or NHC ligand

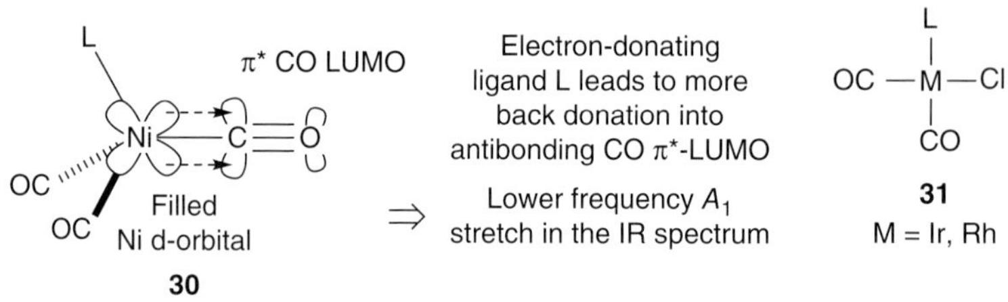

Figure 1.16 The measurement of TEP values.

leads to an increase in electron density at the metal center, which in turn results in an increased degree of electron donation into the π* antibonding orbital of a C≡O spectator ligand. The overall effect of this phenomenon is a decrease in the effective bond order of the carbonyl ligand, which is reflected in a lower frequency A_1 vibration in the complex' infrared (IR) spectrum. Simple comparison of the IR spectra of a series of complexes bearing different NHC ligands therefore allows for an indirect evaluation of their relative electron-donating ability and facilitates a direct comparison with phosphines (Figure 1.16).

There are, however, several limitations associated with the use of TEP values. Firstly, the weakening of the carbonyl ligand in the model complexes is dominated by π-electron considerations, whereas the binding of NHC ligands is mostly (albeit not exclusively) through σ-donation. As a result, the TEP is arguably not best suited for evaluating the overall electron-donating ability of these compounds, and the assumed inverse relationship between ligand binding strength and C≡O bond strength may not always be valid. The range of TEP values encountered with different NHCs is also rather small ($\approx$10 cm^{-1} cf. $\approx$60 cm^{-1} for phosphines) and, as such, care must be taken to ensure the reliability of the measurements. In particular, a suitably high-resolution IR spectrometer must be employed and a common solvent (usually hexane or dichloromethane) should be used. When interpreting TEP values, it is also important to take the steric properties of the ligands into consideration. Bulky NHCs, for example, can lead to distortions in the geometry of the complex leading to changes in the orbital overlap. The trans-influence of other ligands present in the complex can also lead to differences in orbital overlap and, therefore, TEP values. A further limitation arises from the identity of the complex itself. In initial studies, TEP values were measured using tetrahedral nickel(0) complexes of the form $[Ni(CO)_3(L)]$ (**30**) obtained upon ligand exchange with $Ni(CO)_4$. These species are, however, highly toxic and careful handling is required. Because of this limitation, the IR spectra are now usually measured using less toxic square planar iridium(I) or rhodium(I) complexes of the form $[MCl(CO_2)(L)]$ (M = Ir, Rh, **31**). In order to standardize values obtained from the different complexes, correlation equations have been derived (Eqs. (1.1) and (1.2)) that convert the wavenumbers obtained from the Ir [66] or Rh [67] complexes (as an average of the two C≡O stretching frequencies, ν_{CO}^{av}) into TEP values. Care must always be taken, however, only to compare values for which measurements were taken under the same conditions using a common solvent, regardless of which complex was used. In addition to these experimental approaches, TEP values have also been calculated using computational methods [68]. These studies have not only succeeded in accurately reproducing experimentally obtained values for many NHCs but also allowed for the prediction of the electronic properties of hitherto uninvestigated derivatives. As such, the donor properties of new potential NHC ligands or organocatalysts can now be evaluated without having to synthesize them. The TEP values quoted in this section and displayed in Figures 1.17 and 1.18 were measured experimentally using either the Ni, Ir, or Rh complexes in dichloromethane or chloroform and were standardized using Eqs. (1.1) and (1.2) by Nelson and Nolan in their seminal 2013 review on the electronic properties of NHCs [62].

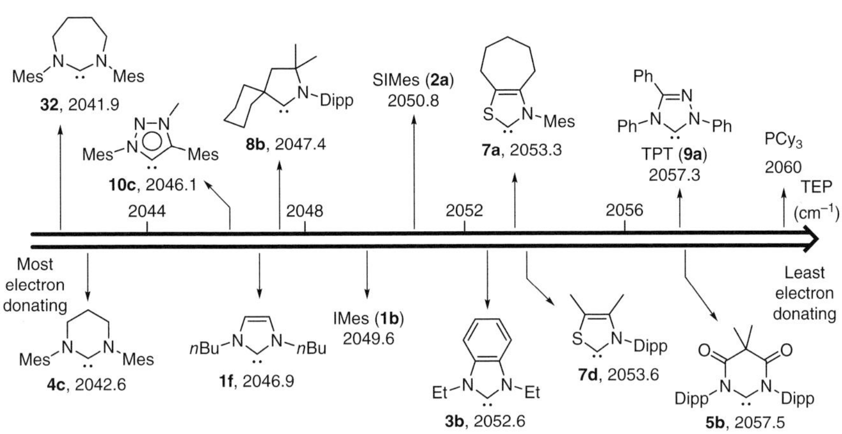

Figure 1.17 Scale showing the range of TEP values measured with different NHCs.

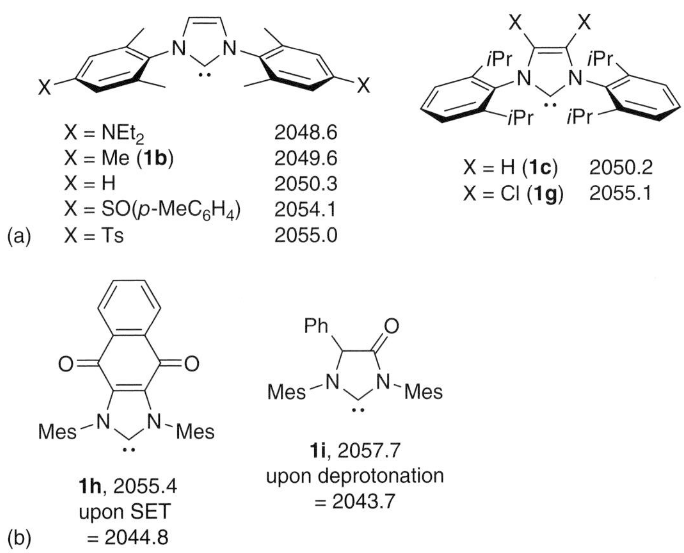

Figure 1.18 (a) Effect of *N*-aryl and backbone substituents on the TEP values of imidazolylidenes [62, 69]. (b) Examples of switchable NHCs (all values in cm^{-1}).

$$\text{Ir to Ni}: \text{TEP} = 0.847\, \nu_{\text{CO}}^{\text{av/Ir}} + 336\,\text{cm}^{-1} \tag{1.1}$$

$$\text{Rh to Ir}: \nu_{\text{CO}}^{\text{av/Ir}} = 0.8695\, \nu_{\text{CO}}^{\text{av/Rh}} + 250.7\,\text{cm}^{-1} \tag{1.2}$$

A scale showing TEP values obtained for various NHCs is displayed in Figure 1.17. The first conclusion one can make from these values is that NHCs are significantly more electron donating than phosphines. Typical imidazolylidenes and imidazolinylidene commonly used as ligands in organometallic chemistry have TEP values in the range of ≈2045–2055 cm^{-1}, lower than even Lewis-basic phosphines such as PCy_3 (2060 cm^{-1}) and substantially lower than common triaryl derivatives (e.g. PPh_3 = 2070 cm^{-1}). Comparing the TEP values obtained for different NHCs, similar trends to those seen with the pK_a values of azolium salts can be observed. The greatest factor affecting the TEP values is the class of heterocycle with the identity, number, and location of the ring heteroatoms having a large influence. For the five-membered NHC series, moving from one electronegative ring nitrogen in CAAC **8a** (TEP = 2047.4 cm^{-1}) to two nitrogens in the imidazolylidene IMes (**1b**, 2049.6 cm^{-1}) and three nitrogens in TPT (**9a**, TEP = 2057.3 cm^{-1}), a significant increase in the TEP values reflecting a supposed decreasing electron-donating ability is observed. Thiazolylidene NHCs such as **7a** (2053.3 cm^{-1}) and **7d** (2053.6 cm^{-1}) also have higher TEP values than the corresponding imidazolylidenes. Abnormal (mesoionic) NHCs are particularly electron rich and lead to the lowest TEP values. The NHC ring size has been found to greatly affect the TEP value. In the series of saturated *N*-mesityl-substituted five-, six-, and seven-membered NHCs **2a**, **4c**, and **32**, the increasing ring size resulted in a decrease in the TEP value (**2a** = 2050.8 cm^{-1}, **4c** = 2042.6 cm^{-1}, **32** = 2041.9 cm^{-1}). The NHC ring size inherently affects the ground-state orbital hybridization and consequently the stabilization of the carbene center, which in turn influences its donating ability. An interesting case concerns the comparison of data for imidazolylidenes and their saturated imidazolinylidene analogs. In general, the TEP values obtained for imidazolylidenes are slightly lower than those of the corresponding identically substituted imidazolinylidenes, suggesting that the unsaturated NHCs are stronger donors. This interpretation, however, must be treated with caution as imidazolinylidenes are considered to be more π-accepting than imidazolylidenes (*vide infra*) and may better compete for the π-electron density at the metal center, reducing Ni to C≡O back-bonding. A more electron-rich nickel center bearing a strongly σ-donating NHC may also lead to a decrease in σ-electron donation from the slightly antibonding C≡O lone pair, which can result in an increase in the C≡O stretching frequency. Density functional theory (DFT) calculations by Nolan and coworkers on $[Ni(CO)_3(NHC)]$ complexes with IMes (**1b**) and SIMes (**2a**) show that the binding energy of the C≡O ligand to the nickel center is greater with the imidazolylidene, suggesting increased Ni to C≡O back-bonding [70]. These results illustrate the complexity of bonding in metal carbonyl complexes and highlight how care should be taken when inferring the donating ability of NHCs from TEP values.

The electron richness of the nitrogen substituents and ring backbone also influence the TEP values. In general, NHCs bearing alkyl *N*-substituents have lower C≡O stretching frequencies than *N*-aryl derivatives (**1f**, = 2046.9 cm^{-1}, cf. IMes, **1b** = 2049.6 cm^{-1}). In a 2007 study by Plenio and coworkers, a series of imidazolylidenes bearing a range of electronically diverse functional groups at the *para*-position of the aryl nitrogen substituents were synthesized, and their TEP values obtained from the iridium(I) complexes [IrCl$(CO)_2$(NHC)] were systematically analyzed [69]. A clear correlation was observed with strongly electron-withdrawing groups such as *para*-toluenesulfonyl, leading to high TEP values of close to 2060 cm^{-1} and more electron-donating substituents leading to low TEP values of 2052.2 cm^{-1}. Backbone substitution also plays a role as exemplified by the higher TEP value obtained for the 4,5-dichloro-substituted NHC (**1g**, 2055.1 cm^{-1}) compared to the simple hydrogen-bearing derivative IPr (**1c**, 2050.2 cm^{-1}). Comparatively, high TEP values are also observed with DACs (**5**) and other NHCs where significant electron density is delocalized into carbonyl groups present in the ring structure. An interesting series of NHCs are derivatives that undergo changes in their electronic properties under different conditions. An example of such a switchable NHC is the redox-active imidazolylidene **1h** developed by Bielawski and coworkers which exhibits a TEP value of 2055.4 cm^{-1} in its diketone form [71]. Single-electron reduction of this species to the corresponding radical anion led to a decrease in the TEP to 2044.8 cm^{-1}, reflecting the predicted increased electron-donating ability of the reduced NHC. Similarly, the electronic properties of the carbonyl-containing imidazolylidene **1i** reported by Glorius and coworkers was found to be highly dependent on the pH [72]. Deprotonation of this species to afford the corresponding enolate led to a decrease in the TEP value of 14 cm^{-1}.

1.4.3 NMR Measurements

A number of techniques for investigating NHCs have been developed using NMR spectroscopy. The most straightforward of these involves simple measurement of the ^{13}C NMR spectrum of the free NHC [17a, 73, 74]. The carbene carbon atom is highly deshielded, and the ^{13}C NMR signals are consequently highly diagnostic, appearing in an area of the spectrum where few other peaks are normally found (between $\delta \approx 200$–330 ppm). The significant increase in shielding observed upon complexation to a metal or other center means that ^{13}C NMR is also a useful technique for assessing adduct formation. The class of heterocycle has a great effect on the observed chemical shifts for the carbene carbon (Figure 1.19). Imidazolylidenes as well as triazolylidenes typically exhibit resonances at the lower end of the range ($\delta \approx 210$–220 ppm), whereas the carbene carbon peaks of imidazolinylidenes are more downfield-shifted, appearing at

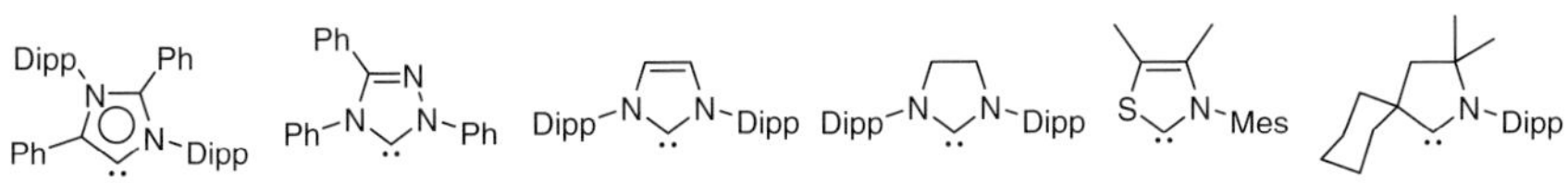

10a δ_C = 201.9 (in THF-d_8)

TPT (**9a**) δ_C = 214.6 (in C_6D_6)

IPr (**1c**) δ_C = 220.6 (in C_6D_6)

SIPr (**2b**) δ_C = 244.0 (in C_6D_6)

7d δ_C = 251.6 (in THF-d_8)

8b δ_C = 309.4 (in THF-d_8)

Figure 1.19 ^{13}C NMR shifts of the carbene carbon in selected free NHCs.

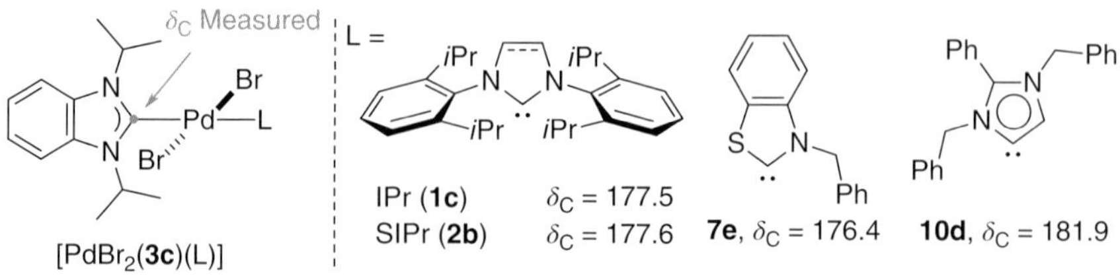

Figure 1.20 ^{13}C NMR shifts of the carbene carbon of benzimidazolylidene **3c** in palladium(II) complexes $[PdBr_2(\mathbf{3c})(L)]$ measured by Huynh and coworkers.

$\delta \approx 235$–260 ppm. The corresponding resonances in thiazolylidenes are similarly found at more downfield chemical shifts whereas the carbene carbon atom in CAACs is exceptionally deshielded, appearing at δ values greater than 300 ppm. Abnormal carbenes such as **10a**, on the other hand, exhibit resonances at the opposite end of the spectrum at around 200 ppm.

A systematic method for comparing the electronic properties both of different NHCs and of NHCs and related ligands including phosphines, amines, and isocyanides was developed by Huynh et al. [75]. Palladium(II) complexes of the form $[PdBr_2(\mathbf{3c})(L)]$ were synthesized and the ^{13}C NMR shift corresponding to the carbene carbon of the benzimidazolylidene spectator ligand **3c** was used as a gauge for assessing the electron-donating ability of the ligand of interest situated trans to it (Figure 1.20). The more electron donating the ligand, the better it competes with the benzimidazolylidene **3c**, leading a more "free carbene"-like, downfield-shifted ^{13}C NMR signal. As expected, NHCs were found to be the most electron-donating class of ligand tested with abnormal carbenes **10**, leading to the most downfield-shifted resonances and triazolylidenes **9** being the least strongly binding ligands. This approach has more recently been conducted using analogous linear gold(I) complexes and a good correlation between the data obtained using either metal was obtained [76].

In 2013, Bertrand and coworkers conducted a systematic study of NHCs using ^{31}P NMR spectroscopy of carbene–phosphine adducts prepared from $PhPCl_2$ [46a, 77]. A related method using ^{77}Se NMR on analogous NHC–selenium adducts was also developed by the Ganter group [46b, 78]. Both of these species can be considered as existing between the two extreme resonance forms **A** and **B** shown in Figure 1.21. In form **A**, the carbene–metal interaction is exclusively of σ character with no back donation of π-electron density. In form **B**, on the other hand, a formal double bond is present. For the phosphinidene adducts, the ^{31}P NMR chemical shifts observed between these two extremes span some 130 ppm ($\delta_P \approx -60$ to 70 ppm), whereas an even larger range of δ_{Se} values spanning nearly 800 ppm is observed for different NHC–selenium adducts. As such, these methods provide a sensitive probe for evaluating the π-accepting properties of NHCs relatively independently of their σ-donating abilities. Further studies by Nolan and coworkers in 2015 supported this interpretation by correlating experimentally determined δ_{Se} and δ_P values with calculated energetic parameters associated with π-back-bonding into the empty carbene p-orbital [79]. The π-accepting ability of various NHCs as determined from the δ_P or δ_{Se} chemical shift values in phosphorus or selenium adducts is shown in Figure 1.21. As

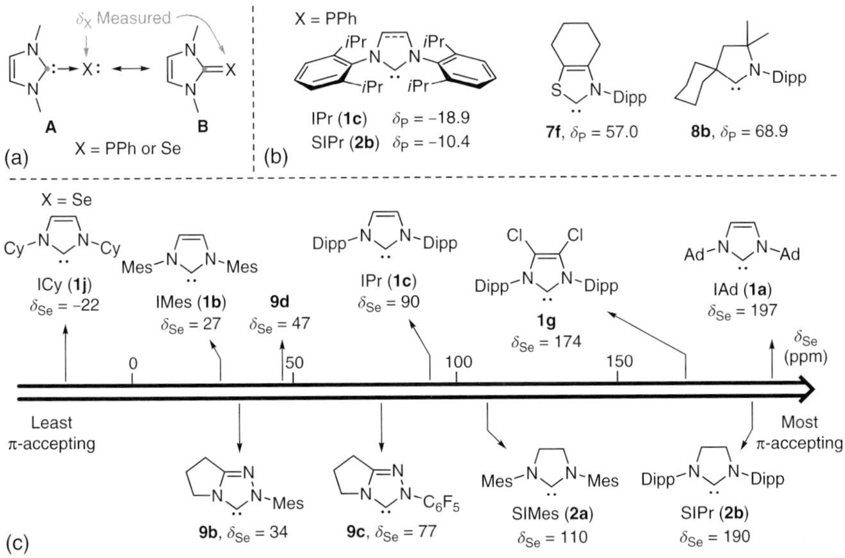

Figure 1.21 (a) Extreme resonance forms **A** and **B** of NHC–phosphorus and NHC–selenium adducts. (b) ^{31}P NMR shifts of NHC–phosphorus adducts for selected NHCs (measured in C_6D_6). (c) Scale showing ^{77}Se NMR shifts for selected NHC–selenium adducts (measured in $CDCl_3$).

expected, CAACs, which have only one π-donating ring nitrogen adjacent to the carbene center, were found to be the most π-accepting NHC class as signified by the highly downfield-shifted δ_P resonances of their phosphine adducts. Among NHC classes measured as the selenium adducts, DACs that feature carbonyl groups that compete with the carbene p-orbital for nitrogen π-electron density were accordingly shown to be highly π-accepting. Other common classes of NHCs, however, may exhibit a wide range of π-accepting abilities depending on the substitution pattern. For imidazolylidene and imidazolinylidene NHCs bearing identical nitrogen and backbone substituents, the saturated derivatives were found to more π-accepting (IMes–selenium adduct δ_{Se} in $CDCl_3 = 27$ ppm cf., SIMes–selenium adduct = 110 ppm). Significant variation in the δ_P or δ_{Se} chemical shift values, however, can be observed even with minor differences in the nitrogen substituents. For example, in the triazolylidene series, a δ_{Se} value of 34 ppm was recorded in $CDCl_3$ for the selenium adduct of TMes (**9b**), whereas the corresponding pentafluorophenyl-substituted derivative **9c** was found to be much more π-accepting with a δ_{Se} of 77 ppm recorded under the same conditions. As a specific metric for the π-accepting character of NHCs, this approach provides a deeper insight both into the individual factors affecting the binding of different NHC ligands in complexes and, by extension, into their relative reactivity as organocatalysts. Taken in combination with TEP or other parameters that provide information on the overall donor ability, the relative contribution of σ-donation can also be inferred. Furthermore, recent studies by Ganter and coworkers also suggest that $^1J_{Se-C}$ coupling constants determined by NMR spectroscopy on NHC–selenium adducts can provide direct information on the σ-donating ability of NHCs [78].

1.4.4 Nucleophilicity and Lewis Basicity

The first step in any organocatalytic reaction involves the addition of the NHC to a carbon electrophile. The nucleophilicity of the carbene carbon is therefore directly related to the NHC's reactivity as an organocatalyst. In 2011, Mayr and coworkers measured the nucleophilicity of three NHCs (IMes **1b**, SIMes **2a**, and TPT **9a**) by studying the kinetics of adduct formation with a selection of benzhydrylium electrophiles [80]. Triazolylidene was found to be significantly less nucleophilic than the imidazole-based NHCs, whereas the saturated imidazolinylidene SIMes reacted at a slightly faster rate than the unsaturated analog IMes. DFT calculations also revealed the methyl cation affinities (MCAs) of each NHC as a means of assessing their Lewis basicities (Figure 1.22). Using this arguably more relevant metric for organocatalysis, which is related to the thermodynamic equilibrium constant of adduct formation, a similar trend was observed (MCA for IMes, **1b** = 767.2 kJ mol^{-1}, SIMes, **2a**, 768.9 kJ mol^{-1}, TPT, **9a** = 712.2 kJ mol^{-1}). In comparison to PPh_3 and 4-(dimethylamino)pyridine (DMAP), all three NHCs were significantly more Lewis basic (MCA for PPh_3 = 618.4 kJ mol^{-1}, DMAP = 581.2 kJ mol^{-1}).

1.4.5 Electrochemical Methods

The electronic properties of metal complexes can also be investigated using electrochemical methods. In comparison to the TEP or Huynh's NMR approach, which measures IR or NMR spectra of spectator CO or benzimidazolylidene ligands, respectively, measuring the redox potential of the overall complex provides arguably more direct information on the electron density at the metal center and therefore on the donating ability of its constituent ligands. Redox potentials are typically measured for the Ru^{II}/Ru^{III} couple in ruthenium complexes bearing one or more 2,2′-bipyridine (bpy) ligands in addition to the ligand of interest, and the data are deconvoluted into Lever electronic parameter (LEP) values specific to the investigated ligand [81]. To date, however, limited LEP values have been obtained for NHCs, and electrochemical data are somewhat limited to studies on iridium and rhodium complexes of the form [MCl(COD)(L)] (COD = 1,5-cyclooctadiene). A survey of imidazolylidene and imidazolinylidene ligands bearing electronically diverse *para*-groups on the *N*-aryl substituents was conducted by Plenio and coworkers (Figure 1.23) [69, 82]. Notable differences in the measured $E_{1/2}$ values were recorded even for these relatively remote electronic variations with more strongly electron-donating groups leading to a more facile oxidation of the metal.

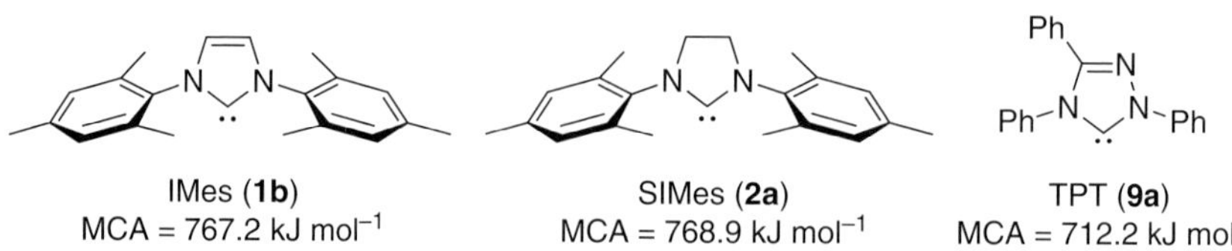

Figure 1.22 Calculations of the MCA of IMes (**1b**), SIMes (**2a**), and TPT (**9a**) by Mayr and coworkers.

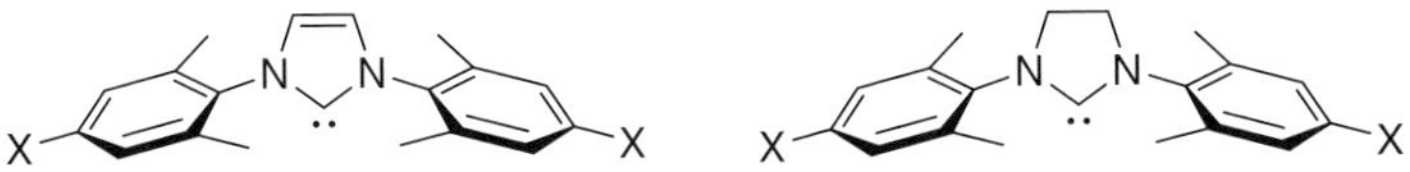

	$E_{1/2}$ (V)		$E_{1/2}$ (V)
X = NEt_2	0.648	X = NEt_2	0.591
X = Me (**1b**)	0.765	X = Me (**2a**)	0.735
X = H	0.786	X = H	0.759
X = SO(*p*-MeC_6H_4)	0.870	X = SO(*p*-MeC_6H_4)	0.846
X = Ts	0.920	X = Ts	0.910

Figure 1.23 Electrochemical measurements on iridium complexes of the form [IrCl(COD)(L)] with different imidazolylidene and imidazolinylidene NHCs (measured in 0.1 M NBu_4PF_6, referenced to $FcMe_8$, scan rate: 100 mV s^{-1}).

Saturated imidazolinylidene NHCs were again found to be more electron donating than their unsaturated imidazolylidene analogs using this method. A good correlation was observed between the electrochemical data and the corresponding TEP values while a comparison with the analogous iridium complex [IrCl(COD)(PCy_3)] again confirmed the strong donating ability of NHCs relative to phosphines.

1.4.6 Computational Methods

Alongside the experimental approaches, significant efforts have been devoted to examining the electronic properties of NHCs using computational methods. As discussed above, calculated TEP values have been obtained for many derivatives with good correlations with experimentally determined values being achieved [68]. A number of research groups have instead focused on other metrics calculated using DFT models (Figure 1.24). As far back as 2001, Clot and coworkers obtained computed ligand electronic parameters (CEPs) from DFT calculations on a small set of nickel carbonyl complexes bearing simple NHCs as part of a larger study of donor ligands [83]. The CEP values were obtained as wavenumbers in analogy to TEP values and could be correlated with both TEP and LEP data. As with the experimental approaches to quantifying ligand electronic properties, this method again demonstrated the extraordinarily

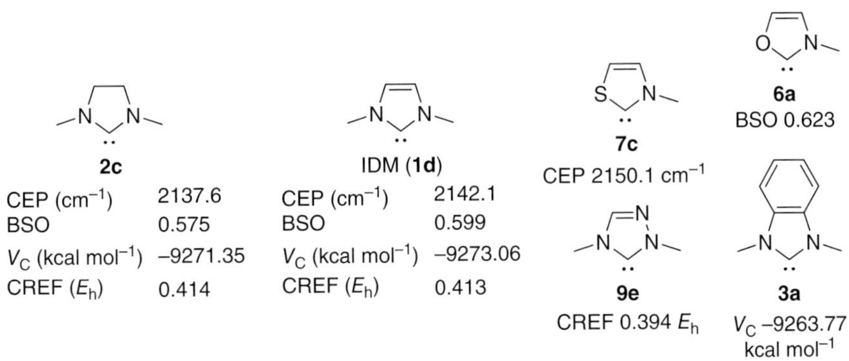

Figure 1.24 Electronic parameters of model NHCs calculated using DFT.

strong donating ability of NHCs compared to phosphines and other ligands. Among the tested NHC substrates, the model imidazolinylidene derivative **2c** had a lower CEP value (2137.6 cm^{-1}) than the corresponding imidazolylidene IDM (**1d**, CEP = 2142.1 cm^{-1}), whereas *N*-methylthiazolylidene (**7c**) was even less donating (CEP = 2150 cm^{-1}). More recently, Cremer and coworkers introduced the metal–ligand electronic parameter (MLEP) as a metric for quantifying the binding strength of ligands to metals [84]. By focusing specifically on the metal–ligand interaction in nickel carbonyl complexes, the MLEP was developed as a more insightful parameter than the TEP, which provides only indirect information on the ligand binding by measuring the IR stretching frequency associated with the spectator carbonyl ligands. The bond strength order (BSO) values thus obtained for a small set of simple NHCs were found to be higher for the aromatic NHCs IDM (**1d**), thiazolylidene **7c**, and, especially, oxazolylidene **6a**, whereas the lowest BSO was calculated for the imidazolinylidene **2c**. Interestingly, calculations of the C≡O bond strength in these complexes as a direct comparison with TEP data revealed a different trend to that expected from Tolman's model. Rather than observing an inverse relationship where increased NHC binding to the metal leads to increased Ni to C≡O back-bonding, stronger C≡O bonds were calculated in complexes containing strong NHC—Ni bonds. These results again emphasize how care should be taken in correlating C≡O stretching frequencies with the ligand properties for NHCs. DFT studies by the groups of Ciancaleoni and Belpassi have shown that the steric influence of bulky NHCs affects the orbital overlap in nickel carbonyl complexes and further complicates the interpretation of TEP values [85].

Rather than calculating NHC-containing metal complexes, DFT studies have also been conducted on free NHCs themselves. HOMO energies, proton affinities [86], and nucleophilicities [87] have been computed while, in 2010, Suresh and coworker calculated molecular electrostatic potentials (MESP, V_C) of a range of imidazolylidenes (**1**), imidazolinylidenes (**2**), and benzimidazolylidenes (**3**) [88]. Good correlations between the V_C values obtained and the corresponding TEP values were observed, while applications of the method for understanding the roles of NHCs in transition metal catalysis have been performed [89]. Recently, a new parameter termed the carbene relative energy of formation (CREF) was introduced by Ramsden and Oziminski, which focuses on DFT-determined energies associated with NHC generation for azolium salts [90]. This metric provides information specifically on the σ-donating properties of the carbene center independently of any π-considerations or steric effects. Comparing the *N*-methyl-substituted NHCs **1d**, **2c**, and **9e**, similar CREF values of 0.413 and 0.414 E_h were obtained for the imidazolylidene and imidazolinylidene derivatives, while the triazolylidene was calculated to be significantly less σ-donating (CREF = 0.394 E_h).

1.5 Quantifying the Steric Properties of NHCs

In addition to knowing the electron richness and donating ability of NHCs, when selecting a potential ligand or organocatalyst, it is also important to consider their steric properties. In the context of organocatalysis, the steric accessibility of the carbene center inherently influences the rate of addition of the NHC to the electrophilic starting material (e.g. an aldehyde), the elimination of the catalyst at the end of the cycle, and all other intermediate steps. The steric influence of chiral NHCs on the reaction intermediates also determines the sense and level of enantioselectivity afforded by the process. The relative size of phosphine ligands can be conveniently described by the Tolman cone angle [65], which is measured by envisaging a cone with an apex at a metal center and an angle defined by the outermost lying portion of the ligand. The uniform steric influence assumed by this metric, however, does not accurately reflect the unsymmetrical nature of NHCs. For this reason, alternative models have been developed principally by the groups of Nolan and Cavallo [63], which allow for a more authentic evaluation of the steric properties of NHCs, while still enabling comparison with phosphines or other related classes of compounds.

1.5.1 Percentage Buried Volume ($\%V_{bur}$)

First introduced in 2003, the percentage buried volume ($\%V_{bur}$) of a ligand is determined by visualizing a complex as a sphere of radius (r) 3.5 Å with the metal at the center and metal–carbene bond length (d) defined as 2.00 or 2.28 Å (Figure 1.25) [47]. The percentage volume of this sphere occupied or "buried" by the ligand is then calculated using standardized atomic radii of the constituent atoms (excluding hydrogen). Suitable data sources for the calculations are X-ray crystal structures either of a metal complex or the uncoordinated ligand (e.g. the free carbene or azolium salt for NHCs) or DFT determined atomic coordinates. There now exists a free Internet tool (SambVca) [91] that calculates $\%V_{bur}$ values from uploaded input files, and because of the large number of crystal structures available for NHC-containing metal complexes, $\%V_{bur}$ values can nowadays be found in the literature for many NHCs. The model can also be readily applied for other classes of ligands such as phosphines and may be used for mono- or poly-coordinating species [92]. It is important, however, only to compare $\%V_{bur}$ values obtained from the same source. The coordination geometry of a complex and the number and type of ancillary ligands can greatly affect the ligand coordination, and unusual or unrepresentative conformations may be adopted to minimize steric clashing. As a result, linear gold(I) complexes of the form [AuCl(L)] have proved useful as standard sources for $\%V_{bur}$ calculations as their uncongested coordination sphere allows the ligands to avoid steric clashing with other ligands. Other inconsistencies in $\%V_{bur}$ values arise from poor-quality X-ray or DFT input data, leading to errors in the atomic coordinates.

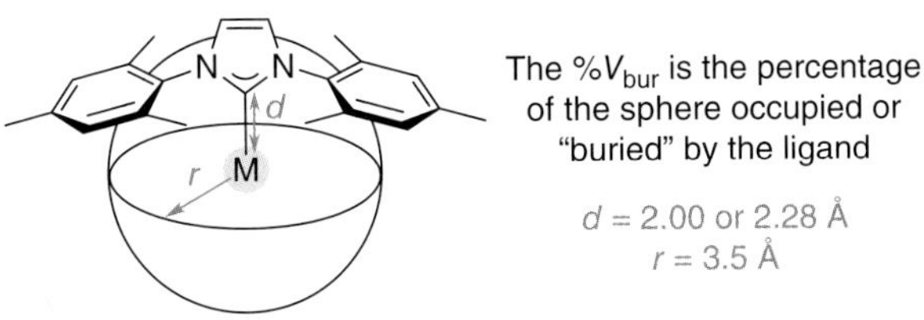

Figure 1.25 Model used to determine the percentage-buried volume ($\%V_{bur}$) of NHCs.

$\%V_{bur}$ values obtained for common classes of NHCs typically range from around 25% up to over 50%. The large number of $\%V_{bur}$ values available has allowed for a systematic evaluation of the effects of different NHC classes and nitrogen and backbone substituents on the steric environment experienced by metals or, in the context of organocatalysis, organic substrates bound to the carbene carbon. A scale showing $\%V_{bur}$ values collated from the literature by Nolan and coworkers [63] from gold(I) chloride complexes (d = 2.00 Å) of a range of imidazolylidene NHCs featuring different nitrogen and backbone substituents is provided in Figure 1.26. As could be expected from their close proximity to the carbene carbon, the nitrogen substituents have a large steric influence. Increasing the size of the *ortho*-substituents in 2,6-disubstituted *N*-aryl imidazolylidenes from methyl in IMes (**1b**) to isopropyl in IPr (**1c**) and diphenylmethyl in IPr* (**1k**) leads to a large increase in $\%V_{bur}$ from 36.5% to 45.4% and 50.4%. $\%V_{bur}$ values calculated for *N*-alkyl-substituted NHCs are in general smaller than for *N*-aryl-substituted derivatives with IDM (**1d**), ICy (**1j**), and IAd (**1a**), having values of 26.3%, 27.5%, and 39.8%, respectively. In the case of ICy (**1j**) and other NHCs bearing comparatively flexible substituents, the $\%V_{bur}$ values are strongly affected by the ability of these groups to adopt conformations that reduce steric crowding. As $\%V_{bur}$ values are determined from single snapshots in a specific complex, special care should be taken when drawing conclusions about the steric influence of these NHCs in different contexts. Indeed, the concept of "flexible steric bulk" brought about through conformation changes in the ligand can be especially significant in catalytic systems as different stages in the mechanistic cycle may favor different steric environments [93]. In contrast to changes in the nitrogen substituents, the backbone substitution pattern has a relatively minor influence on the $\%V_{bur}$ values for most imidazolylidenes. For example, the 2,6-diisopropylphenyl N-substituted species bearing hydrogen (IPr, **1c**), chlorine (**1g**), or methyl substituents at

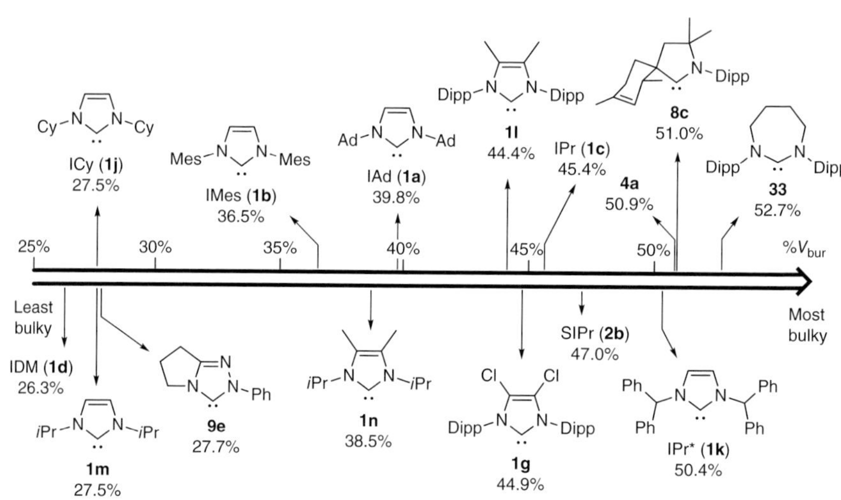

Figure 1.26 Scale showing the variation in $\%V_{bur}$ values for imidazolylidenes with different nitrogen and backbone substituents.

the 4- and 5-positions all have $\%V_{bur}$ values within 1% of each other, with the dimethyl-substituted derivative **1l** actually showing the smallest value of 44.4%. Indeed, backbone substitution is often conducted in order to vary the electronic properties of the NHC without changing the steric properties. Exceptions to this trend, however, can occur when the backbone substituents affect the flexibility or rotational freedom of the *N*-substituents. For example, the $\%V_{bur}$ values observed for *N*-isopropyl-substituted imidazolylidenes increase significantly from 27.5% (for **1m**) to 38.5% (for **1n**) upon backbone substitution with methyl groups. Restricted rotation of the *N*-substituents that leads to an increase in steric bulk closer to the carbene carbon likely explains this observation.

The class of NHC heterocycle also has a major influence on the steric properties at the carbene carbon. Depending on the identity and position of the ring heteroatoms, the carbene carbon experiences inherently different steric environments for different heterocycles. Thiazolylidene or oxazolylidene NHCs, for example, possess only one substituent adjacent to the carbene carbon by virtue of the bivalency of their sulfur or oxygen ring heteroatoms. By contrast, CAACs possess an adjacent sp^3-hybridized, quaternary carbon, which exerts a large steric effect and results in high $\%V_{bur}$ values (e.g. 51.0% for **8c**). The size of the heterocycle also plays an important role as changes to the ring geometry effectively push the nitrogen or other proximal substituents either closer to or further away from the carbene carbon. Increasing the ring size from 5 to 6 and 7 in the series of saturated 2,6-diisopropylphenyl N-substituted NHCs SIPr (**2b**), **4a**, and **33**, for example, results in an increase in $\%V_{bur}$ from 47.0% to 50.9% and 52.7%, reflecting the closer proximity of the bulky *N*-aryl substituents to the carbene carbon in the larger rings. A notable increase in the $\%V_{bur}$ values is also observed upon saturation of the heterocycle with imidazolinylidene derivatives being more sterically demanding than the corresponding imidazolylidene analogs. This effect has been rationalized as resulting from an increase in the ring flexibility, which allows the nitrogen substituents to approach closer to the metal center.

1.5.2 Steric Maps

Although the percentage-buried volume gives a useful insight into the overall steric influence exerted by NHC ligands and organocatalysts, as a single number parameter, it does not provide any information as to the spatial distribution of the steric bulk around the metal or organic substrate bound to the carbene center. As previously discussed, the steric influence of NHCs is highly anisotropic and different environments will be experienced by other moieties in complexes or adducts depending on their position in relation to the NHC. This feature is of particular relevance for enantioselective organocatalysis as the relative position of steric bulk in the chiral NHC-containing intermediates is fundamental for determining the sense and level of enantioselectivity afforded. One strategy to account for the anisotropy of NHCs in percentage-buried volume calculations is to split the NHC–metal sphere into four and determine $\%V_{bur}$ values for each quadrant. A more complete model, however, was introduced by Cavallo and coworkers [94], which instead represents the spatial distribution as a 2D diagram called a steric map. Visualized along the metal–carbene axis with the metal at the center, the steric bulk present throughout the sphere can be easily interpreted in the x, y plane with contours being used to show the extent of steric bulk along the z-axis. The SambVca online tool can now be used to generate steric maps from cif or XYZ files alongside $\%V_{bur}$ values [95].

1.6 Concluding Remarks

Since the initial studies on NHC ligands in the 1960s and the first isolation of a free NHC by Arduengo and coworkers in 1991, these compounds have found multiple uses across many areas of organic and inorganic chemistry. Nature itself has chosen an NHC in the form of vitamin B_1 to perform organocatalytic reactions *in vivo*, and over the last decades, NHCs have opened up new classes of often enantioselective organic transformations. As a background to the more specific chapters that follow, this introductory chapter provides a concise overview of the general properties of NHCs and briefly summarizes some of their major applications as ligands for metals and nonmetals. We hope this will serve as a general foundation for newcomers to NHCs and place the developments in organocatalysis described in the subsequent chapters into a wider context. Through a survey of the different methods used to quantify and their electronic and steric profiles, this overview should also allow for informed analogies between the organocatalytic reactivity of different NHCs and their fundamental properties to be made.

References

1 Arduengo, A.J., Harlow, R.L., and Kline, M. (1991). *J. Am. Chem. Soc.* 113: 361–363.

2 (a) Hopkinson, M.N., Richter, C., Schedler, M., and Glorius, F. (2014). *Nature* 510: 485–496. (b) Dröge, T. and Glorius, F. (2010). *Angew. Chem. Int. Ed.* 49: 6940–6952. (c) Bourissou, D., Guerret, O., Gabbaï, F.P., and Bertrand, G. (2000). *Chem. Rev.* 100: 39–92. (d) de Frémont, P., Marion, N., and Nolan, S.P. (2009). *Coord. Chem. Rev.* 253: 862–892. (e) Herrmann, W.A. and Köcher, C. (1997). *Angew. Chem. Int. Ed. Engl.* 36: 2162–2187.

3 (a) Díez-González, S., Marion, N., and Nolan, S.P. (2009). *Chem. Rev.* 109: 3612–3676. (b) Hahn, F.E. and Jahnke, M.C. (2008). *Angew. Chem. Int. Ed.* 47: 3122–3172. (c) Bellemin-Laponnaz, S. and Dagorne, S. (2014). *Chem. Rev.* 114: 8747–8774. (d) Herrmann, W.A. (2002). *Angew. Chem. Int. Ed.* 41: 1290–1309. (e) Crabtree, R.H. (2005). *J. Organomet. Chem.* 690: 5451–5457. (f) Samojłowicz, C., Bieniek, M., and Grela, K. (2009). *Chem. Rev.* 109: 3708–3742. (g) Marion, N. and Nolan, S.P. (2008). *Chem. Soc. Rev.* 37: 1776–1782. (h) Glorius, F. (2007). *N-Heterocyclic Carbenes in Transition Metal Catalysis*. Springer. (i) Kantchev, E.A.B., O'Brien, C.J., and Organ, M.G. (2007). *Angew. Chem. Int. Ed.* 46: 2768–2813. (j) Fortman, G.C. and Nolan, S.P. (2011). *Chem. Soc. Rev.* 40: 5151–5169. (k) Würtz, S. and Glorius, F. (2008). *Acc. Chem. Res.* 41: 1523–1533. (l) Valente, C., Çalimsiz, S., Hoi, K.H. et al. (2012). *Angew. Chem. Int. Ed.* 51: 3314–3332. (m) Bézier, D., Sortais, J.-B., and Darcel, C. (2013). *Adv. Synth. Catal.* 355: 19–33. (n) Ritleng, V.,

Henrion, M., and Chetcuti, M.J. (2016). *ACS Catal.* 6: 890–906. (o) Henrion, M., Ritleng, V., and Chetcuti, M.J. (2015). *ACS Catal.* 5: 1283–1302. (p) Zhang, D. and Zi, G. (2015). *Chem. Soc. Rev.* 44: 1898–1921. (q) Nolan, S.P. (2006). *N-Heterocyclic Carbenes in Synthesis*. Wiley. (r) Wang, Z., Jiang, L., Mohamed, D.K.B. et al. (2015). *Coord. Chem. Rev.* 293–294: 292–326. (s) Arnold, P.L. and Casely, I.J. (2009). *Chem. Rev.* 109: 3599–3611. (t) Wang, F., Liu, L.-J., Wang, W. et al. (2012). *Coord. Chem. Rev.* 256: 804–853. (u) Vougioukalakis, G.C. and Grubbs, R.H. (2010). *Chem. Rev.* 110: 1746–1787. (v) Crabtree, R.H. (2013). *Coord. Chem. Rev.* 257: 755–766. (w) Lazreg, F., Nahra, F., and Cazin, C.S.J. (2015). *Coord. Chem. Rev.* 293–294: 48–79.

4 (a) Hindi, K.M., Panzner, M.J., Tessier, C.A. et al. (2009). *Chem. Rev.* 109: 3859–3884. (b) Liu, W. and Gust, R. (2016). *Coord. Chem. Rev.* 329: 191–213. (c) Zou, T., Lum, C.T., Lok, C.-N. et al. (2015). *Chem. Soc. Rev.* 44: 8786–8801.

5 Curran, D.P., Solovyev, A., Makhlouf Brahmi, M. et al. (2011). *Angew. Chem. Int. Ed.* 50: 10294–10317.

6 Martin, D., Soleilhavoup, M., and Bertrand, G. (2011). *Chem. Sci.* 2: 389–399.

7 (a) Enders, D., Niemeier, O., and Henseler, A. (2007). *Chem. Rev.* 107: 5606–5655. (b) Flanigan, D.M., Romanov-Michailidis, F., White, N.A., and Rovis, T. (2015). *Chem. Rev.* 115: 9307–9387.

8 Ukai, T., Tanaka, R., and Dokawa, T.A. (1943). *J. Pharm. Soc. Jpn.* 63: 296–300.

9 (a) Bugaut, X. and Glorius, F. (2012). *Chem. Soc. Rev.* 41: 3511–3522. (b) Nair, V., Menon, R.S., Biju, A.T. et al. (2011). *Chem. Soc. Rev.* 40: 5336–5346. (c) Yetra, S.R., Patra, A., and Biju, A.T. (2015). *Synthesis* 47: 1357–1378. (d) Menon, R.S., Biju, A.T., and Nair, V. (2016). *Beilstein J. Org. Chem.* 12: 444–461.

10 (a) Vora, H.U., Wheeler, P., and Rovis, T. (2012). *Adv. Synth. Catal.* 354: 1617–1639. (b) Knappke, C.E.I., Imami, A., and Jacobi von Wangelin, A. (2012). *ChemCatChem* 4: 937–941. (c) Ryan, S.J., Candish, L., and Lupton, D.W. (2013). *Chem. Soc. Rev.* 42: 4906–4917. (d) Reyes, E., Uria, U., Carrillo, L., and Vicario, J.L. (2017). *Synthesis* 49: 451–471. (e) Wang, M.H. and Scheidt, K.A. (2016). *Angew. Chem. Int. Ed.* 55: 14912–14922.

11 Arduengo, A.J., Goerlich, J.R., and Marshall, W.J. (1995). *J. Am. Chem. Soc.* 117: 11027–11028.

12 Moerdyk, J.P., Schilter, D., and Bielawski, C.W. (2016). *Acc. Chem. Res.* 49: 1458–1468.

13 Lavallo, V., Canac, Y., Präsang, C. et al. (2005). *Angew. Chem. Int. Ed.* 44: 5705–5709.

14 (a) Soleilhavoup, M. and Bertrand, G. (2015). *Acc. Chem. Res.* 48: 256–266. (b) Rao, B., Tang, H., Zeng, X. et al. (2015). *Angew. Chem. Int. Ed.* 54: 14915–14919.

15 Martin, C.D., Soleilhavoup, M., and Bertrand, G. (2013). *Chem. Sci.* 4: 3020–3030.

16 Enders, D., Breuer, K., Raabe, G. et al. (1995). *Angew. Chem. Int. Ed. Engl.* 34: 1021–1023.

17 (a) Aldeco-Perez, E., Rosenthal, A.J., Donnadieu, B. et al. (2009). *Science* 326: 556–559. (b) Ghadwal, R.S. (2016). *Dalton Trans.* 45: 16081–16095.
18 Schuster, O., Yang, L., Raubenheimer, H.G., and Albrecht, M. (2009). *Chem. Rev.* 109: 3445–3478.
19 Melaimi, M., Soleilhavoup, M., and Bertrand, G. (2010). *Angew. Chem. Int. Ed.* 49: 8810–8849.
20 Vignolle, J., Cattoën, X., and Bourissou, D. (2009). *Chem. Rev.* 109: 3333–3384.
21 (a) Igau, A., Grutzmacher, H., Baceiredo, A., and Bertrand, G. (1988). *J. Am. Chem. Soc.* 110: 6463–6466. (b) Kato, T., Gornitzka, H., Baceiredo, A. et al. (2000). *J. Am. Chem. Soc.* 122: 998–999. (c) Igau, A., Baceiredo, A., Trinquier, G., and Bertrand, G. (1989). *Angew. Chem. Int. Ed. Engl.* 28: 621–622.
22 Lavallo, V., Canac, Y., Donnadieu, B. et al. (2006). *Science* 312: 722–724.
23 Dyker, C.A., Lavallo, V., Donnadieu, B., and Bertrand, G. (2008). *Angew. Chem. Int. Ed.* 47: 3206–3209.
24 Carter, E.A. and Goddard, W.A. (1986). *J. Phys. Chem.* 90: 998–1001.
25 Runyon, J.W., Steinhof, O., Dias, H.V.R. et al. (2011). *Aust. J. Chem.* 64: 1165–1172.
26 Heinemann, C., Müller, T., Apeloig, Y., and Schwarz, H. (1996). *J. Am. Chem. Soc.* 118: 2023–2038.
27 Boehme, C. and Frenking, G. (1996). *J. Am. Chem. Soc.* 118: 2039–2046.
28 Arduengo, A.J., Dias, H.V.R., Harlow, R.L., and Kline, M. (1992). *J. Am. Chem. Soc.* 114: 5530–5534.
29 Wanzlick, H.W. and Schönherr, H.J. (1968). *Angew. Chem. Int. Ed. Engl.* 7: 141–142.
30 Öfele, K. (1968). *J. Organomet. Chem.* 12: P42–P43.
31 Cardin, D.J., Cetinkaya, B., and Lappert, M.F. (1972). *Chem. Rev.* 72: 545–574.
32 Sanford, M.S., Love, J.A., and Grubbs, R.H. (2001). *J. Am. Chem. Soc.* 123: 6543–6554.
33 Gaillard, S., Slawin, A.M.Z., and Nolan, S.P. (2010). *Chem. Commun.* 46: 2742–2744.
34 Ezugwu, C.I., Kabir, N.A., Yusubov, M., and Verpoort, F. (2016). *Coord. Chem. Rev.* 307 (Part 2): 188–210.
35 Mercs, L. and Albrecht, M. (2010). *Chem. Soc. Rev.* 39: 1903–1912.
36 (a) Zhukhovitskiy, A.V., MacLeod, M.J., and Johnson, J.A. (2015). *Chem. Rev.* 115: 11503–11532. (b) Wang, G., Rühling, A., Amirjalayer, S. et al. (2017). *Nat. Chem.* 9: 152–156.
37 Zhong, R., Lindhorst, A.C., Groche, F.J., and Kühn, F.E. (2017). *Chem. Rev.* 117: 1970–2058.
38 (a) Schaper, L.-A., Hock, S.J., Herrmann, W.A., and Kühn, F.E. (2013). *Angew. Chem. Int. Ed.* 52: 270–289. (b) Levin, E., Ivry, E., Diesendruck, C.E., and Lemcoff, N.G. (2015). *Chem. Rev.* 115: 4607–4692.
39 Wang, Y., Xie, Y., Wei, P. et al. (2008). *J. Am. Chem. Soc.* 130: 14970–14971.
40 Wang, Y., Xie, Y., Wei, P. et al. (2008). *Science* 321: 1069–1071.
41 Wang, Y., Chen, M., Xie, Y. et al. (2015). *Nat. Chem.* 7: 509–513.
42 Mahoney, J.K., Martin, D., Moore, C.E. et al. (2013). *J. Am. Chem. Soc.* 135: 18766–18769.

43 Ueng, S.-H., Makhlouf Brahmi, M., Derat, É. et al. (2008). *J. Am. Chem. Soc.* 130: 10082–10083.
44 Moerdyk, J.P. and Bielawski, C.W. (2012). *Nat. Chem.* 4: 275–280.
45 (a) Díez-González, S. and Nolan, S.P. (2007). *Coord. Chem. Rev.* 251: 874–883. (b) Jacobsen, H., Correa, A., Poater, A. et al. (2009). *Coord. Chem. Rev.* 253: 687–703.
46 (a) Back, O., Henry-Ellinger, M., Martin, C.D. et al. (2013). *Angew. Chem. Int. Ed.* 52: 2939–2943. (b) Liske, A., Verlinden, K., Buhl, H. et al. (2013). *Organometallics* 32: 5269–5272.
47 Hillier, A.C., Sommer, W.J., Yong, B.S. et al. (2003). *Organometallics* 22: 4322–4326.
48 Zhao, Y. and Truhlar, D.G. (2007). *Org. Lett.* 9: 1967–1970.
49 Kuhn, N. and Kratz, T. (1993). *Synthesis* 561–562.
50 Nyce, G.W., Csihony, S., Waymouth, R.M., and Hedrick, J.L. (2004). *Chem. Eur. J.* 10: 4073–4079.
51 Bantu, B., Pawar, G.M., Decker, U. et al. (2009). *Chem. Eur. J.* 15: 3103–3109.
52 Otto, M., Conejero, S., Canac, Y. et al. (2004). *J. Am. Chem. Soc.* 126: 1016–1017.
53 Fürstner, A., Seidel, G., Kremzow, D., and Lehmann, C.W. (2003). *Organometallics* 22: 907–909.
54 Benhamou, L., Chardon, E., Lavigne, G. et al. (2011). *Chem. Rev.* 111: 2705–2733.
55 Arduengo, A.J. III, Krafczyk, R., Schmutzler, R. et al. (1999). *Tetrahedron* 55: 14523–14534.
56 (a) Benhamou, L., Cesar, V., Gornitzka, H. et al. (2009). *Chem. Commun.* 4720–4722. (b) Kuhn, K.M. and Grubbs, R.H. (2008). *Org. Lett.* 10: 2075–2077. (c) César, V., Lugan, N., and Lavigne, G. (2008). *J. Am. Chem. Soc.* 130: 11286–11287. (d) Jazzar, R., Liang, H., Donnadieu, B., and Bertrand, G. (2006). *J. Organomet. Chem.* 691: 3201–3205. (e) Bertogg, A., Camponovo, F., and Togni, A. (2005). *Eur. J. Inorg. Chem.* 347–356.
57 Binobaid, A., Iglesias, M., Beetstra, D.J. et al. (2009). *Dalton Trans.* 7099–7112.
58 Sheehan, J.C. and Hara, T. (1974). *J. Org. Chem.* 39: 1196–1199.
59 Lebeuf, R., Hirano, K., and Glorius, F. (2008). *Org. Lett.* 10: 4243–4246.
60 (a) Vora, H.U., Lathrop, S.P., Reynolds, N.T. et al. (2010). *Org. Synth.* 87: 350. (b) Bode, J.W., Struble, J.T., and Lian, Y. (2010). *Org. Synth.* 87: 362.
61 (a) Enders, D., Breuer, K., Kallfass, U., and Balensiefer, T. (2003). *Synthesis* 1292–1295. (b) Enders, D., Breuer, K., Raabe, G. et al. (1995). *Angew. Chem. Int. Ed. Engl.* 34: 1021–1023.
62 Nelson, D.J. and Nolan, S.P. (2013). *Chem. Soc. Rev.* 42: 6723–6753.
63 Gomez-Suarez, A., Nelson, D.J., and Nolan, S.P. (2017). *Chem. Commun.* 53: 2650–2660.
64 (a) Higgins, E.M., Sherwood, J.A., Lindsay, A.G. et al. (2011). *Chem. Commun.* 47: 1559–1561. (b) Amyes, T.L., Diver, S.T., Richard, J.P. et al. (2004). *J. Am. Chem. Soc.* 126: 4366–4374. (c) Massey, R.S., Collett, C.J., Lindsay, A.G. et al. (2012). *J. Am. Chem. Soc.* 134: 20421–20432.
65 Tolman, C.A. (1977). *Chem. Rev.* 77: 313–348.

66 Kelly, R.A. III, Clavier, H., Giudice, S. et al. (2008). *Organometallics* 27: 202–210.
67 Wolf, S. and Plenio, H. (2009). *J. Organomet. Chem.* 694: 1487–1492.
68 (a) Gusev, D.G. (2009). *Organometallics* 28: 763–770. (b) Shi, Q., Thatcher, R.J., Slattery, J. et al. (2009). *Chem. Eur. J.* 15: 11346–11360. (c) Tonner, R. and Frenking, G. (2009). *Organometallics* 28: 3901–3905. (d) Gusev, D.G. (2009). *Organometallics* 28: 6458–6461.
69 Leuthäußer, S., Schwarz, D., and Plenio, H. (2007). *Chem. Eur. J.* 13: 7195–7203.
70 Dorta, R., Stevens, E.D., Scott, N.M. et al. (2005). *J. Am. Chem. Soc.* 127: 2485–2495.
71 Rosen, E.L., Varnado, C.D., Tennyson, A.G. et al. (2009). *Organometallics* 28: 6695–6706.
72 Biju, A.T., Hirano, K., Fröhlich, R., and Glorius, F. (2009). *Chem. Asian J.* 4: 1786–1789.
73 Tapu, D., Dixon, D.A., and Roe, C. (2009). *Chem. Rev.* 109: 3385–3407.
74 Arduengo, A.J., Goerlich, J.R., and Marshall, W.J. (1997). *Liebigs Ann.* 1997: 365–374.
75 Huynh, H.V., Han, Y., Jothibasu, R., and Yang, J.A. (2009). *Organometallics* 28: 5395–5404.
76 (a) Guo, S., Sivaram, H., Yuan, D., and Huynh, H.V. (2013). *Organometallics* 32: 3685–3696. (b) Meier, M., Tan, T.T.Y., Hahn, F.E., and Huynh, H.V. (2017). *Organometallics* 36: 275–284.
77 Rodrigues, R.R., Dorsey, C.L., Arceneaux, C.A., and Hudnall, T.W. (2014). *Chem. Commun.* 50: 162–164.
78 Verlinden, K., Buhl, H., Frank, W., and Ganter, C. (2015). *Eur. J. Inorg. Chem.* 2416–2425.
79 Vummaleti, S.V.C., Nelson, D.J., Poater, A. et al. (2015). *Chem. Sci.* 6: 1895–1904.
80 Maji, B., Breugst, M., and Mayr, H. (2011). *Angew. Chem. Int. Ed.* 50: 6915–6919.
81 Lever, A.B.P. (1990). *Inorg. Chem.* 29: 1271–1285.
82 Vorfalt, T., Leuthäußer, S., and Plenio, H. (2009). *Angew. Chem. Int. Ed.* 48: 5191–5194.
83 Perrin, L., Clot, E., Eisenstein, O. et al. (2001). *Inorg. Chem.* 40: 5806–5811.
84 Setiawan, D., Kalescky, R., Kraka, E., and Cremer, D. (2016). *Inorg. Chem.* 55: 2332–2344.
85 Ciancaleoni, G., Scafuri, N., Bistoni, G. et al. (2014). *Inorg. Chem.* 53: 9907–9916.
86 (a) Tukov, A.A., Normand, A.T., and Nechaev, M.S. (2009). *Dalton Trans.* 7015–7028. (b) Bernhammer, J.C., Frison, G., and Huynh, H.V. (2013). *Chem. Eur. J.* 19: 12892–12905.
87 Rezaee, N., Ahmadi, A., and Kassaee, M.Z. (2016). *RSC Adv.* 6: 13224–13233.
88 Mathew, J. and Suresh, C.H. (2010). *Inorg. Chem.* 49: 4665–4669.
89 Mathew, J. and Suresh, C.H. (2011). *Organometallics* 30: 3106–3112.
90 Ramsden, C.A. and Oziminski, W.P. (2016). *J. Org. Chem.* 81: 10295–10301.

91 Poater, A., Cosenza, B., Correa, A. et al. (2009). *Eur. J. Inorg. Chem.* 2009: 1759–1766.

92 Clavier, H. and Nolan, S.P. (2010). *Chem. Commun.* 46: 841–861.

93 Altenhoff, G., Goddard, R., Lehmann, C.W., and Glorius, F. (2003). *Angew. Chem. Int. Ed.* 42: 3690–3693.

94 Poater, A., Ragone, F., Mariz, R. et al. (2010). *Chem. Eur. J.* 16: 14348–14353.

95 Falivene, L., Credendino, R., Poater, A. et al. (2016). *Organometallics* 35: 2286–2293.

2

Benzoin Reaction

Steven M. Langdon, Karnjit Parmar, Myron M.D. Wilde, and Michel Gravel

University of Saskatchewan, Department of Chemistry, 110 Science Place, Saskatoon, SK S7N 5C9, Canada

2.1 Background and Mechanism

The benzoin reaction has its origins in the 1832 report of Wöhler and Liebig demonstrating that cyanide may be used to produce a then-undetermined compound, named benzoin [1]. With the advent of structural theory, this was shown to be a coupling of two molecules of benzaldehyde to form an α-hydroxy ketone. In 1903, Lapworth expanded on this work by proposing a mechanism for the transformation [2]. This proposal would later form the basis for mechanistic proposals using *N*-heterocyclic carbenes (NHCs).

A variety of catalytic and stoichiometric approaches to the benzoin reaction were subsequently developed [3]. Most notably, in 1943, Ukai et al. demonstrated that in the presence of a base, thiazolium salts could be used to catalyze the homo-benzoin reaction of aromatic and heteroaromatic aldehydes, effectively initiating research in the field of NHC organocatalysis [4]. Fifteen years after Ukai's report, Ronald Breslow combined Ukai's work, Lapworth's proposed mechanism, and a series of mechanistic experiments to propose the currently accepted mechanism for the NHC-catalyzed benzoin reaction (Scheme 2.1) [5]. Deprotonation of the salt (**I**) generates a carbene/ylide (**II**). This new species is nucleophilic and attacks a molecule of aldehyde to generate a tetrahedral intermediate (**III**) analogous to that proposed by Lapworth. Protonation of the alkoxide followed by a subsequent deprotonation forms a resonance-stabilized hydroxyenamine (**V**) with the resulting reversal of polarity (*umpolung*) [6] on the former carbonyl carbon. This species has since come to be referred to as the "Breslow intermediate" (BI) and attacks another aldehyde molecule to form an unstable intermediate (**VI**), which collapses to eject the benzoin product.

Several alternative mechanisms have been proposed for the NHC-catalyzed benzoin reaction, although these have not gained widespread acceptance [7]. Additionally, aspects of Breslow's mechanism have proven contentious since its proposal. Formation of the Breslow intermediate may proceed through a protonation–deprotonation sequence as initially proposed, involve a 1,2-proton shift [8], or an additional epoxide intermediate [9]. Protonated tetrahedral intermediates (as **IV**) have been observed and/or isolated under several reaction

N-Heterocyclic Carbenes in Organocatalysis, First Edition. Edited by Akkattu T. Biju.
© 2019 Wiley-VCH Verlag GmbH & Co. KGaA. Published 2019 by Wiley-VCH Verlag GmbH & Co. KGaA.

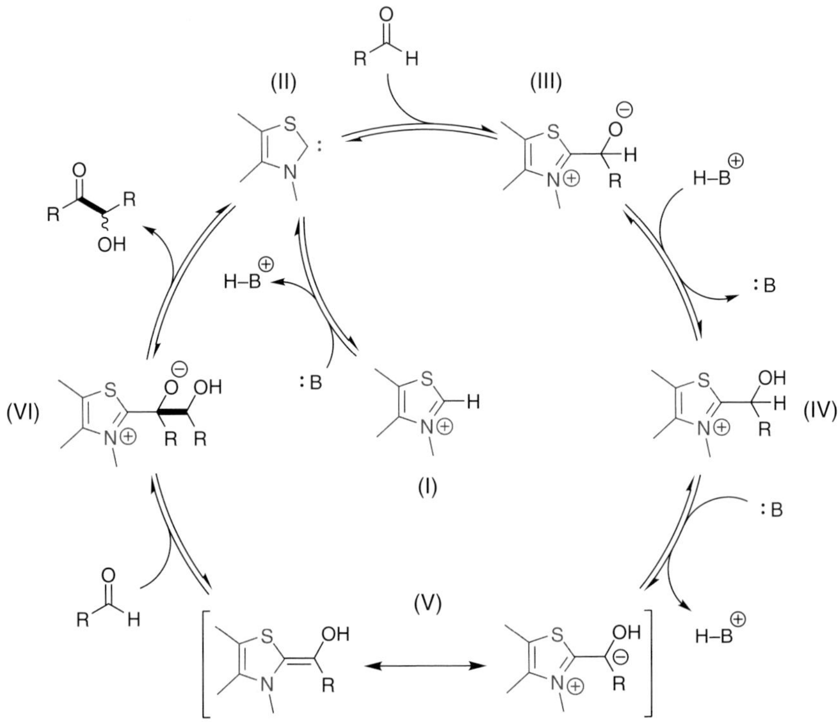

Scheme 2.1 Proposed mechanism of the NHC-catalyzed benzoin reaction.

conditions, supporting their presence in the reaction pathway [7a, 10]. Recent work has resulted in the isolation and characterization of a series of Breslow intermediate, or related, species (Figure 2.1), providing strong support for the proposed reaction pathway [11]. These compounds show a high sensitivity to the presence of oxygen, emphasizing the need for anaerobic conditions in many benzoin reactions. Evidence suggesting that the carbon–carbon bond formation step may proceed through sequential single-electron transfer reactions compounds this concern [7e].

A variety of experiments have confirmed that in many cases, each step of the NHC-catalyzed benzoin reaction is reversible, particularly with aromatic and heteroaromatic aldehydes [10, 12]. However, the use of aliphatic aldehydes results, in some instances, in reactions that are effectively irreversible after the

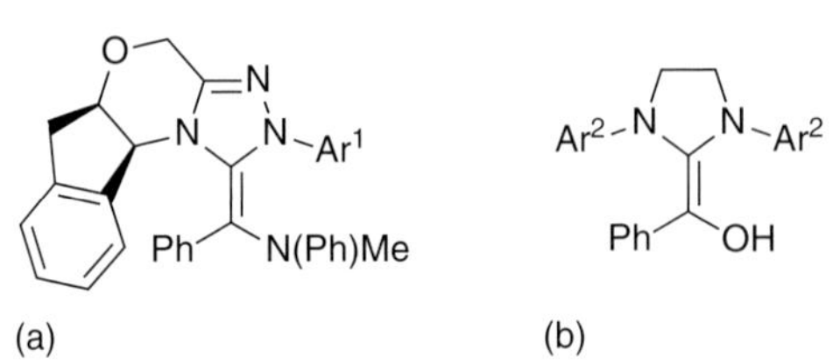

Figure 2.1 Examples of isolated (aza-)Breslow intermediates. (a) A triazolylidene-derived aza-Breslow intermediate, Ar^1 = 2,4,6-tribromophenyl; (b) a set of imidazolidinylidene-derived Breslow intermediates, Ar^2 = 2,6-bis(2-propyl)phenyl, 2,4,6-trimethylphenyl.

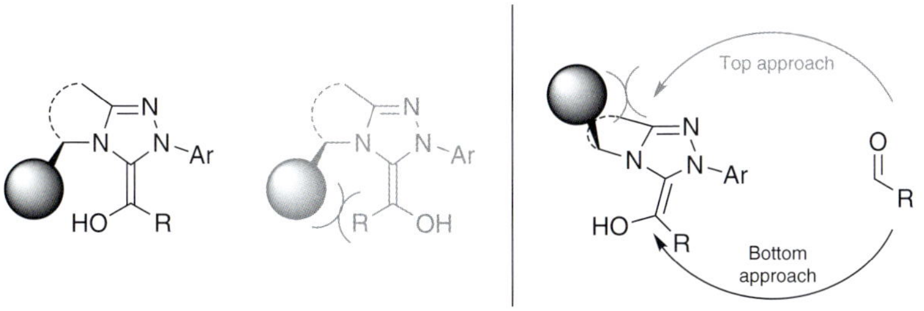

Figure 2.2 Simplified models for enantioinduction in triazolylidene-catalyzed benzoin reactions.

carbon–carbon bond forming step [10]. The homo-benzoin reaction has been variously reported as first or second order with respect to the aldehyde [7b, 12b, 13]. Based on these reports, it is likely that the rate-limiting step depends on the choice of solvent, base/buffer, NHC, and the nature of the substrate.

A substantial body of work has focused on attempts to induce enantioselectivity in the thiazolylidene-catalyzed homo-benzoin reaction of aromatic aldehydes (*vide infra*, Section 2.3) with generally modest results by modern standards. The introduction of triazolylidene species in the mid-1990s revitalized interest in the homo-benzoin reaction as substantially higher enantioselectivities were observed [14]. It has been suggested that triazolylidene-derived Breslow intermediates are better suited to preferentially form the *E*-isomer (Figure 2.2) [3a]; control of the *E*/*Z* geometry of the Breslow intermediate is desirable as it limits the number of possible transition states. The combination of this preference with restricted rotation of the catalyst's substituents and the consequent restriction of the possible approaches for carbon–carbon bond formation is suggested as the driving force for the increased enantioselectivity [15].

Dudding and Houk explored the origins of enantioselectivity in these reactions computationally using a variation of ONIOM (*n*-layered integrated molecular orbital method) with a combination of B3LYP/6-31G(d) and AM1 [16]. The energies of the modeled transition states confirm that the reaction pathway proceeding from the *E*-isomer of the Breslow intermediate is preferred. However, the authors suggest two additional factors in achieving enantioselectivity: the presence of a hydrogen bond between the Breslow intermediate and the incoming electrophile and a π–π interaction between the intermediate's triazolium core and the aromatic substituent of the aldehyde aide in organizing and stabilizing the lowest energy transition states (Figure 2.3).

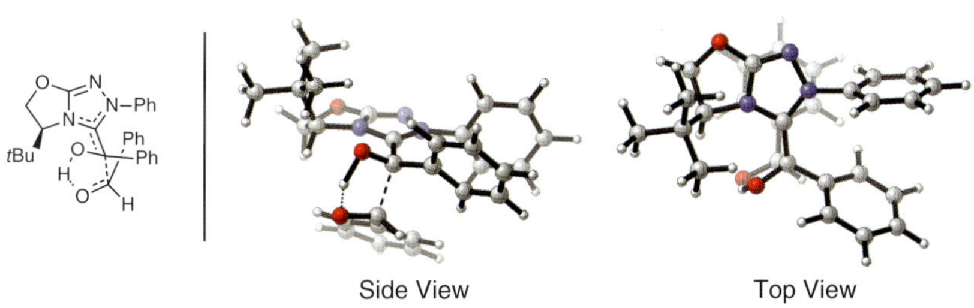

Figure 2.3 Proposed lowest energy transition state.

2.2 Standard Conditions and Substrate Scope

In general, thiazolium and triazolium salts are used as precatalysts for NHC-catalyzed benzoin reactions. Imidazolylidene catalysts are employed infrequently and are generally considered better suited for enolate/homo-enolate reactivity with enals. Common solvents include dichloromethane, tetrahydrofuran, and toluene, although more polar/protic solvents such as alcohols or *N,N*-dimethylformamide are occasionally used, particularly in thiazolium-based systems. Non-nucleophilic bases such as tertiary amines, carbonates, carboxylates, and 1,8-diazabicyclo[5.4.0]undec-7-ene (DBU) are preferred, although alkoxide salts such as potassium *tert*-butoxide may be used with some precatalysts. Selection of base is often dependent on the precatalyst salt as the acidity of azolium species is variable [17].

Aromatic aldehydes are generally well tolerated in the NHC-catalyzed homo-benzoin reaction. In addition to generally requiring extended reaction times, highly electron-rich aromatic substrates may prove problematic (Scheme 2.2a) [18]. Aromatic substrates with electron-withdrawing groups are

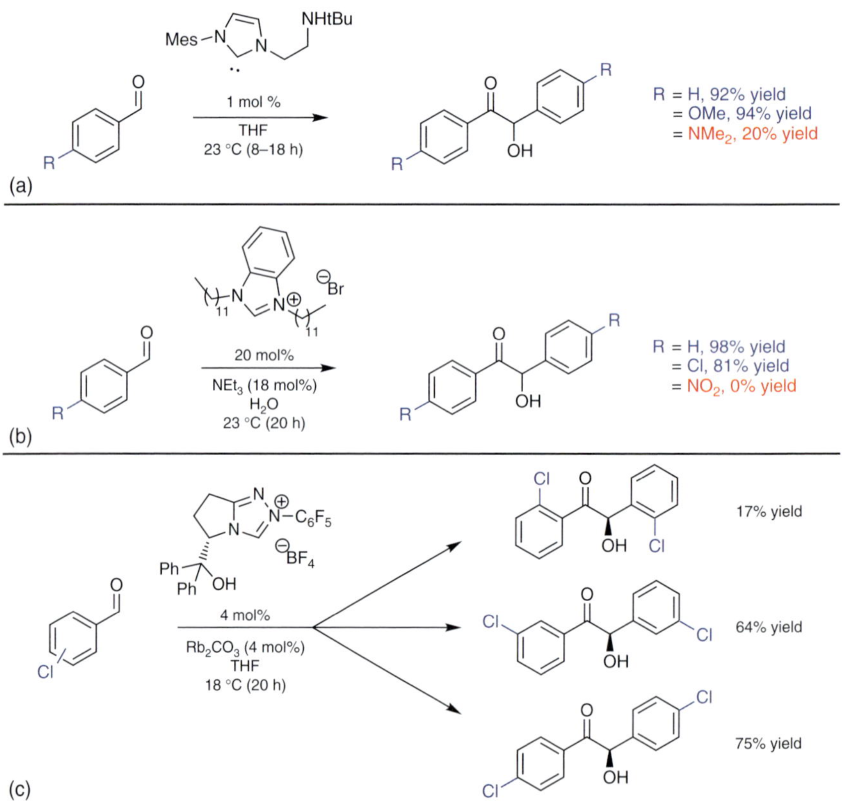

Scheme 2.2 NHC-catalyzed homo-benzoin reactions of aromatic aldehydes highlighting substrate limitations. (a) Electron-rich aldehydes, (b) Electron-poor aldehydes, (c) *o*-, *m*-, *p*-substituted aldehydes.

commonly used, although reactions with highly electron-poor aromatic species are susceptible to redox processes (Scheme 2.3b) [19]. *ortho*-Substitution significantly reduces the reactivity of aromatic substrates (Scheme 2.3c), although moderate yields are possible when unhindered/highly reactive NHCs are used [20].

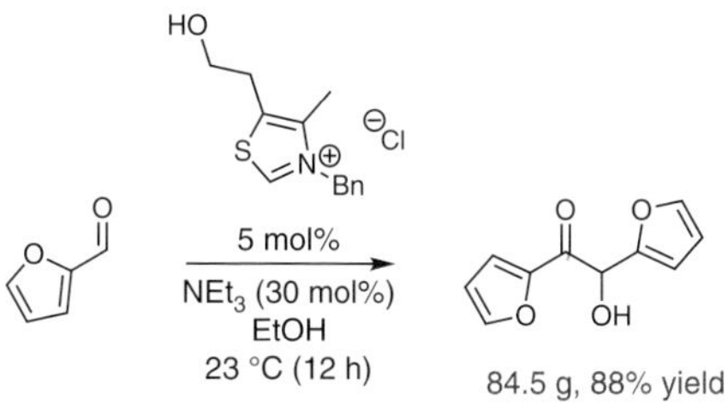

Scheme 2.3 Large-scale NHC-catalyzed homo-benzoin reaction of 2-furaldehyde.

NHC-catalyzed homo-benzoin reactions of heteroaromatic aldehydes, particularly 2-furaldehyde, are often high yielding, including on large scale (Scheme 2.3) [21]. Substituting sulfur (i.e. thiophene-2-carboxaldehyde) does not adversely affect reactivity, although these substrates are used infrequently [22]. Conversely, the nitrogenous equivalent (i.e. pyrrole-2-carboxaldehyde) and related compounds have yet to be shown as effective substrates [3a], although 2-pyridinecarboxaldehyde displays high reactivity [23]. It is noteworthy that the overwhelming majority of reactive heteroaromatic substrates has the aldehyde functionality at the 2-position.

Aliphatic aldehydes (Scheme 2.4a) [21] and formaldehyde (Scheme 2.4c) [24] can be used in NHC-catalyzed homo-benzoin reactions, although they may require slightly modified reaction conditions. α-Substituted aliphatic aldehydes typically afford substantially reduced reactivity (Scheme 2.4b), whereas β-substitution has a lesser effect [25]. The reduction in the reactivity from α-substitution can be overcome by tethering the aldehydes, forcing an intramolecular homo-benzoin reaction [26]. Ester, ether, acetal, and protected amine functional groups are tolerated. Alkenes, provided they are not conjugated (or readily isomerized into conjugation) with the aldehyde moiety, are also tolerated [25]. Phenolic substrates may be used, although poor yields are obtained [27].

2.3 Enantioselective Homo-benzoin Reactions

Beginning in 1966 with Sheehan and Hunneman's report, a substantial effort was made to find conditions for highly enantioselective NHC-catalyzed homo-benzoin reactions [28]. Initial efforts focused on thiazolylidene catalysts that produced only modest results (Scheme 2.5) [13a, 15a–c, 28, 29]. Poor yields and low enantioselectivity were obtained in almost all attempts. The most notable advance was achieved by Leeper and coworkers (**5** and **6**), who hypothesized that tethering the *N*-substituent by forming a bicyclic system would restrict rotation and effectively increase steric interactions at the active

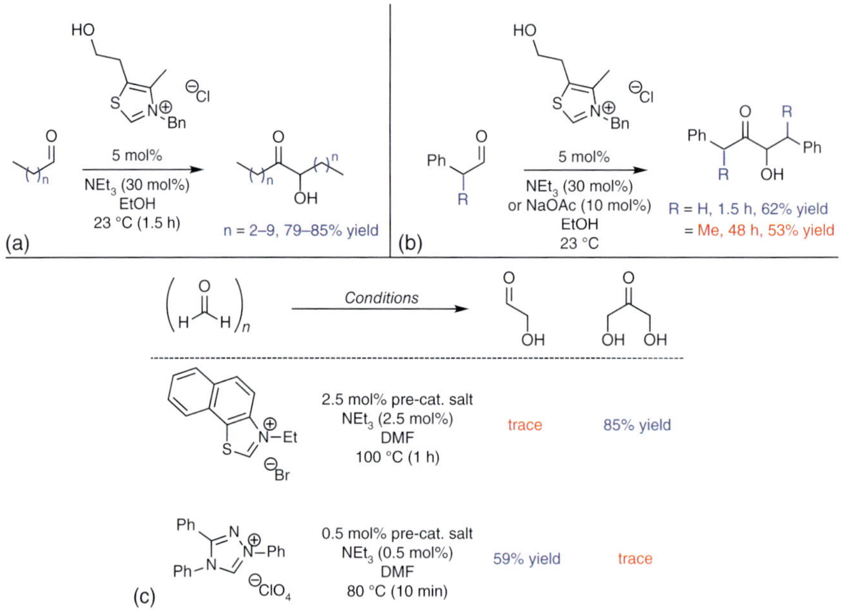

Scheme 2.4 NHC-catalyzed homo-benzoin reactions of aliphatic aldehydes highlighting substrate limitations. (a) Unbranched aliphatic aldehydes, (b) α-Substituted aldehydes, (c) Formaldehyde.

site. Although low enantioselectivities were achieved with their precatalyst, the rationale behind their catalyst design would later be incorporated into triazolylidenes catalysts.

The first triazolylidene NHC catalyst used in an enantioselective homo-benzoin reaction afforded improved yields and substantially improved enantioselectivity (Scheme 2.6, **12**) [14]. Combining the proven efficacy of triazolylidenes with their bicyclic catalyst design, Leeper's group reported the seminal morpholinone-derived **13** in 1998 [29c]. Variations of this design became the focus of the majority of subsequent work, culminating in precatalysts bearing substituents capable of hydrogen bonding, thought to aid in rigidifying/ordering the carbon–carbon bond formation step [20, 30]. Care must be taken with these reactions as even small variations in reaction conditions (i.e. use of K_2CO_3 vs. Cs_2CO_3 vs. Rb_2CO_3 as the base) [20] can affect the reproducibility and have a subtle influence on yields and/or enantioselectivities.

2.4 Cross-benzoin Reactions

2.4.1 Intramolecular Cross-benzoin Reactions

Suzuki and coworkers reported the first intramolecular cross-benzoin reactions between aldehydes and ketones, using thiazolylidenes as catalysts in the synthesis of preanthraquinones (Scheme 2.7) [31]. Reactions afford high diastereoselectivities while maintaining high yields.

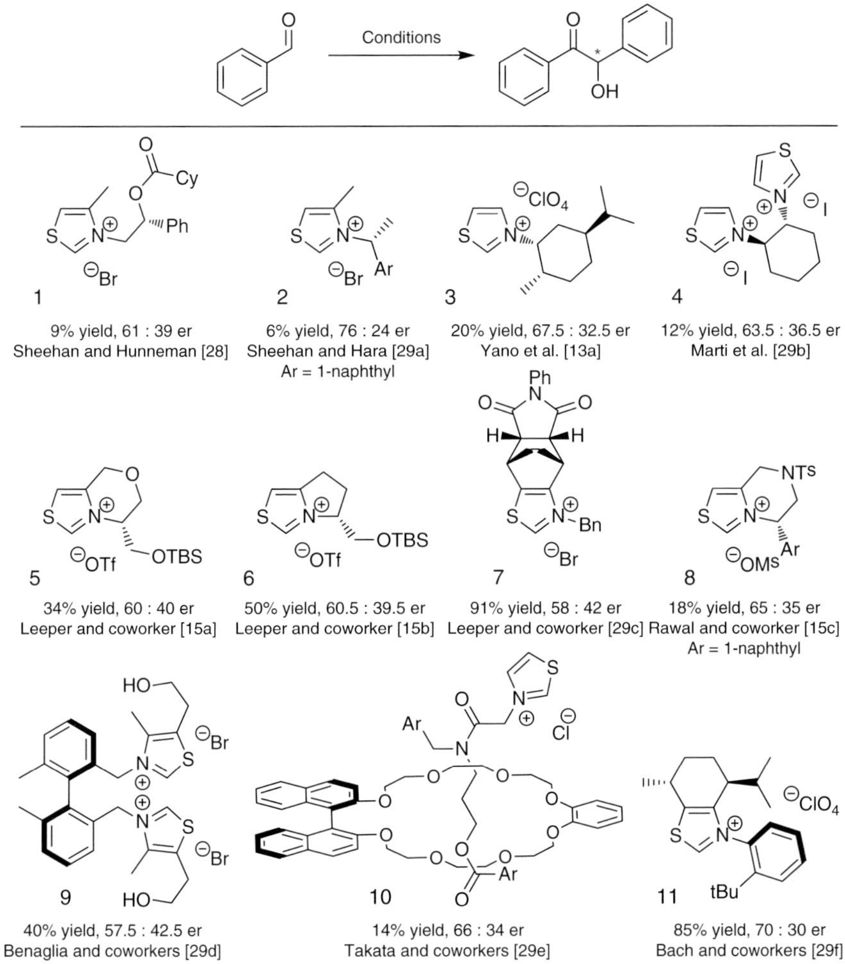

Scheme 2.5 Select thiazolium salt precatalysts for the enantioselective homo-benzoin reaction.

More general studies of intramolecular cross-benzoin reactions were later reported simultaneously by Enders and Suzuki who explored the substrate scope [32]. Several trends were noted with respect to the substituents on the aldehyde and ketone moieties. The coupling of aromatic aldehydes with aliphatic ketones is possible in the formation of five- and six-membered rings but fails for larger systems (Scheme 2.8a). The use of aliphatic aldehydes provides variable degrees of success, with results heavily dependent on the nature of the ketone (Scheme 2.8b). Methyl ketones react favorably as acceptors, whereas longer tethers ($n = 2$) or phenyl-substituted ketones result in intermolecular product formation. Intermolecular reactions between aldehydes and ketones failed under the optimized reaction conditions. Note that competing aldol pathways can be suppressed by using substoichiometric quantities of the base.

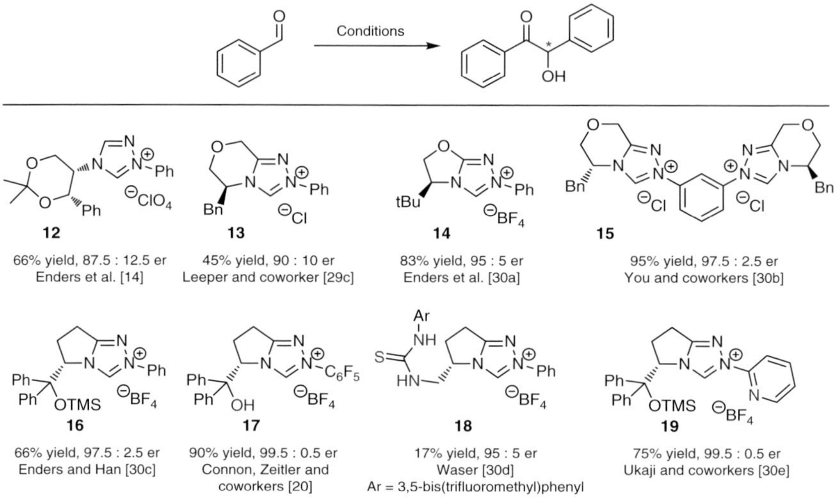

Scheme 2.6 Select triazolium salt precatalysts for the enantioselective homo-benzoin reaction highlighting hydrogen-bonding substituents.

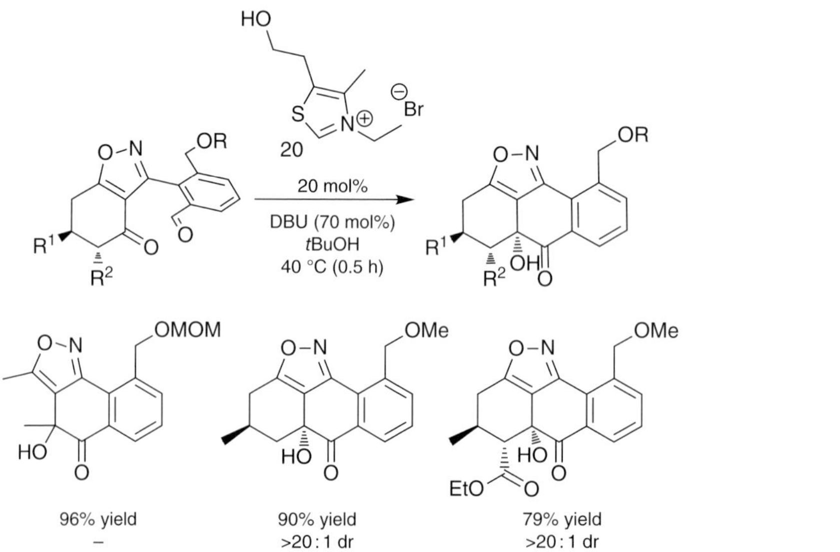

Scheme 2.7 First intramolecular cross-benzoin reaction between an aldehyde and a ketone.

The potential for enantioselectivity was examined by Enders and Suzuki using chiral triazolium salts [33]. Oxazolidinone-derived triazolylidenes (as **14**), which perform well in homo-benzoin reactions between aldehydes, afford poor yields in these reactions [33a]. Use of aminoindanol-derived catalysts affords excellent yields and enantioselectivities in aromatic aldehyde–aliphatic ketone couplings (Scheme 2.9) [33c]. A sole example of cyclization involving an aliphatic aldehyde is provided, which afforded a low yield of the cross-benzoin product. However, a variety of preanthraquinones, chromanones, and other carbocycles could

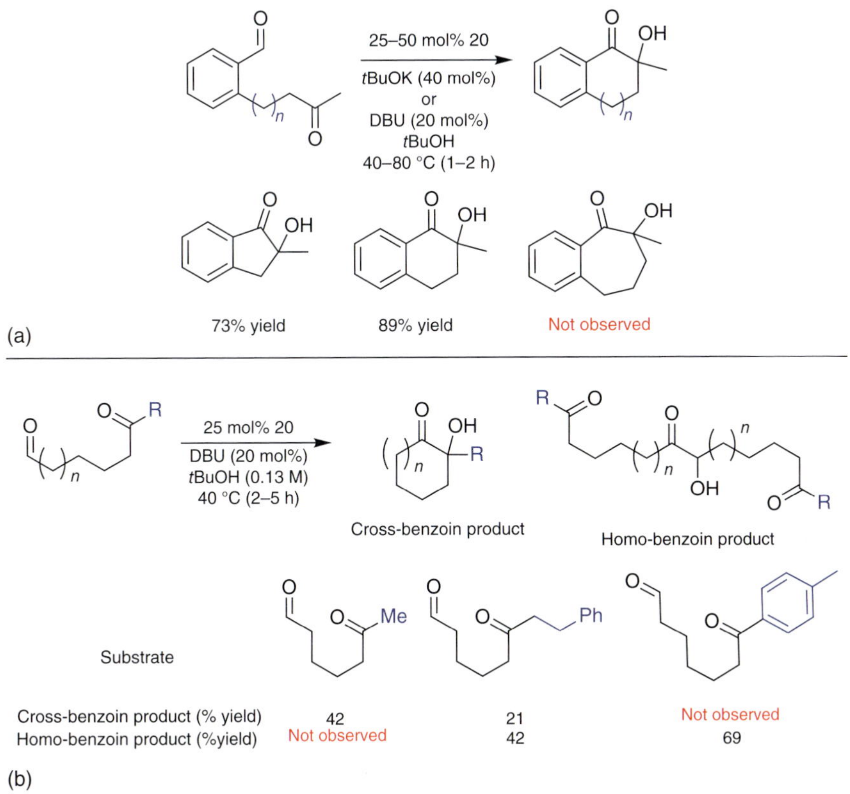

Scheme 2.8 Intramolecular cross-benzoin reactions of aromatic/aliphatic aldehydes and aliphatic/aromatic ketones. (a) Aromatic aldehydes, (b) Aliphatic aldehydes.

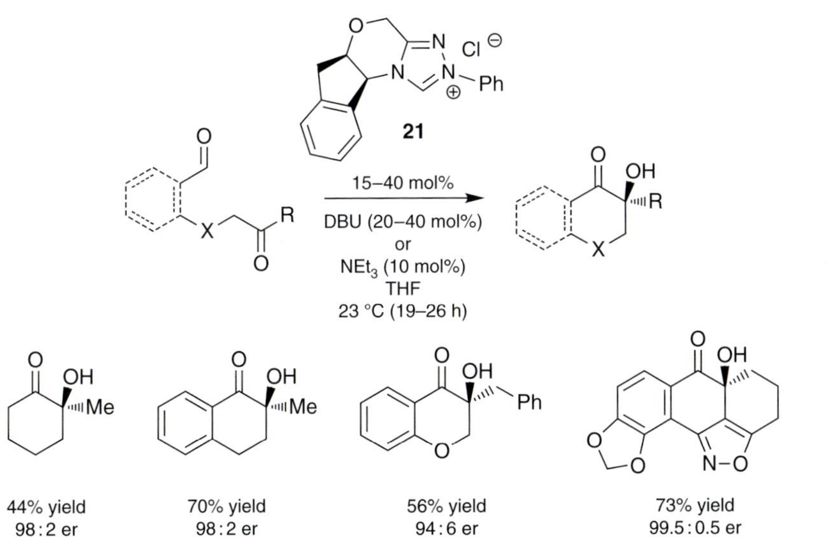

Scheme 2.9 Enantioselective intramolecular cross-benzoin reactions of aliphatic/aromatic aldehydes and aliphatic ketones.

be synthesized efficiently when aromatic aldehydes were utilized. The formation of five-membered rings with aromatic aldehydes resulted in products with lower enantiomeric ratios under conditions reported by Enders. The utility of this methodology was subsequently demonstrated in the total synthesis of sappanone [34]. Similar methodologies containing *N*-tosyl substituents on the carbocycle were also reported [35].

Ema, Sakai, and coworkers subsequently expanded the substrate scope for the intramolecular cross-benzoin reaction between aliphatic aldehydes and aliphatic ketones (Scheme 2.10) [36]. Changing the size of the formed ring does not affect the high (>20 : 1) diastereoselectivity. Aminoindanol-derived triazolylidenes again impart moderate-to-high enantioselectivities for these reactions [37].

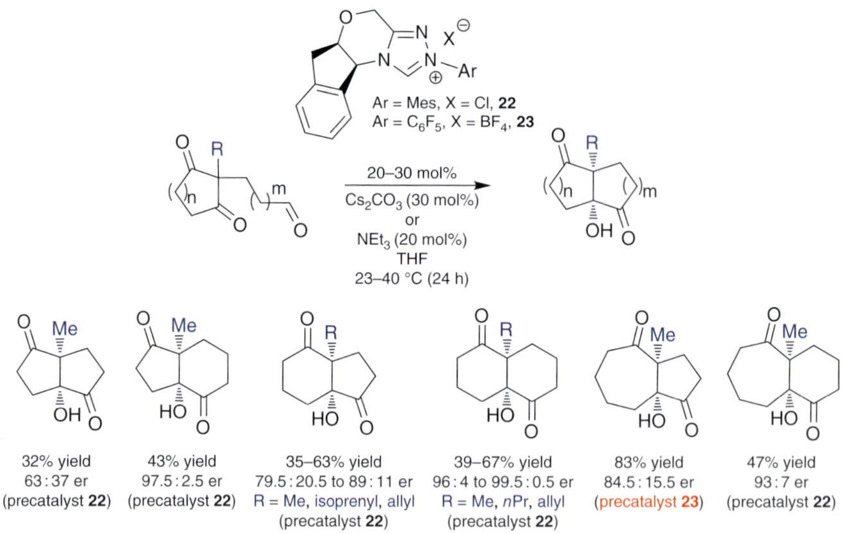

Scheme 2.10 Desymmetrization via cross-benzoin reactions of aliphatic aldehydes with aliphatic ketones.

Variation of the substituent at the quaternary center was well tolerated, and enantioselectivities are heavily dependent on the size of the newly formed ring, rather than substituents. In general, the formation of six-membered rings affords higher enantioselectivities than for the corresponding five-membered products. Switching from precatalyst **22** to the one containing an *N*-pentafluorophenyl substituent (**23**) was necessary to impart enantioselectivity in the formation of a hydroazulene derivative. Curiously, a positive temperature dependence on enantioselectivity was observed for some substrates.

Fang and coworkers further expanded these reactions to include acyclic aliphatic and aromatic diketones [38]. In general, these desymmetrizations proceed with good enantioselectivity and in high yield without the corresponding formation of any minor diastereomer. Use of similar substrates in cross-benzoin reactions may also proceed through a dynamic kinetic resolution, affording high enantio- and diastereoselectivities (Scheme 2.11) [39]. Use of an electron-poor triazolylidene catalyst is required to minimize the formation of

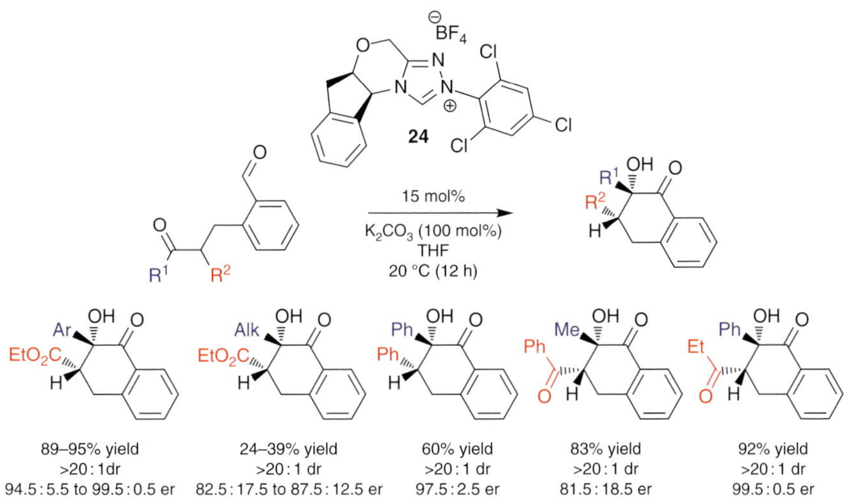

Scheme 2.11 Dynamic kinetic resolution in intramolecular cross-benzoin reactions of aromatic aldehydes and aliphatic or aromatic ketones.

aldol side products. Products are typically obtained as a single diastereomer, with enantioselectivities highly dependent on the ketone substituent. β-Carbonyl substitution is unnecessary with α-phenyl-substituted substrates. The drastic change in regioselectivity observed for unsymmetrical diketones containing either methyl or ethyl substituents is especially notable.

Intramolecular cross-benzoin reactions between two aldehydes have also been attempted with limited success. Reactions between unsymmetrical aliphatic aldehydes are poorly chemoselective [40], whereas macrocyclizations between *ortho*-substituted aromatic aldehydes and aliphatic aldehydes produce moderate yields [41].

2.4.2 Intermolecular Cross-benzoin Reactions

Enders and Henseler found that trifluoromethyl ketones were sufficiently electrophilic to behave as electrophiles in intermolecular cross-benzoin reactions [42]. Use of either electron-poor or electron-rich benzaldehyde derivatives affords moderate-to-good yields of cross-benzoin products. Additionally, heteroaromatic aldehydes such as 2-furaldehyde and thiophene-2-carboxaldehyde are exceptionally competent substrates. Moderate enantioselectivities (c. 90 : 10 er) can be achieved for this transformation using chiral triazolylidene catalysts [43].

The groups of Connon and coworkers showed that α-ketoesters are also effective electrophiles for intermolecular cross-benzoin reactions (Scheme 2.12) [44]. The substrate scope and functional group tolerance of the reaction is quite extensive; use of aliphatic aldehydes substituted with distal ethers, amides, or esters, as well as aromatic and heteroaromatic aldehydes, all afford good yields. Conversely, increasing the steric bulk on the ketone has a detrimental effect on reaction yields.

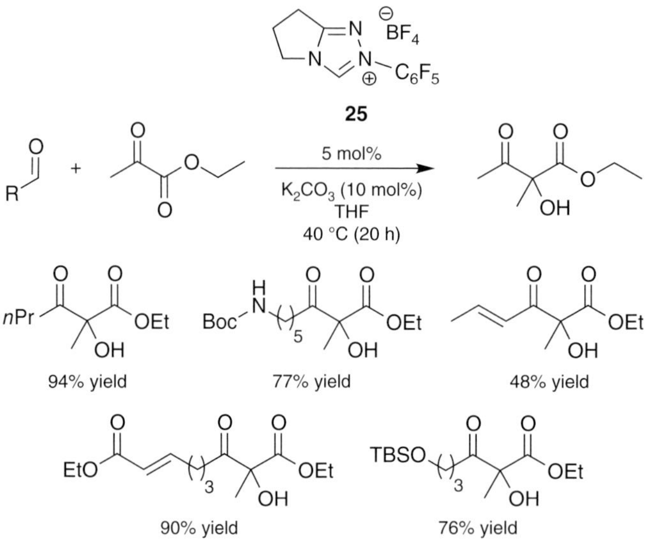

Scheme 2.12 Intermolecular cross-benzoin reactions of aldehydes with α-ketoesters.

Notably, the presence of an unsaturated ester does not lead to an intramolecular Stetter reaction. Large-scale (multigram) reactions were also demonstrated.

An enantioselective variation of the reaction was explored by Gravel and coworkers [45]. Use of morpholinone-derived triazolylidene catalysts affords moderate-to-high enantioselectivities with aryl-substituted α-ketoesters (Scheme 2.13). Although aliphatic aldehydes were found to be good substrates for this reaction, the use of aromatic and heteroaromatic aldehydes leads to low yields. The products were also shown to undergo a highly diastereoselective hydroxyl-directed zinc borohydride reduction to furnish α,β-dihydroxy esters.

Cross-benzoin reactions between aldehydes possess the additional challenge of chemoselectivity, as four possible benzoin products may be formed. Early work demonstrated that the use of hindered *ortho*-substituted aromatic aldehydes with α-substituted aliphatic aldehydes affords products chemoselectively. Chemoselective couplings of *ortho*-substituted aromatic aldehydes with unhindered aromatic aldehydes [46], and of heteroaromatic aldehydes with α-substituted aliphatic aldehydes [47], have been reported. The reaction was further expanded by Connon, Zeitler, and coworkers who discovered that steric requirements are less stringent with the use of triazolylidene catalysts, reporting chemoselective cross-benzoin reactions between *ortho*-substituted aromatic aldehydes and unbranched aliphatic aldehydes [12a].

Expansion of the fused ring on the triazolylidene catalyst (precatalyst **27**) allowed Gravel and coworkers to eliminate the need for additional steric bulk on the substrates while maintaining excellent chemoselectivity (Scheme 2.14) [10, 48]. A variety of substitution patterns on the aromatic aldehyde as well as branching on the aliphatic aldehyde were well tolerated. Use of a chiral catalyst (precatalyst **26**) afforded the product with an enantiomeric ratio of 70 : 30. The observed chemoselectivity when employing **27** was later found to be due to a

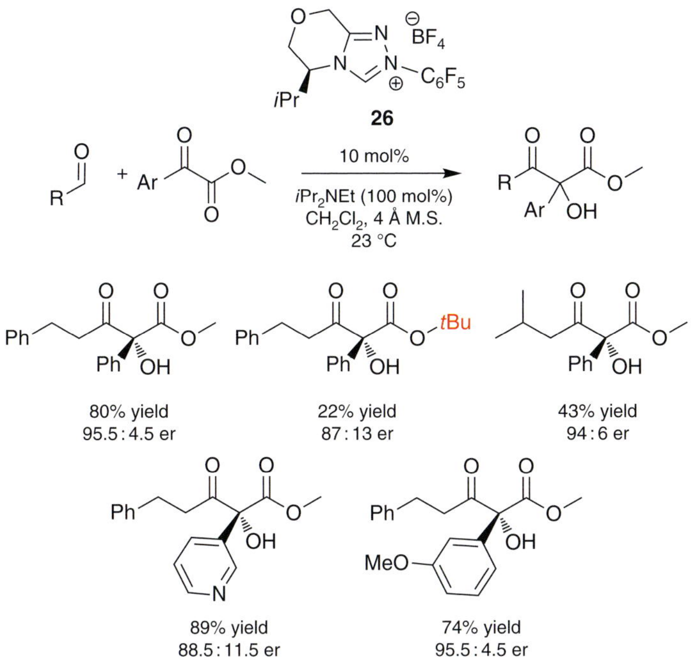

Scheme 2.13 Enantioselective intermolecular cross-benzoin reactions of aldehydes with α-ketoesters.

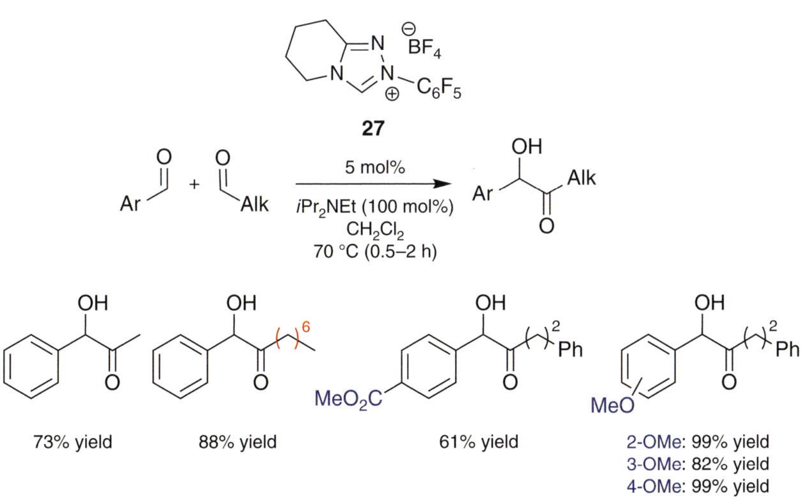

Scheme 2.14 Chemoselective intermolecular cross-benzoin reactions of unhindered aromatic and aliphatic aldehydes.

kinetic bias toward the favored cross-benzoin product. It is notable that the major product is also thermodynamically favored, and under slightly modified conditions, **25** can also provide comparable chemoselectivity.

Yang and coworkers could modulate chemoselectivity through the use of either thiazolylidene or triazolylidene catalysts (Scheme 2.15) [49]. The authors suggest that the observed switch in chemoselectivity is thermodynamically derived. It is also suggested that the formation of the Breslow intermediate with acetaldehyde is favored with precatalyst **25** because of its enhanced steric bulk. The exact reason for such a drastic reversal of selectivity is unclear, and the effect of a large excess of acetaldehyde cannot be deconvoluted without further investigation.

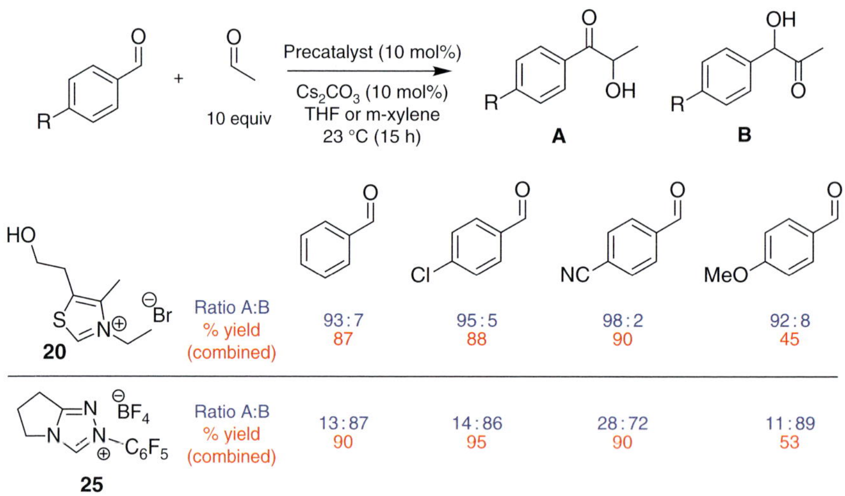

Scheme 2.15 Catalyst-dependent chemoselectivity switch in an intermolecular cross-benzoin reaction of aromatic and aliphatic aldehydes.

Chemoselective cross-benzoin reactions between various aldehydes and paraformaldehyde (hydroxymethylation) have also been demonstrated [50]. A wide variety of both aromatic and aliphatic aldehydes may be utilized with good yields and chemoselectivities. Selectivity in these reactions is the result of thermodynamic control.

α-Amino aldehydes are notable because of their ability to undergo a chemo- and diastereoselective cross-benzoin reaction (Scheme 2.16) [51]. Both potential diastereomers of the product can be obtained through changes in the *N*-protecting group; Boc monoprotection affords Cram-chelate-like selectivities, presumably because of N—H—O hydrogen bonding, while doubly protected α-amino aldehydes afford products predicted by the Felkin-Anh model. Aliphatic aldehydes were compatible only with mono-protected amino aldehydes, whereas heteroaromatic aldehydes showed universal compatibility. The authors showed that no erosion of the enantiomeric ratio was occurring under the reaction conditions. The utility of the methodology was showcased in a total synthesis of D-*arabino*-phytosphingosine.

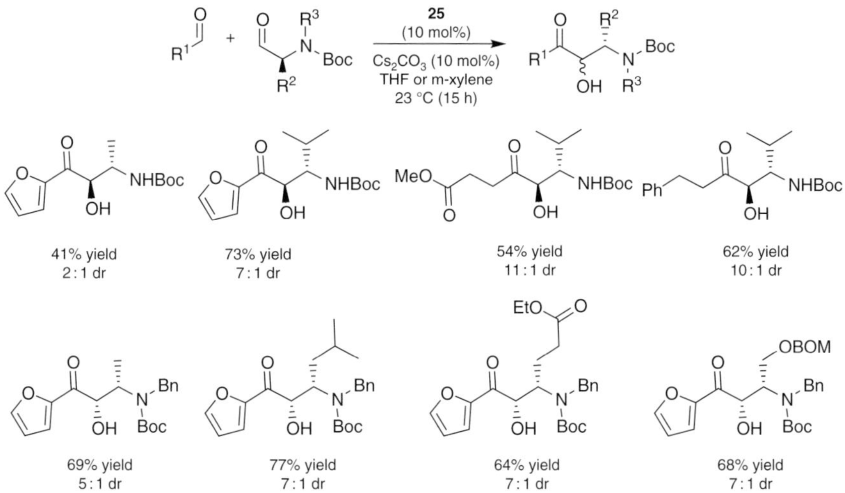

Scheme 2.16 Chemo- and diastereoselective cross-benzoin reactions of α-amino aldehydes.

2.5 Aza-benzoin Reactions

An aza-benzoin reaction arises when the electrophile in the carbon–carbon bond-forming step (**V** to **VI**, Scheme 2.1) is replaced by an imine. In most reported cases, the imine partner bears an electron-withdrawing group on the nitrogen. Chemoselectivity is not an obstacle, either because carbene catalysts have a higher affinity for aldehydes or because formation of a carbene–imine adduct is reversible [11c]. Notably, examples of NHC-catalyzed aza-benzoin reactions are limited, and generalizations on reactivity are difficult at this point.

2.5.1 Aza-benzoin Reactions of Aldimines

The first aza-benzoin reaction was reported by López-Calahorra and coworkers [52]. They discovered that the iminium salts generated from paraformaldehyde and morpholine or piperidine could react with aromatic aldehydes in the presence of a preformed thiazolylidene to generate α-amino ketones (Scheme 2.17).

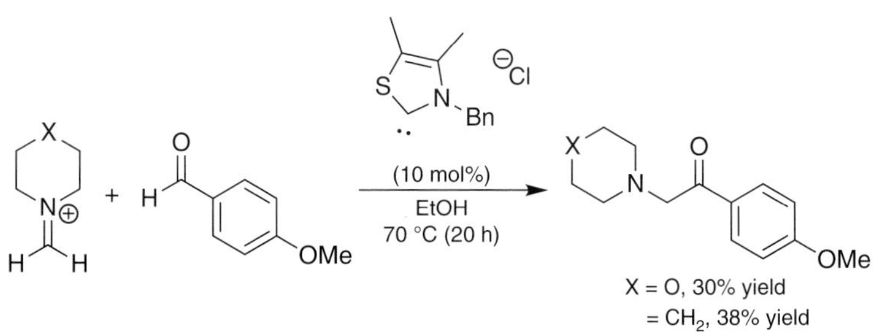

Scheme 2.17 Aza-benzoin reactions of formiminium ions.

In a seminal paper, Murry, Reider, and coworkers reported that acyl imines and Boc-imines generated in situ from their sulfonic acid adducts are excellent substrates for this transformation (Scheme 2.18). The use of aromatic, heteroaromatic, and aliphatic aldehydes as well as enals is well tolerated. However, imine partners are limited to those derived from aromatic aldehydes or formaldehyde. Crossover experiments showed that the reaction is irreversible and under kinetic control under standard reaction conditions [53]. Later studies showed that phosphinoylimines can be employed either by using acylsilanes as the reaction partner [54] or by using a bis(amino)cyclopropenium catalyst [55]. Both strategies have the advantage of avoiding formation of undesired homo-benzoin products.

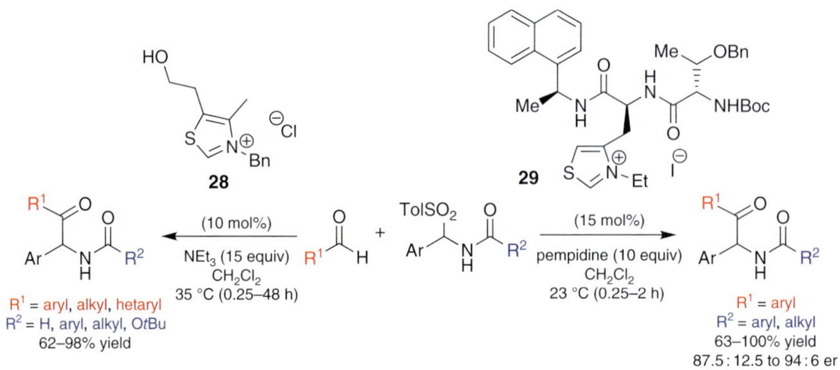

Scheme 2.18 Aza-benzoin reactions of imine-sulfonic acid adducts.

Miller and coworkers developed the first enantioselective variant by employing thiazolium-bearing peptide **29** (Scheme 2.18) [56]. The use of a hindered tertiary amine as the base was essential to avoid racemization of the products. Good yields were obtained only when using electron-poor benzaldehyde derivatives. A directing hydrogen bond between the peptide backbone and the imine nitrogen was postulated in the transition state. The effect of an alternative hydrogen bond to that between the Breslow intermediate and the electrophile [16] (imine) in aza-benzoin reactions is unclear at the moment.

The Rovis group reported highly enantioselective aza-benzoin reactions of aliphatic aldehydes using triazolium catalyst **30** (Scheme 2.19) [57]. α- and/or β-substitution on the aldehyde considerably lowers the yield. Acetic acid formed during the reaction ensures the reversibility of imine–carbene adduct formation, whereas product racemization could be suppressed by lowering the reaction temperature. Additionally, You and coworkers showed that less electrophilic *N*-aryl imine substrates can undergo aza-benzoin reactions under appropriate conditions [58]. The highest yields were obtained using a thiazolium salt. The reaction is tolerant to a number of aromatic and heteroaromatic aldehydes. Both benzoin and aza-benzoin pathways were found reversible through crossover experiments.

In an elegant combination of carbene and photoredox catalysis, the group of Rovis showed that *N*-aryl iminium ions formed from in situ oxidation of isoquinolines are viable reaction partners (Scheme 2.20) [59]. Enantioselectivity was

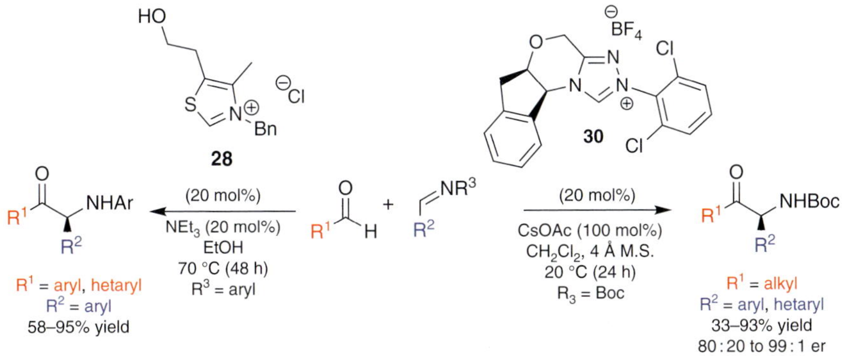

Scheme 2.19 Aza-benzoin reactions of aldehydes with imines.

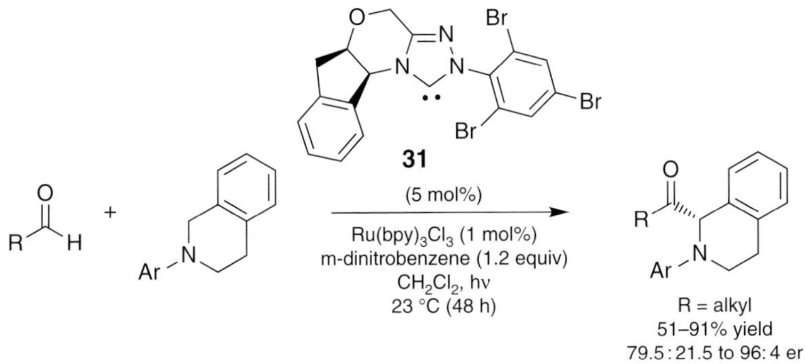

Scheme 2.20 Dual photoredox and NHC catalysis.

highly dependent on the nature of the *N*-aryl substituent on the catalyst, with triazolylidene **31** featuring a tribromophenyl moiety being optimal. Although oxygen is assumed to be the terminal oxidant, the presence of a weak oxidant (*m*-dinitrobenzene) was necessary to obtain good yields.

2.5.2 Aza-benzoin Reactions of Ketimines

Ketimines are much less reactive than aldimines, but they can nevertheless undergo aza-benzoin reactions, provided they bear additional electron-withdrawing substituents. Although the scope is more limited as a result, the products obtained from ketimines cannot racemize via enolization. The first report in this area came from the group of Enders who demonstrated the ability of *N*-aryl trifluoromethylimines to undergo the desired transformation with furfural as the reaction partner [60]. Ye and coworkers later disclosed an enantioselective variant using *N*-Boc trifluoromethylimines and enals in the presence of triazolium **32** (Scheme 2.21) [61]. Ynals may also be used although yields are low. α-Imino-esters and α-imino-nitriles were also shown to be suitable ketimine partners. The groups of Song and Chi subsequently reported isatin-derived ketimines to be competent reaction partners,

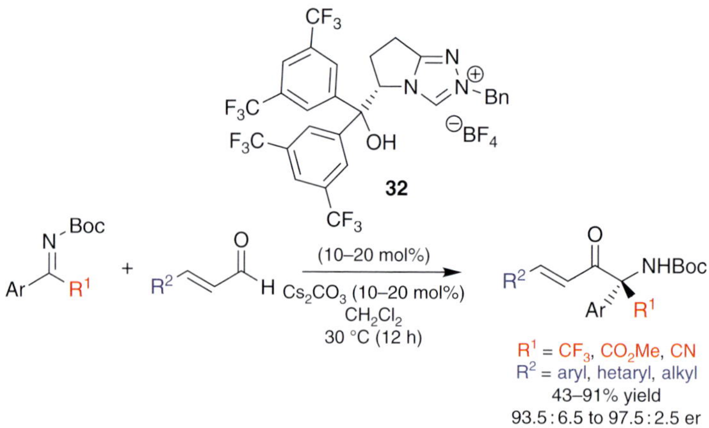

Scheme 2.21 Enantioselective aza-benzoin reactions of ketimines.

again in the reaction with enals [62]. For these substrates, an unhindered electron-withdrawing *N*-substituent on the triazolium salt was crucial in avoiding unwanted homoenolate pathways.

References

1 Wöhler, F. and Liebig, J. (1832). *J. Ann. Pharm.* 3: 249–282.
2 Lapworth, A. (1903). *J. Chem. Soc. Trans.* 83: 995–1005.
3 (a) Bugaut, X. (2014). Benzoin and aza-benzoin. In: *Comprehensive Organic Synthesis II*, vol. 1 (ed. I. Marek and P. Knochel), 424–470. Elsevier. (b) Thai, K., Sánchez-Larios, E., and Gravel, M. (2013). *Comprehensive Enantioselective Organocatalysis: Catalysts, Reactions, and Applications*. Weinheim, Germany: Wiley-VCH. (c) Menon, R.S., Biju, A.T., and Nair, V. (2016). *Beilstein J. Org. Chem.* 12: 444–461. (d) Enders, D., Niemeier, O., and Henseler, A. (2007). *Chem. Rev.* 107: 5606–5655. (e) Flanigan, D.M., Romanov-Michailidis, F., White, N.A., and Rovis, T. (2015). *Chem. Rev.* 115: 9307–9387. (f) Moore, J.L. and Rovis, T. (2009). *Asymmetric Organocatalysis* 291: 77–144. (g) Phillips, E.M., Chan, A., and Scheidt, K.A. (2009). *Aldrichim Acta* 42: 55–66. (h) Chiang, P.-C. and Bode, J.W. (2011). *N-Heterocyclic Carbenes: From Laboratory Curiosities to Efficient Synthetic Tools*. Cambridge: Royal Society of Chemistry. (i) Campbell, C.D., Ling, K.B., and Smith, A.D. (2011). *N-Heterocyclic Carbenes in Transition Metal Catalysis and Organocatalysis*, vol. 32. Dordrecht, Germany: Springer. (j) Bugaut, X. and Glorius, F. (2012). *Chem. Soc. Rev.* 41: 3511–3522. (k) O'Bryan, E.A. and Scheidt, K.A. (2014). Acyloin coupling reactions. In: *Comprehensive Organic Synthesis II*, 2e, vol. 3 (ed. I. Marek and P. Knochel), 621–655. Elsevier. (l) Hopkinson, M.N., Richter, C., Schedler, M., and Glorius, F. (2014). *Nature* 510: 485–496.
4 Ukai, T., Tanaka, R., and Dokawa, T.J. (1943). *Pharm. Soc. Jpn.* 63: 296–300.
5 Breslow, R. (1958). *J. Am. Chem. Soc.* 80: 3719–3726.

6 (a) Wittig, G., Davis, P., and Koenig, G. (1951). *Chem. Ber. Recl.* 84: 627–632. (b) Seebach, D. (1979). *Angew. Chem. Int. Ed. Engl.* 18: 239–258.
7 (a) Chen, Y.T., Barletta, G.L., Haghjoo, K. et al. (1994). *J. Org. Chem.* 59: 7714–7722. (b) Breslow, R. and Kim, R. (1994). *Tetrahedron Lett.* 35: 699–702. (c) Breslow, R. and Schmuck, C. (1996). *Tetrahedron Lett.* 37: 8241–8242. (d) Lemal, D.M., Lovald, R.A., and Kawano, K.I. (1964). *J. Am. Chem. Soc.* 86: 2518–2519. (e) Rehbein, J., Ruser, S.M., and Phan, J. (2015). *Chem. Sci.* 6: 6013–6018.
8 Holloczki, O., Kelemen, Z., and Nyulaszi, L. (2012). *J. Org. Chem.* 77: 6014–6022.
9 Gronert, S. (2007). *Org. Lett.* 9: 3065–3068.
10 Langdon, S.M., Legault, C.Y., and Gravel, M. (2015). *J. Org. Chem.* 80: 3597–3610.
11 (a) Berkessel, A., Elfert, S., Etzenbach-Effers, K., and Teles, J.H. (2010). *Angew. Chem. Int. Ed.* 49: 7120–7124. (b) Berkessel, A., Yatham, V.R., Elfert, S., and Neudorfl, J.M. (2013). *Angew. Chem. Int. Ed.* 52: 11158–11162. (c) DiRocco, D.A., Oberg, K.M., and Rovis, T. (2012). *J. Am. Chem. Soc.* 134: 6143–6145.
12 (a) O'Toole, S.E., Rose, C.A., Gundala, S. et al. (2011). *J. Org. Chem.* 76: 347–357. (b) White, M.J. and Leeper, F.J. (2001). *J. Org. Chem.* 66: 5124–5131.
13 (a) Yano, Y., Tamura, Y., and Tagaki, W. (1980). *Bull. Chem. Soc. Jpn.* 53: 740–744. (b) Breslow, R. and Kool, E. (1988). *Tetrahedron Lett.* 29: 1635–1638. (c) Lopez-Calahorra, F. and Rubires, R. (1995). *Tetrahedron* 51: 9713–9728. (d) Diederich, F. and Lutter, H.D. (1989). *J. Am. Chem. Soc.* 111: 8438–8446. (e) Vandenberg, H.J., Challa, G., and Pandit, U.K. (1989). *J. Mol. Catal.* 51: 1–12.
14 Enders, D., Breuer, K., and Teles, J.H. (1996). *Helv. Chim. Acta* 79: 1217–1221.
15 (a) Knight, R.L. and Leeper, F.J. (1997). *Tetrahedron Lett.* 38: 3611–3614. (b) Gerhard, A.U. and Leeper, F.J. (1997). *Tetrahedron Lett.* 38: 3615–3618. (c) Dvorak, C.A. and Rawal, V.H. (1998). *Tetrahedron Lett.* 39: 2925–2928. (d) Vora, H.U. and Rovis, T. (2011). *Aldrichim Acta* 44: 3–11. (e) Rovis, T. (2008). *Chem. Lett.* 37: 2–7.
16 Dudding, T. and Houk, K.N. (2004). *Proc. Natl. Acad. Sci. U.S.A.* 101: 5770–5775.
17 Massey, R.S., Collett, C.J., Lindsay, A.G. et al. (2012). *J. Am. Chem. Soc.* 134: 20421–20432.
18 Li, C.Y., Kuo, Y.Y., Tsai, J.H. et al. (2011). *Chem. Asian J.* 6: 1520–1524.
19 Iwamoto, K., Hamaya, M., Hashimoto, N. et al. (2006). *Tetrahedron Lett.* 47: 7175–7177.
20 Baragwanath, L., Rose, C.A., Zeitler, K., and Connon, S.J. (2009). *J. Org. Chem.* 74: 9214–9217.
21 Stetter, H. and Kuhlmann, H. (1984). *Org. Synth.* 62: 170.
22 Iwamoto, K., Kimura, H., Oike, M., and Sato, M. (2008). *Org. Biomol. Chem.* 6: 912–915.

23 (a) Miyashita, A., Suzuki, Y., Iwamoto, K., and Higashino, T. (1994). *Chem. Pharm. Bull.* 42: 2633–2635. (b) DiRocco, D.A. and Rovis, T. (2011). *J. Am. Chem. Soc.* 133: 10402–10405.

24 Teles, J.H., Melder, J.P., Ebel, K. et al. (1996). *Helv. Chim. Acta* 79: 61–83.

25 Stetter, H. and Ramsch, R.Y. (1981). *Synthesis* 477–478.

26 Stockton, K.P., Greatrex, B.W., and Taylor, D.K. (2014). *J. Org. Chem.* 79: 5088–5096.

27 Zhang, X., Fu, J., Zhan, T.-G. et al. (2013). *Tetrahedron Lett.* 54: 5039–5042.

28 Sheehan, J.C. and Hunneman, D.H. (1966). *J. Am. Chem. Soc.* 88: 3666.

29 (a) Sheehan, J.C. and Hara, T. (1974). *J. Org. Chem.* 39: 1196–1199. (b) Marti, J., Castells, J., and Lopez-Calahorra, F. (1993). *Tetrahedron Lett.* 34: 521–524. (c) Knight, R.L. and Leeper, F.J. (1998). *J. Chem. Soc., Perkin Trans. 1* 1891–1893. (d) Orlandi, S., Caporale, M., Benaglia, M., and Annunziata, R. (2003). *Tetrahedron: Asymmetry* 14: 3827–3830. (e) Tachibana, Y., Kihara, N., and Takata, T. (2004). *J. Am. Chem. Soc.* 126: 3438–3439. (f) Pesch, J., Harms, K., and Bach, T. (2004). *Eur. J. Org. Chem.* 2025–2035.

30 (a) Enders, D. and Kallfass, U. (2002). *Angew. Chem. Int. Ed.* 41: 1743–1745. (b) Ma, Y.J., Wei, S.P., Wu, J. et al. (2008). *Adv. Synth. Catal.* 350: 2645–2651. (c) Enders, D. and Han, J.W. (2008). *Tetrahedron: Asymmetry* 19: 1367–1371. (d) Brand, J.P., Siles, J.I.O., and Waser, J. (2010). *Synlett* 881–884. (e) Soeta, T., Tabatake, Y., Inomata, K., and Ukaji, Y. (2012). *Tetrahedron* 68: 894–899.

31 Hachisu, Y., Bode, J.W., and Suzuki, K. (2003). *J. Am. Chem. Soc.* 125: 8432–8433.

32 (a) Enders, D. and Niemeier, O. (2004). *Synlett* 2111–2114. (b) Hachisu, Y., Bode, J.W., and Suzuki, K. (2004). *Adv. Synth. Catal.* 346: 1097–1100.

33 (a) Enders, D., Niemeier, O., and Balensiefer, T. (2006). *Angew. Chem. Int. Ed.* 45: 1463–1467. (b) Enders, D., Niemeier, O., and Raabe, G. (2006). *Synlett* 2431–2434. (c) Takikawa, H., Hachisu, Y., Bode, J.W., and Suzuki, K. (2006). *Angew. Chem. Int. Ed.* 45: 3492–3494.

34 Takikawa, H. and Suzuki, K. (2007). *Org. Lett.* 9: 2713–2716.

35 Jia, M.Q. and You, S.L. (2013). *ACS Catal.* 3: 622–624.

36 Ema, T., Oue, Y., Akihara, K. et al. (2009). *Org. Lett.* 11: 4866–4869.

37 Ema, T., Akihara, K., Obayashi, R., and Sakai, T. (2012). *Adv. Synth. Catal.* 354: 3283–3290.

38 Li, Y., Yang, S., Wen, G. et al. (2016). *J. Org. Chem.* 81: 2763–2769.

39 Zhang, G.X., Yang, S., Zhang, X.Y. et al. (2016). *J. Am. Chem. Soc.* 138: 7932–7938.

40 Cookson, R.C. and Lane, R.M. (1976). *J. Chem. Soc., Chem. Commun.* 804–805.

41 Mennen, S.M. and Miller, S.J. (2007). *J. Org. Chem.* 72: 5260–5269.

42 Enders, D. and Henseler, A. (2009). *Adv. Synth. Catal.* 351: 1749–1752.

43 Enders, D., Grossmann, A., Fronert, J., and Raabe, G. (2010). *Chem. Commun.* 46: 6282–6284.

44 Rose, C.A., Gundala, S., Fagan, C.L. et al. (2012). *Chem. Sci.* 3: 735–740.

45 Thai, K., Langdon, S.M., Bilodeau, F., and Gravel, M. (2013). *Org. Lett.* 15: 2214–2217.

46 Piel, I., Pawelczyk, M.D., Hirano, K. et al. (2011). *Eur. J. Org. Chem.* 5475–5484.

47 Stetter, H. and Dambkes, G. (1977). *Synthesis* 403–404.

48 Langdon, S.M., Wilde, M.M.D., Thai, K., and Gravel, M. (2014). *J. Am. Chem. Soc.* 136: 7539–7542.

49 Jin, M.Y., Kim, S.M., Han, H. et al. (2011). *Org. Lett.* 13: 880–883.

50 Kuhl, N. and Glorius, F. (2011). *Chem. Commun.* 47: 573–575.

51 (a) Haghshenas, P. and Gravel, M. (2016). *Org. Lett.* 18: 4518–4521. (b) Haghshenas, P., Quail, J.W., and Gravel, M. (2016). *J. Org. Chem.* 81: 12075–12083.

52 Castells, J., López-Calahorra, F., Bassedas, M., and Urrios, P. (1988). *Synthesis* 314–315.

53 Murry, J.A., Frantz, D.E., Soheili, A. et al. (2001). *J. Am. Chem. Soc.* 123: 9696–9697.

54 Mattson, A.E. and Scheidt, K.A. (2004). *Org. Lett.* 6: 4363–4366.

55 Wilde, M.M.D. and Gravel, M. (2014). *Org. Lett.* 16: 5308–5311.

56 Mennen, S.M., Gipson, J.D., Kim, Y.R., and Miller, S.J. (2005). *J. Am. Chem. Soc.* 127: 1654–1655.

57 DiRocco, D.A. and Rovis, T. (2012). *Angew. Chem. Int. Ed.* 51: 5904–5906.

58 Li, G.Q., Dai, L.X., and You, S.L. (2007). *Chem. Commun.* 852–854.

59 DiRocco, D.A. and Rovis, T. (2012). *J. Am. Chem. Soc.* 134: 8094–8097.

60 Enders, D., Henseler, A., and Lowins, S. (2009). *Synthesis* 4125–4128.

61 Sun, L.H., Liang, Z.Q., Jia, W.Q., and Ye, S. (2013). *Angew. Chem. Int. Ed.* 52: 5803–5806.

62 Xu, J.F., Mou, C.L., Zhu, T.S. et al. (2014). *Org. Lett.* 16: 3272–3275.

3

N-Heterocyclic Carbene-catalyzed Stetter Reaction and Related Chemistry

Santigopal Mondal[1], *Santhivardhana R. Yetra*[1], *and Akkattu T. Biju*[2]

[1] *CSIR-National Chemical Laboratory (CSIR-NCL), Organic Chemistry Division, Dr. Homi Bhabha Road, Pune 411008, India*
[2] *Indian Institute of Science, Department of Organic Chemistry, Bangalore 560012, India*

3.1 Introduction

The N-heterocyclic carbene (NHC)-organocatalysis has emerged as a one of the most important field in organic synthesis for the formation of carbon–carbon and carbon–heteroatom bonds in an unconventional manner. In 1958, the mechanistic rationalization for the thiazolium salt-catalyzed benzoin condensation was provided by Breslow [1], and almost three decades later, the isolation of free carbenes was reported by the Bertrand and coworkers [2] and Arduengo groups [3]. These seminal discoveries resulted in the exponential growth of research in NHC-organocatalysis.

Carbonyl carbons are electrophilic in nature. However, the reversal of polarity of carbonyls is a challenging task and needs a strong base and harsh condition. The traditional methods for the generation of acyl anion equivalent involve the use of dithianes or protected cyanohydrin derivatives [4]. However, the recent advancement in umpolung of carbonyl functionality using catalytic and mild reaction condition has drawn much attention. In 1973, Stetter and Schreckenberg developed the first cyanide ion-catalyzed reaction of aldehydes with Michael acceptors such as α,β-unsaturated carboxylic esters, ketones, and nitriles proceeding via the umpolung strategy [5]. Three years later, Stetter demonstrated the thiazolium salt-catalyzed highly selective conjugate addition of a wide variety of aliphatic, aromatic, and heteroaromatic aldehydes with a broad range of Michael acceptors for the synthesis of 1,4-bifunctional compounds (Scheme 3.1) [6]. Later, this reaction was called Stetter reaction [7] and became one of the most important tool for the synthesis of 1,4-bifunctional compounds, which are the precursors for valuable heterocycles such as furans, pyrroles, and thiophenes.

N-Heterocyclic Carbenes in Organocatalysis, First Edition. Edited by Akkattu T. Biju.
© 2019 Wiley-VCH Verlag GmbH & Co. KGaA. Published 2019 by Wiley-VCH Verlag GmbH & Co. KGaA.

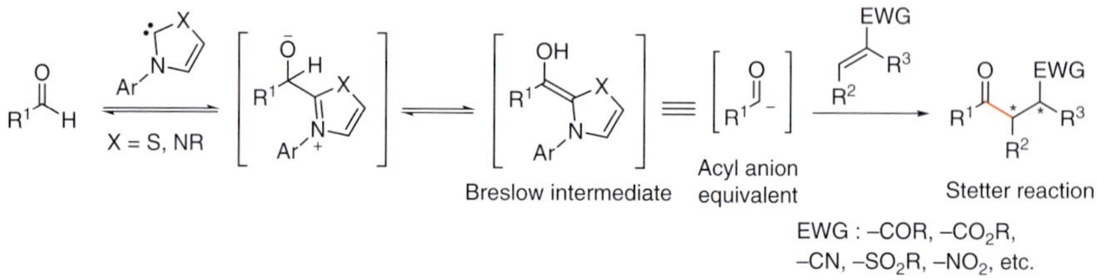

Scheme 3.1 NHC-catalyzed Stetter reaction.

3.2 Proposed Mechanism of the Stetter Reaction

Similar to the benzoin reaction mechanism proposed by Breslow [1], the thiazolium-catalyzed Stetter reaction also proceeds via the key enaminol intermediate. Interestingly, in most cases, the benzoin reaction is found to be reversible, but in Stetter reaction, the addition of Breslow intermediate to Michael acceptors is irreversible in nature. According to the proposed mechanism of NHC-catalyzed Stetter reaction (Scheme 3.2), the free carbene **2** is generated from the azolium precursor **1** upon treatment with base. The carbene **2** adds to the aldehyde generating the nucleophilic Breslow intermediate **4** via the tetrahedral intermediate **3**. Up to the generation of **4**, the steps are considered to be reversible. However, a recent mechanistic and kinetic study by the Rovis group showed that the proton transfer from the tetrahedral intermediate **3** to the Breslow intermediate **4** is the first irreversible step [8]. The conjugated addition of Breslow intermediate **4** to the Michael acceptor generates the intermediate **5**, which on proton transfer and subsequent release

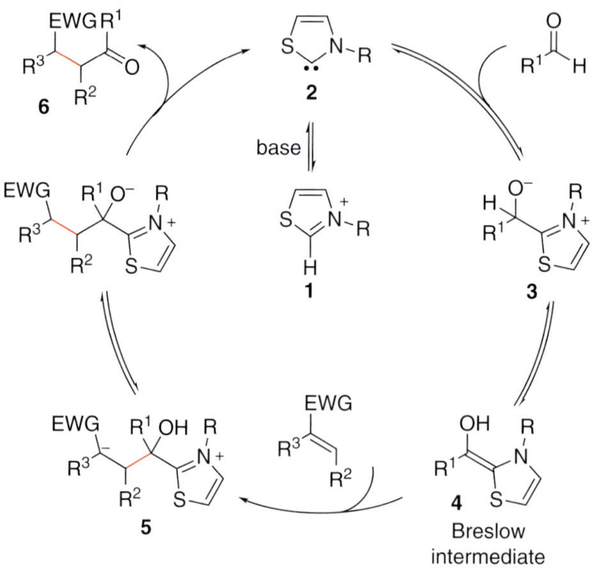

Scheme 3.2 Proposed mechanism of Stetter reaction.

of free carbene affords the desired Stetter product **6**. Moreover, this method is applied for the enantioselective synthesis of 1,4-bifunctional compounds using chiral NHCs, and the Breslow intermediate plays a central role for the product enantioselectivity. Additionally, density functional theory (DFT) calculations showed a stepwise mode of addition of Breslow intermediate to the Michael acceptor, in which the carbon–carbon coupling is the rate-determining step, which is followed by the final proton transfer [9].

3.3 Intramolecular Stetter Reaction

Almost two decades after Stetter's initial contribution, Ciganek reported the carbene generated from thiazolium salt **9** catalyzed intramolecular cyclization of formylaryloxy crotonates **7** for the synthesis of benzannulated pyranones **8** (Scheme 3.3) [10]. The reactivity of the substrate was enhanced because of the entropic factor and reaction took place in the absence of base and in dimethyl formamide (DMF) solvent under heating conditions. Later, this highly facile intramolecular Stetter reaction became a standard reaction for the development of various chiral NHC catalysts. In 1996, the first enantioselective NHC-catalyzed intramolecular Stetter reaction was developed by Enders et al. [11]. The use of chiral triazolium salt **10** as a catalyst precursor, various aromatic aldehydes **7**, with tethered Michael acceptor underwent smooth cyclization reaction affording several 4-chromanones in moderate yield (22–73%) and enantioselectivity (41–74%). A highly enantioselective (up to 97%) intramolecular Stetter reaction of substrate of the type **7** using the aminoindanol-derived chiral triazolium salt **11** was disclosed in 2002 by Rovis and coworkers. The reaction is found to be quite general with a broad range of substrates [12].

In the later stage, various thiazolium- and triazolium-based NHC catalysts have been developed for the intramolecular Stetter reaction (Scheme 3.3). The NHC derived from an axially chiral thiazolium salt **12** with an *N*-aryl substituent was developed by Bach and coworkers [13], and thiazolium salts **13** and **14** based on a peptide backbone developed by Miller group [14] afforded the product in moderate yield and enantioselectivity. A highly efficient 1,2,3-triazolo-linked azolium precatalyst **15** was examined in this annulation reaction by the Zeitler group, which afforded the chromanone product in high yields [15]. Recently, the You group disclosed the synthesis of chiral triazolium salts **16** and **17** from readily available diphenyl ethylene diamine (DPEN) as well as D-camphor, which furnished the chromanone derivatives **8** in excellent yields and up to 97% ee [16]. Very recently, a highly efficient, (−)-β-pinene-derived triazolium salt **18** was developed by the Rafiński group and applied for this reaction, which afforded the desired product in excellent yield and enantioselectivity [17]. This intramolecular Stetter reaction resulted in moderate yield and enantioselectivity when performed using the bis(hydroxyaryl)diamide containing chiral thiazolium salt **19** developed by Shibasaki and coworkers [18]. Notably, Zhang and coworkers developed the C_2 symmetric chiral cyclophane dihydroimidazolium salt **20** for intramolecular Stetter reaction for the synthesis of chromanones in moderate yield and ee values [19].

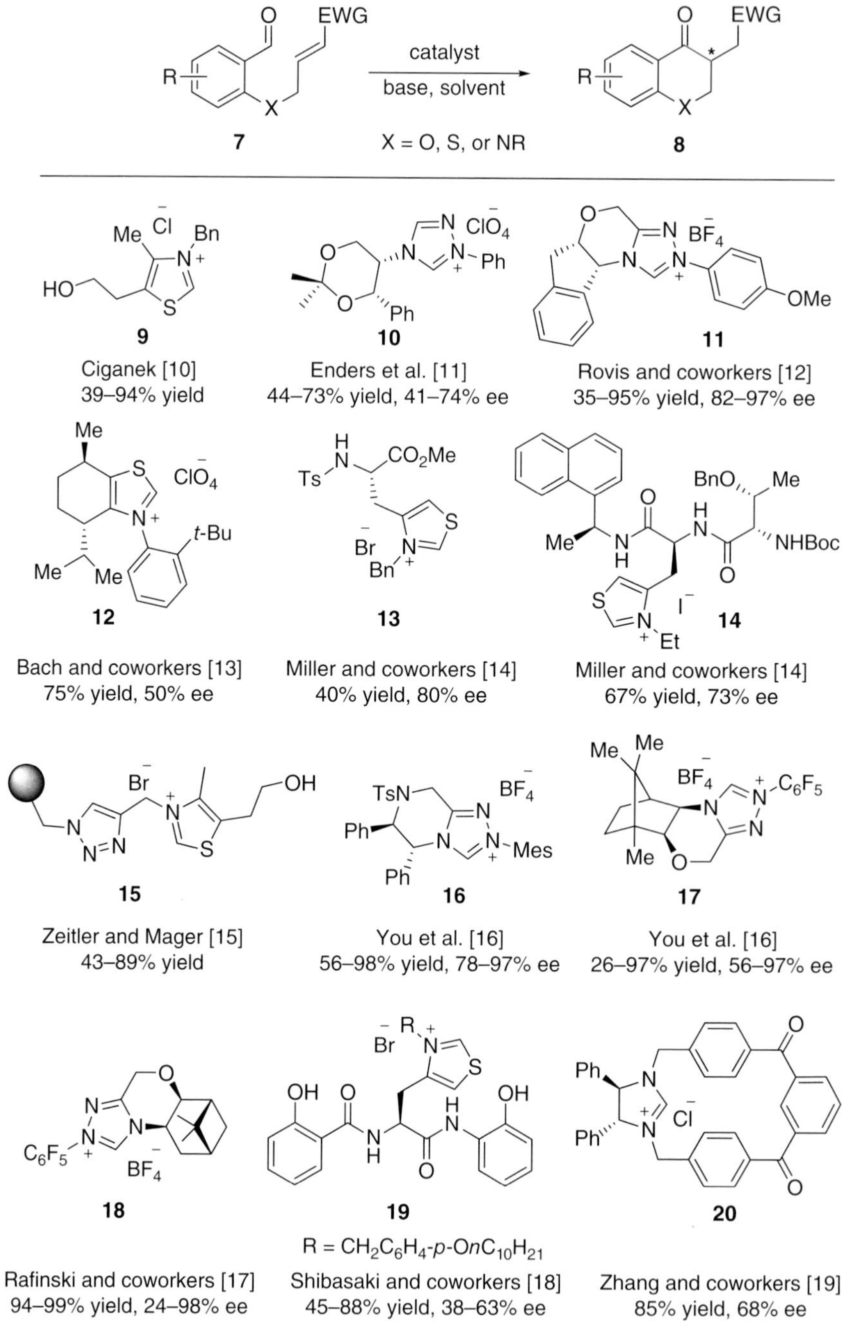

Scheme 3.3 Catalysts developed for intramolecular Stetter reaction.

A highly enantioselective and diastereoselective intramolecular Stetter reaction of salicylaldehyde-derived substrate **21** containing a tethered α,β-disubstituted Michael acceptor was developed by the Rovis group in 2005 for the synthesis of chromanones **22** (Scheme 3.4) [20]. The NHC **23** was the key for this transformation, and the reaction involves enantioselective construction of C—C bond and the diastereoselective protonation. A deuterium labeling experiment suggested that the use of external base for the carbene generation diminished the diastereoselectivity of the product. The high diastereoselectivity was obtained by the removal of base before the reaction. Additionally, the aliphatic substrates underwent smooth annulation reaction in good yield and stereoselectivity.

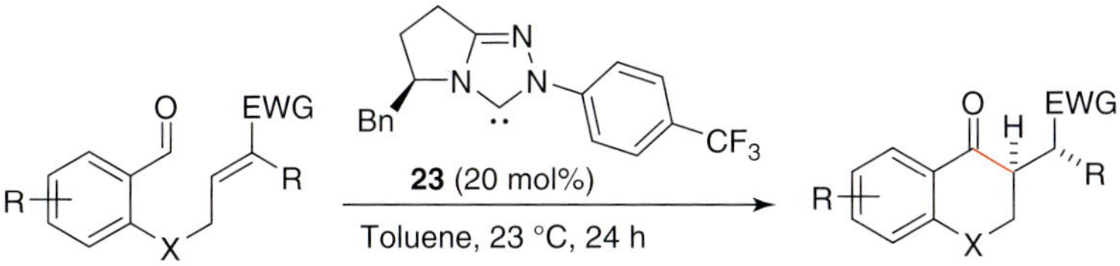

21
X = O, S, or NR

22

Scheme 3.4 Asymmetric intramolecular Stetter reaction of α,β-disubstituted Michael acceptor.

The use of vinylphosphonate as Michael acceptor in intramolecular Stetter was uncovered recently by the Rovis group. Employing carbene generated from aminoindanol-derived pentafluoroaryl-substituted triazolium salt **25**, the addition of Breslow intermediate to the moderately electron-poor Michael acceptor afforded enantioenriched products **24** containing phosphorous group in high yield (Scheme 3.5) [21].

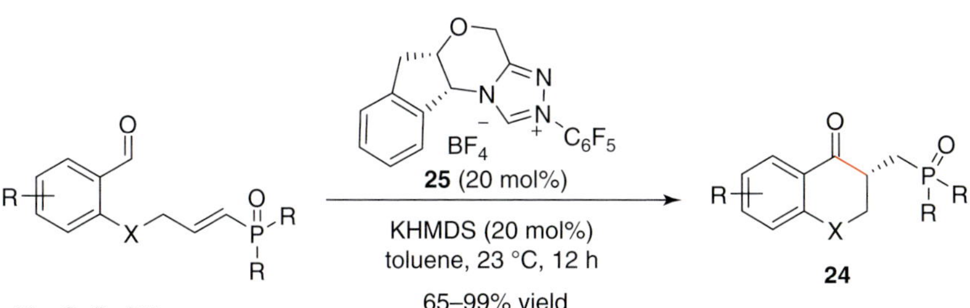

X = O, S, CH_2

24

Scheme 3.5 Asymmetric intramolecular Stetter reaction of vinylphosphonates.

The intramolecular addition of catalytically generated nucleophilic Breslow intermediates to activated alkynes was described recently by the Liu group for the synthesis of chromones **26** [22]. The cheap and commercially available NHC precursor **9** was used for this transformation, and the expected product was formed in good-to-excellent yields (Scheme 3.6).

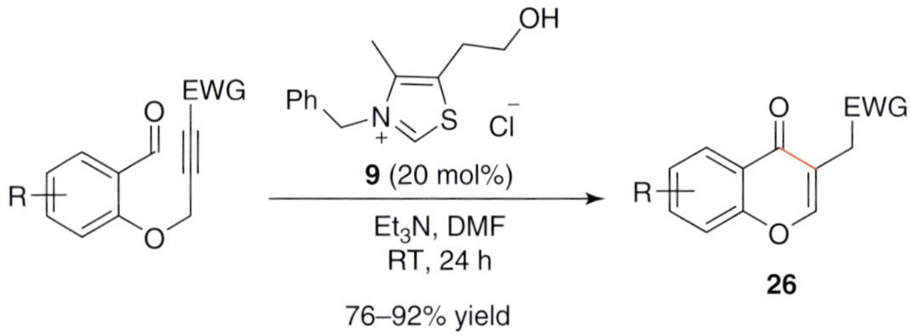

Scheme 3.6 Intramolecular Stetter reaction of activated alkynes.

The successful extension of the intramolecular Stetter reaction using a tethered vinylogous Michael acceptor **27** for the synthesis of chromanone **28** was demonstrated in 2013 by the McErlean group (Scheme 3.7) [23]. Various aromatic and aliphatic aldehydes containing heteroatom-tethered vinylogous Michael acceptor worked well in the presence of triazolium salt **29** as a NHC precursor. Additionally, an enantioselective synthesis of chromanone was described employing aminoindanol-derived pentafluoroaryl-substituted triazolium salt **25** and the desired product obtained in high enantioselectivity (up to 96%).

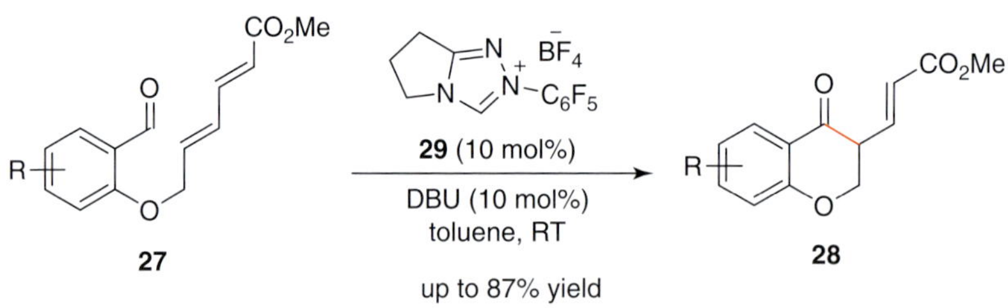

Scheme 3.7 Intramolecular Stetter reaction of vinylogous Michael acceptor.

Ren and coworkers by a detailed experimental and computational study found that the intramolecular cyclization of 4-(2-formylphenoxy) but-2-enoate **30** depends on the counterion on the imidazolium salt. The more basic counteranion generates the higher concentration of free carbene and vice versa. A series of imidazolium salts were applied for this cyclization reaction and found that at a relatively low concentration, carbene shows nucleophilic reactivity to form intramolecular Stetter product **31**. On the other hand, at a higher concentration, carbene behaves as a base and resulted in the cyclized product **32** (Scheme 3.8) [24].

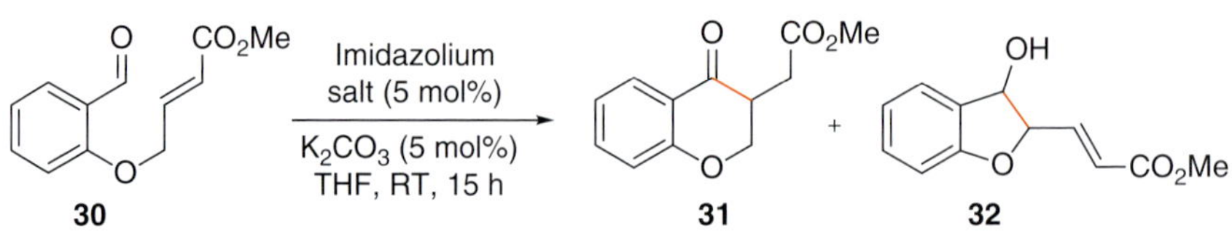

Scheme 3.8 Role of the counteranion of an imidazolium salt in Stetter reaction.

Recently, the Smith group studied the role of *N*-aryl substituent on triazolium-derived NHCs in the intramolecular Stetter reaction. A series of tetrahedral intermediate **34** were synthesized by the addition of carbene to aldehyde **33** and found that the formation of **34** was reversible under the reaction condition, followed by the slow cyclization step (Scheme 3.9). A kinetic study on this reaction revealed that pentafluoroaryl-substituted triazolium precatalysts are the best for intramolecular Stetter reaction using Et_3N as a base [25].

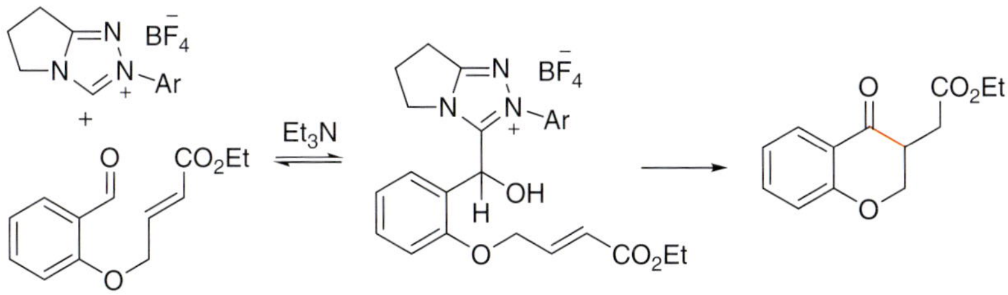

33 **34**

Scheme 3.9 Isolation of aldehyde–NHC adducts in Stetter reaction.

An intramolecular Stetter reaction of β,β-disubstituted Michael acceptor **35** was successfully demonstrated by the Rovis group [26]. The NHCs generated from the triazolium salts **25** or **37** as a catalyst precursors were found to be the best for this transformation and the 1,4-dicarbonyl compounds **36** were formed in high yield (up to 96%) and enantioselectivity (up to 99%) with a quaternary stereocenter. A variety of heteroatom-tethered aldehydes (aromatic or aliphatic) and the Michael acceptors were well tolerated under the reaction condition (Scheme 3.10). Later, the same research group described a one-pot Michael addition–Stetter reaction sequence between salicylaldehyde and activated alkynes [27]. A variety of salicylaldehyde derivatives and alkynes worked smoothly, and the benzofuranone derivatives formed in good yield (up to 87%) and excellent enantioselectivity (up to 94%).

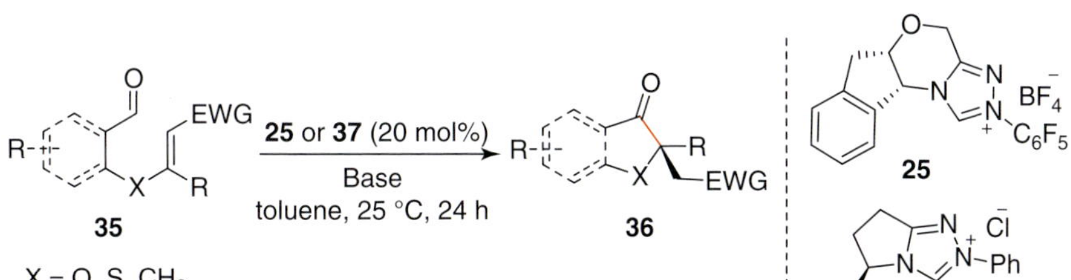

35 **36** **25** **37**

Scheme 3.10 Asymmetric intramolecular Stetter reaction of β,β-disubstituted Michael acceptors.

The asymmetric synthesis of spirocyclic compounds via intramolecular Stetter reaction was described by the Rovis group using aldehyde **38** containing *N*-benzyl-substituted maleimide moiety (Scheme 3.11) [28]. Carbene generated from electron-deficient chiral triazolium salt **25** by potassium hexamethyl disilazide (KHMDS) as the base was used for the synthesis of spirofuranone products **39**, obtained in 80% yield and 99% enantioselectivity. A valuable application of this methodology has been shown by converting the Stetter product to spirocyclic core of the antibiotic FD-838.

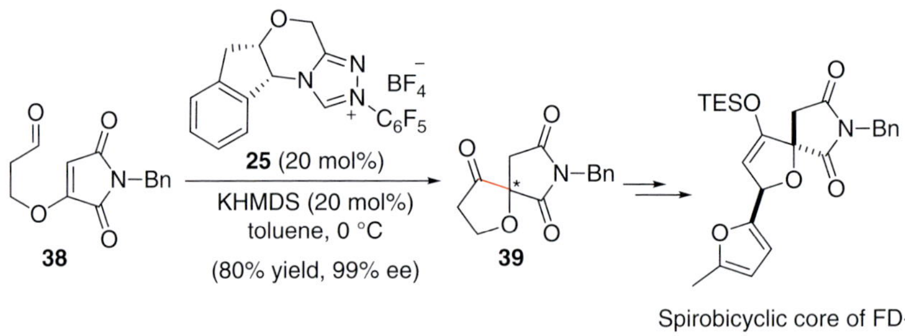

Scheme 3.11 Application of intramolecular Stetter reaction to the spiro-bicyclic core of FD-838.

A similar approach for the synthesis of spirofuranone lactam core was developed by the Rovis group by the intramolecular Stetter reaction of enal **40** with a tethered malemide group. The major difficulty in this reaction is the olefin geometry and the α-substitution of the enal. This key issue was solved by the photoinduced isomerization of the olefin (preferably *Z* for this reaction) and the proper choice of the catalyst precursor **42**. Under the optimized reaction conditions, the desired spirocyclic products **41** were formed in 62% yield and 95% enantioselectivity. Additionally, this process was applied to the total synthesis of (−)-cephamycin A (Scheme 3.12) [29].

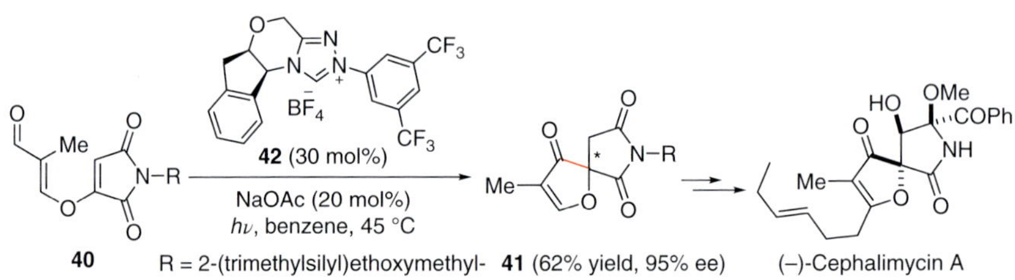

Scheme 3.12 Photoisomerization–Stetter reaction cascade.

In 2006, the Rovis group employed intramolecular Stetter reaction for the desymmetrization of cyclohexadienone **43** (Scheme 3.13). A broad range of cyclohexadienone substrate containing α and β substitution was smoothly converted to the bicyclic product **44** with three contiguous stereocenters in <5-minute reaction time. Using 4-methoxyaryl substituted chiral triazolium salt **11** as a carbene precursor, the highly diastereoselective hydrobenzofuran product was formed in good yield and excellent enantioselectivity [30].

A similar type of desymmetrization strategy was applied to the spirocyclohexadienones **45** by the You group for the enantioselective synthesis of

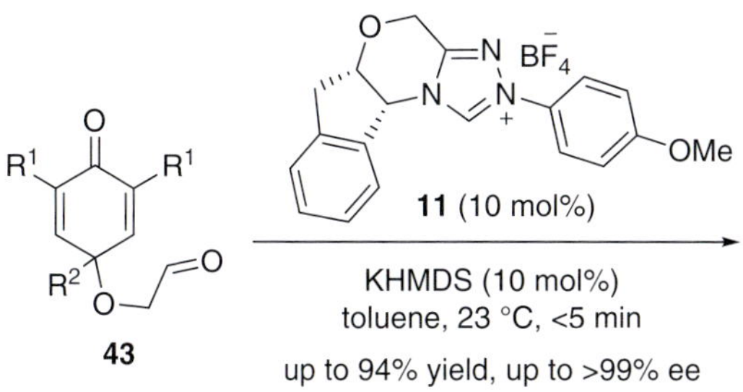

Scheme 3.13 Desymmetrization of cyclohexadienones via intramolecular Stetter reaction.

tricyclic compounds **46** (Scheme 3.14) [31]. This process involves the use of camphor-derived triazolium salt **17** as a catalyst precursor and *i*-Pr_2NEt as a base. A variety of cyclohexadienone derivative underwent smooth conversion to the desire product in good yield and enantioselectivity. An α-substitution in the cyclohexadienone moiety furnished the unsatisfactory result of only 9% yield under the optimized reaction condition. Later, this issue was solved by the use of aminoindanol-derived pentafluoroaryl-substituted triazolium salt ***ent*-25**, and the tricyclic product was formed with three contiguous stereocenter in excellent yield (up to 95%) and enantioselectivity (up to >99%) [32].

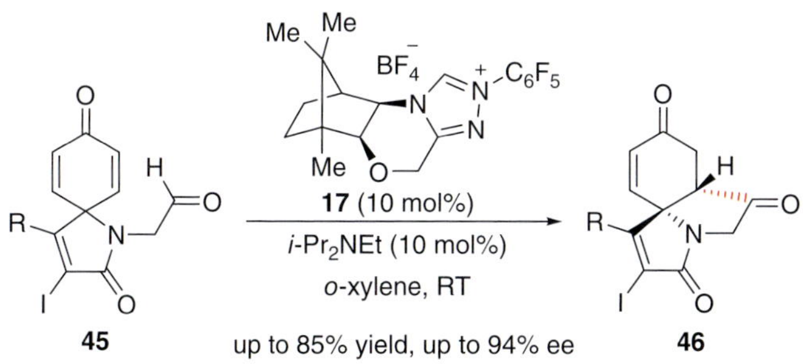

Scheme 3.14 Desymmetrization of spirocyclohexadienones via intramolecular Stetter reaction.

Furthermore, the You group demonstrated an efficient construction of tricyclic carbocycles **48** by an intramolecular Stetter reaction of cyclohexadienones **47** (Scheme 3.15) [33]. Employing camphor-derived triazolium salt **17** as a carbene precursor and KOAc as the base, the tricyclic products were obtained in high yield and enantioselectivity.

Very recently, the NHC-catalyzed intramolecular Stetter reaction was described by Fang and coworkers for the desymmetrization of acyclic 1,4-dienes **49** (Scheme 3.16) [34]. With an appropriate choice of 2,4,6-trichloroaryl substituted chiral triazolium salt **51** as a carbene precursor, a wide variety of 1,4-diene containing tethered aromatic or aliphatic aldehyde underwent smooth annulation reaction. The desired α-tetralone derivatives **50** was formed with two contiguous stereocenter in high yield, enantioselectivity, and moderate diastereoselectivity.

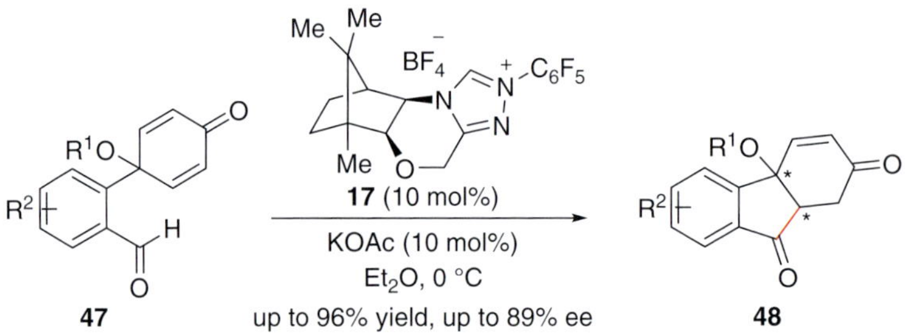

Scheme 3.15 Desymmetrization of cyclohexadienones via intramolecular Stetter reaction.

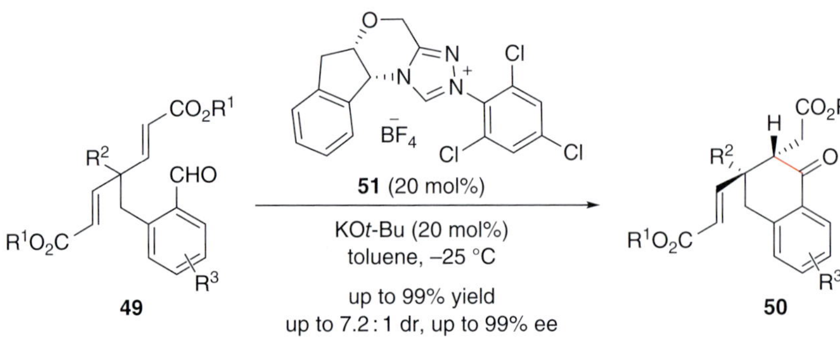

Scheme 3.16 Desymmetrization of functionalized 1,4-dienes.

3.4 Intermolecular Stetter Reaction

Since the first observation of NHC-catalyzed intermolecular Stetter reaction in 1976, the enantioselective version of this reaction continued as a challenge for a long time. Unlike the asymmetric intramolecular Stetter reaction, which was developed earlier in a highly efficient manner, the enantioselective intermolecular Stetter reaction faced difficulties may be because of the choice of proper catalyst and, in some cases, the formation of stable adduct between NHC and the Michael acceptor [35]. In 1989, Enders' group reported the first enantioselective Stetter reaction. The reaction between *n*-butanal and chalcone in the presence of NHC generated from the chiral thiazolium salt **53** using K_2CO_3 as a base, the desired 1,4-diketone **52** was formed in 39% ee but in a low yield of 4% (Scheme 3.17) [36].

In 2008, Enders' group reported the general and broad scope, enantioselective intermolecular Stetter reaction of aromatic aldehydes with chalcones (Scheme 3.18) [37]. Employing a new chiral triazolium salt **55** as carbene precursor, a variety of 1,4-diketones **54** were formed in moderate-to-good yields and moderate-to-good enantioselectivity. It is noteworthy that up to 99% ee values of the products could be obtained by single recrystallization. The key to success for the high level of enantioselectivity was explained by the proposed transition-state diagram. It is anticipated that the bulky silyl group in the catalyst moiety block the *Si*-face of the Breslow intermediate and the 1,4-addition to the chalcone would occur at its less hindered *Re*-face to furnish the product in high ee values. Subsequently, the same research group demonstrated the

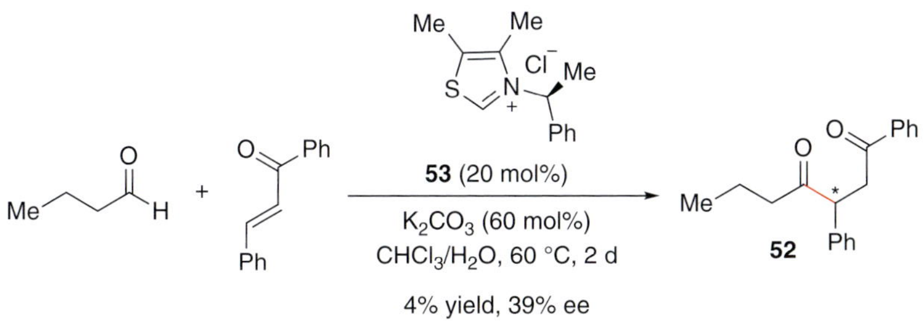

Scheme 3.17 First asymmetric intermolecular Stetter reaction of *n*-butanal with chalcone.

enantioselective intermolecular Stetter reaction of heteroaromatic aldehydes with arylidene malonates as a Michael acceptor. Using NHC generated from the chiral triazolium salt **55**, the ketomalonate products were obtained in 84–98% yield and up to 78% ee [38].

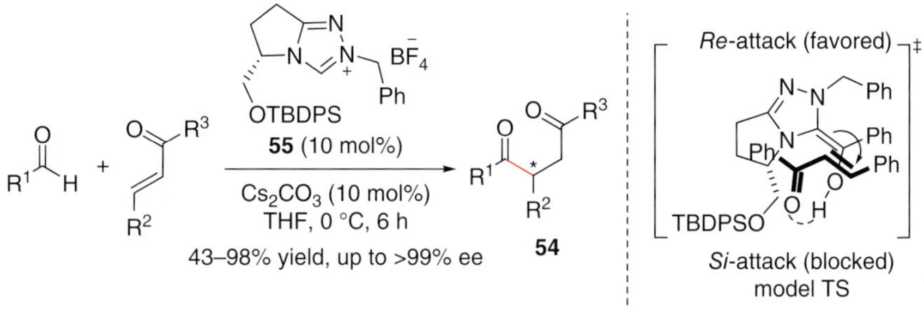

Scheme 3.18 Asymmetric intermolecular Stetter reaction of aldehyde with chalcone.

A highly enantioselective intermolecular Stetter reaction was described independently by the Rovis group using glyoxamides **56** as an aldehyde component and using electrophilic Michael acceptor as alkylidenemalonate **57** (Scheme 3.19) [39]. A variety of glyoxamide derivatives were tested for this reaction, and it was found that the morpholine-derived glyoxamide derivative **56** was the best aldehyde component for this reaction in terms of yield and enantioselectivity. A wide range of β-alkyl substituted alkylidenemalonates were well tolerated in the presence of phenylalanine-derived NHC precatalyst **59** and the desired product was obtained in high yield and enantioselectivity.

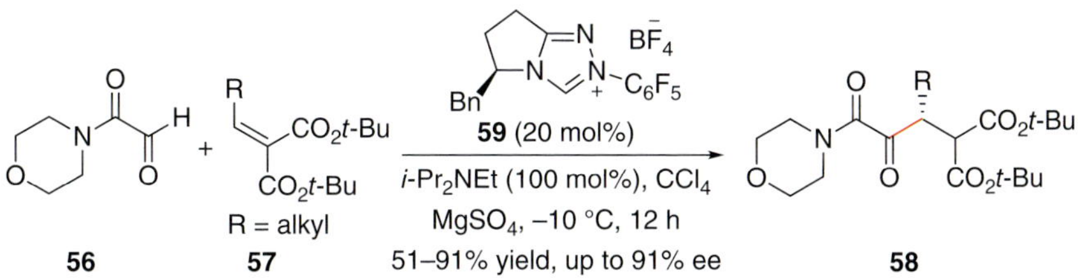

Scheme 3.19 Asymmetric intermolecular Stetter reaction of glyoxamides with alkylidenemalonates.

Subsequently, a similar type of intermolecular Stetter reaction was developed by the Rovis group using alkylidene β-ketoamide **60** as the Michael acceptor (Scheme 3.20) [40]. The key steps involve the addition of nucleophilic Breslow intermediate to Michael acceptor **60** followed by diastereoselective protonation in the presence of chiral triazolium salt **59** as a catalyst precursor. Additionally, some useful synthetic transformations of the 1,4-dicarbonyl product **61** have been shown demonstrating the utility of the method.

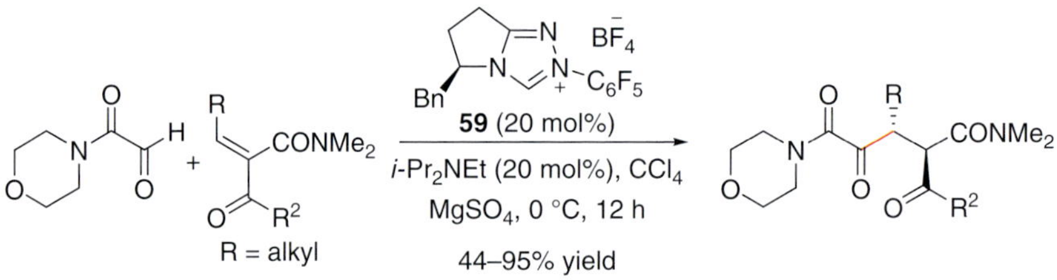

56 **60** **61**

Scheme 3.20 Asymmetric intermolecular Stetter reaction of glyoxamides with alkylidene ketoamides.

A highly efficient and enantioselective synthesis of α-amino acid derivatives via an intermolecular Stetter reaction was disclosed in 2011 by the Glorius group. Employing *N*-acylamido acrylate **62** as the Michael acceptor, a broad range of aldehydes worked well under the optimized reaction conditions (Scheme 3.21). In the presence of carbene generated from morpholine-derived chiral triazolium salt **64**, the desired α-amino acid derivatives **63** were obtained in high yield and enantioselectivity [41]. The key steps involved in this reaction are the addition of the Breslow intermediate to the Michael acceptor and an enantioselective protonation.

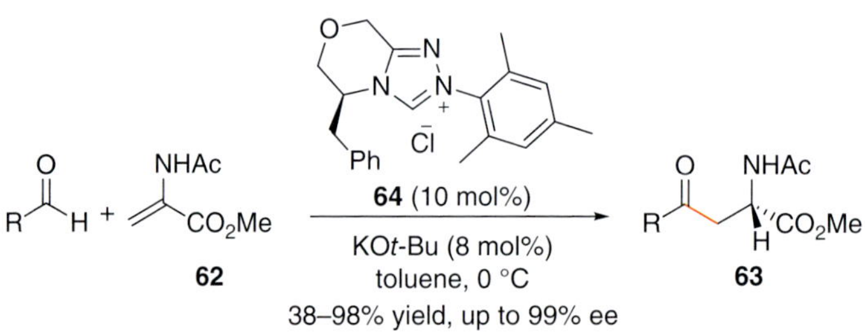

62 **63**

Scheme 3.21 Asymmetric intermolecular Stetter reaction of *N*-acylamido acrylates.

Moreover, two possible modes of addition of acyl anion to the Michael acceptor were proposed (Figure 3.1). The stepwise addition of chiral Breslow intermediate to the Michael acceptor **62** from the *Si*-face, as shown in Figure 3.1, was supported by a hydrogen bond between the enol hydrogen and the carbonyl oxygen of the Michael acceptor. A concerted mechanism was also suggested, which proceeds via cyclic five-membered transition state **66**. Additionally, a detailed computational study reveals that the use of KO*t*-Bu is crucial for

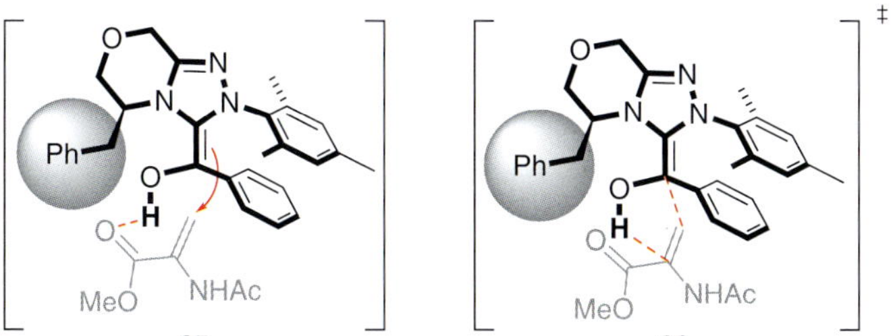

Figure 3.1 Stepwise and concerted pathways in enantioselective intermolecular Stetter reaction involving *N*-acylamido acrylate.

enantioselectivity, and *t*-BuOH played an important role in intermolecular proton transfer step [42].

A comparatively less electrophilic Michael acceptor such as acrylates **67** was employed as a coupling partner in the enantioselective intermolecular Stetter reaction by Glorius and coworkers. A new class of triazolium salt **69** containing electron-rich 2,6-dimethoxyphenyl substitution was used as the catalyst precursor (Scheme 3.22). A variety of electronically different (hetero)aromatic and aliphatic aldehydes and several acrylates underwent smooth Stetter reaction under the optimized reaction conditions to afford the desired α-chiral γ-ketoesters **68** in high yield and enantioselectivity [43].

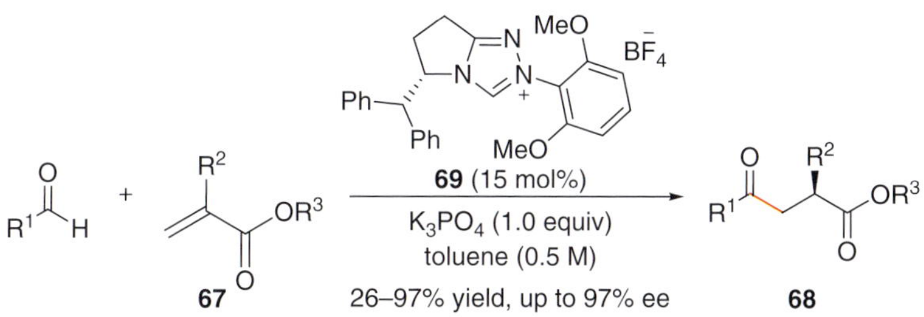

Scheme 3.22 Asymmetric intermolecular Stetter reaction using α,β-unsaturated esters.

An enantioselective Stetter reaction of α,β-unsaturated aldehydes with α-acyl chalcones **70** was uncovered in 2011 by the Chi group (Scheme 3.23) [44]. The carbene generated from the chiral aminoindanol-derived triazolium salt **72** was efficient in catalyzing the addition of enals to modified chalcone **70**, and the desired triketones **71** were obtained in high yield and enantioselectivity. Moreover, a broad range of β-alkyl, β-aryl, and β,β-disubstituted enals were found to be tolerable under the present reaction conditions. It may be noted that under slightly different reaction conditions, the same reagents and catalyst afforded the Diels–Alder adducts by the [4+2] cycloaddition of the NHC-bound enolate generated from enal and **72** with the modified chalcones [45].

A new batch of chiral triazolium salts were synthesized by the Rovis group and successfully applied for the intermolecular Stetter reaction using nitroalkenes **74**

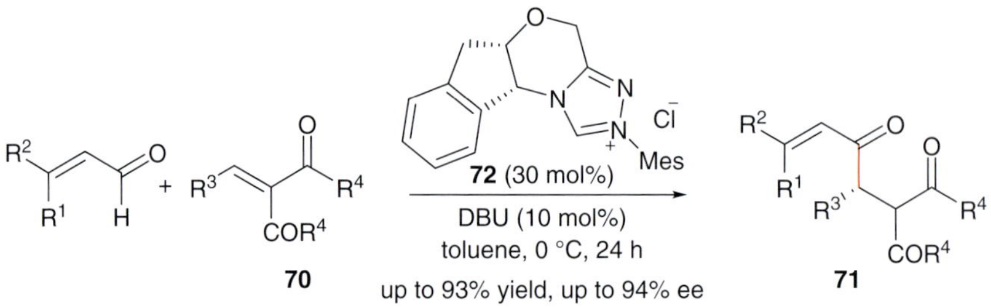

Scheme 3.23 Asymmetric intermolecular Stetter reaction of α,β-unsaturated aldehydes with modified chalcones.

as a Michael acceptor. Highly enantioselective β-nitro ketone products **75** were obtained employing carbene generated from fluoro- and isopropyl-substituted triazolium salt **76** (Scheme 3.24) [46]. Notably, the use of heteroaromatic aldehyde **73** was necessary for the high level of reactivity and selectivity. For a better understanding of the role of fluorine atom in the triazolium moiety, an X-ray crystallographic analysis was performed, and it was found that various stereoelectronic factor dominates over steric effect. Additionally, a quantum mechanical calculation supported the role of fluorine atom in **76** on the reactivity and selectivity [47].

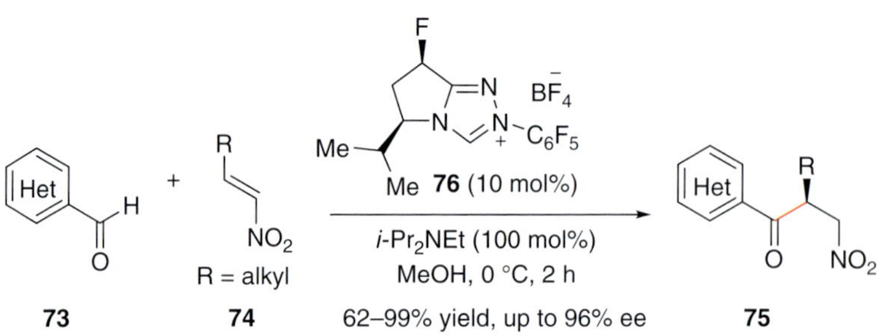

Scheme 3.24 Asymmetric intermolecular Stetter reaction of heteroaromatic aldehydes with nitroalkenes.

Subsequently, the Rovis group disclosed an extension of the asymmetric intermolecular Stetter reaction employing α,β-unsaturated aldehydes as the aldehyde components along with nitro olefins as the Michel acceptors. In the presence of fluorinated chiral triazolium salt **76** as a NHC precatalyst, and catechol as a bifunctional Brønsted acid, the desired Stetter product **77** was formed in good-to-excellent yield and enantioselectivity (Scheme 3.25) [48]. Interestingly, under these reaction conditions, the homoenolate-derived product was not observed. The use of 1.0 equiv of catechol was found to be essential for the reactivity, and the reaction proceeded with a low catalyst loading. The activation of initially generated tetrahedral intermediate via hydrogen-bonding interaction using catechol was proposed (Scheme 3.26). Through a series of isotope labeling experiments, it was concluded that the turnover-limiting step in this transformation is the initial proton transfer step.

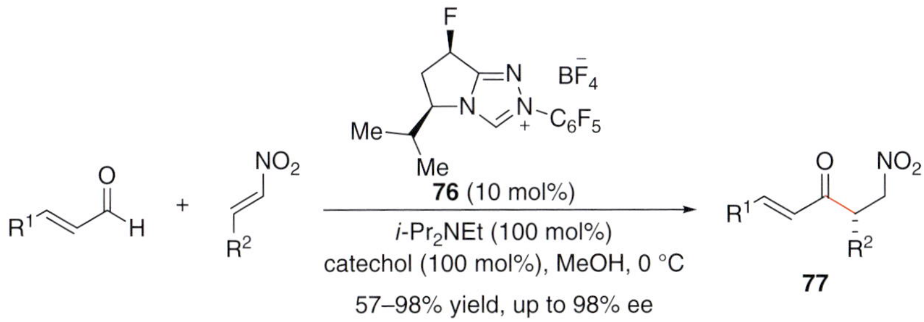

Scheme 3.25 Asymmetric intermolecular Stetter reaction of α,β-unsaturated aldehydes with nitro olefins.

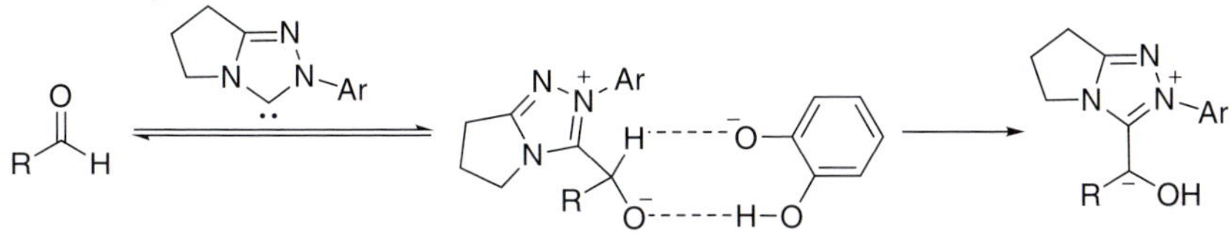

Scheme 3.26 Proposed mode of activation.

An enantioselective intermolecular Stetter reaction employing enolizable aldehydes was described by the Rovis group using β-nitro styrenes as the coupling partner (Scheme 3.27). A wide variety of aliphatic aldehydes and β-nitro styrenes underwent smooth carbon–carbon coupling using carbene generated from chiral triazolium salt **79**. Notably, the *trans*-fluorinated geometry in the pyrrolidine moiety of the carbene dramatically improved the reactivity and selectivity. By computational study, a reasonable explanation for the high enantioselectivity of the Stetter product **78** was provided [49].

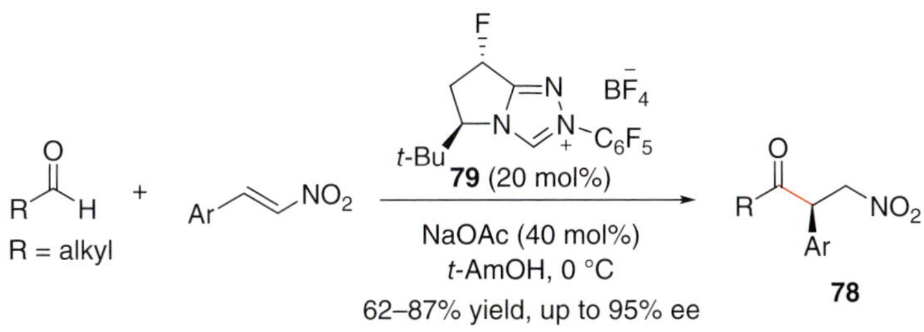

Scheme 3.27 Asymmetric intermolecular Stetter reaction of enolizable aldehydes with β-nitro styrenes.

Liu and coworkers uncovered the NHC-catalyzed intermolecular Stetter reaction for the β-selective *C*-glycosidation of 2-nitroglucals **80** (Scheme 3.28). The reaction of acyl anions, generated from heteroaromatic aldehydes, to the anomeric carbon of 2-nitroglucal resulted in the formation of the desired Stetter product **81** in high yields [50]. Moreover, in the presence of Cs_2CO_3 as the base, the reaction resulted in the formation of the nitro-eliminated *C*-glycosides.

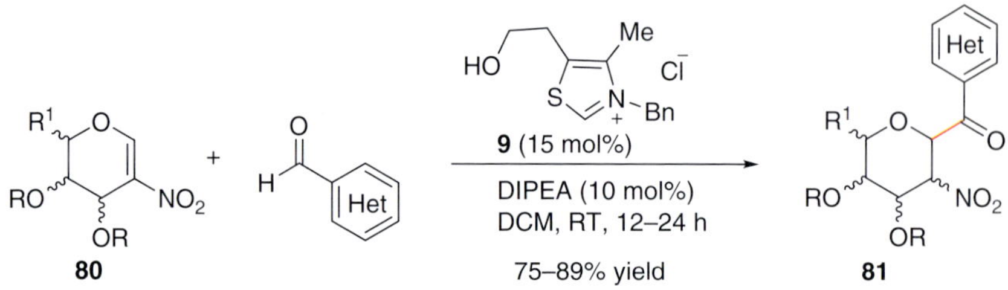

Scheme 3.28 NHC-catalyzed *C*-glycosylation using Stetter reaction.

The NHC generated from the fluorinated chiral triazolium salt **76** was further utilized by the Gravel group for an enantioselective intermolecular Stetter reaction using β,γ-unsaturated α-ketoesters **82** as Michael acceptors. This reaction condition was found to be suitable for a broad range of (hetero)aromatic aldehydes as well as Michael acceptors and the desired 1,2,5-tricarbonyl compounds **83** were formed in high yield and ee values (Scheme 3.29) [51]. A synthetically useful and stereoselective transformation of the chiral α,δ-diketoester product was also shown.

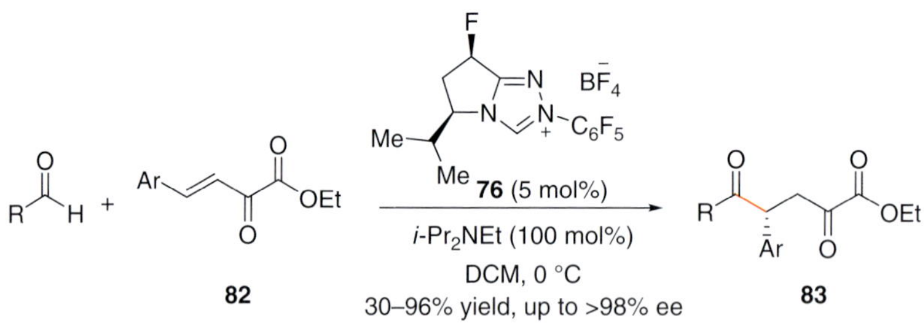

Scheme 3.29 Asymmetric intermolecular Stetter reaction using β,γ-unsaturated α-ketoesters as Michael acceptors.

In 1978, Stetter and Bender for the first time introduced vinyl sulfone as a Michael acceptor in the intermolecular Stetter reaction [52]. However, instead of the expected γ-keto sulfones **84**, the reaction resulted in a 1 : 1 mixture of symmetric 1,4-diketones **85** and γ-disulfone **86**. Recently, the NHC-catalyzed highly selective intermolecular Stetter reaction using vinyl sulfones as Michael acceptors was reported by the Biju group for the selective synthesis of γ-ketosulfones **84** (Scheme 3.30) [53]. The reaction was efficiently catalyzed by the carbene generated from thiazolium salt **87**, and a variety of γ-ketosulfones were formed in good yields and selectivity. A mechanistic study revealed that the reaction proceeds via the reversible generation of benzoin as the intermediate in this reaction, and the nucleophilic Breslow intermediate can add to the vinyl sulfone in a stepwise manner or in a concerted pathway.

Subsequently, the Biju group developed the intermolecular Stetter reaction using moderately electron-poor vinyl phosphonates as the Michael acceptor for the synthesis of γ-keto phosphonates **88** (Scheme 3.31) [54]. Considering

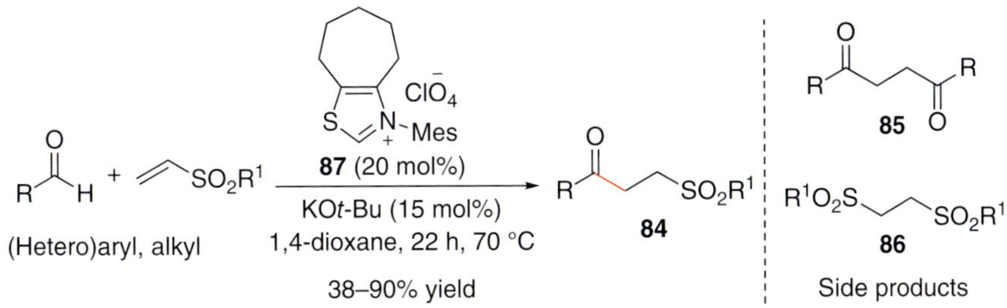

Scheme 3.30 Intermolecular Stetter reaction using vinyl sulfones.

the pharmaceutical and agrochemical importance of γ-ketophosphonates, this method was found to be an efficient and alternative approach compared to the traditional phospha-Michael reaction. A wide range of γ-keto phosphonates were synthesized by the generation of acyl anion via the umpolung of aldehyde under NHC derived from imidazolium salt **89** and their subsequent addition with vinyl phosphonates. Along with the vinyl phosphonates, the α-substituted vinyl phosphonates also tolerated under the reaction conditions.

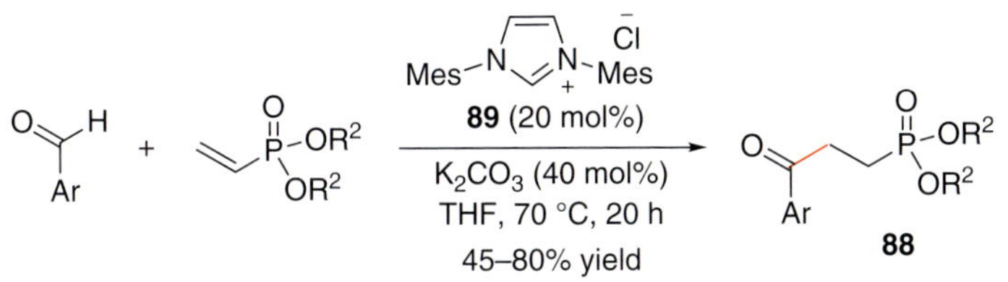

Scheme 3.31 Intermolecular Stetter reaction using vinyl phosphonates.

In 2011, the Yang group demonstrated the Stetter reaction of acetaldehyde. The generation of acyl anion from acetaldehyde was considered to be challenging. The intermolecular Stetter reaction of acetaldehyde with chalcones as Michael acceptors in the presence of carbene generated from the aminoindanol-derived pentafluoro-aryl-substituted chiral triazolium salt **25** furnished the 1,4-diketo product **90** in moderate-to-good yield and moderate enantioselectivity (Scheme 3.32) [55].

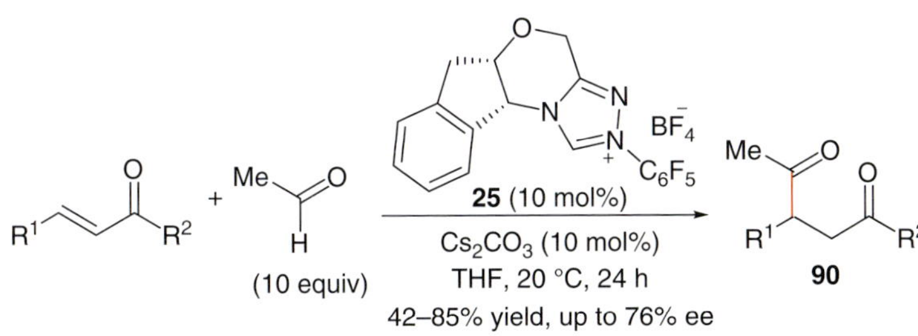

Scheme 3.32 Synthetic utility of acetaldehyde in Stetter reaction.

One of the difficulties in carrying out intermolecular Stetter reaction is the self-condensation of the donor aldehyde to form the benzoin side product.

In 2004, the Scheidt group solved this issue by employing acylsilanes **91** as the acyl anion precursor in a sila-Stetter reaction. The reaction was smoothly catalyzed by the carbene generated from thiazolium salt **93**, and the various electronically different substitutions on either side of α,β-unsaturated ketones were well-tolerated (Scheme 3.33) [56]. Under the optimized reaction conditions, the desired 1,4-diketone products **92** were formed in moderate-to-good yield.

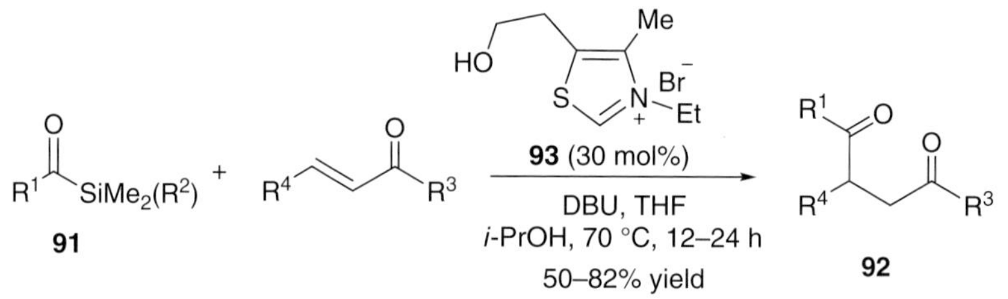

Scheme 3.33 NHC-catalyzed sila-Stetter reaction.

The proposed mechanism of this reaction involves the initial attack of the carbene to the acylsilanes followed by a Brook rearrangement to generate silyl-protected Breslow intermediate **94**, which is in equilibrium with intermediate **95**. In the presence of an added alcohol, enaminol intermediate **96** was formed by the desilylation of intermediate **94**. The conjugated addition of Breslow intermediate **96** to the Michael acceptor followed by the removal of catalyst furnished the 1,4-diketone product **92** (Scheme 3.34). In a related context, nucleophilic acylation of the nitroalkenes was developed for the synthesis of β-nitroketones by the Scheidt group [57].

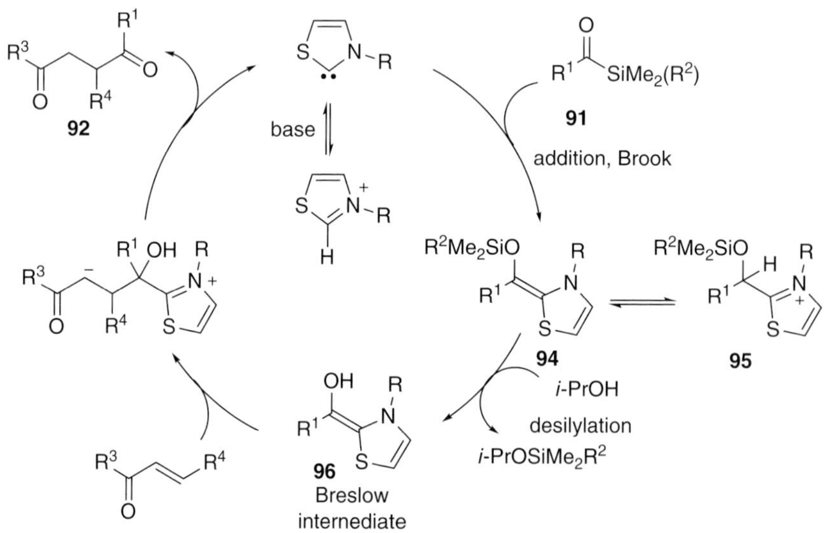

Scheme 3.34 Proposed mechanism of sila-Stetter reaction.

The NHC-catalyzed conjugated addition of in situ generated acyl anions under an aqueous medium was demonstrated by Scheidt group in 2005 [58]. The Breslow intermediate was generated in a biomimetic manner by the decarboxylation of pyruvates in the presence of commercially available thiazolium salt **9** as the precatalyst (Scheme 3.35). Various α-ketoesters and β-aryl substituted unsaturated 2-acyl imidazoles **97** were well tolerated under the mild reaction conditions, and the desired 1,4-diketone product **98** was formed in good-to-excellent yield.

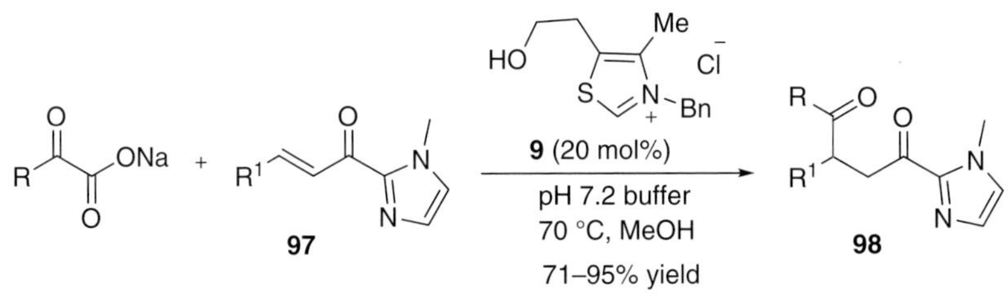

Scheme 3.35 Stetter reaction in aqueous medium.

In an effort to find an efficient, bench-stable, and alternative acyl anion precursor, 2,3-butanedione **99** or cyclohexane-1,2-dione were successfully employed in NHC-catalyzed intermolecular Stetter reaction (Scheme 3.36). A variety of enone derivatives underwent smooth conversion to 1,4-diketo product **100** in the presence of carbene generated from thiazolium salt **9** in moderate-to-good yield [59]. Moreover, under microwave conditions, the reaction furnished the desired product in lesser reaction time.

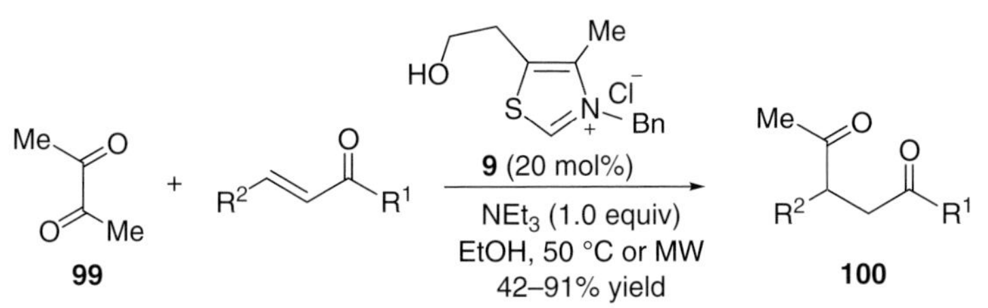

Scheme 3.36 Use of biacetyl for acyl anion generation in Stetter reaction.

Mechanistically, the reaction proceeds via the initial attack of the carbene to the α-diketone **99** to form the tetrahedral intermediate **101**. The attack of EtOH to the intermediate **101** followed by the elimination of the ethyl acetate generates nucleophilic Breslow intermediate **102**. The addition of Breslow intermediate to the chalcone derivative forms the 1,4-dicarbonyl product **100** following the usual Stetter mechanism (Scheme 3.37). Moreover, a similar type of intermolecular Stetter reaction of biacetyl and chalcones was developed using environmentally benign solvent system and a thiazolium salt precatalyst, which are immobilized on silica and monolithic polystyrene [60].

In 2014, Takaki and coworkers reported a Stetter-type double acylation of enone using diaryl 1,2-diones as the acyl anion source. The reaction proceeds

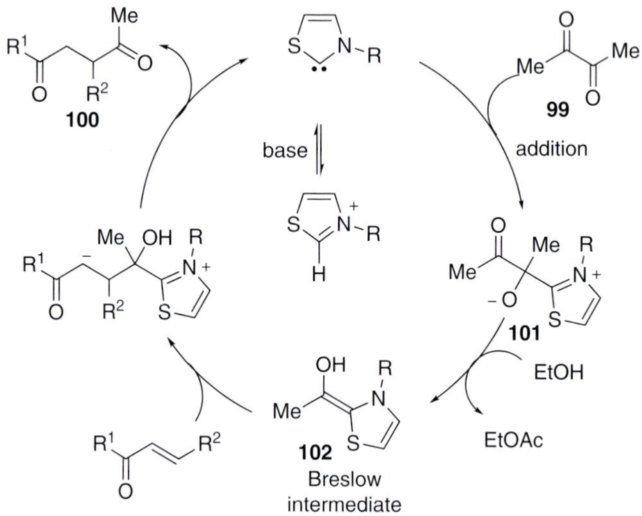

Scheme 3.37 Proposed reaction pathway for the Stetter reaction of linear α-diketones.

through the benzoyl-protected Breslow intermediate **104** in the presence of NHC generated from the thiazolium salt **87**, and the desired triketone products **103** was formed in moderate-to-good yield (Scheme 3.38) [61]. Disappointingly, reactions using aliphatic 1,2-diketones and cyclic 1,2-diketones resulted in the formation of the expected product in low yields.

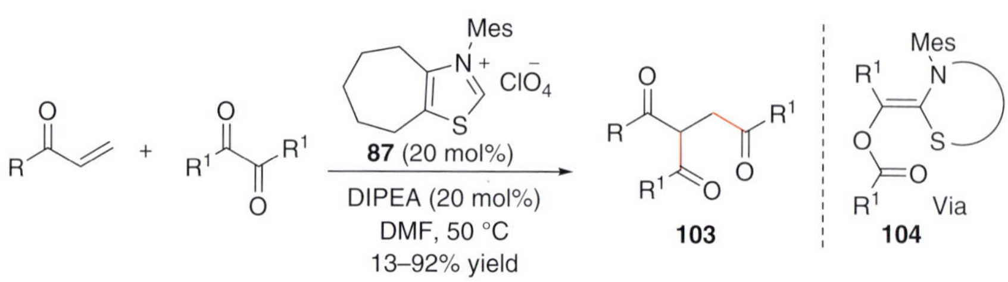

Scheme 3.38 NHC-catalyzed double acylation of enones with diaryl 1,2-diones.

In 2013, the Chi group reported the generation of the one-carbon nucleophile (formyl anion equivalent) by the activation of the carbohydrates for intermolecular Stetter reaction. The one-carbon nucleophile was formed via a retro-benzoin-type process and catalytic C—C bond cleavage of carbohydrates. A variety of β-formyl ketones **105** were synthesized using biomass-based carbohydrates and α,β-unsaturated ketones under microwave reaction conditions in the presence of NHC generated from the thiazolium precatalyst **106** (Scheme 3.39). The renewable nature of carbohydrate and the first application of formaldehyde-derived acyl anions are the noteworthy features of this reaction [62].

The use of bis(amino)cyclopropenium salt **108** as a carbene precursor for intermolecular Stetter reaction was disclosed by the Gravel group in 2013. The less

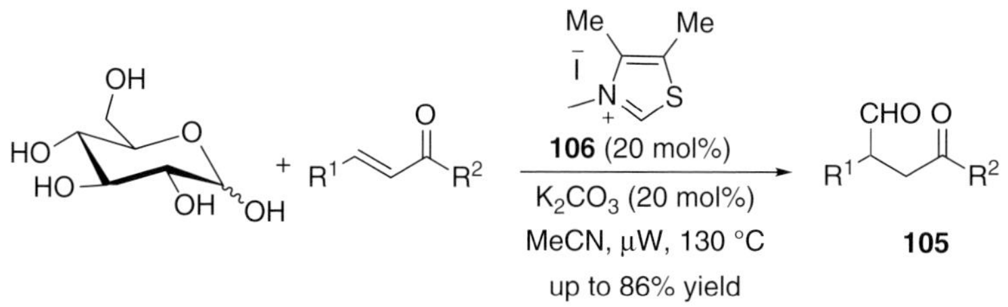

Scheme 3.39 Use of carbohydrates as formaldehyde equivalents in Stetter reaction.

bulky nature is the main advantages of this catalyst and the reaction does not proceed via the benzoin intermediate. Under the optimized reaction conditions, various electronically different aldehydes worked well and the 1,4-diketone products **107** were formed in high yields (Scheme 3.40). Additionally, the use of a chiral bis(amino)cyclopropenium salt-derived carbene resulted in the formation of the chiral 1,4-diketone product in up to 36% ee [63].

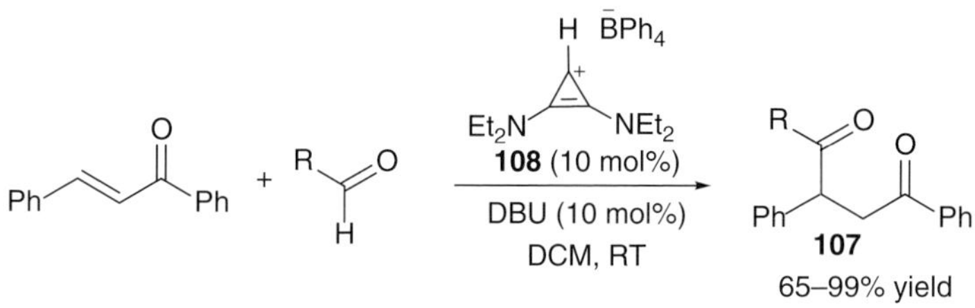

Scheme 3.40 Diamino cyclopropenylidene-catalyzed intermolecular Stetter reaction.

The enantioselective enzyme-catalyzed intermolecular Stetter reaction was uncovered by the research group of Müller. The thiamine diphosphate-dependent enzyme PigD generated the acyl anion intermediate (Breslow intermediate) via the decarboxylation of the pyruvate, and their subsequent reaction with various enones furnished the 1,4-diketone products in moderate yields and excellent enantioselectivity (Scheme 3.41) [64].

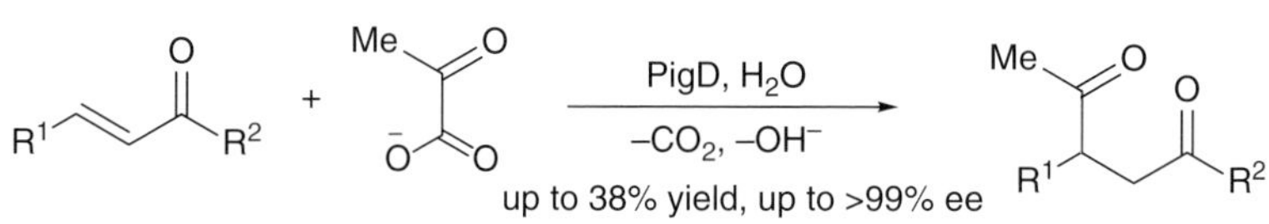

Scheme 3.41 Enzymatic intermolecular Stetter reaction.

3.5 Cascade Processes Involving Stetter Reaction

NHC-catalyzed cascade processes involving intermolecular Stetter reaction was reported by the Gravel group [65]. The addition of catalytically generated acyl anions to the dienones **109** followed by an intramolecular Michael addition afforded the indane derivatives **110**. In the presence of NHC generated from the thiazolium salt **93**, various aldehydes and differently substituted dienones furnished the expected product in good yield and diastereoselectivity (Scheme 3.42).

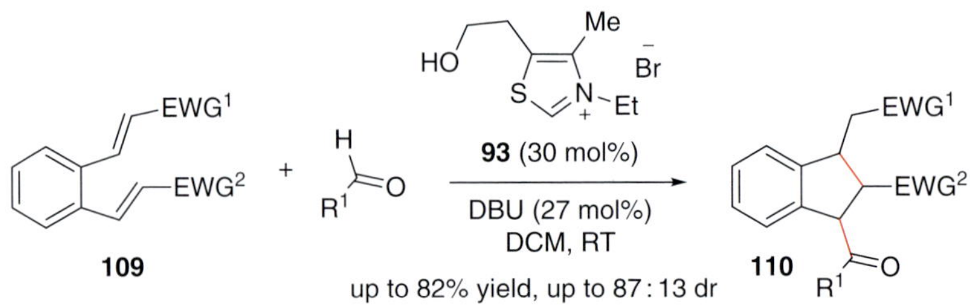

Scheme 3.42 NHC-catalyzed domino Stetter–Michael cascade for indane synthesis.

Subsequently, the Gravel group uncovered the Stetter–aldol–Michael and Stetter–aldol–aldol cascade reaction for a diastereoselective synthesis of spiro-bis-indanes **112** [66]. This efficient cascade reaction proceeds smoothly in the presence of carbene generated from thiazolium salt **93** and forges three new carbon–carbon bonds with the construction of a spiro all-carbon quaternary center in high yield and diastereoselectivity (Scheme 3.43).

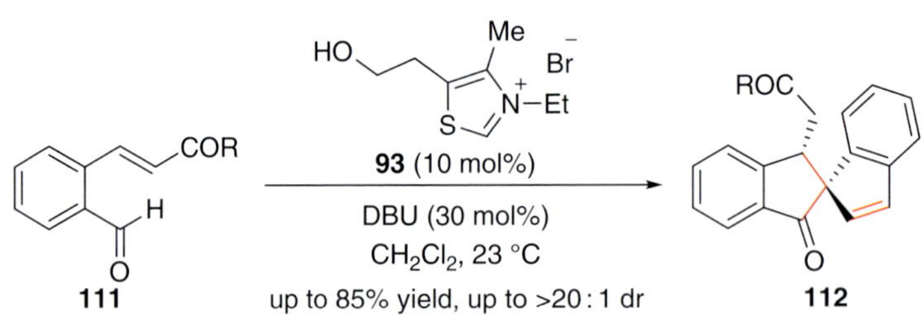

Scheme 3.43 NHC-catalyzed Stetter–aldol–Michael cascade.

The proposed pathway for this cascade reaction involves the initial generation of nucleophilic Breslow intermediate **113** from 2-formyl chalcone **111** and their addition to another molecule of **111** to generate the enolate intermediate **114**. An intramolecular aldol reaction of enolate **114** followed by the regeneration of the free NHC resulted in the formation of the indanol intermediate **115**. In the presence of a base, the intermediate **115** undergoes an intramolecular Michael addition with the enone moiety, which was followed by dehydration to furnish the desired spiro product **112** (Scheme 3.44).

The NHC-catalyzed organocascade reaction involving intermolecular Stetter–aldol sequence was described by Ye and coworkers for the diastereoselective synthesis of 3-substituted 4-hydroxy tetralones **116** (Scheme 3.45) [67]. A wide variety of *trans*-selective tetralone derivatives were synthesized in good yields in the presence of carbene generated from thiazolium salt **9**. Notably, the *cis*-selective product was obtained when the reaction was performed in a stepwise manner. Moreover, using 1,2-diactivated Michael acceptors as

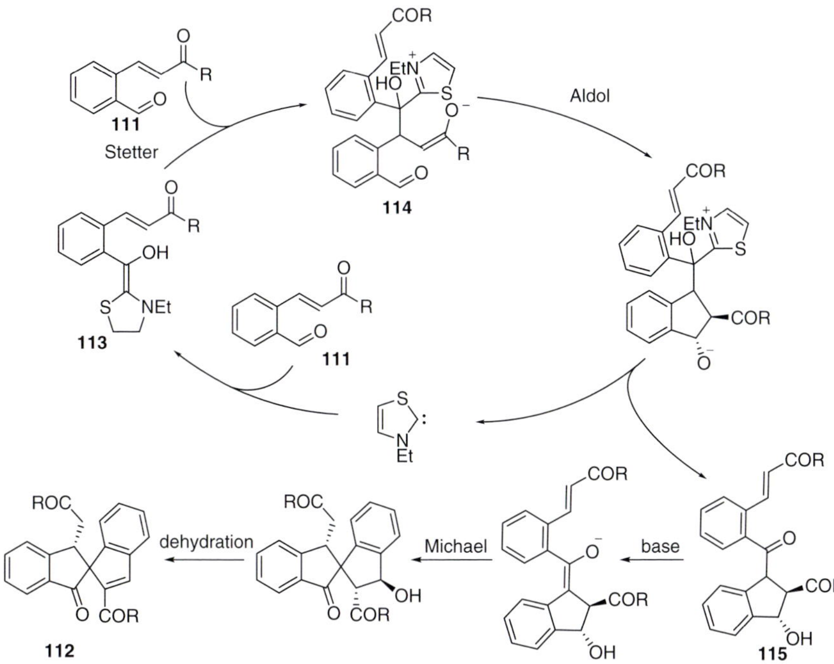

Scheme 3.44 Proposed mechanism for NHC-catalyzed Stetter–aldol–Michael cascade reaction.

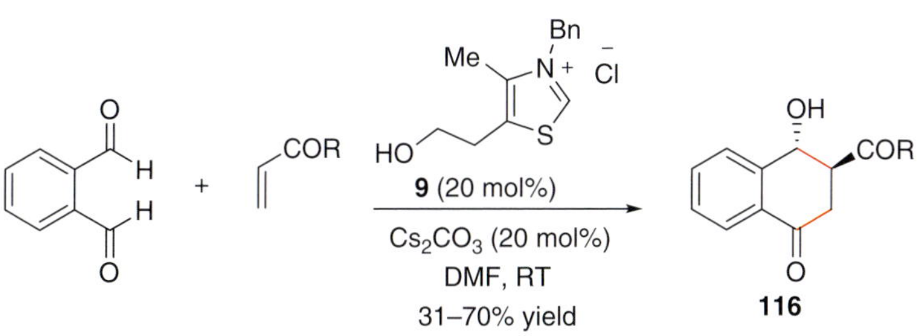

Scheme 3.45 NHC-catalyzed domino Stetter–aldol cascade for 4-hydroxy tetralone synthesis.

a coupling partner, the reaction afforded the synthesis of 2,2-disubstituted 3-hydroxyindanones via a formal [4+1] annulation [68].

An organocatalyzed [1+2+2] annulation reaction via sequential NHC-catalyzed Stetter reaction and chiral secondary amine-catalyzed Michael–aldol condensation for the stereoselective synthesis of fully functionalized cyclopentenes **118** were demonstrated by Liao and coworkers [69]. The Stetter reaction between aromatic aldehydes and nitroalkenes afforded the β-nitroketones **117**, which react with α,β-unsaturated aldehydes through a chiral secondary amine (**119**)-catalyzed Michael-aldol condensation cascade, and generates the expected product **118** in high yield and enantioselectivity (Scheme 3.46). Similarly, when the reaction was carried out using heteroaromatic aldehydes such as picolin aldehyde, a fully substituted cyclopentanol product was formed through sequential Stetter and Michael–aldol reaction pathway [70].

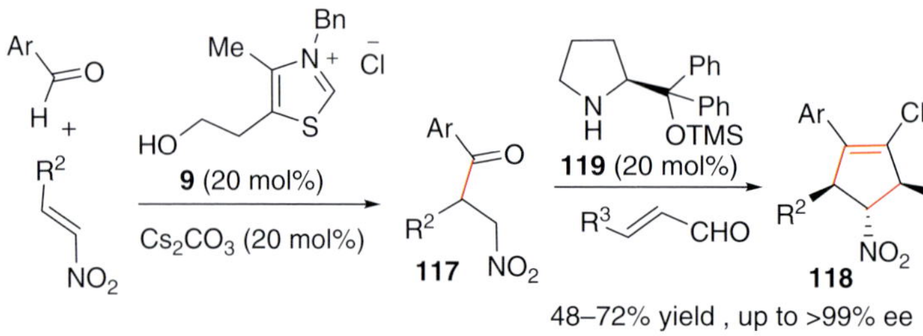

Scheme 3.46 Sequential Stetter and Michael–aldol condensation cascade.

3.6 NHC-catalyzed Hydroacylation Reactions

The NHC-catalyzed umpolung of aldehydes via the generation of nucleophilic Breslow intermediates and their interception with aldehydes, ketones (benzoin), imines (aza-benzoin), and Michael acceptors (Stetter) have been well explored. Moreover, the umpolung concept has been extended to Michael acceptors such as methacrylates for intermolecular reactions [71]. Recently, in the NHC-catalyzed addition of acyl anions to unactivated carbon–carbon multiple bonds, the hydroacylation reactions have drawn more attention (Scheme 3.47) [72]. It may be noted that the related transition metal-catalyzed hydroacylation reactions to unactivated carbon–carbon bonds have been developed earlier [73].

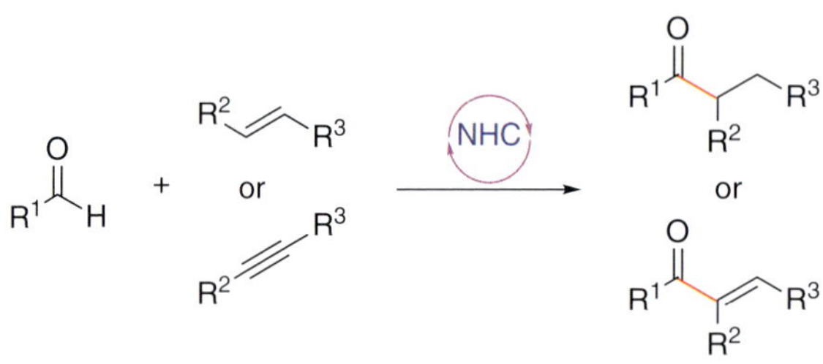

Scheme 3.47 Hydroacylation of unactivated olefins.

In 2008, She and coworkers reported the NHC-catalyzed intramolecular addition of acyl anions to enol ethers for the synthesis of benzofuranones **120** [74]. The carbene generated from commercially available thiazolium salt **121** was efficient in catalyzing the reaction, and the desired product was obtained in excellent yield (Scheme 3.48). With the available information, it was unclear whether the reaction proceeds via a concerted pathway or through an oxonium intermediate.

In 2009, Glorius and coworkers developed the NHC-organocatalyzed intramolecular hydroacylation of unactivated carbon–carbon double bonds. In the presence of carbene generated from thiazolium salt **87** containing seven-membered ring back bone, and 1,8-diaza bicycle [5.4.0] undec-7-ene

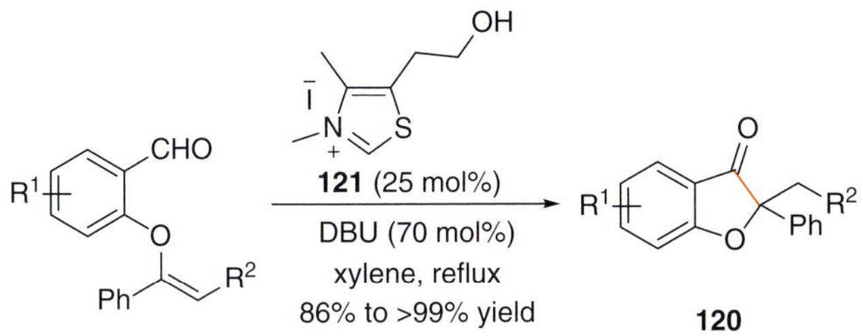

Scheme 3.48 NHC-catalyzed intramolecular hydroacylation of enol ethers.

(DBU) as base under heating conditions, the intramolecular cyclization of 2-allyloxy benzaldehydes **122** furnished the corresponding chromanones **123** (Scheme 3.49) [75]. A broad range of substrates with electronically different substitution on aryl ring did not affect the reaction outcome, and the corresponding product was formed in good-to-excellent yields. Additionally, the chromanone moiety containing all-carbon quaternary center was also synthesized using this methodology.

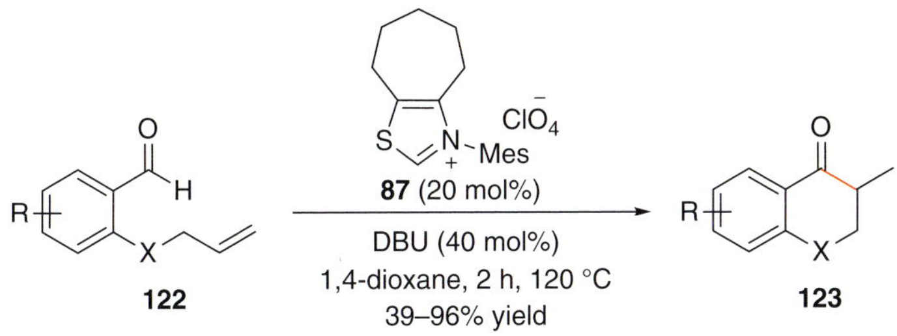

Scheme 3.49 NHC-catalyzed hydroacylation of unactivated double bonds.

A tentative reaction pathway of the NHC-catalyzed hydroacylation reaction is presented in Scheme 3.50. The addition of carbene derived from thiazolium salt **87** to aldehyde **122** forms the nucleophilic Breslow intermediate **124**. This acyl anion equivalent **124** adds intramolecularly to the tethered unactivated alkene via a concerted transition state **125** to form the cyclic zwitterionic tetrahedral intermediate **126**. This intermediate **126** was converted to the desired product **123** by the release of the free carbene. Interestingly, the transition state **125** looks like the Conia-ene-type transition state (based on the polarity of the reacting alkenes) or the reverse-Cope-type elimination (based on the five-membered transition state). Additionally, the DFT calculations have been performed to get more insight into the essential step of the reaction, [76]. These studies supported a concerted but highly asynchronous transition state in the intramolecular hydroacylation of unactivated olefins. Moreover, the recent electron localization function (ELF) analysis indicated that the NHC-catalyzed hydroacylation reactions are mechanistically dissimilar from intramolecular Stetter reaction [77].

Subsequently, a similar type of hydroacylation reaction was demonstrated for the enantioselective construction of the chromanone derivatives **127** [76]. Using

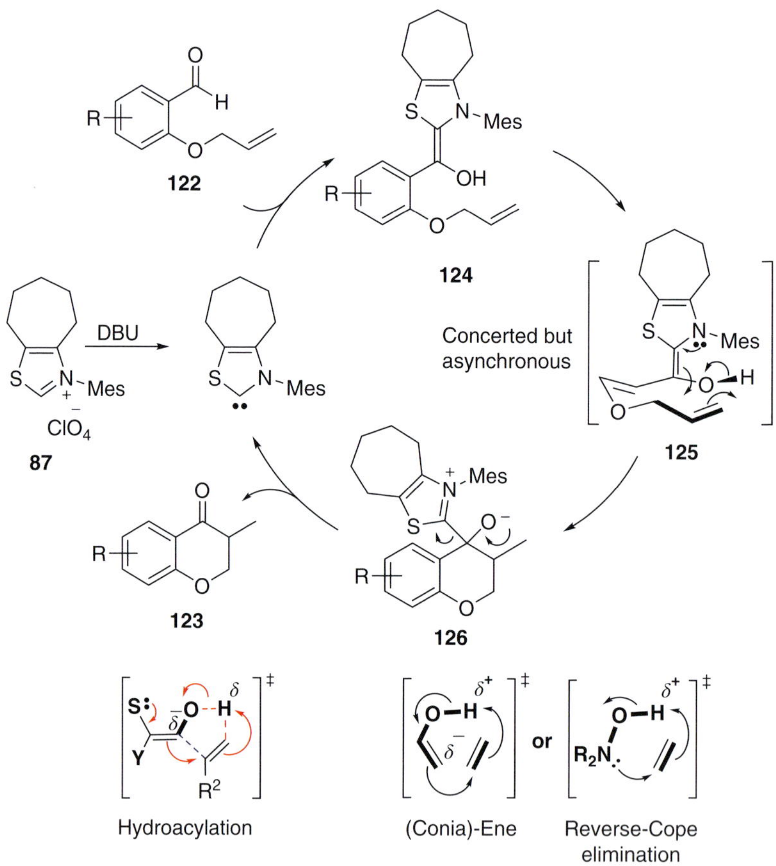

Scheme 3.50 Proposed mechanism of the hydroacylation of unactivated olefins.

morpholine-derived chiral triazolium salt **64** as the NHC precursor, the chromanone derivatives **127** containing a quaternary stereocenter were formed in a high yield and enantioselectivity (Scheme 3.51). Additionally, the high level of ee value and the reaction mechanism were supported by DFT studies.

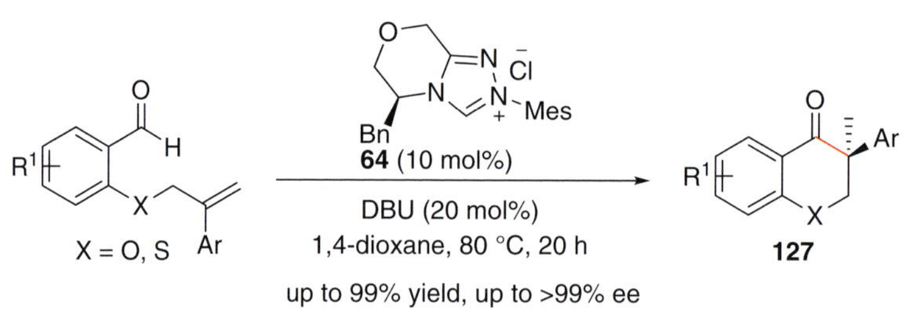

Scheme 3.51 NHC-catalyzed enantioselective hydroacylation of unactivated double bonds.

In 2011, the Glorius group employed cyclopropenes as electron–neutral olefins in the NHC-catalyzed intermolecular hydroacylation reactions [78]. The NHC generated from the triazolium salt **129** was efficient in catalyzing the reaction between aromatic aldehydes and cyclopropenes to afford the acyl cylopropanes **128** in excellent diastereoselectivity and good yield (Scheme 3.52). A deuterium labeling experiment showed that the product formation took place via a concerted *syn* hydroacylation mechanism. Moreover, DFT calculations reveal the important role of base in hydroacylation of cyclopropenes, which promotes the proton transfer step leading to the generation of the Breslow intermediate [79]. Subsequently, a new class of electron-rich chiral triazolium salts was synthesized by the Glorius group and utilized for intermolecular hydroacylation of cyclopropenes [80]. A highly enantioselective and diastereoselective intermolecular hydroacylation of aromatic aldehydes to the electron-neutral cyclopropenes were described, which was efficiently catalyzed by the NHC generated from the chiral triazolium salt **130**. Moreover, a detailed kinetic study revealed that the installation of 2,6-dimethoxy phenyl group in the catalyst precursor enhanced the reactivity and selectivity in intermolecular hydroacylation reactions.

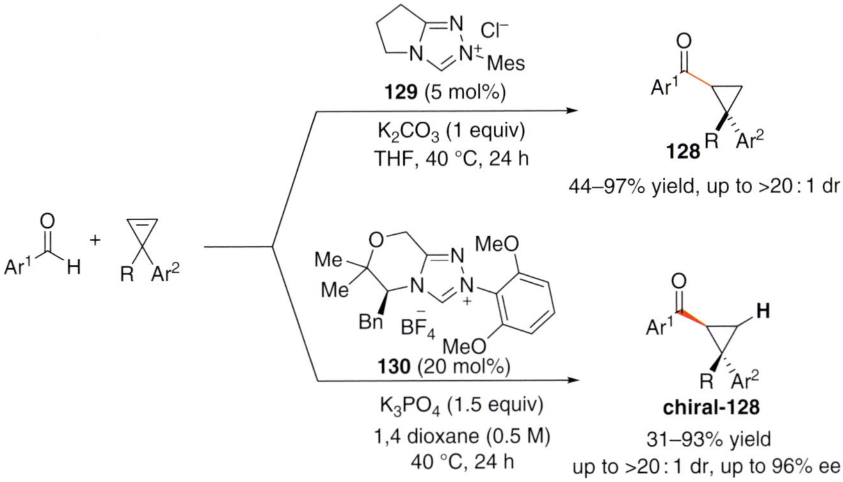

Scheme 3.52 NHC-catalyzed intermolecular hydroacylation of cyclopropenes.

Although the NHC-catalyzed intramolecular hydroacylation reactions are well explored, the intermolecular hydroacylation of electron-poor olefins remained as a challenge. In 2013, The Glorius group disclosed the NHC-catalyzed intermolecular hydroacylation of styrenes (Scheme 3.53) [81]. In the presence of carbene generated from electron-rich triazolium salt **131**, a wide variety of (hetero)aromatic aldehydes as well as electron-poor and electron-rich styrenes were well tolerated under the mild reaction condition, and a mixture of linear and branched hydroacylated products were formed in good yields. The nucleophilic Breslow intermediate generated from aldehydes and NHCs can add to styrenes in three possible modes (cationic, anionic, and concerted) resulting in the formation of different regioisomeric products.

Zhou and coworkers reported the NHC-catalyzed intramolecular acylation of allylic electrophiles **132** for the synthesis of α,β-unsaturated

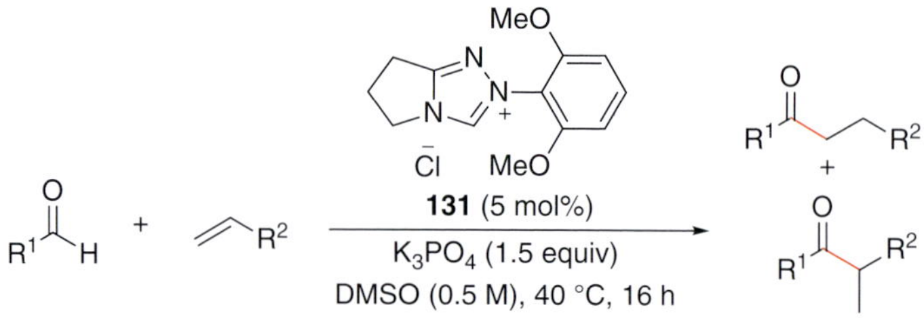

Scheme 3.53 NHC-catalyzed intermolecular hydroacylation of styrenes.

chromanones **133** (Scheme 3.54) [82]. Employing carbene generated from the bulky mesityl-substituted thiazolium salt **87**, the addition of acyl anion to tethered allyl bromide proceeded through S_N2' pathway. Additional mechanistic investigation also supported the formation of the product via the S_N2' pathway and ruled out the possibility of addition–elimination mechanism. Later, an enantioselective version of this intramolecular S_N2' reaction was developed using chiral NHCs [83].

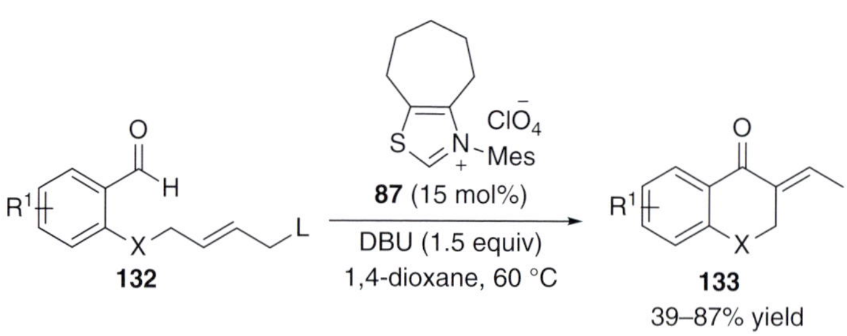

Scheme 3.54 NHC-catalyzed intramolecular acylation of allyl electrophiles.

The NHC-catalyzed intramolecular hydroacylation of allenes were uncovered recently for the synthesis of chromones **135** (Scheme 3.55) [84]. The generation of Breslow intermediate from allenyl aldehydes **134**, followed by the addition to allene moiety, afforded the desired product in moderate-to-good yield. It may be mentioned in this context that a cascade reaction involving the tertiary amine-catalyzed *N*-allylation of indole 2-carboxaldehydes followed by the NHC-catalyzed hydroacylation of the resultant allylic amines for the synthesis of pyrroloindolones has been demonstrated recently [85].

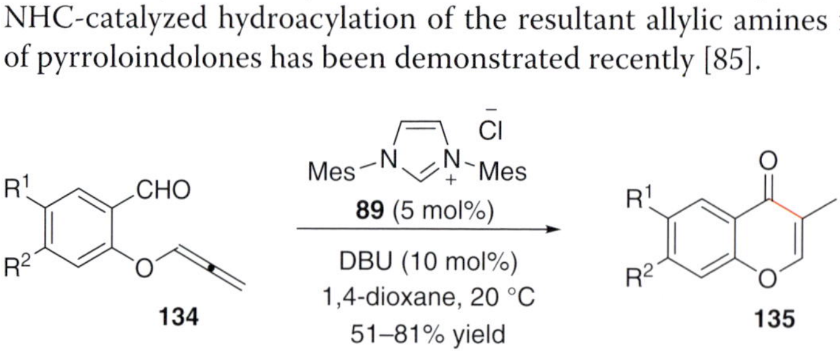

Scheme 3.55 NHC-catalyzed hydroacylation of allenyl aldehydes.

Apart from the NHC-catalyzed hydroacylation of carbon–carbon double bonds, the unactivated carbon–carbon triple bond can also be used as a coupling partner in hydroacylation reactions. In 2010, the Glorius group reported the intramolecular hydroacylation of triple bonds by using aldehydes bearing a propargyl moiety **136** for the synthesis of arylidene chromanones **137** as a single isomer (Scheme 3.56) [86]. Various electronically different substitutions in both aromatic rings are well tolerated under the optimized conditions, and the desired product was formed in an excellent yield. Also, this protocol is applicable for the synthesis of quinolin-4-ones. Similarly, related NHC-catalyzed intramolecular hydroacylation reactions of alkynal phosphonates leading to the synthesis of chromone phosphonates have been developed recently [87].

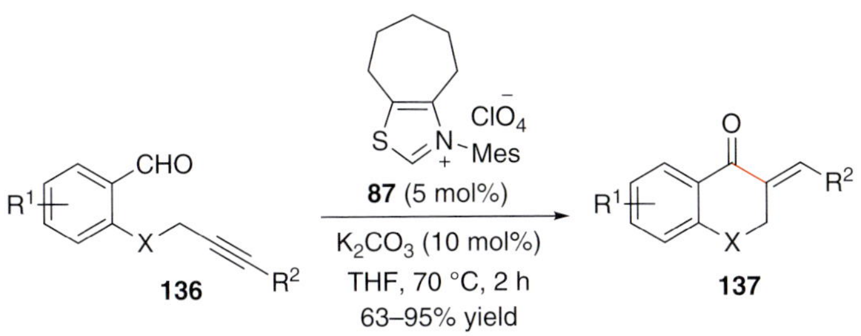

Scheme 3.56 NHC-catalyzed hydroacylation of unactivated internal alkynes.

Moreover, this NHC-catalyzed hydroacylation strategy of triple bonds was applied for an organocascade reaction. The reaction of propargyloxy aldehydes **138** with another molecule of aldehyde furnished the chromanone **139** with a 1,4-diketone motif via an intramolecular hydroacylation reaction of terminal alkynes followed by intermolecular Stetter reaction (Scheme 3.57) [86]. The use of mild base, low catalyst loading (5 mol%), and a broad functional group tolerance are the noteworthy features of this highly selective cascade reaction. Additionally, this methodology was applied for a one-pot synthesis of benzopyranopyrrole derivative through a hydroacylation–Stetter cascade reaction followed by the reaction with aniline derivative.

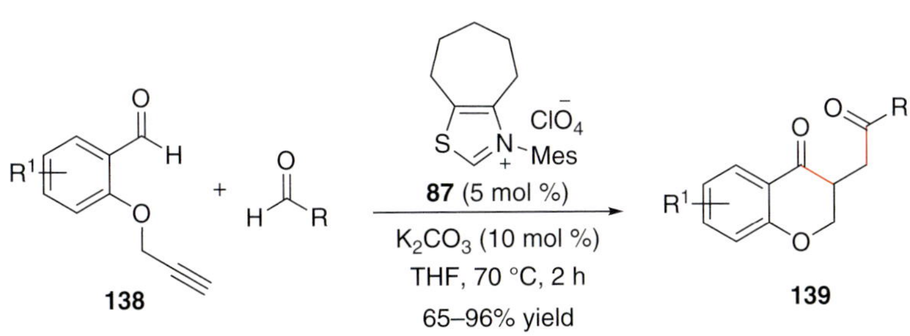

Scheme 3.57 NHC-catalyzed hydroacylation–Stetter cascade.

A further investigation on this organocascade reaction found that the chromanone **139** formed under the NHC-catalyzed hydroacylation–Stetter cascade reaction was very sensitive to the base used and the reaction temperature. When the reaction was performed using 20 mol% DBU at 80 °C for a longer reaction time, the chromanone **139** underwent a retro-Michael addition to generate the phenol **141**, which, on 1,3-H shift followed by intramolecular oxa-Michael reaction, furnished the 2,2-disubstituted benzofuranone derivative **140** (Scheme 3.58) [88]. A variety of electronically different substituents on both reacting partners were well tolerated under the intramolecular hydroacylation–Stetter–rearrangement cascade pathway and desired products are formed in high yield.

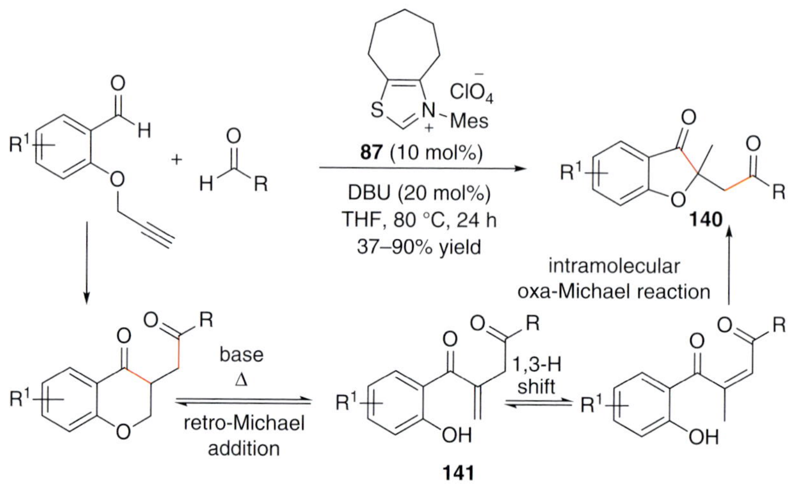

Scheme 3.58 NHC/base-catalyzed hydroacylation–Stetter–rearrangement cascade.

Furthermore, the Glorius group demonstrated the NHC-catalyzed intermolecular hydroacylation reaction of arynes, which are highly electrophilic intermediates in organic synthesis [89]. A formal insertion of arynes into the C_{formyl}—H bond of aldehydes afforded various aryl ketones **143** in the presence of carbene generated from thiazolium salt **87**. Notably, the fluoride-induced generation of aryne from 2-trimethylsilyaryl triflate **142** using 2.0 equiv each of KF and 18-crown-6 was efficiently combined with nucleophilic carbene catalysis for the first time (Scheme 3.59). A broad range of aldehydes and aryne precursors were well tolerated and the desired product formed in high yield.

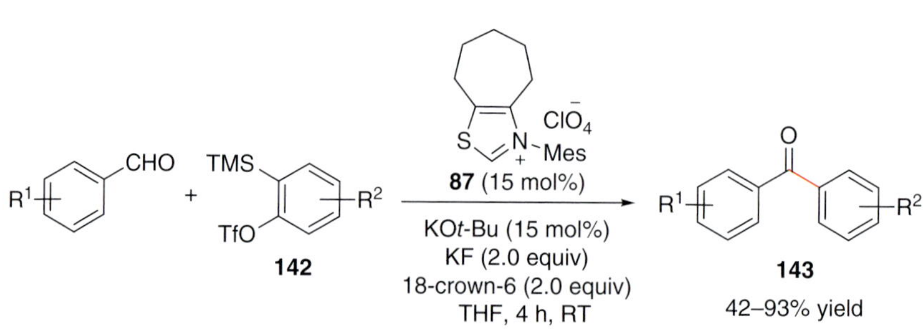

Scheme 3.59 NHC-catalyzed intermolecular hydroacylation of arynes.

Very recently, the hydroacylation–Stetter cascade reaction was described for the efficient synthesis of bisbenzopyrones **145** from alkynyl bisbenzaldehydes **144** (Scheme 3.60) [90]. The reaction proceeds through the intramolecular hydroacylation of one aldehyde with alkyne followed by intramolecular Stetter reaction to obtain the desired product. A variety of symmetrical and unsymmetrical substitution on alkynyl bisbenzaldehyde **144** was well tolerated in the presence of carbene generated from **87** under heating condition.

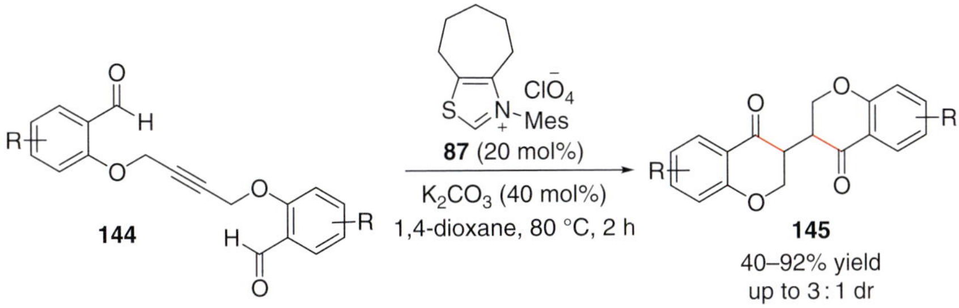

Scheme 3.60 NHC-catalyzed intramolecular hydroacylation–Stetter reaction cascade.

3.7 Conclusion

In summary, remarkable developments in the NHC-catalyzed Stetter reaction leading to the unconventional access to 1,4-diketones or related molecules are presented in this chapter. Advances in catalyst development, various intramolecular Stetter reactions for the synthesis of chromanone and furanone scaffolds, enantioselective Stetter reaction, and the developments in the intermolecular version of this reaction have been discussed. Moreover, advances in the addition of catalytically generated Breslow intermediates to electron-poor olefins (hydroacylation reactions) are also described. The products of the Stetter reaction could be easily functionalized to synthesize valuable heterocycles such as furans, pyrroles, thiophenes, etc. Future direction of research in this area will be probably directed to the use of β,β-disubstituted Michael acceptors for the acyl anion addition, thereby leading to the construction of ketones having an all-carbon quaternary stereocenter at the α-position. A rational design of the new NHC catalyst may solve this challenging problem. It is reasonable to believe that the simplicity and sophistication of the NHC-catalyzed Stetter reaction will inspire a broad range of organic chemists to explore new possibilities and imaginative applications of this precious carbon–carbon coupling reaction.

References

1 Breslow, R. (1958). *J. Am. Chem. Soc.* 80: 3719–3726.
2 Igau, A., Grutzmacher, H., Baceiredo, A., and Bertrand, G. (1988). *J. Am. Chem. Soc.* 110: 6463–6466.
3 Arduengo, A.J. III, Harlow, R.L., and Kline, M.A. (1991). *J. Am. Chem. Soc.* 113: 361–363.

4 For review, see: (a)Albright, J.D. (1983). *Tetrahedron* 39: 3207–3233. (b) Aitken, R.A. and Thomas, A.W. (2001). *Adv. Heterocycl. Chem.* 79: 89–109. (c) Seebach, D. (1979). *Angew. Chem. Int. Ed. Engl.* 18: 239–336.
5 Stetter, H. and Schreckenberg, M. (1973). *Angew. Chem. Int. Ed. Engl.* 85: 89; (1973). *Angew. Chem. Int. Ed. Engl.* 12: 81.
6 Stetter, H. (1976). *Angew. Chem. Int. Ed. Engl.* 88: 695–704; (1976). *Angew. Chem. Int. Ed. Engl.* 15: 639–647.
7 For reviews on Stetter reaction, see: (a)Read de Alaniz, J. and Rovis, T. (2009). *Synlett* 1189–1207. (b) Rovis, T. (2008). *Chem. Lett.* 37: 2–7. (c) Yetra, S.R., Patra, A., and Biju, A.T. (2015). *Synthesis* 47: 1357–1378. For a recent highlight, see: (d)Christmann, M. (2005). *Angew. Chem. Int. Ed.* 117: 2688–2690; (2005). *Angew. Chem. Int. Ed.* 44: 2632–2634. See also: (e)Read de Alaniz, J., Kerr, M.S., Moore, J.L., and Rovis, R. (2008). *J. Org. Chem.* (73): 2033–2040.
8 Moore, J.L., Silvestri, A.P., Read de Alaniz, J. et al. (2011). *Org. Lett.* 13: 1742–1745.
9 Hawkes, K.J. and Yates, B.F. (2008). *Eur. J. Org. Chem.* 5563–5570.
10 Ciganek, E. (1995). *Synthesis* 1311–1314.
11 Enders, D., Breuer, K., Runsink, J., and Teles, J.H. (1996). *Helv. Chim. Acta* 79: 1899–1902.
12 (a) Kerr, M.S., Read de Alaniz, J., and Rovis, T. (2002). *J. Am. Chem. Soc.* 124: 10298–10299. (b) Kerr, M.S., Read de Alaniz, J., and Rovis, T. (2005). *J. Org. Chem.* 70: 5725–5728.
13 Pesch, J., Harms, K., and Bach, T. (2004). *Eur. J. Org. Chem.* 2025–2035.
14 Mennen, S.M., Blank, J.T., Tran-Dubé, M.B. et al. (2005). *Chem. Commun.* (0): 195–197.
15 Zeitler, K. and Mager, I. (2007). *Adv. Synth. Catal.* 349: 185–1857.
16 (a) Jia, M.-Q., Li, Y., Rong, Z.-Q., and You, S.-L. (2011). *Org. Biomol. Chem.* 9: 2072–2074. (b) Rong, Z.-Q., Li, Y., Yang, G.-Q., and You, S.-L. (2011). *Synlett* 1033–1037.
17 Rafiński, Z., Kozakiewicz, A., and Rafinska, K. (2014). *ACS Catal.* 4: 1404–1408.
18 Bao, Y., Kumagai, N., and Shibasaki, M. (2014). *Tetrahedron: Asymmetry* 25: 1401–1408.
19 Lu, T., Gu, L., Kang, Q., and Zhang, Y. (2012). *Tetrahedron Lett.* 53: 6602–6605.
20 Read de Alaniz, J. and Rovis, T. (2005). *J. Am. Chem. Soc.* 127: 6284–6289.
21 Cullen, S.C. and Rovis, T. (2008). *Org. Lett.* 10: 3141–3144.
22 Vedachalam, S., Wong, Q.-L., Maji, B. et al. (2011). *Adv. Synth. Catal.* 353: 219 225.
23 Law, K.R. and McErlean, C.S.P. (2013). *Chem. Eur. J.* 19: 15852–15855.
24 Wei, S., Wei, X.-G., Su, X. et al. (2011). *Chem. Eur. J.* 17: 5965–5971.
25 Collett, C.J., Massey, R.S., Maguire, O.R. et al. (2013). *Chem. Sci.* 4: 1514–1522.
26 (a) Kerr, M.S. and Rovis, T. (2004). *J. Am. Chem. Soc.* 126: 8876–8877. (b) Reynolds, N.T. and Rovis, T. (2005). *Tetrahedron* 61: 6368–6378. (c) Moore, J.L., Kerr, M.S., and Rovis, T. (2006). *Tetrahedron* 62: 11477–11482.

27 Filloux, C.M., Lathrop, S.P., and Rovis, T. (2010). *Proc. Natl. Acad. Sci. U.S.A.* 107: 20666–20671.
28 Orellana, A. and Rovis, T. (2008). *Chem. Commun.* 730–732.
29 Lathrop, S. and Rovis, T. (2013). *Chem. Sci.* 4: 1668–1673.
30 (a) Liu, Q. and Rovis, T. (2006). *J. Am. Chem. Soc.* 128: 2552–2553. (b) Liu, Q. and Rovis, T. (2007). *Org. Process Res. Dev.* 11: 598–604.
31 Jia, M.-Q. and You, S.-L. (2012). *Chem. Commun.* 48: 6363–6365.
32 Jia, M.-Q., Liu, C., and You, S.-L. (2012). *J. Org. Chem.* 77: 10996–11001.
33 Jia, M.-Q. and You, S.-L. (2013). *Synlett* 24: 1201–1204.
34 Lin, Q., Li, Y., Das, D.K. et al. (2016). *Chem. Commun.* 52: 6459–6462.
35 Enders, D., Breuer, K., Runsink, J., and Teles, J.H. (1996). *Liebigs Ann. Chem.* 2019–2028.
36 (a) Tiebes, J. (1990). Diploma thesis. Aachen, Germany: RWTH Aachen; (b) Enders, D. (1993). *Stereoselective Synthesis*, 63–90. Heidelberg, Germany: Springer-Verlag. (c) Enders, D., Bockstiegel, B., Dyker, H. et al. (1993). *Dechema-Monographies*, vol. 129, 209–223. Weinheim, Germany: VCH.
37 Enders, D., Han, J., and Henseler, A. (2008). *Chem. Commun.* 3989–3991.
38 Enders, D. and Han, J. (2008). *Synthesis* 3864–3868.
39 Liu, Q., Perreault, S., and Rovis, T. (2008). *J. Am. Chem. Soc.* 130: 14066–14067.
40 Liu, Q. and Rovis, T. (2009). *Org. Lett.* 11: 2856–2859.
41 Jousseaume, T., Wurz, N.E., and Glorius, F. (2011). *Angew. Chem. Int. Ed.* 123: 1446–1450; (2011). *Angew. Chem. Int. Ed.* 50: 1410–1414.
42 Kuniyil, R. and Sunoj, R.B. (2013). *Org. Lett.* 15: 5040–5043.
43 Wurz, N.E., Daniliuc, C.G., and Glorius, F. (2012). *Chem. Eur. J.* 18: 16297–16301.
44 Fang, X., Chen, X., Lv, H., and Chi, Y.R. (2011). *Angew. Chem. Int. Ed.* 123: 11986–11989; (2011). *Angew. Chem. Int. Ed.* 50: 11782–11785.
45 Fang, X., Chen, X., and Chi, Y.R. (2011). *Org. Lett.* 13: 4708–4711.
46 DiRocco, D.A., Oberg, K.M., Dalton, D.M., and Rovis, T. (2009). *J. Am. Chem. Soc.* 131: 10872–10874.
47 Um, J.M., DiRocco, D.A., Noey, E.L. et al. (2011). *J. Am. Chem. Soc.* 133: 11249–11254.
48 DiRocco, D.A. and Rovis, T. (2011). *J. Am. Chem. Soc.* 133: 10402–10405.
49 DiRocco, D.A., Noey, E.L., Houk, K.N., and Rovis, T. (2012). *Angew. Chem. Int. Ed.* 124: 2441–2444; *Angew. Chem. Int. Ed.* 51: 2391–2394.
50 Vedachalam, S., Tan, S.M., Teo, H.P. et al. (2012). *Org. Lett.* 14: 174–177.
51 Sánchez-Larios, E., Thai, K., Bilodeau, F., and Gravel, M. (2011). *Org. Lett.* 13: 4942–4945.
52 Stetter, H. and Bender, H.-J. (1978). *Angew. Chem. Int. Ed. Engl.* 90: 130; (1978). *Angew. Chem. Int. Ed. Engl.* 17: 131.
53 Bhunia, A., Yetra, S.R., Bhojgude, S.S., and Biju, A.T. (2012). *Org. Lett.* 14: 2830–2833.
54 Patra, A., Bhunia, A., and Biju, A.T. (2014). *Org. Lett.* 16: 4798–4801.
55 Kim, S.M., Jin, M.Y., Kim, M.J. et al. (2011). *Org. Biomol. Chem.* 9: 2069–2071.

56 (a) Mattson, A.E., Bharadwaj, A.R., and Scheidt, K.A. (2004). *J. Am. Chem. Soc.* 126: 2314–2315. (b) Mattson, A.E., Bharadwaj, A.R., Zuhl, A.M., and Scheidt, K.A. (2006). *J. Org. Chem.* 71: 5715–5724.
57 Mattson, A.E., Zuhl, A.M., Reynolds, T.E., and Scheidt, K.A. (2006). *J. Am. Chem. Soc.* 128: 4932–4933.
58 Myers, M.C., Bharadwaj, A.R., Milgram, B.C., and Scheidt, K.A. (2005). *J. Am. Chem. Soc.* 127: 14675–14680.
59 Bortolini, O., Fantin, G., Fogagnolo, M. et al. (2011). *Org. Biomol. Chem.* 9: 8437–8444.
60 Bortolini, O., Cavazzini, A., Dambruoso, P. et al. (2013). *Green Chem.* 15: 2981–2992.
61 Takaki, K., Ohno, A., Hino, M. et al. (2014). *Chem. Commun.* 50: 12285–12288.
62 Zhang, J., Xing, C., Tiwari, B., and Chi, Y.R. (2013). *J. Am. Chem. Soc.* 135: 8113–8116.
63 (a) Wilde, M.M.D. and Gravel, M. (2013). *Angew. Chem. Int. Ed.* 125: 12883–12886; (2013). *Angew. Chem. Int. Ed.* 52: 12651–12654. For a highlight on this work, see: (b)Connon, S.J. (2014). *Angew. Chem. Int. Ed.* 126: 1225–1227; (2014). *Angew. Chem. Int. Ed.* 53: 1203–1205.
64 (a) Dresen, C., Richter, M., Pohl, M. et al. (2010). *Angew. Chem.* 122: 6750–6753; (2010). *Angew. Chem. Int. Ed.* 49: 6600–6603. (b) Beigia, C., Waltzera, S., Zareib, M., and Müller, M. (2014). *J. Biotechnol.* 191: 64–68.
65 Sanchez-Larios, E. and Gravel, M. (2009). *J. Org. Chem.* 74: 7536–7539.
66 Sanchez-Larios, E., Holmes, J.M., Daschner, C.L., and Gravel, M. (2010). *Org. Lett.* 12: 5772–5775.
67 Sun, F.-G., Huang, X.-L., and Ye, S. (2010). *J. Org. Chem.* 75: 273–276.
68 Sun, F.-G. and Ye, S. (2011). *Synlett* 1005–1009.
69 Hong, B.-C., Dange, N.S., Hsu, C.-S., and Liao, J.-H. (2010). *Org. Lett.* 12: 4812–4815.
70 Hong, B.C., Dange, N.S., Hsu, C.-S. et al. (2011). *Org. Lett.* 13: 1338–1341.
71 For initial reports, see: (a)Katada, A., Ichioka, K., Takagi, K., and Suzuki, M. (2011). *Org. Lett.* 13: 3722–3725. (b) Biju, A.T., Padmanaban, M., Wurz, N.E., and Glorius, F. (2011). *Angew. Chem. Int. Ed.* 123: 8562–8565; (2011). *Angew. Chem. Int. Ed.* 50: 8412–8415.
72 For a review, see: Biju, A.T., Kuhl, N., and Glorius, F. (2011). *Acc. Chem. Res.* 44: 1182–1195.
73 Willis, M.C. (2010). *Chem. Rev.* 110: 725–748.
74 He, J., Tang, S., Liu, J. et al. (2008). *Tetrahedron* 64: 8797–8800.
75 Hirano, K., Biju, A.T., Piel, I., and Glorius, F. (2009). *J. Am. Chem. Soc.* 131: 14190–14191.
76 Piel, I., Steinmetz, M., Hirano, K. et al. (2011). *Angew. Chem. Int. Ed.* 123: 5087–5091; (2011). *Angew. Chem. Int. Ed.* 50: 4983–4987.
77 Domingo, L.R., Saéz, J.A., and Arnó, M. (2012). *RSC Adv.* 2: 7127–7134.
78 Bugaut, X., Liu, F., and Glorius, F. (2011). *J. Am. Chem. Soc.* 133: 8130–8133.
79 Ajitha, M.J. and Suresh, C.H. (2013). *Tetrahedron Lett.* 54: 7144–7146.
80 (a) Liu, F., Bugaut, X., Schedler, M. et al. (2011). *Angew. Chem. Int. Ed.* 123, 12834–12839; *Angew. Chem. Int. Ed.* 50: 12626–12630. (b) Schedler,

M., Fröhlich, R., Daniliuc, C.-G., and Glorius, F. (2012). *Eur. J. Org. Chem.* 4164–4171.

81 Schedler, M., Wang, D.S., and Glorius, F. (2013). *Angew. Chem. Int. Ed.* 125: 2645–2649; (2013). *Angew. Chem. Int. Ed.* 52: 2585–2589.

82 Zhao, M., Yang, H., Li, M.M. et al. (2014). *Org. Lett.* 16: 2904–2907.

83 Zhao, M., Chen, J., Yang, H., and Zhou, L. (2017). *Chem. Eur. J.* 23: 2783–2787.

84 Alcaide, B., Almendros, P., Fernandez, I. et al. (2015). *Chem. Eur. J.* 21: 1533–1541.

85 Lu, H., Lin, J.-B., Liu, J.-B., and Xu, P.-F. (2014). *Chem. Eur. J.* 20: 11659–11663.

86 Biju, A.T., Wurz, N.E., and Glorius, F. (2010). *J. Am. Chem. Soc.* 132: 5970–5971.

87 Wang, Z., Yu, Z., Wang, Y., and She, D. (2012). *Synthesis* 44: 1559–1568.

88 (a) Padmanaban, M., Biju, A.T., and Glorius, F. (2011). *Org. Lett.* 13: 5624–5627. For a related one-pot protocol, see: (b)Franz, J.F., Fuchs, P.J.W., and Zeitler, K. (2011). *Tetrahedron Lett.* 52: 6952–6956.

89 Biju, A.T. and Glorius, F. (2010). *Angew. Chem. Int. Ed.* 122: 9955–9958; (2010). *Angew. Chem. Int. Ed.* 49: 9761–9764.

90 Zhao, M., Liu, J.-L., Liu, H.-F. et al. (2018). *Org. Lett.* 20: 2676–2679.

4

N-Heterocyclic Carbene (NHC)-Mediated Generation and Reactions of Homoenolates

Vijay Nair[1], Rajeev S. Menon[2], and Jagadeesh Krishnan[1]

[1] *National Institute for Interdisciplinary Science and Technology (CSIR-NIIST), Chemical Science and Technology Division, Pappanamcode, Trivandrum, 695019, India*
[2] *Central University of Haryana, Department of Chemistry, Jant-Pali, Mahendergarh 123031, India*

4.1 Homoenolates – An Introduction

Carbonyl chemistry is widely recognized as the backbone of organic synthesis because of the versatility of this functional group. The chemistry of carbonyl group pivots mainly around three aspects: the electrophilicity of the carbonyl carbon, acidity of protons of α-carbon, and the influence of the carbonyl group on adjacent unsaturation. These reactivity patterns are well established and routinely applied in laboratories all around the world. Majority of the conventional carbon–carbon bond-forming reactions involve a carbonyl-derived component either as the electrophile or as the nucleophile (usually enolate). Even many modern methods in organic synthesis are built on the foundations of carbonyl chemistry. For example, modern organocatalysis is mainly centered on the *enamine activation* of carbonyl compounds originally described in the pioneering contributions of Stork et al. [1].

The formation of a new bond at the α-carbon of a carbonyl group is invariably achieved by the generation of enolate or its synthetic equivalent and its reaction with an electrophile (e.g. Aldol reaction). Similarly, bond formation at the β-carbon relies on conjugate addition of nucleophiles to an α,β-unsaturated carbonyl system (Michael addition). A rather contraintuitive approach of generating a negatively polarized β-carbon opens up myriads of new synthetic opportunities. The reactive intermediate in this approach would be a homolog of an enolate and thus may be termed as *homoenolate* [2]. A general definition of a homoenolate may be advanced as *a species possessing a nucleophilic carbon β to a carbonyl group or a carbonyl surrogate.*

The first instance of a *homoenolate occurrence* was described in a 1962 report by Nickon and Lambert [3]. They explained the base-catalyzed racemization of (+)-camphenilone **1** by invoking the intermediacy of a homoenolate **2** that is generated by deprotonation at the β-carbon (note that the solitary α-hydrogen in **1** is on a more hindered, bridgehead carbon). The homoenolate **2** engages in an intramolecular nucleophilic addition to the ketone to produce a symmetric cyclopropanolate **3**. The latter collapses via two equally probable carbon–carbon bond

N-Heterocyclic Carbenes in Organocatalysis, First Edition. Edited by Akkattu T. Biju.
© 2019 Wiley-VCH Verlag GmbH & Co. KGaA. Published 2019 by Wiley-VCH Verlag GmbH & Co. KGaA.

cleavage pathways to eventually afford a racemic sample of the starting ketone (Scheme 4.1).

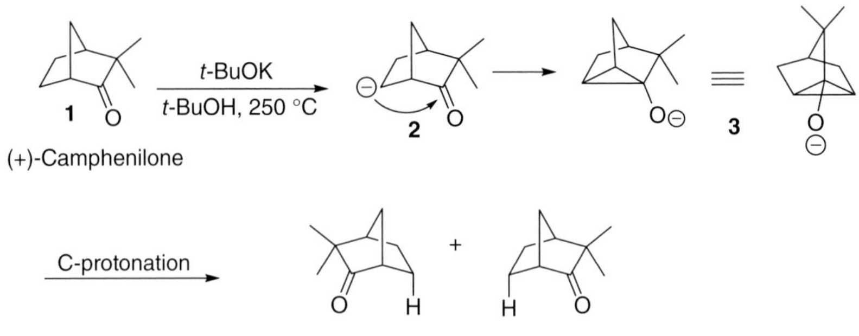

Scheme 4.1 Base-catalyzed racemization of (+)-camphenilone via a homoenolate intermediate.

It is obvious that the generation of homoenolates would allow direct reactions of electrophiles at the β-carbon of carbonyl groups; however, the development of suitable and convenient experimental protocols poses serious challenges. Indirect methods involving cyclopropanone derivative **4** (Nakamura) [4a], acetal-containing Grignard reagent **5** (Helquist) [4b], or 2-alkenyl-1-metallocarbamates **6** generated by directed asymmetric metallation (Hoppe) [4c] managed to circumvent the challenges to some extent (Figure 4.1); however, the synthetic potential of homoenolates remained largely untapped.

The chemistry of homoenolates re-emerged strongly in the dawn of this century along with the advent of organocatalysis by nucleophilic heterocyclic carbenes (NHCs). A brief description of the events that paved way for this renaissance of homoenolate chemistry is presented below as it is essential for appreciating the results that follow.

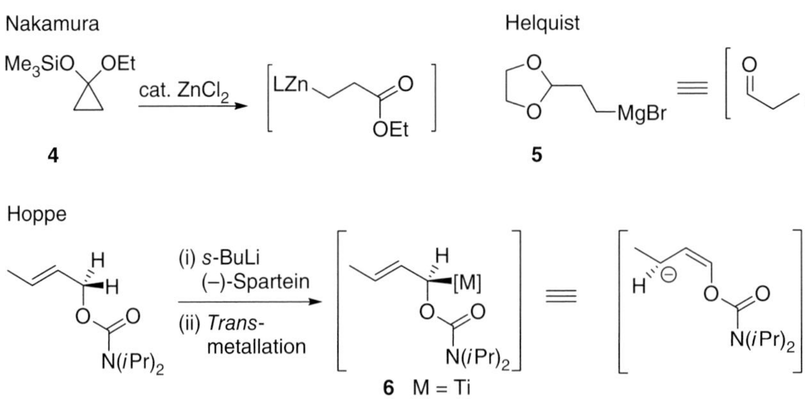

Figure 4.1 Examples of some metal–homoenolate reagents.

4.2 N-Heterocyclic Carbenes (NHCs)

The conceptual origins of NHC can be traced back to the investigations by Breslow on the mechanism of biochemical reactions catalyzed by the coenzyme thiamine. Breslow identified the thiazolium moiety present in the coenzyme as the seat of its catalytic activity. In his seminal report in 1958, Breslow described that the thiazolium ring on deprotonation is converted into a thiazolylidene species (ylide), which functions as a nucleophilic catalyst that mediates reactions such as pyruvic acid decarboxylation and acetoin condensation [5]. More importantly, he demonstrated that a similar thiazolylidene species **7** can catalyze the benzoin condensation of aldehydes. His proposal hinges on the formation of an enol intermediate **8** from the thiazolylidene and aldehyde. By virtue of conjugation with the nitrogen and sulfur atoms in the thiazolylidene, the double bond in the enol **8** is rendered nucleophilic at the (originally) acyl carbon. This reversal of polarity (or *umpolung*) of the aldehyde carbon turns it into an *acyl anion equivalent*, and this allows for a straightforward mechanistic rationalization of the benzoin condensation (Scheme 4.2). The enol **8** has since been aptly named "the Breslow intermediate."

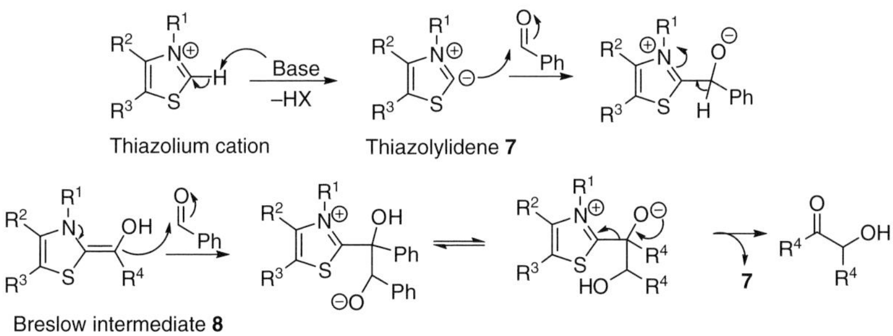

Scheme 4.2 Breslow's postulate on thiazolium-mediated benzoin condensation.

As already stated, Breslow depicted the catalytically active species of the above transformation as a thiazolylidene (**7**, an ylide). He stopped just a resonance form short of what would have been a revolutionary idea for that era, *a heterocyclic carbene* **9** (Figure 4.2). The equivalence of the ylide and the carbene structures were soon recognized, and such compounds were collectively categorized as NHCs. Synthesis and isolation of stable, crystalline NHCs by Arduengo et al. in 1991 proved their existence and structure beyond doubt [6]. Presently, the utility of

Figure 4.2 Thiazolylidene: ylide and carbene forms.

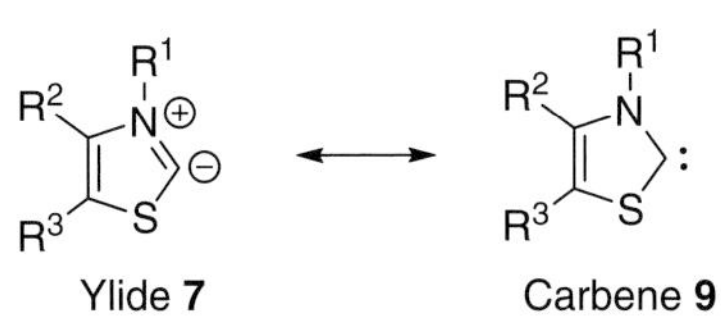

NHCs as catalysts, reagents, and ligands in important organometallic reagents is well established [7].

The importance of Breslow's postulate cannot be overstated as the vast majority of NHC-mediated reactions proceeds via the formation of Breslow intermediates of different structural attributes. The first experimental evidence for the formation of Breslow intermediate, however, was obtained much later in 2003 by Nair et al. [8]. A three-component reaction of an NHC **10**, aldehyde, and dimethylacetylene dicarboxylate (DMAD) afforded the 2-oxy maleate **11** as one of the products (Scheme 4.3). The formation of **11** may be viewed as a result of the interception of the Breslow intermediate **12** (at the oxygen end of enol) by the highly electrophilic DMAD via a hetero-Michael addition. Later, Breslow intermediate was also observed in solution and characterized using NMR spectroscopy [9].

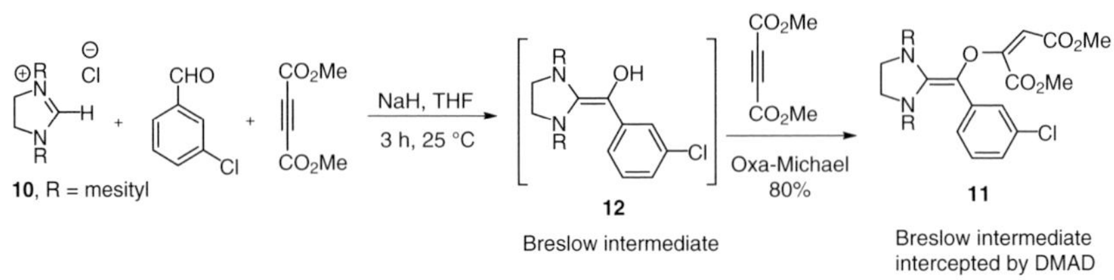

Scheme 4.3 Nair's interception of Breslow intermediate.

4.3 NHC-Derived Homoenolates – The Beginning

In the same way as the interaction of an NHC and an aldehyde leads to the formation of Breslow intermediate, the union of an NHC and an α,β-unsaturated aldehyde would furnish the analogous Breslow intermediate **13** with extended conjugation. This scenario is depicted in Scheme 4.4 taking cinnamaldehyde as a representative example.

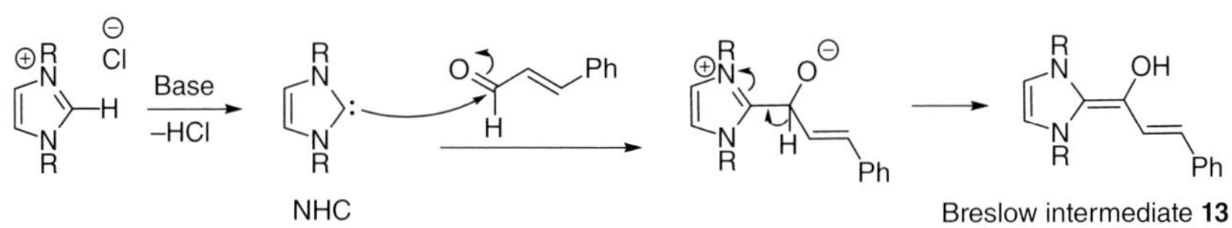

Scheme 4.4 Formation of Breslow intermediate with extended conjugation from cinnamaldehyde.

By virtue of conjugation, the extended Breslow intermediate **13** would be nucleophilic at the benzylic position (originally β-carbon of cinnamaldehyde). The reactive intermediates **13A–13B** shown in Scheme 4.5 clearly satisfy the conditions stated in the definition of *homoenolates* (vide supra). In principle, the extended Breslow intermediate **13** may thus be considered as a homoenolate.

The reactions of enals expected from the above assumptions were experimentally demonstrated independently by the groups of Glorius and coworker [10a] and Bode and coworkers [10b] in 2004. They found that α,β-unsaturated

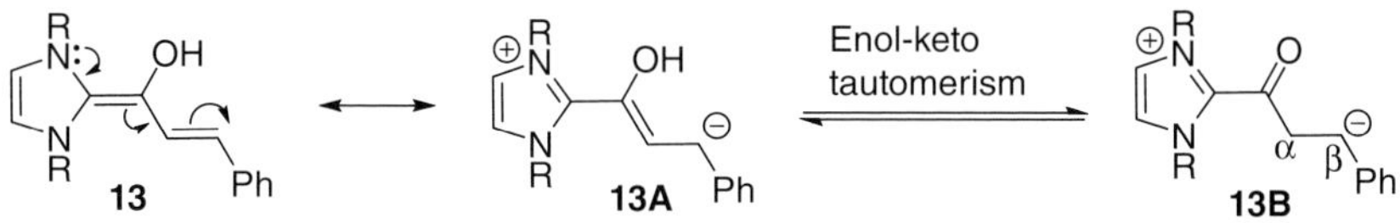

Scheme 4.5 Homoenolate formation from enals and NHC.

aldehyde **14** and aldehyde **15** react in the presence of NHC catalyst **16** to afford γ-butyrolactones **17**. The reaction is initiated by the formation of the homoenolate **13** and its addition to the second aldehyde to form an alkoxide **18**. The latter forms the lactone ring via an acyl displacement of the NHC unit (Scheme 4.6).

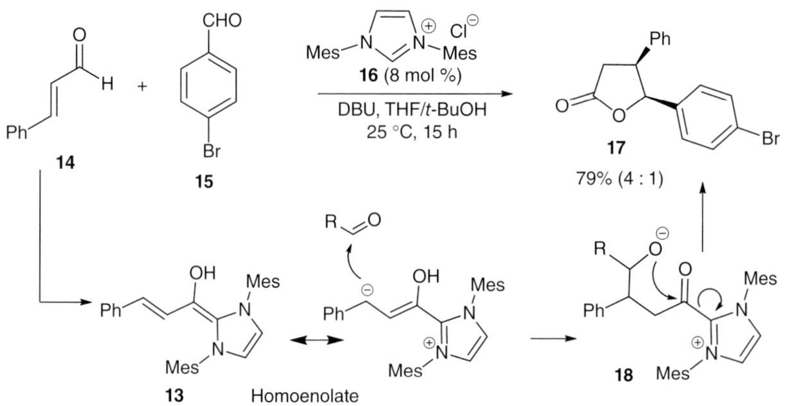

Scheme 4.6 NHC-catalyzed synthesis of γ-lactones via homoenolate intermediates.

When a second aldehyde is not available for the interception of the homoenolate, a homodimerization of the enal is observed, leading to the formation of the corresponding γ-lactone **19** (Scheme 4.7) [10a].

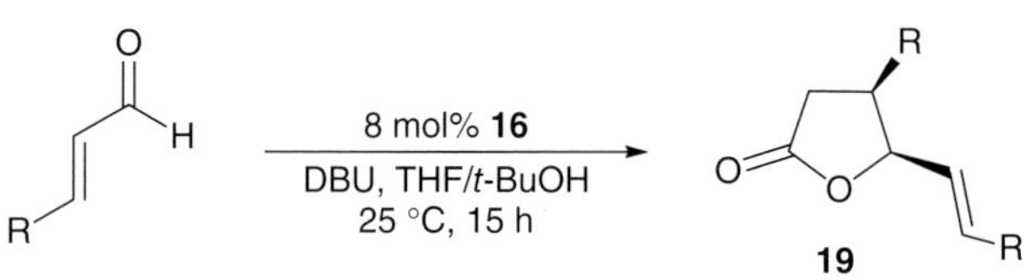

Scheme 4.7 NHC-catalyzed homodimerization of enals.

These landmark reports by Glorius and Bode heralded a new era in homoenolate chemistry. Reactive homoenolates could now be generated from readily available enals via organocatalysis by NHCs under mild conditions. Modulation of homoenolate reactivity could be explored with confidence as a sumptuous array of NHC catalysts of varying electronic and steric demands was available. Additionally, NHC catalysis offered plenty of opportunities for the development of asymmetric transformations. It is not surprising that the Glorius–Bode reports were followed by vigorous activity in this area, and as a result, a variety of novel transformations were developed [11].

4.4 Mechanistic Pathways Available for NHC-Homoenolates

The extended conjugation in NHC-derived homoenolates leads to manifestation of diverse modes of reactivity. A broad classification of reactions into four major mechanistic pathways is provided below.

A: Homoenolate: This pathway is characterized by the initial reaction of an electrophile X at the β-position. The enol **20** then tautomerizes to the keto form to afford the acyl azolium **21**. Final reaction of a nucleophile Z with the acyl azolium **21** affords the product **22** and releases the NHC catalyst. The NHC-homoenolate functions as a 1,3-dipole-like synthon **23** in this pathway (Scheme 4.8). The Glorius–Bode γ-lactone formation (Scheme 4.6) is an example for this type of reactivity.

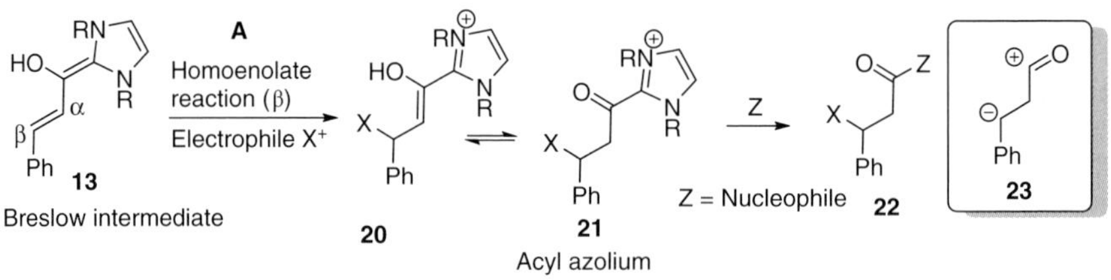

Scheme 4.8 Homoenolate-type reactivity of Breslow intermediate **13**.

B: Homoenolate–enolate: The initial step here is the same as that of the previous pathway; however, the enol intermediate **20** further reacts with a second electrophile Y and a bond is formed at the α-position. The acyl azolium **24** then reacts with a nucleophile Z and affords the final product **25**. In this pathway, two electrophiles (X and Y) sequentially react with the NHC-homoenolate at the β- and α-positions, and the NHC-homoenolate may be viewed as the synthon **26** (Scheme 4.9).

C: Enolate: The initial electrophile in this pathway is a proton, and it gets attached to the β-carbon. The β-protonated homoenolate **27** possesses an enol moiety that can react with an electrophile Y to install a C—Y bond at the α-position. The 1,2-dipolar synthon **28** represents this type of reactivity of homoenolates (Scheme 4.10).

D: Acyl anion: The acyl anion-type reactivity is in principle a feasible pathway for the Breslow intermediate; however, such reactions are seldom observed. The apparent paucity of this type of reactions is presumably the result of severe steric crowding near the acyl carbon caused by the bulky NHC substituents (Scheme 4.11).

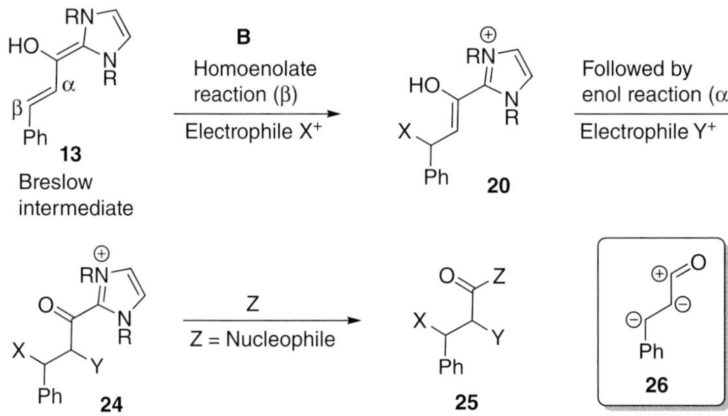

Scheme 4.9 Homoenolate–enolate reaction pathway of Breslow intermediate **13**.

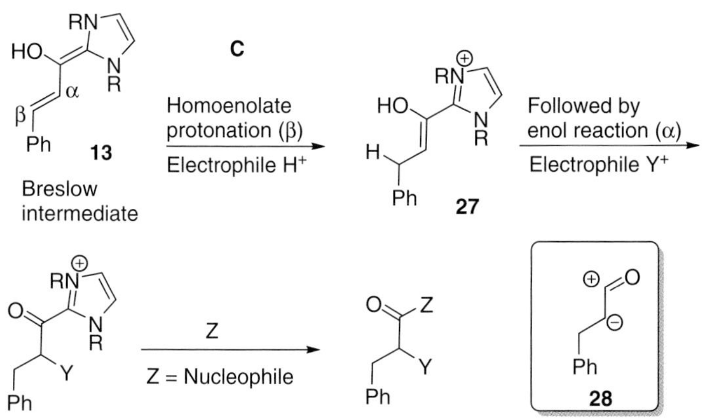

Scheme 4.10 Enolate-type reactivity of Breslow intermediate **13**.

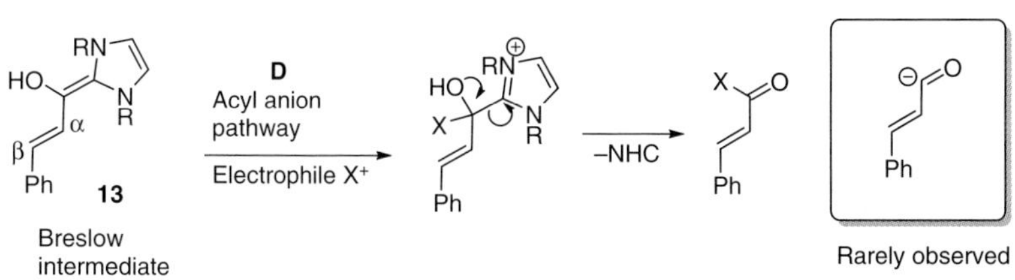

Scheme 4.11 Acyl anion-type reactivity of Breslow intermediate **13**.

In view of the large number of various types of reactions that proceed via the intermediacy of NHC-homoenolates, their classification based on the type of electrophilic partner as well as the major mechanistic pathway is deemed appropriate for elaborate discussion. The details of such reactions are presented in the following sections under various categories. It may be noted that the categorization is attempted here for the convenience of discussion, and some overlap among sections is inevitable.

4.5 Reaction of NHC-Homoenolates with Ketones and Ketimines

The Glorius–Bode γ-lactone synthesis from enals and aldehydes could be extended to only a few reactive ketones such as 1,2-diones. Work from Nair's laboratory showed that both acyclic and cyclic 1,2-diones including isatin derivatives underwent γ-lactonization on treatment with imidazolium precatalyst **16** and enals. A representative example is depicted in Scheme 4.12 [12]. The reaction mechanism is presumably analogous to that of the γ-lactonization of aldehydes (see Scheme 4.6).

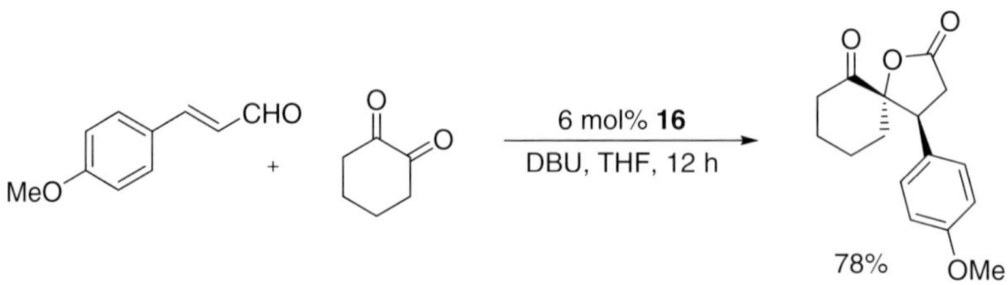

Scheme 4.12 Homoenolate reaction with cyclic 1,2-diones.

Scheidt and coworkers reported that cooperative catalysis by chiral triazolium NHC **29** and lithium chloride was essential for successful enantioselective homoenolate cyclization of enals and isatins **30** (Scheme 4.13). Enals endowed with β-alkyl substituents, however, reacted poorly in the presence of LiCl. This method was used as the key step in the asymmetric total synthesis of the anticancer marine natural product maremycin B [13]. Independent study by Ye and coworkers suggested that hydrogen-bonding interaction of NHC with the isatins may be playing a key role in asymmetric induction [14].

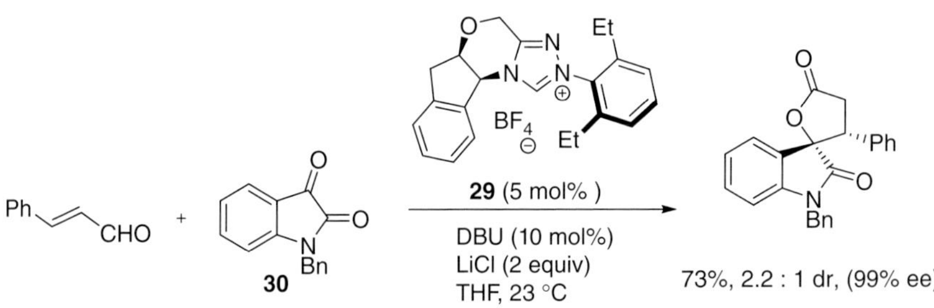

Scheme 4.13 NHC–LiCl cooperative catalysis for asymmetric spirolactone synthesis.

Cooperative catalysis using NHC **31** and Brønsted acid has allowed the union of sterically demanding β-disubstituted enals and isatins **30** to afford spirocyclic oxindoles **32** endowed with two congested and contiguous quaternary stereocenters (Scheme 4.14). The Brønsted acid engages in hydrogen bonding to activate the reaction partners and bring them together [15].

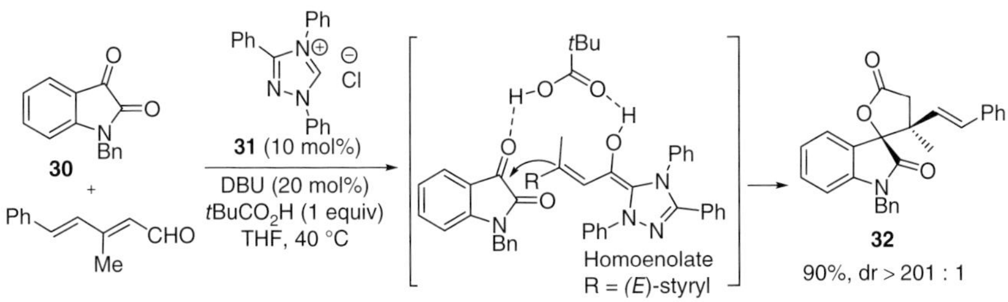

Scheme 4.14 NHC–Brønsted acid cooperative catalysis.

It has been shown that NHCs can be successfully used in combination with metal salts or complexes to promote novel reactions. This cooperative catalysis, however, warrants a judicious selection of both the catalysts. Scheidt demonstrated that Mg- and Ti-based Lewis acids are generally tolerant toward NHCs. When used in combination, the NHC activates the enal to produce a homoenolate (nucleophile), whereas the Lewis acid coordinates with the electrophile rendering it more reactive. Therefore, a wider substrate scope is permitted. This approach permits the use of either chiral NHCs and/or chiral ligands on Lewis acids, thereby allowing chemists to exploit the benefits of both traditional asymmetric catalysis and asymmetric organocatalysis.

The NHC-catalyzed annulation of enals with hydrazones **33** furnished desired cyclic products in moderate yields and stereoselectivity. Scheidt reported that the use of $Mg(Ot\text{-}Bu)_2$ as a cocatalyst improved both these aspects significantly (Scheme 4.15). The chiral γ-lactams **34** thus obtained were further converted into a valuable pyroglutamic acid derivative and a 3-phenyl proline derivative [16].

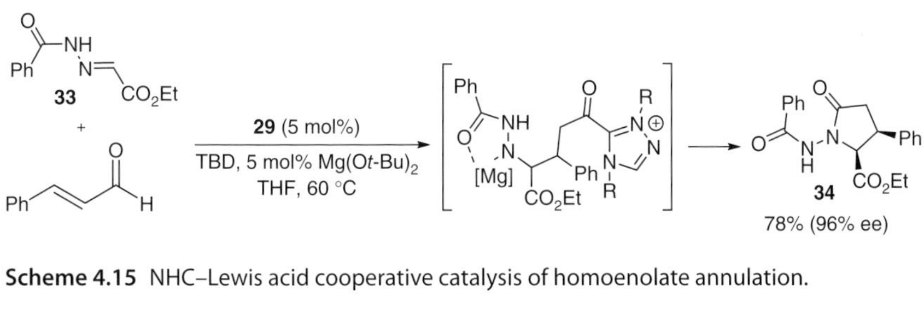

Scheme 4.15 NHC–Lewis acid cooperative catalysis of homoenolate annulation.

N-Boc imines **35** derived from isatins also underwent facile spiroannulation with homoenolate in the presence of triazole precatalyst **36** and base to produce corresponding spiro-cycloadducts **37** in excellent enantiomeric excess (Scheme 4.16) [17].

Johnson and coworkers reported dynamic kinetic resolution of racemic β-chloro-α-keto esters **38** via an intermolecular asymmetric NHC-homoenolate annulation reaction (Scheme 4.17). The triazolium **39**-derived NHC promotes homoenolate cyclization of the reactive keto group of **38** with enal. At the same time, rapid epimerization of the halogen-bearing chiral center is promoted by the base. This leads to the accumulation of a single enantiomer product **40** via

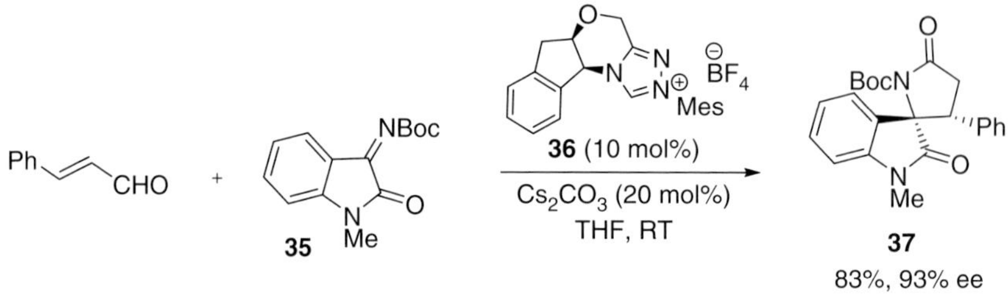

Scheme 4.16 Enantioselective spiroannulation of isatin-imines by homoenolates.

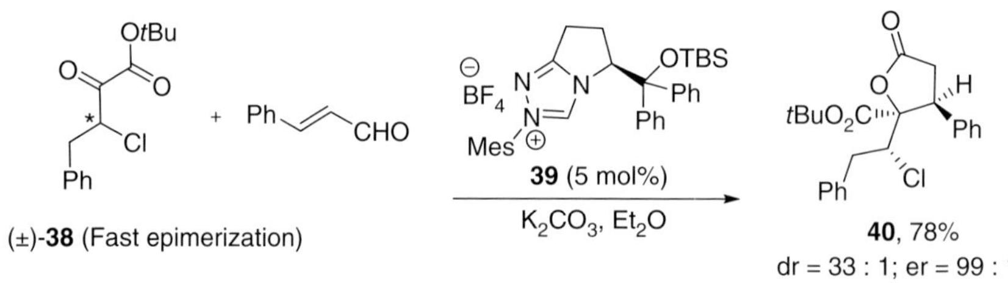

Scheme 4.17 NHC-catalyzed intermolecular dynamic kinetic resolution via homoenolate reaction.

dynamic kinetic resolution. It may be noted that three contiguous stereocenters are installed in one operation to furnish a highly functionalized product [18].

Snyder reported an elegant application of NHC-titanium isopropoxide cooperative catalysis for the synthesis of Securinega alkaloids. The spiro-butenolide moiety and the bicyclic scaffold of the target natural product were constructed by an intramolecular spirocyclization of an ynal and ketone. The low solubility of the catalyst **41** in toluene ensures that the concentration of reactive species is low, and the competing decomposition routes for the labile ynal **42** are suppressed. The butenolide **43** was readily converted into the natural product 3-deshydroxy-secu'amamine A **44** in two steps (Scheme 4.18) [19].

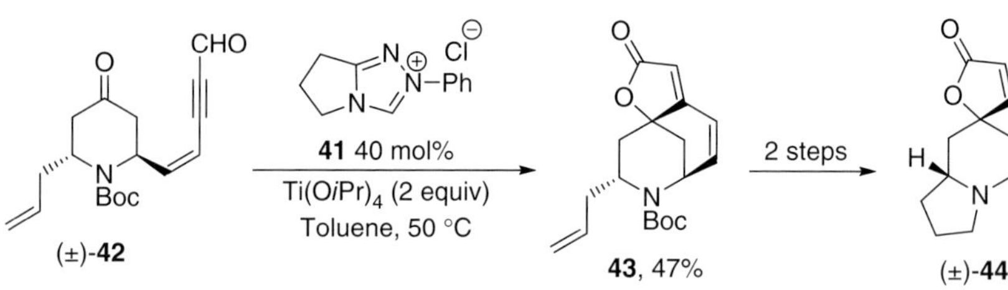

Scheme 4.18 NHC–Lewis acid cooperative catalysis in Securinega alkaloid synthesis.

Ynals can also afford homoenolate-type reactive intermediates on exposure to NHCs and an acidic cocatalyst. She and Scheidt independently reported that ynals and 1,2-diones may be assembled together in a [3+2] annulation mediated by NHCs to afford butenolides. She's approach [20] involved the combination of the chiral triazolium NHC **45** and LiCl, whereas Scheidt and coworker [21] employed a chiral imidazolium NHC **46** and a chiral Brønsted acid **47** for mediating the reaction (Scheme 4.19).

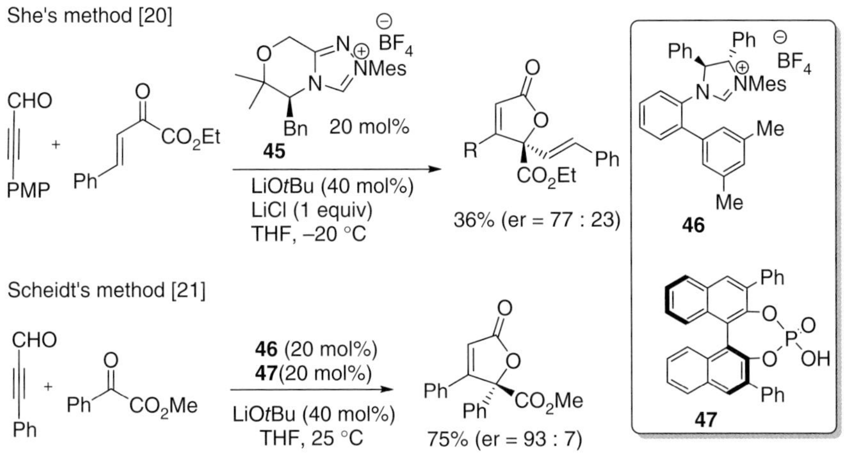

Scheme 4.19 Asymmetric synthesis of butenolides from ynals and 1,2-diones.

Interestingly, the reactions of benzofuran-2,3-dione **48** with enals, when catalyzed by two different NHCs, led to slightly different outcomes. The triazolium salt **31** catalyzed the formation of spirocyclic products **49** [22a], whereas the use of thiazolium salt **50** led to the formation of two spirocyclic products **49** and **51** (Scheme 4.20) [22b].

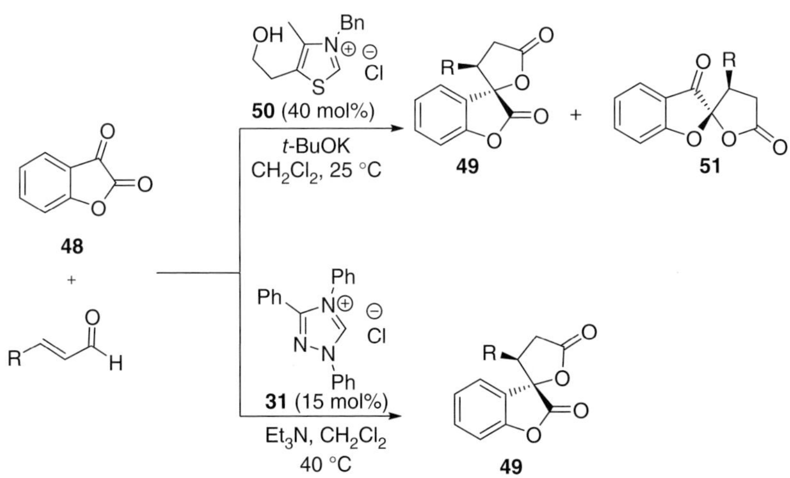

Scheme 4.20 NHC-catalyzed spirocyclization of benzofurandiones and enals.

Acylphosphonates **52** readily undergo [3+2] annulation with chiral NHC-bound homoenolates to afford enantioenriched γ-lactones **53**. The chiral

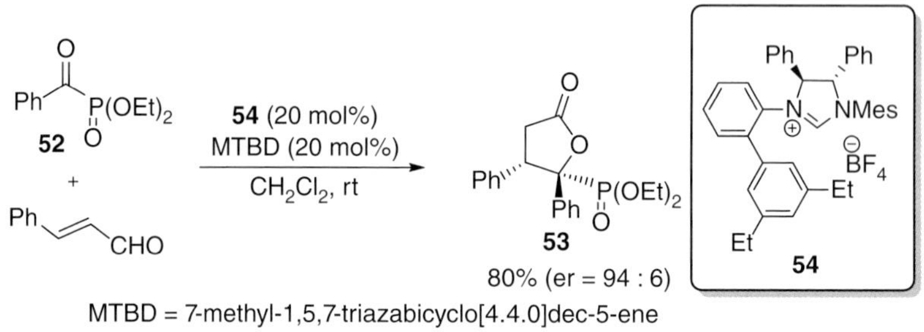

Scheme 4.21 [3+2] Cycloaddition of homoenolates and acylphosphonates.

imidazolium catalyst **54** was rationally designed on the basis of computational modeling of the transition state (Scheme 4.21) [23].

Bode and coworker used an intramolecular [3+2] cyclization of NHC-homoenolate annulation with a ketone as a key step in the formal synthesis of the anticancer therapeutic agent salinosporamide A. The desired stereoisomer **55** was, however, obtained as the minor product from the keto-enal **56** (Scheme 4.22) [24]. This constituted a formal total synthesis of salinosporamide A, as the γ-lactone **55** has previously been converted to salinosporamide A in eight steps.

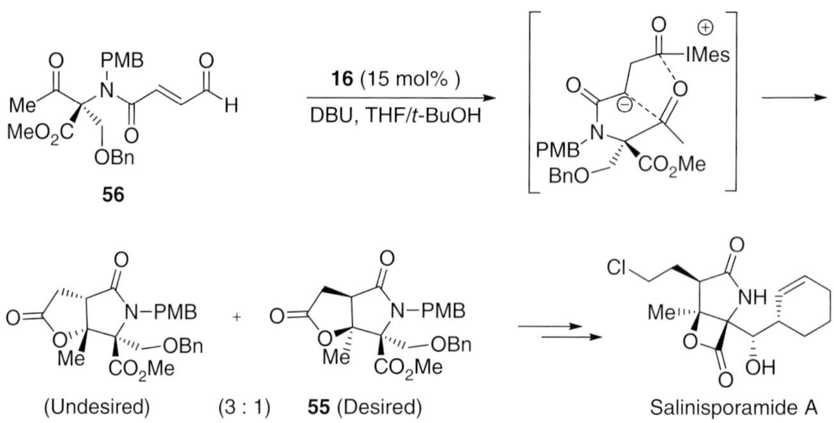

Scheme 4.22 Homoenolate annulation in formal total synthesis of salinosporamide A.

Interestingly, the reaction of tropone **57** with homoenolates afforded bicyclic δ-lacones **58** unlike other ketones (Scheme 4.23). A plausible mechanism may involve the conjugate addition of the homoenolate to tropone followed ring closure to afford the δ-lactone **58** as depicted in Scheme 4.23 [25].

Rovis and coworkers reported that unactivated α,β-unsaturated imines **59** can be made to react with homoenolates by careful selection of the catalyst and base. He used the sodium salt of *o*-chlorobenzoic acid **60** as a base to generate the NHC from the pentafluorophenyl triazolium salt **61**. The conjugate acid, *o*-chlorobenzoic acid, sets up the reaction by bringing the reaction partners together via hydrogen bonding. The relatively less electron-rich NHC formed

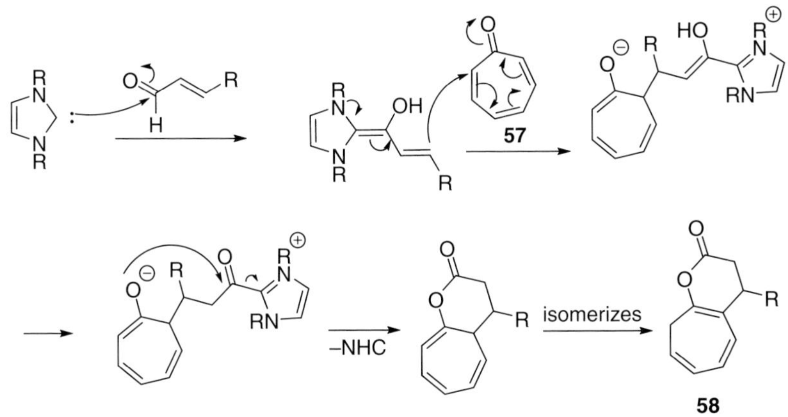

Scheme 4.23 Formation of δ-lactone from NHC-homoenolate and tropone.

from **61** is not affected by the Brønsted acid either. The γ-lactams **62** are formed under these conditions in an excellent enantioselectivity (Scheme 4.24) [26].

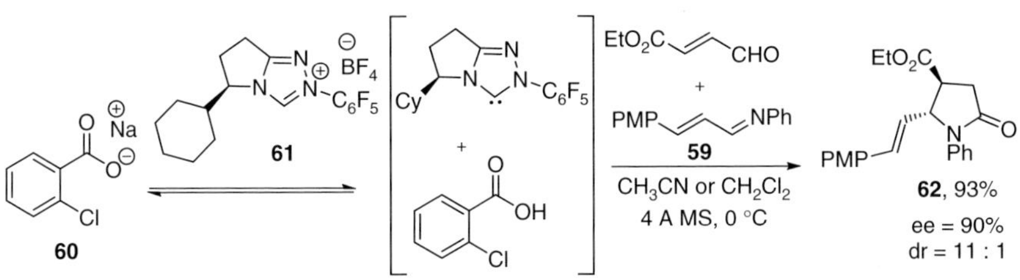

Scheme 4.24 Asymmetric synthesis of γ-lactams from NHC-homoenolate and unsaturated imines.

Interestingly, the combination of the achiral NHC **16** and the chiral Ti(IV)-based Lewis acid **63** promoted a stereoselective homodimerization to afford the γ-lactone **64** (Scheme 4.25). Although the reasons behind the change in outcome remains unclear, these results demonstrate that stereocontrol of reactions may be achieved by using either of the catalysts in enantiomerically pure forms [27].

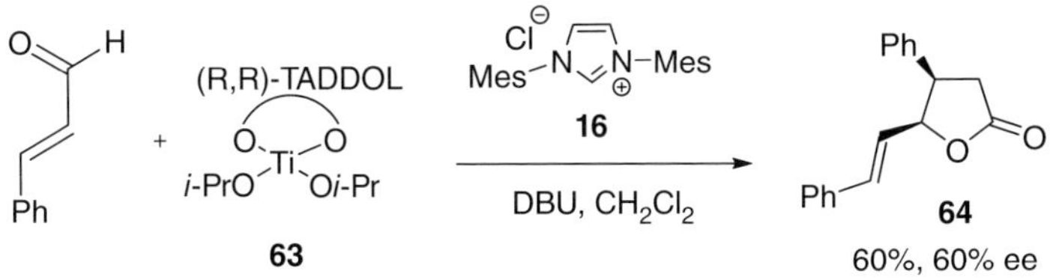

Scheme 4.25 Homodimerization of enals cocatalyzed by achiral NHC and chiral Lewis acid.

Yu and coworkers has shown that carbohydrate-derived cyclic nitrones **65** react readily with homoenolates derived from NHC **66** to form γ-hydroxy amino esters **67**. The latter may be used further in the synthesis of hydroxylated pyrrolizidines and indolizidines (Scheme 4.26) [28].

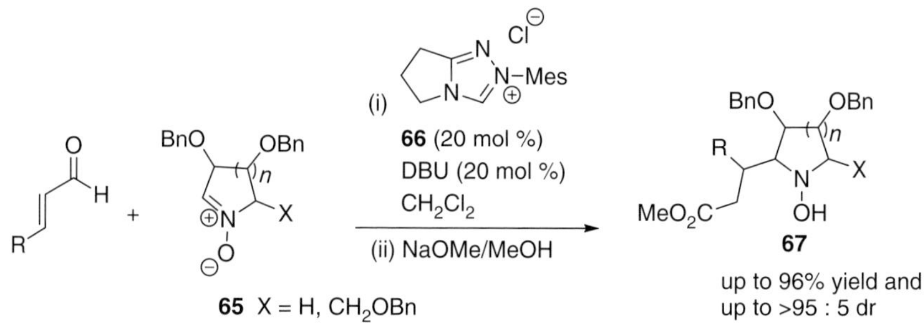

Scheme 4.26 Addition of homoenolates to carbohydrate-derived cyclic nitrones.

4.6 Reaction of NHC-Homoenolates with Michael Acceptors

The carbon–carbon bond-forming event in the Glorius–Bode γ-lactone synthesis is a 1,2-addition of NHC-homoenolate to the α,β-unsaturated carbonyl group. Another important carbon–carbon bond-forming reaction of the homoenolates, namely the Michael addition, was developed by Nair and colleagues in 2004. Their investigations revealed that α,β-unsaturated ketones such as chalcones on exposure to imidazolium-derived homoenolates afforded 1,3,4-trisubstituted cyclopentenes **68** in excellent diastereoselectivity (Scheme 4.27). The cyclization is most likely initiated by a Michael addition of the homoenolate and enone, followed by internal proton shift and an intramolecular aldol reaction. Subsequent intramolecular acyl displacement of the NHC moiety by the alkoxide of **69** results in the formation of a transient β-lactone **70** (the presence of which was detected by IR spectroscopic monitoring of the reaction mixture). Extrusion of CO_2 from the fleeting β-lactone generated the cyclopentene **68** [29].

Subsequently, a number of modifications of this cyclopentene synthesis were developed. This included an enantioselective variant developed by Bode and coworkers that employed a chiral triazolium-based catalyst **71** (Scheme 4.28) [30]. It was proposed by the authors that the reaction proceeds via the cross-benzoin addition of the homoenolate **72** with enone followed by oxy-Cope rearrangement and cyclization. However, subsequent DFT calculations indicate that the proposed cross-benzoin reaction is highly unlikely due to the high energy demand [31].

Bode and coworker also demonstrated that the chiral NHC-bound homoenolate can react with *N*-sulfonyl ketimines **73** to afford bicyclic β-lactams **74** (Scheme 4.29) [32].

The pioneering studies by Bode and Glorius showed that NHCs can catalyze the homodimerization of α,β-unsaturated aldehydes to afford γ-lactones. This reaction pathway can be altered by employing NHC–Lewis acid cooperative

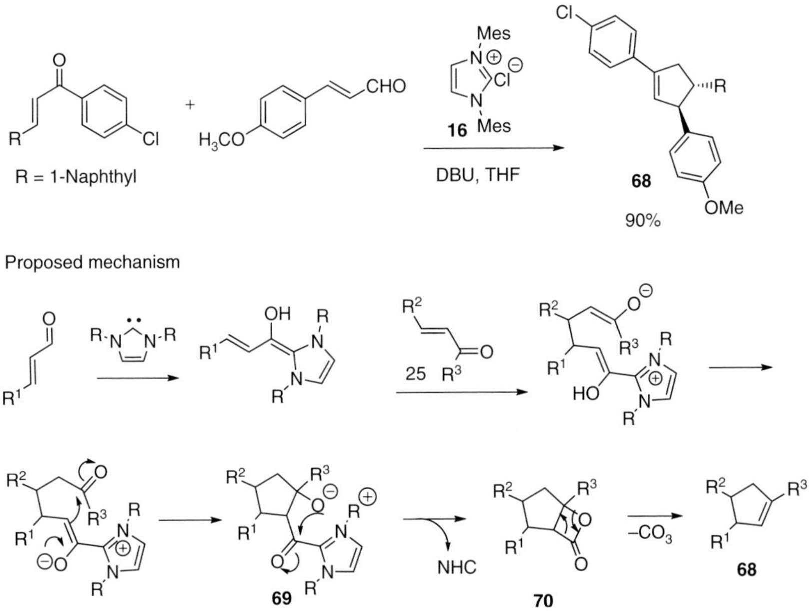

Scheme 4.27 Diastereoselective formation of cyclopentenes from homoenolates and chalcones.

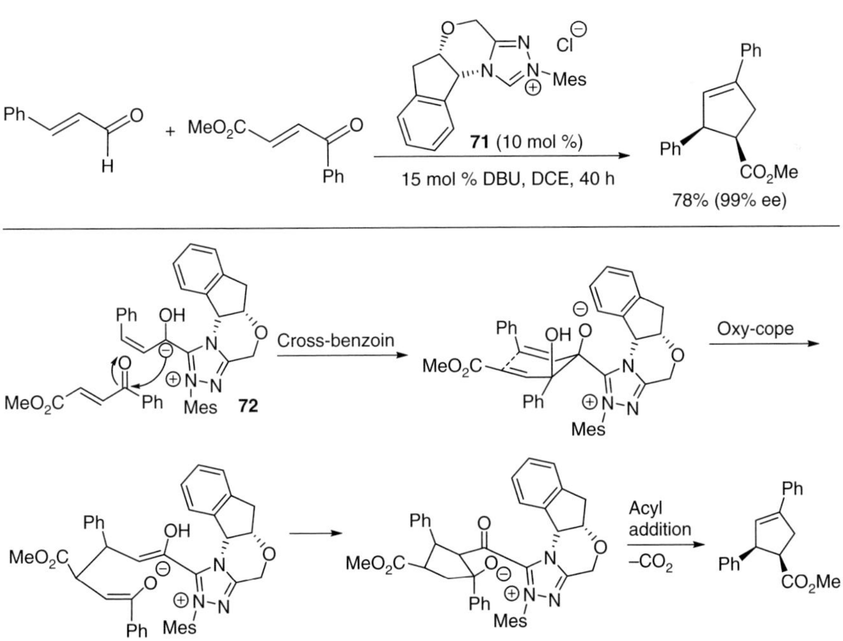

Scheme 4.28 Enantioselective cyclopentene synthesis via homoenolates.

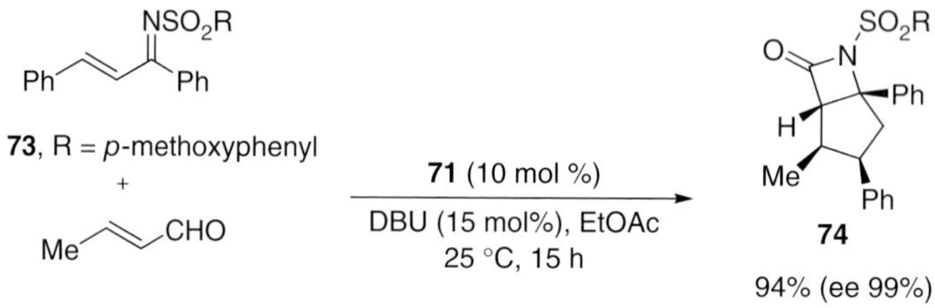

Scheme 4.29 Enantioselective β-lactam synthesis via homoenolates.

catalysis. Scheidt's group developed a stereoselective homodimerization of enals mediated by the combination of the chiral triazolium NHC **75**, titanium isopropoxide, and isopropanol culminating in the formation of *cis*-cyclopentene derivatives **76** (Scheme 4.30) [33].

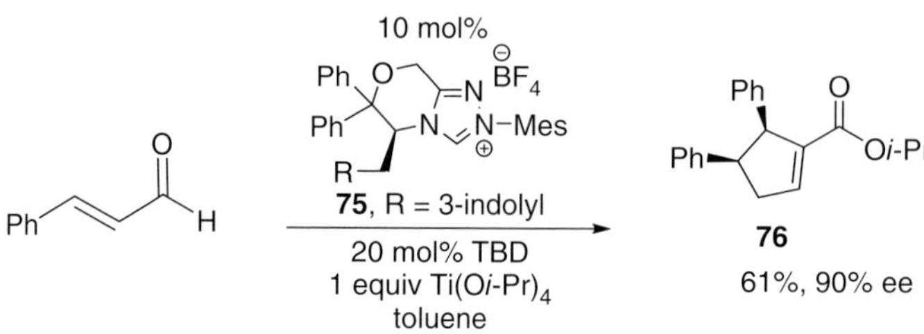

Scheme 4.30 Homodimerization of enals under NHC–Lewis acid cooperative catalysis.

The cyclopentene formation from chalcones and homoenolates has also been exploited for the construction of macrocyclic frameworks. Nair reported that benzimidazole-NHC **77** efficiently promoted the macrocyclization of chalcones in **78** chained to enals via oxygen-containing tethers. Cyclic ethers **79** ranging from ring sizes 10–13 were generated efficiently via this method. Similarly, a designer "semicrown" **80** was also synthesized in 45% yield (Scheme 4.31) [34].

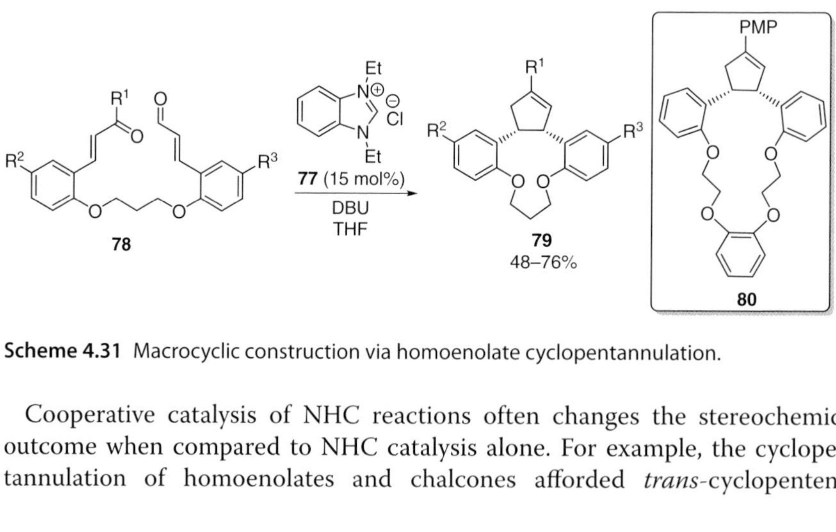

Scheme 4.31 Macrocyclic construction via homoenolate cyclopentannulation.

Cooperative catalysis of NHC reactions often changes the stereochemical outcome when compared to NHC catalysis alone. For example, the cyclopentannulation of homoenolates and chalcones afforded *trans*-cyclopentenes under NHC catalysis, whereas the combination of chiral triazolium NHC **29**, $Ti(Oi\text{-}Pr)_4$ (Lewis acid), and isopropanol (additive) promoted the formation of the corresponding *cis*-isomer **81** as the major product (Scheme 4.32). The titanium Lewis acid coordinates with both the homoenolate and the chalcone. This coordination enhances the electrophilicity of chalcone, places the β-carbons in vicinity to promote the Michael process, and controls the facial selectivity [28].

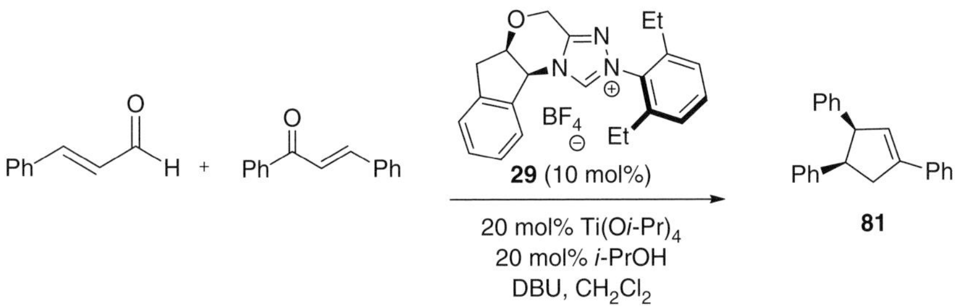

Scheme 4.32 NHC–Lewis acid cooperative catalysis for cyclopentene synthesis.

Reactions of NHC-homoenolates with cross-conjugated enones are highly sensitive toward structural changes in the enone partner. The examples depicted in Scheme 4.33, reported by Nair et al., illustrate this feature clearly. Acyclic dienones such as **82** afforded a separable mixture of cyclopentenes **83** (major product) and cyclcopenatanones **84**. Dibenzylidine cyclopentanones **85**, on the other hand, led to the stereoselective formation of spirocyclopentanone products **86** under identical reaction conditions. The corresponding dibenzylidine cyclohexanones **87**, however, furnished cyclopentene derivatives **88** exclusively (Scheme 4.33) [35].

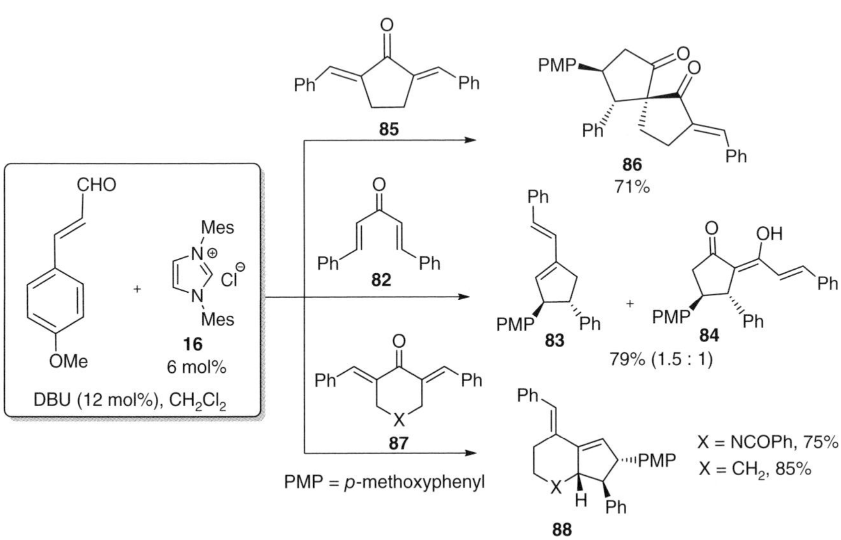

Scheme 4.33 Reactions of NHC-homoenolate with cross-conjugated dienones.

The different results in these reactions may be attributed to the difference in reactivity of the enolates formed after the addition of the homoenolates to the dienones in each case. The cyclopentene-enolate **89** is presumably very reactive, and it rapidly undergoes C-acylation to displace the NHC catalyst to afford the spirocyclic product **86**. A similar enolate **90**, formed from the six-membered dienone **87**, is less nucleophilic, and it may isomerize to form the enolate **91** that endures an Aldol cyclization to deliver the cyclopentene derivative **88**. It is reasonable to assume that the enolate **92** (formed from acyclic dienone **82**) has a mixed reactivity profile and therefore further reacts to form both cyclopentene and cyclopentanone derivatives **83** and **84**, respectively (Scheme 4.34) [35].

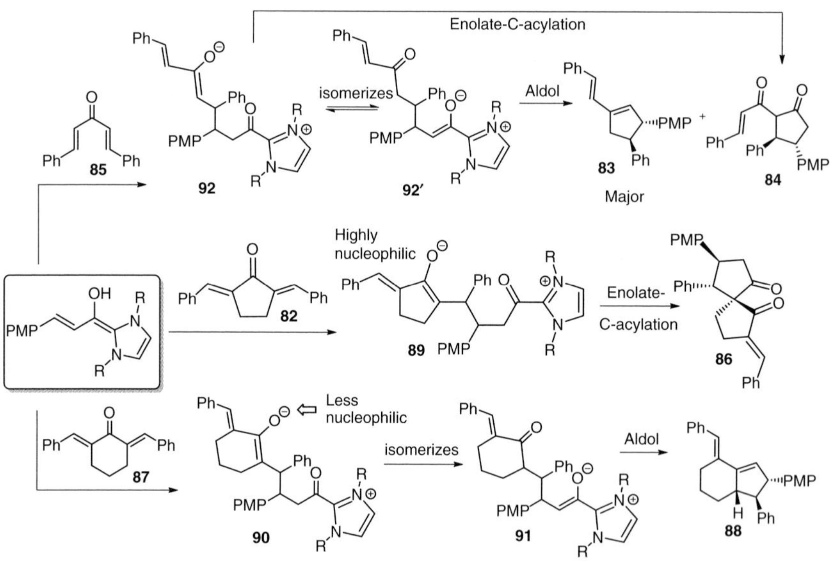

Scheme 4.34 Plausible mechanistic pathways for homoenolate-dienone reactions.

Enders developed an asymmetric synthesis of 3,4-disubstituted adipic acid derivatives via NHC-catalyzed [3+2] cycloaddition of enals and masked cinnamates **93**. The [3+2] cycloaddition afforded chiral cyclopentanones **94** bearing a masked carboxylic acid group in moderate-to-good yields with high stereoselectivities. Further synthetic modifications of the cyclopentanone furnished 3,4-disubstituted adipic esters **95**, which are precursors of pharmaceutically important gababutins (Scheme 4.35) [36].

Scheidt and coworkers had demonstrated that NHC-homoenolates can also be generated in protic solvents [37]. Nair et al. observed that when homoenolates were exposed to chalcones in a protic solvent such as methanol, cyclopenetene formation did not occur. Instead, the imidazolium salt **16** catalyzed the cyclization of enals and chalcones to afford a single diastereomer of β-hydroxycyclopentane carboxylate **96** as the major product. Minor amounts of acyclic δ-ketoester **97** were also formed in this reaction (Scheme 4.36) [38]. The major product **96** contains four contiguous stereocenters out of which one is

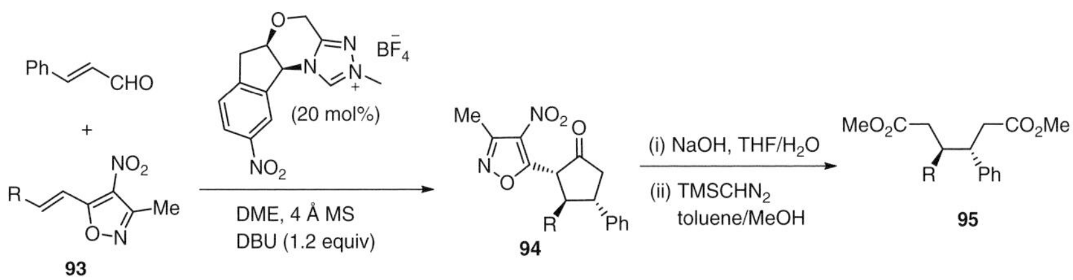

Scheme 4.35 NHC-catalyzed asymmetric synthesis of 3,4-disubstituted adipic acid derivatives.

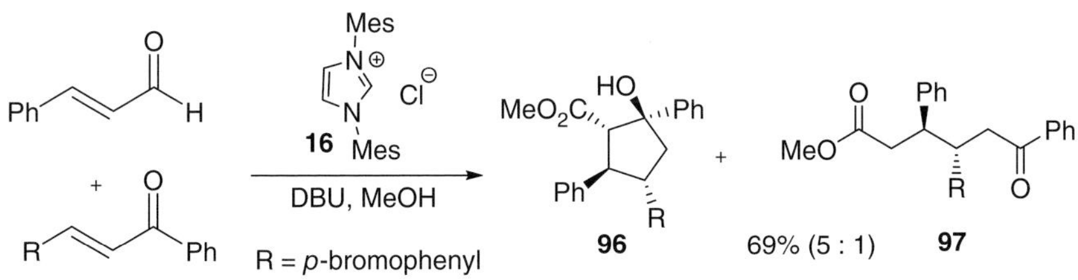

Scheme 4.36 Reaction of NHC-homoenolates with chalcones in a protic solvent.

quaternary. Interestingly, the relative stereochemistry of the product is opposite to that observed in the formation of cyclopentenes. This difference in outcome may be attributed to the methanol-assisted facile proton transfer.

NHC-catalyzed stereoselective Michael/aldol/lactamization cascade reaction of indole-derived enones **98** with enals afforded functionalized cyclopenta[3,4]pyrrolo[1,2-*a*]indol-4-ones **99**. The products incorporate four contiguous stereocenters and were formed in good yield, high diastereoselectivity, and excellent enantioselectivity (Scheme 4.37) [39].

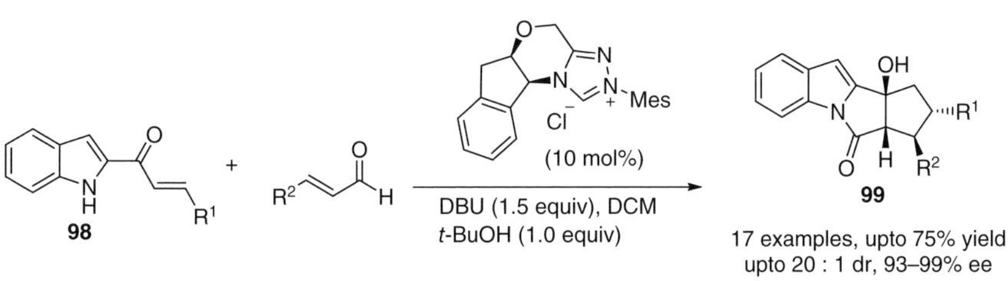

Scheme 4.37 NHC-catalyzed stereoselective Michael/aldol/lactamization cascade reaction.

Michael acceptors devoid of carbonyl groups also react efficiently with homoenolates. For example, Nair reported that a Michael addition of homoenolates to β-nitrostyrenes **100** can be efficiently catalyzed by imidazolidinium catalyst **10**. A highly stereoselective formation of the anti-isomer of δ-nitroesters **101** was observed in this case. Presumably, a *trans*-Michael addition leads to the formation of the observed products (Scheme 4.38). The multifunctional acyclic molecule **101** may be employed as the precursor for the construction of δ-lactams and δ-amino acid derivatives [40].

Liu and coworkers independently showed that such Michael addition reactions may be carried out in an enantioselective manner by employing chiral

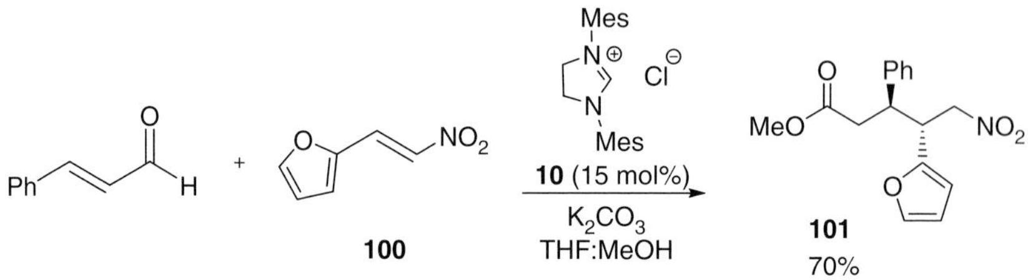

Scheme 4.38 Diastereoselective Michael addition of homoenolates to β-nitrostyrenes.

triazolium-derived NHC catalysts. The *anti*-δ-nitro ester **102** is formed selectively from nitroenynes **103** (or nitrodienes) in the presence of catalyst **29** (Scheme 4.39) [41].

Scheme 4.39 Chiral NHC-catalyzed Michael addition of homoenolates to nitroenynes.

Rovis and coworkers used the more sterically demanding triazolium catalyst **104** to efficiently construct the syn isomers of δ-nitro ester **105**. The bulky chiral NHC precatalyst **104** ensures that the undesired acyl anion addition is minimized. Additionally, this method is also compatible with aliphatic nitroolefins (Scheme 4.40) [42].

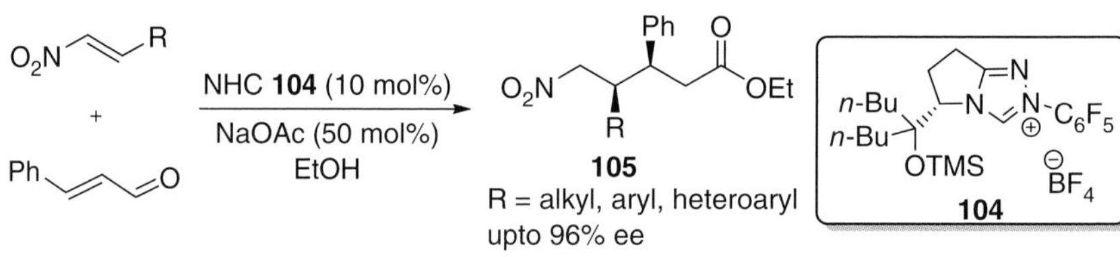

Scheme 4.40 Enantioselective Michael addition of homoenolates to nitroolefins.

The sensitivity of NHC-homoenolate reactivity was demonstrated by Bode in a divergent reaction of cinnamaldehydes and α′-hydroxyenones. The chiral triazolium catalyst **71** promoted the formation of the cyclopentane-fused β-lactone derivative **106**, whereas the structurally similar imidazolium catalyst **107** furnished the γ-lactone **108**. Both the products are formed in an excellent enantiomeric purity. The triazolium species **71** is a better leaving group than the imidazolium **107** and therefore favors the formation of the β-lactone product. The long-lived imidazolium intermediate **109**, on the other hand, undergoes a retro-aldol–aldol reaction sequence to form an epimeric cyclopentanol. The latter is readily poised to cyclize to afford a more stable γ-lactone **108** (Scheme 4.41) [43]. Importantly, the starting material (α′-hydroxy enone **110**) does not react with NHC to produce homoenolates (see Scheme 4.70 for such reactions).

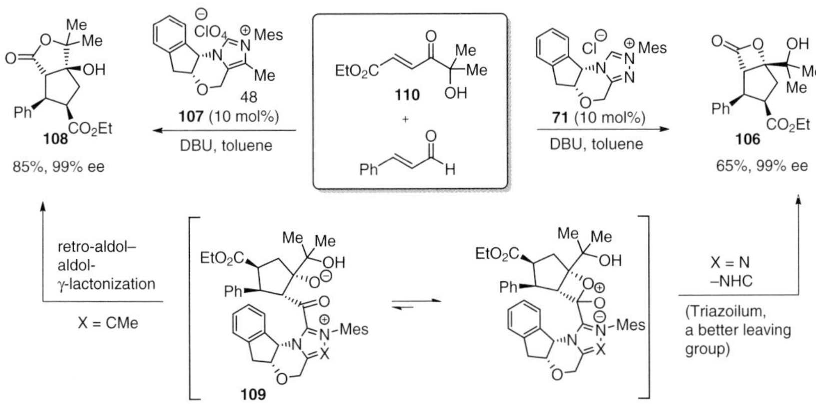

Scheme 4.41 Enantioselective and divergent synthesis of cyclopentanols.

Oxindole-derived α,β-unsaturated imines **111** constitute important building blocks for the construction of valuable spirocyclic products via homoenolate annulations. The example depicted in Scheme 4.41 illustrates their utility. The use of triazole-derived NHC **112** allows to fine-tune the reaction to afford either the β-lactam product **113** or the spiro-cyclopentene **114** by changing the reaction temperature (Scheme 4.42) [44].

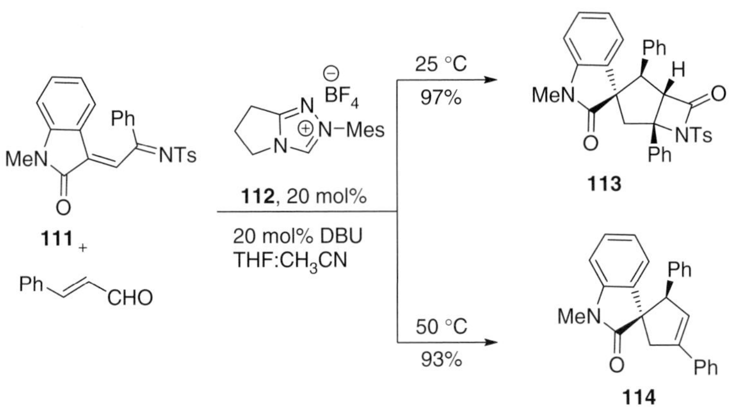

Scheme 4.42 Oxindole-containing spirocycle construction via homoenolates.

Glorius and coworkers reported the asymmetric construction of spirocyclopentanones **115** from aza-aurones **116** by homoenolate annulation catalyzed by the aminoindanol–triazolium catalyst **36** (Scheme 4.43) [45].

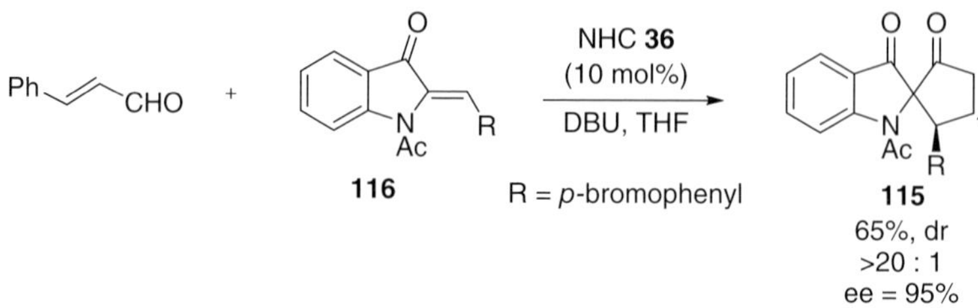

Scheme 4.43 Asymmetric synthesis of spiro-cyclopentanones from azaaurones.

Zhao and coworkers reported that aurones **117** undergo stereoselective [4+3] cyclization with NHC-homoenolates to afford ε-lactones **118** (Scheme 4.44). The formation of five-membered spirocyclic products were suppressed by the use of a Lewis acid catalyst, $Ti(O\textit{i}\text{-}Pr)_4$ [46].

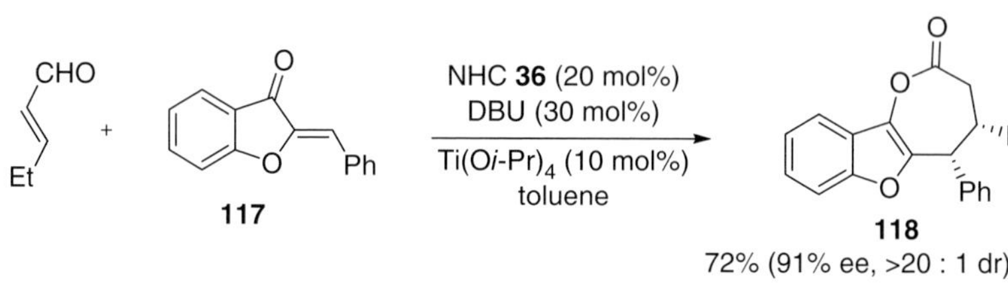

Scheme 4.44 Synthesis of seven-membered lactones via homoenolate annulation.

Related exocyclic enones, viz., 2-aroylidene benzofuran-3-ones **119**, afforded cyclopentene-fused spirobenzofuran-3-ones **120** after annulation with NHC-homoenolates as shown in Scheme 4.45 [47].

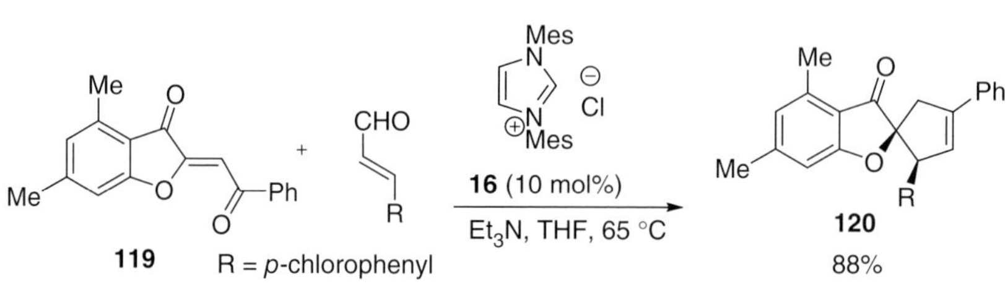

Scheme 4.45 Spirocycle synthesis from NHC-homoenolates and enones.

Scheidt and coworkers reported an asymmetric cyclization reaction of homoenolates and *o*-quinomethides to afford 2-benzoxepinones **121** in high enantioselectivity. Importantly, these two transient intermediates are generated in situ and successfully assembled. In addition to the NHC catalyst **29**, tetrabutyl ammonium acetate, cesium fluoride, and 18-crown-6 are employed as promoters/catalysts in this double-Lewis-base-mediated annulation reaction (Scheme 4.46) [48].

Stable, preformed *o*-quinomethides **122** also react similarly with homoenolates as reported independently by Ye and coworkers (Scheme 4.47) [49].

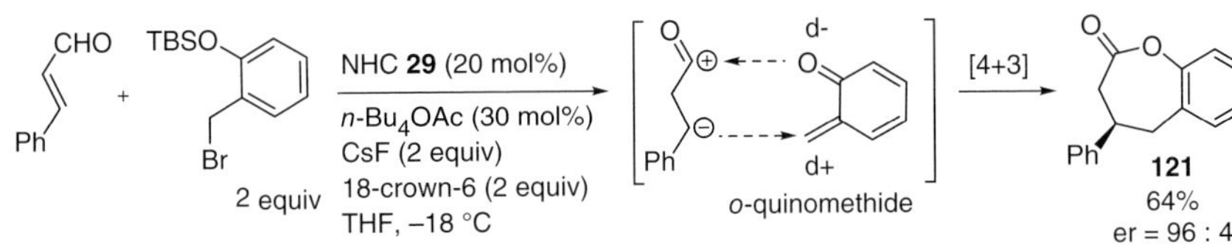

Scheme 4.46 Scheidt's [4+3] annulation of homoenolates and *o*-quinomethides.

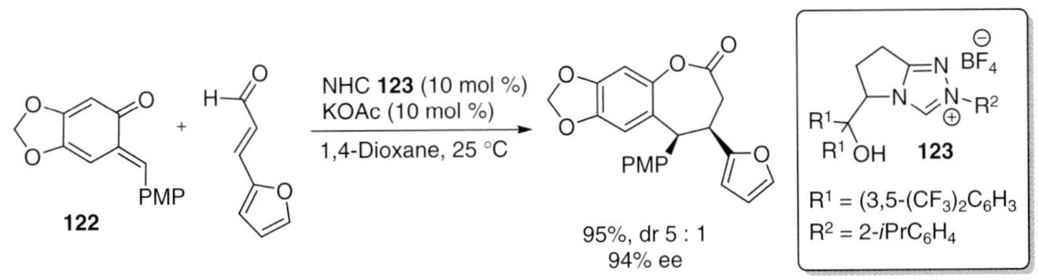

Scheme 4.47 Cyclization reaction of stable *o*-quinomethides and homoenolates.

Qi and coworkers reported a novel and convenient strategy for the enantioselective synthesis of γ-lactam derivatives **124** via NHC-catalyzed formal [3+2] annulation of enals with 2-aminoacrylates **125**. This activation mode provides a complementary approach for the synthesis of various γ-lactam derivatives. In this process, two contiguous stereocenters are installed including a quaternary carbon center (Scheme 4.48) [50].

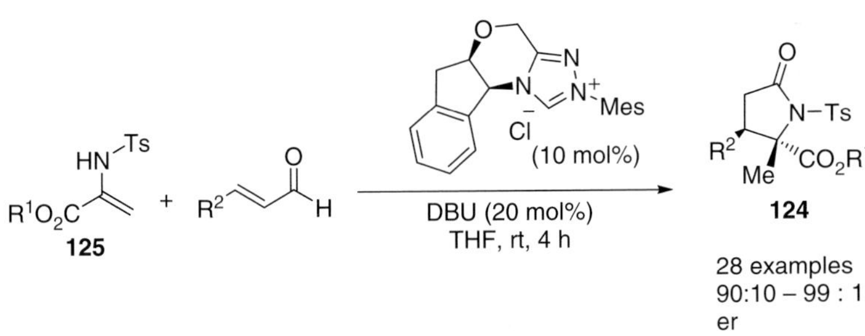

Scheme 4.48 NHC-catalyzed enantioselective synthesis of γ-lactam derivatives.

4.7 β-Protonation of Homoenolates and Subsequent Reactions

As described earlier, the Breslow intermediate formed from conjugated enals can exhibit a variety of reactivity patterns depending on the reaction conditions. In the presence of a proton source, the homoenolate can protonate at the β-carbon to generate an enol (see Scheme 4.10). Subsequent reaction of such enols with suitable electrophiles results in the formation of a new bond at the α-carbon. Alternatively, the keto form of this enol (acyl azolium) may interact with nucleophiles to undergo acyl displacement and NHC regeneration. A few selected examples of these types of reactions are presented in the following passages.

Chan and Scheidt reported that enals may be converted to saturated esters by mediation of NHCs. The β-protonation of homoenolate is effected by phenol here. The resultant acyl azolium then reacts with benzyl alcohol to furnish the corresponding ester **126** (Scheme 4.49) [51].

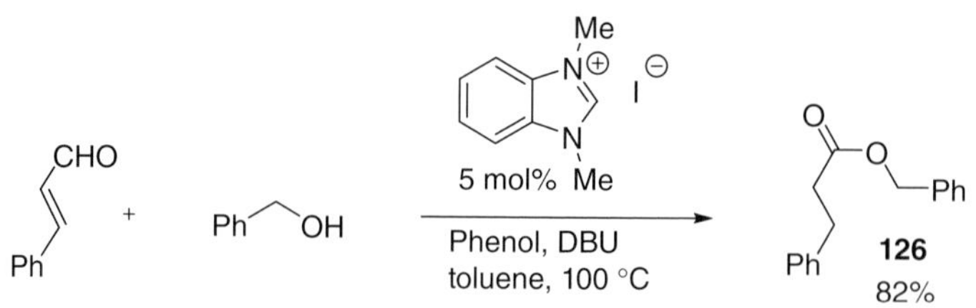

Scheme 4.49 NHC-catalyzed redox esterification of enals.

Studies by Sohn and Bode revealed that the β-protonation may occur even without the addition of external proton source [52]. Conjugate acids of mild bases that are used for the generation of NHC may occasionally act as proton sources and promote similar redox esterification of unsaturated aldehydes. It was also observed that this reactivity is facilitated further by the use of less nucleophilic carbenes.

Glorius and coworkers reported that the enolate-type reactivity can be promoted by using a mild base such as triethylamine in a nonpolar solvent at elevated temperature. The reaction of unsaturated aldehyde **127** and trifluromethyl ketone **128** afforded the β-lactone **129** as a mixture of diastereomers (Scheme 4.50) [53].

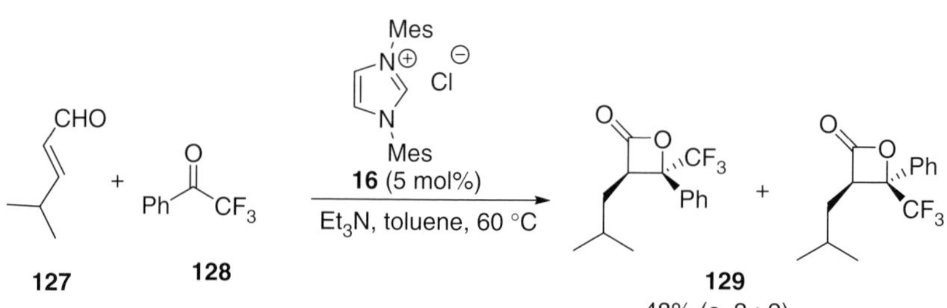

Scheme 4.50 NHC-catalyzed formation of β-lactone from enals and ketones.

The annulation of enolate equivalent (**28**, Scheme 4.10) to Michael acceptors could potentially lead to the formation of six-membered rings in a formal [4+2] cycloaddition. Such a transformation was developed by Bode and colleagues. Here, α,β-unsaturated imines and NHC-bound enals underwent an enantioselective formal Diels–Alder cycloaddition to afford dihydropyridinone derivatives **130** (Scheme 4.51) [54].

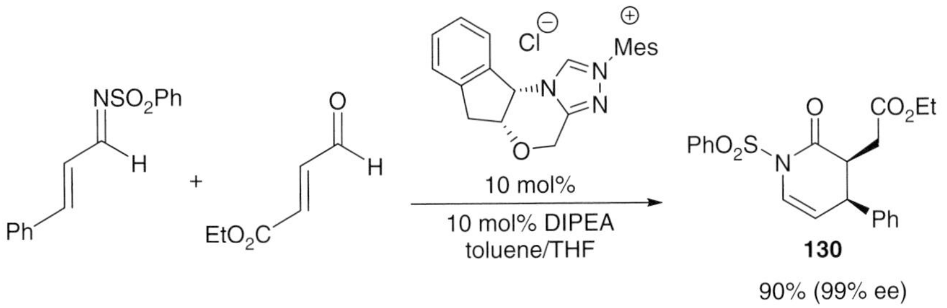

Scheme 4.51 NHC-catalyzed enantioselective formal Diels–Alder cycloaddition.

Highly diastereo- and enantioselective intramolecular Michael reaction of enolate equivalents and chalcones was reported by Scheidt and coworkers. The initially formed dihydropyranone **131** is cleaved by transesterificaion with methanol to afford a 1,5-dicarbonyl compound **132** (Scheme 4.52) [55].

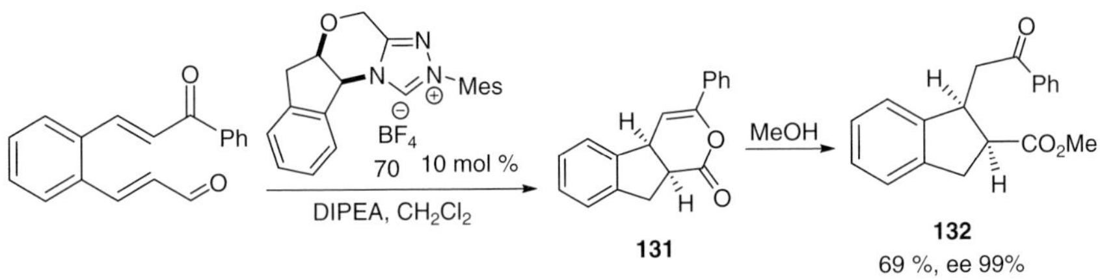

Scheme 4.52 Intramolecular Michael addition of enolate equivalents.

Dihydropyranone derivatives **133** were isolated as the final products in another similar intermolecular reaction. It may be noted that the overall transformation is equivalent to a formal inverse electron demand hetero-Diels–Alder reaction between a chiral ester enolate (derived from the starting enal) and the enone **134** (Scheme 4.53) [56].

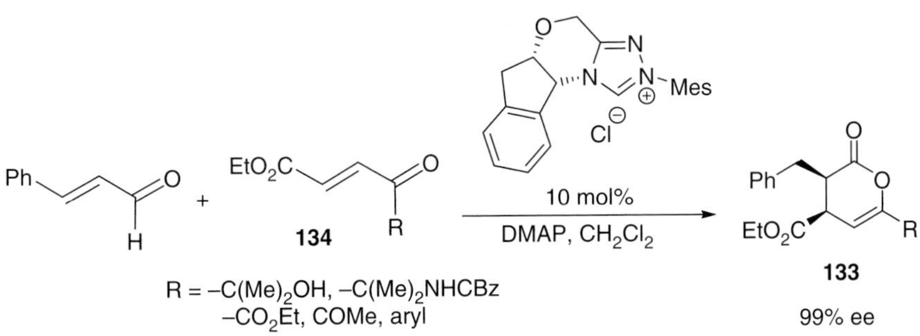

Scheme 4.53 Enantioselective formal Diels–Alder reactions of NHC enolates.

These enolate equivalents can also engage in Aldol-type reactions. An intramolecular aldol clyclization was used by Scheidt and coworkers to efficiently desymmetrize the 1,3-diketone **135**. The β-lactone intermediate could be isolated in one case; however, with the rest of the 1,3-diketones in this study, enantiomerically pure cyclopentenes **136** were obtained as the final products (Scheme 4.54) [57].

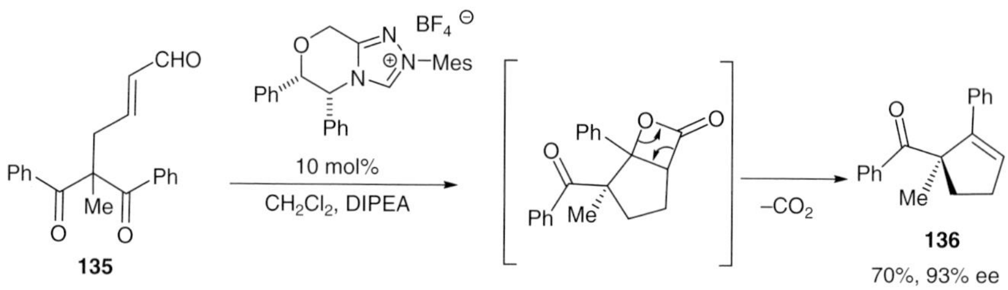

Scheme 4.54 Enantioselective intramolecular aldol reaction of NHC-enolates.

This approach was elegantly applied by Scheidt and coworkers in the enantioselective total synthesis of sesquiterpene natural products, bakkenolides I, J, and S. Desymmetrization of the easily accessible cyclic 1,3-diketone **137** was mediated by the chiral triazolium catalyst **138**. The β-lactone **139** could be generated in multigram quantities in excellent enantiomeric purity. The desired bakkenolides were then synthesized from **139** in a few steps (Scheme 4.55) [58].

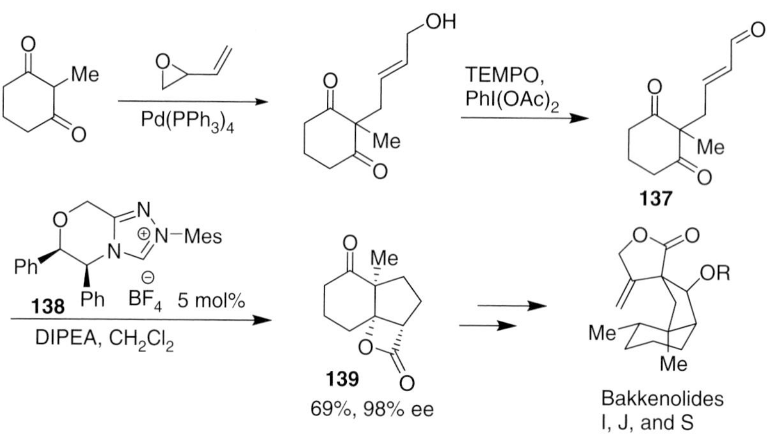

Scheme 4.55 Total synthesis of bakkenolides by NHC-assisted desymmetrization.

Scheidt's group also developed this approach to carry out a dynamic kinetic resolution of β-keto esters [59a]. The enolate **140** generated from chiral triazolium NHC **29** and the enal unit of **141** presumably undergoes rapid epimerization in the presence of base. Intramolecular aldol reaction of **140** then proceeds via a six-membered transition state held together by hydrogen-bonding interaction between the NHC-enol and ketone oxygen. It is proposed that a gauche interaction of the ethyl ester and the aryl group destabilizes the transition state that would lead to the minor product. Highly substituted β-lactones **142** are formed as the final products in excellent enantioselectivity. This method was also applied in the formal enantioselective synthesis of a benzopyran estrogen receptor β-agonist (Scheme 4.56) [59b].

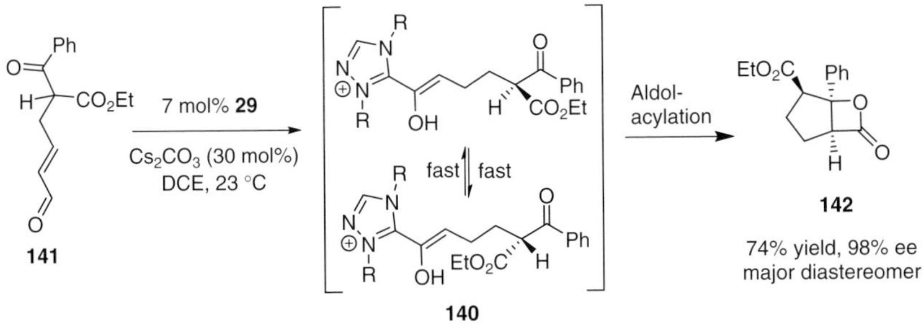

Scheme 4.56 Chiral NHC-catalyzed dynamic kinetic resolution of β-keto esters.

An intriguing structural reorganization of 2*H*-chromene-3-carboxaldehyde **143** catalyzed by the imidazolidinium catalyst **10** also proceeded via an enolate species. Nair et al. reported that **143** rearranges, when exposed to NHC **10** and DBU, to afford 3-methylcoumarin **144**. The reaction presumably involves β-protonation of the Breslow intermediate **145**, a Grob-type fragmentation, subsequent displacement of the acyl azolium by phenoxide **146**, and isomerization (Scheme 4.57) [60].

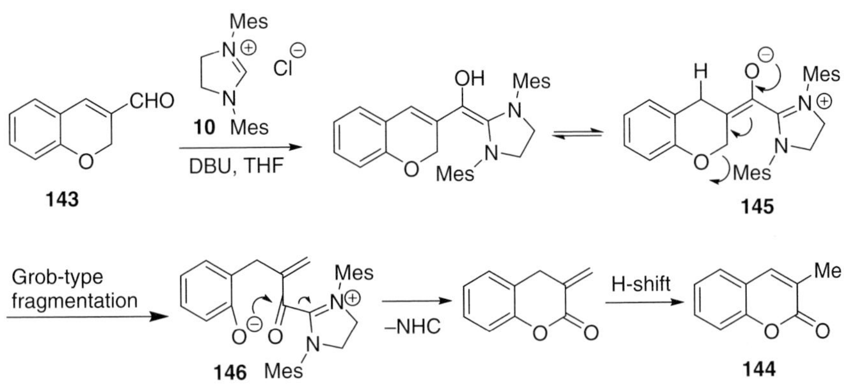

Scheme 4.57 NHC-catalyzed isomerization of a chromene aldehyde to coumarin.

Reaction of chiral NHC-bound enols with alkylidene imidazolidinones **147** led to the formation of enantioenriched bicyclic lactones **148** in good yields. This formal [4+2] cycloaddition reaction is promoted by equimolar amounts of acetic acid. The latter protonates the homoenolate and thereby suppresses the reaction at the β-carbon. Subsequent conjugate addition and cyclization of the enolate with alkylidene imidazolidinones afford the products **148** (Scheme 4.58) [61].

Alkylation of NHC-bound enols with π-allyl palladium complexes was reported by Scheidt and colleagues in 2014. The high affinity of NHCs to bind to transition metal complexes poses a serious challenge to the use of transition metal-based

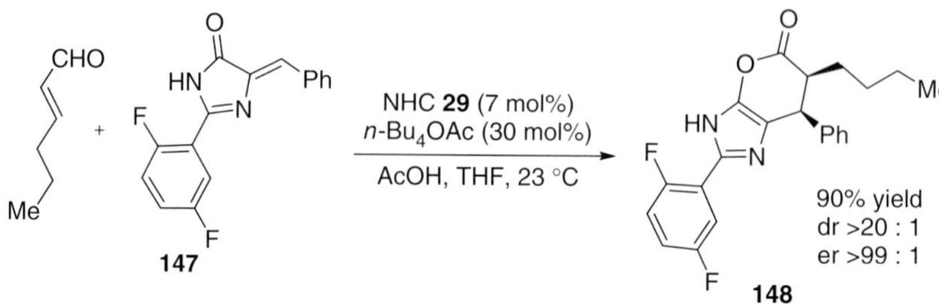

Scheme 4.58 Cyclization of NHC-bound enolates and alkylidene imidazolidinones.

cocatalysts. The substrate **149** incorporated an enal unit as well as the precursor for the π-allyl palladium complex. It was found that an additional source of π-allyl palladium complex (**150**) is essential for the success of the reaction. The enal and NHC interact to generate the corresponding Breslow intermediate, which then undergoes β-protonation to furnish the reactive enol **151**. The latter then reacts with the π-allyl palladium complex generated from its precursors. Finally, the acycl azolium is quenched by lactonization by the phenoxide to afford the allyl dihydrocoumarin **152** (Scheme 4.59) [62]. This constitutes one of the many examples of cooperative catalysis where NHC and a cocatalyst are employed together to promote an otherwise sluggish transformation.

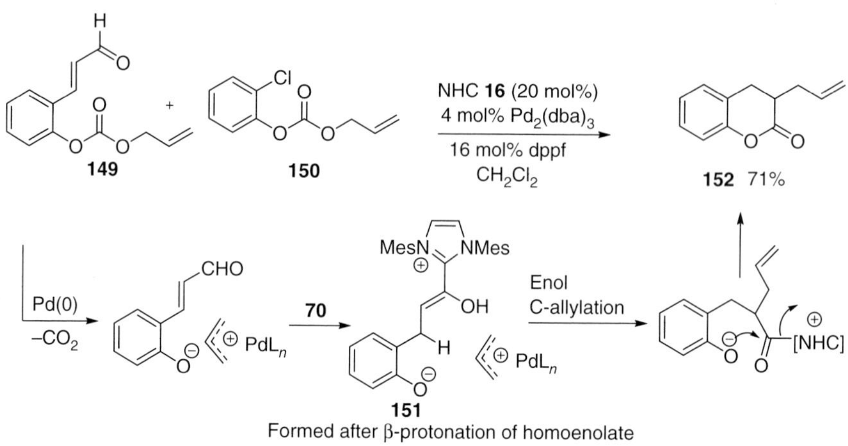

Scheme 4.59 Reaction of NHC-enol and π-allyl palladium complex via cooperative catalysis.

4.8 Homoenolates in Carbon–Nitrogen Bond Formation

Formal [3+2] cycloaddition reactions of homoenolates with 1,3-dipoles such as nitrones and azomethine imines have been reported. Rapid cycloaddition of cinnamaldehyde-derived homoenolate to nitrones was promoted by a triazolium NHC. This method allows rapid synthesis of various γ-amino esters **153** (Scheme 4.60) [63].

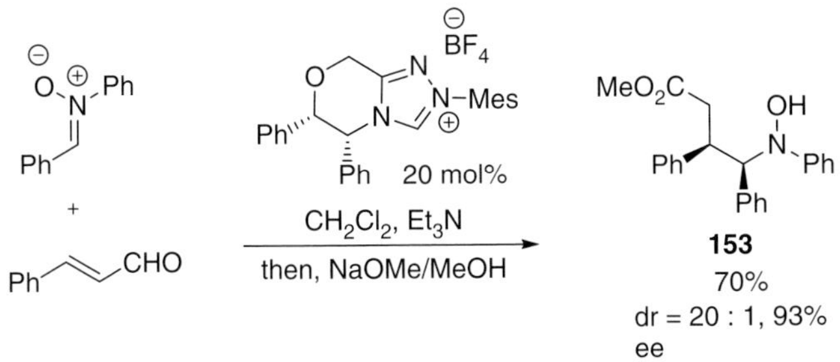

Scheme 4.60 Reaction of homoenolates with nitrone.

The cycloadduct pyridazinone **154** was isolated as the final product in the reaction of homoenolate with azomethine imine **155** (Scheme 4.61) [64].

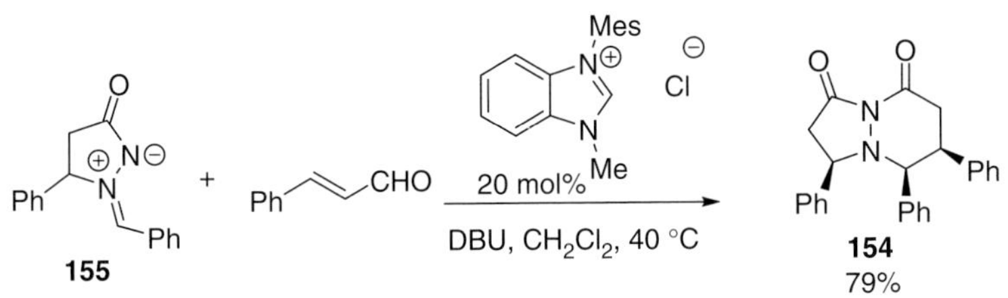

Scheme 4.61 Formal [3+2] cycloaddition of homoenolates and azomethine imines.

Diazenes **156** constitutes another class of nitrogen electrophiles that successfully intercepted homoenolates in a regioselective cyclization reaction. Substituted pyrazolidones **157** were obtained as the products in this cycloaddition (Scheme 4.62) [65].

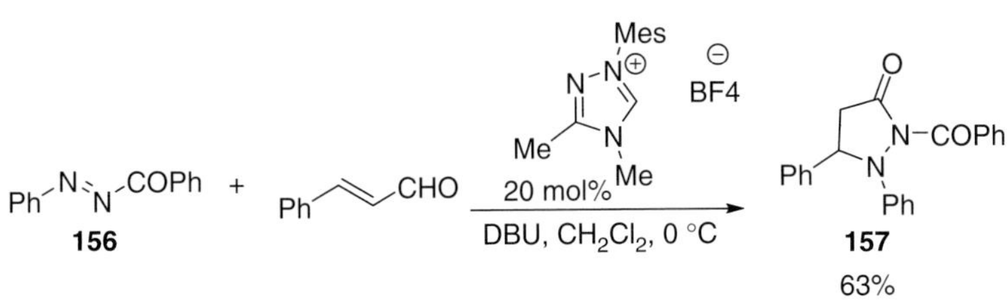

Scheme 4.62 Formal cycloaddition of diazene with homoenolate.

Formation of carbon–nitrogen bond at the β-position of enals can be achieved by the reaction of NHC-homoenolate with nitrosobenzene. The initially formed isoxozolidinone **158** is readily cleaved on exposure to methanol and acid to afford the β-amino acid derivative **159** (Scheme 4.63) [66].

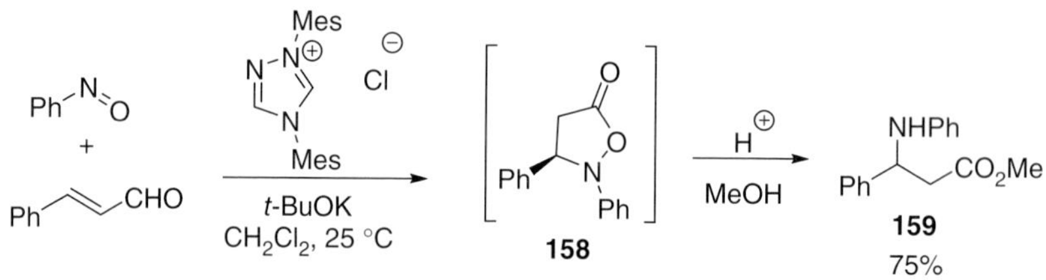

Scheme 4.63 Reaction of homoenolate with nitrosobenzene.

4.9 Domino Reactions of Homoenolates

It is well known that domino/cascade reactions allow generating molecular complexity in a single operation. When homoenolates interact with designed reaction partners that incorporate multiple reactive sites, a number of new bonds are formed. This leads to the formation of structurally diverse and complex scaffolds from relatively simple substrates. A few such examples are described below.

Chi and coworkers reported that the reaction of homoenolates with the benzodi(enone) **160**, in which two Michael acceptors are placed in close proximity, led to the formation of benzotricyclic product **161**. Initially, the homoenolate underwent a Michael addition to the more electron-deficient enone. This was followed by internal proton exchange to generate the enolate **162**, which took part in an intramolecular Michael addition with the second enone moiety. Finally, acylation of the enolate **163** by acyl azolium delivered the product **161** (Scheme 4.64). Despite the availability of numerous competing reaction pathways, the reaction furnished a single product in excellent yield and enantioselectivity. New C—C bonds are formed, with complete regio- and stereocontrol, at all the three carbons of the homoenolate moiety here [67].

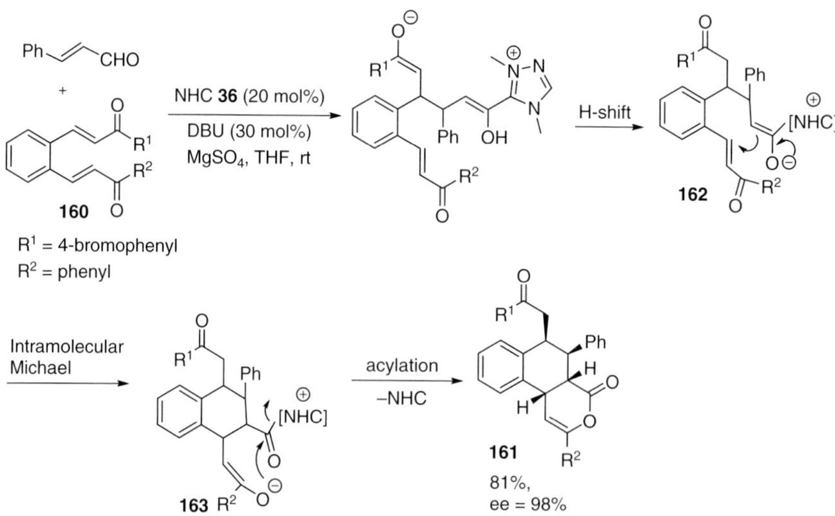

Scheme 4.64 Stereoselective domino cyclization of homoenolates with di(enones).

Biju and coworkers developed a domino reaction of homoenolates and 2′-hydroxy chalcones **164** that led to the diastereoselective formation of cyclopentane-fused coumarins **165**. The imidazolium catalyst **16** presumably promoted the cyclopentannulation of the chalcone and homoenolate to afford the cyclopentanol intermediate **166**. The latter was transformed into a six-membered lactone (instead of the β-lactone that is formed with simple chalcones) via acyl displacement of NHC by the phenol. Subsequent dehydration furnished the coumarin derivative **165** (Scheme 4.65) [68].

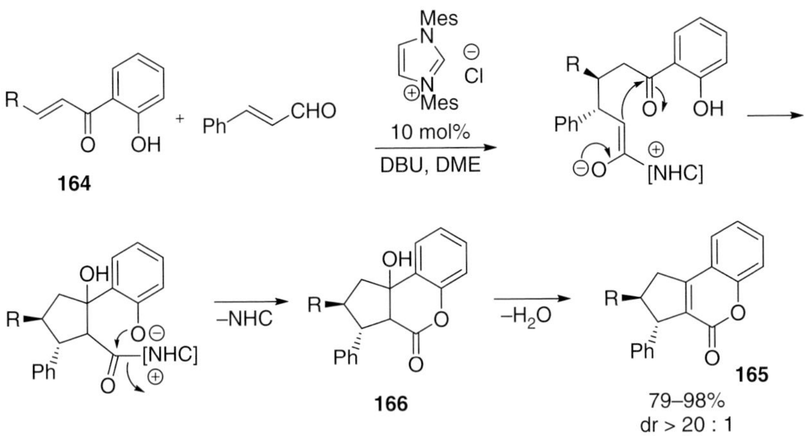

Scheme 4.65 Domino synthesis of cyclopentane-fused coumarins from homoenolates.

Substituted coumarins were obtained as the final products in the intramolecular domino reaction of homoenolates reported by Nair and coworkers. The homoenolate generated from the enal **167** engaged in an intramolecular conjugate addition to the tethered 2-*O*-alkenoate group. The stabilized carbanion **168** underwent elimination of the phenoxide and acyl substitution and double-bond isomerization to form the coumarin derivative **169** (Scheme 4.66) [69].

Azalactone **170** was found to undergo a diastereoselective domino cyclization with homoenolates generated from enals under standard NHC catalysis.

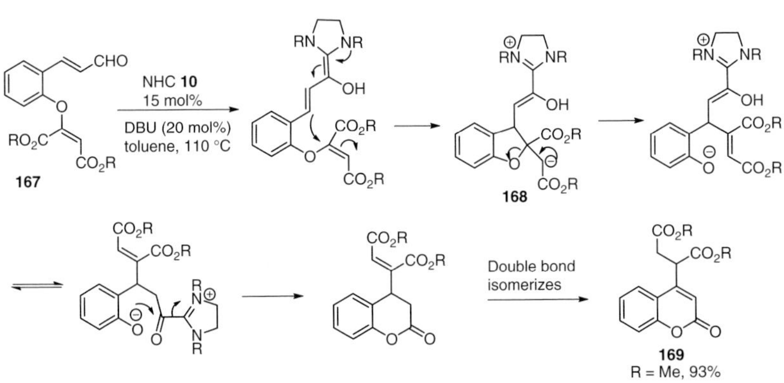

Scheme 4.66 Coumarin formation via intramolecular domino reaction of homoenolates.

Addition of the homoenolate to azalactone carbonyl opens up the heterocyclic ring to reveal an amide anion. The latter then displaces the NHC from the acyl azolium to complete a formal [3+3] cycloaddition to furnish multifunctionalized piperidones **171** (Scheme 4.67) [70].

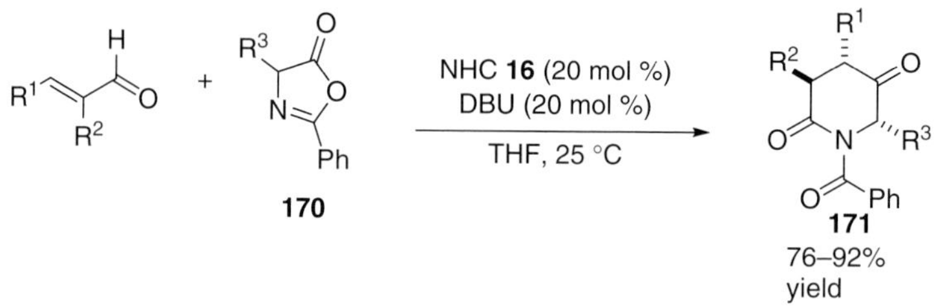

Scheme 4.67 Domino reaction of homoenolates and azalactones.

4.10 New Precursors for Homoenolates

Homoenolates are typically generated by the action of NHC catalysts on α,β-unsaturated aldehydes. Recent studies, however, have shown that other substrates can also be employed for this purpose. In an important discovery in this area, Chi and coworkers demonstrated that saturated carboxylic esters can be used as homoenolate equivalents under NHC catalysis. This not only expanded the scope of NHC-mediated transformations but also opened up new synthetic avenues in the realm of remote sp^3-CH functionalization. Chi and coworkers showed that carboxylic esters of nitrophenol **172** when reacted with the chiral NHC **173** lead to the formation of a reactive intermediate generated by deprotonation at the unactivated β-position. This deprotonation presumably occurs as depicted in Scheme 4.67. This homoenolate intermediate was successfully intercepted with a variety of electrophiles such as enones, trifluoromethyl ketones, and hydrazones. These reactions afforded enantioenriched cyclopentenes **174**, γ-lactones **175**, and γ-lactams **176**, respectively (Scheme 4.68) [71].

The homoenolate generated from esters can partake in domino reactions with suitable substrates. Chi and coworkers showed that esters **172** and amino enones **177** can be combined under chiral NHC catalysis to generate polycyclic quinolinones **178** in excellent enantiopurity. Mechanistic events leading to the formation of quinolinones are depicted in Scheme 4.69 [72].

Carboxylic anhydrides can also be employed as precursors for homoenolates under NHC catalysis. Chi and coworkers reported that the aliphatic anhydride **179** generates the homoenolate intermediate **180**, on exposure to the triazolium-derived NHC **181**. Reactions of this homoenolate with arylidene diketones, chalcones, and isatins afforded the corresponding cyclic products in good yield and stereoselectivity (Scheme 4.70) [73].

Bode has employed α′-hydroxyenones **181** as surrogates for enals in the generation of NHC-homoenolates. They are readily available via the aldol condensation of aromatic aldehydes and 3-hydroxy-3-methylbutanone **182** [74]. The addition

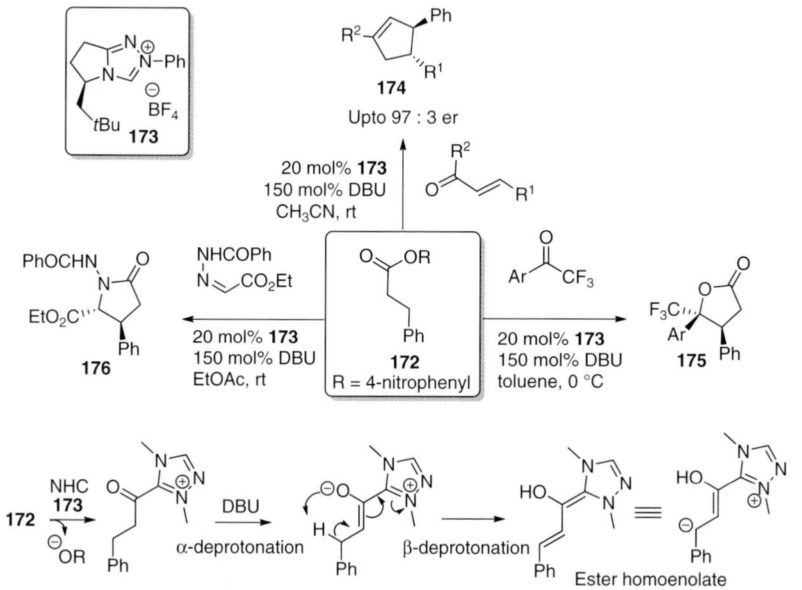

Scheme 4.68 Generation of homoenolates from saturated esters and further reactions.

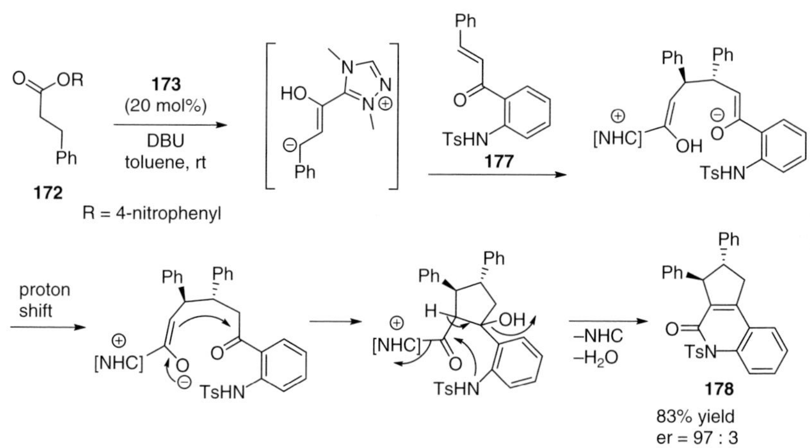

Scheme 4.69 Domino synthesis of quinolinones via NHC-catalyzed sp^3-CH activation.

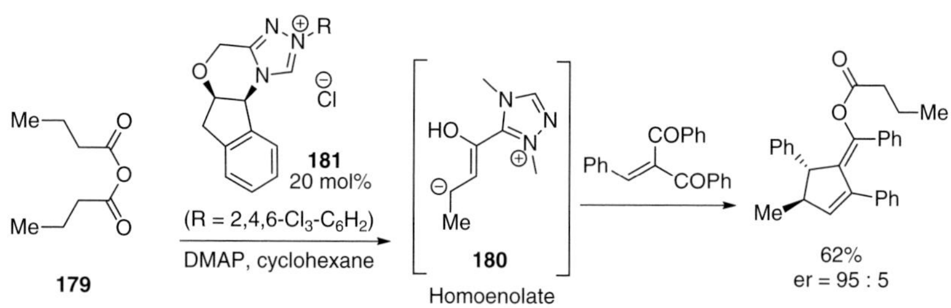

Scheme 4.70 Aliphatic anhydrides as a source of NHC-homoenolates.

of an NHC to the carbonyl group of **181** produces the tetrahedral intermediate **183**. This can undergo a retro-benzoin reaction to generate the homoenolate **184** and acetone (Scheme 4.71).

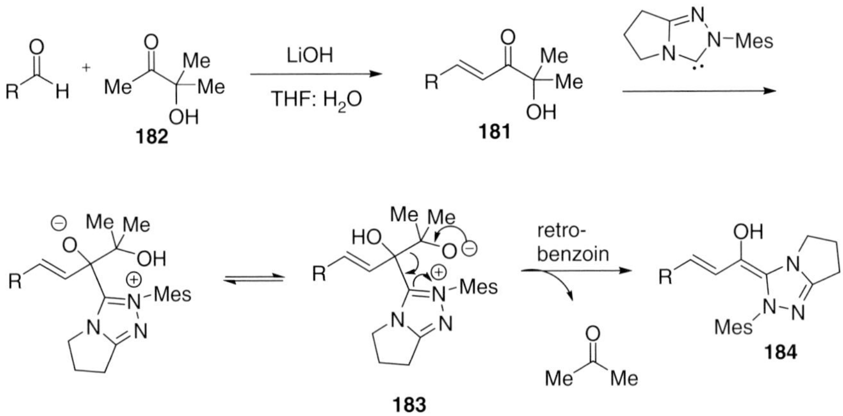

Scheme 4.71 α′-Hydroxyenones as enal surrogates for homoenolate generation.

The homoenolates generated in the above manner interact with electrophiles in the same manner as the ones generated from enals. A variety of NHC-catalyzed transformations were carried out by this method by Bode and coworkers, and a few representative examples are depicted in Scheme 4.72 [75].

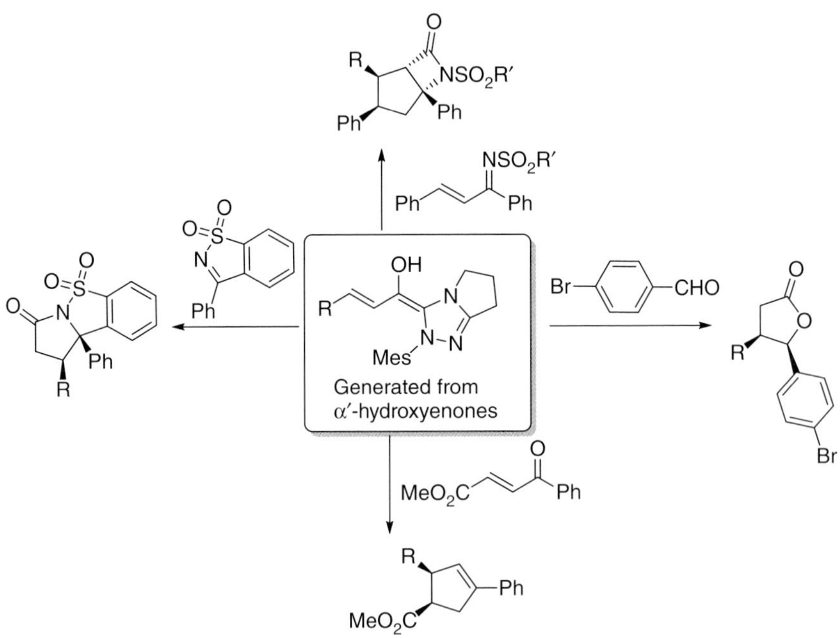

Scheme 4.72 Reactions of NHC-homoenolates derived from α′-hydroxyenones.

Although redox esterification of enals by alcohols to afford saturated esters is readily catalyzed by NHCs (see Scheme 4.49), the analogous amidation reactions generally fail because of competing imine formation. This problem was circumvented by the use of enal surrogates, α′-hydroxyenones. The latter and amines reacted smoothly in the presence of the triazole-derived NHC **66**, Hünig's base, and 1,2,4-triazole (cocatalyst) to furnish the corresponding amides in good yields (Scheme 4.73) [76].

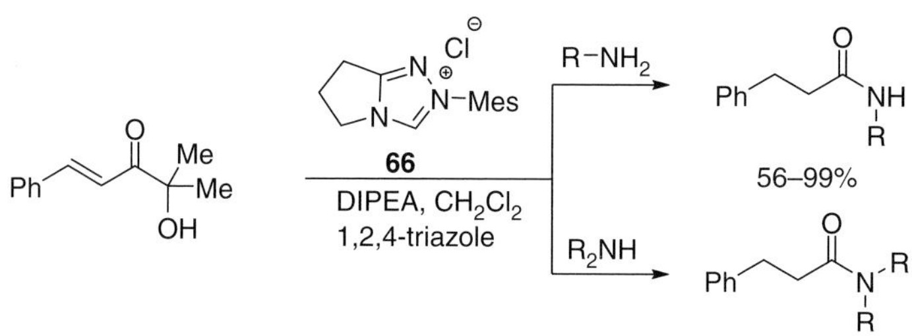

Scheme 4.73 NHC-catalyzed amidation of α′-hydroxyenones.

4.11 Conclusion

The chemistry of homoenolates has advanced a long way since the pioneering work by Glorius and Bode in 2004. This versatile three-carbon synthon has found applications in the stereoselective construction of a number of carbocyclic, heterocyclic, and acyclic molecules. Selective formation of carbon–carbon bonds at each of the carbons of the homoenolate may be achieved by careful selection of the catalyst and reaction conditions. The availability of chiral NHCs of different varieties allows chemists to fine-tune the steric and electronic features of the catalyst for the task at hand. The emergence of cooperative catalysis wherein NHCs are used in combination with metal-based catalysts has broadened the scope of this area significantly. A number of natural products have been synthesized by using NHC-homoenolates as key reactive intermediates. Chi's path-breaking discovery that saturated esters and anhydrides may also be used as precursors of homoenolates have opened up new avenues of exploration. It is presumable that this area will continue to grow and numerous novel, intriguing and useful transformations will be reported in future.

References

1 Stork, G., Brizzolara, A., Landesman, H. et al. (1963). *J. Am. Chem. Soc.* 85: 207–222.
2 Kuwajima, I. and Nakamura, E. (1991). *Comprehensive Organic Synthesis*, vol. 2 (ed. B.M. Trost and I. Fleming), 441–473. New York: Pergamon.
3 Nickon, A. and Lambert, J.L. (1962). *J. Am. Chem. Soc.* 84: 4604–4605.

4 (a) Nakamura, E., Aoki, S., Sekiya, K. et al. (1987). *J. Am. Chem. Soc.* 109: 8056–8066. (b) Bal, S.A., Marfat, A., and Helquist, P. (1982). *J. Org. Chem.* 47: 5045–5050. (c) Hoppe, D. and Hense, T. (1997). *Angew. Chem., Int. Ed. Engl.* 36: 2282–2316.
5 Breslow, R. (1958). *J. Am. Chem. Soc.* 80: 3719–3726.
6 Arduengo, A.J., Harlow, R.L., and Kline, M. (1991). *J. Am. Chem. Soc.* 113: 361–363.
7 (a) For a monograph, see: Nolan, S.P. (2006). *N-Heterocyclic Carbenes in Synthesis*, 1–304. Weinheim: Wiley-VCH. (b) For reviews, see: Flanigan, D.M., Romanov-Michailidis, F., White, N.A., and Rovis, T. (2015). *Chem. Rev.* 115: 9307–9387.
8 Nair, V., Bindu, S., Sreekumar, V., and Rath, N.P. (2003). *Org. Lett.* 5: 665–667.
9 (a) Berkessel, A., Elfert, S., Etzenbach-Effers, K., and Teles, J.H. (2010). *Angew. Chem. Int. Ed.* 49: 7120–7124. (b) Berkessel, A., Yatham, V.R., Elfert, S., and Neudörfl, J.-M. (2013). *Angew. Chem. Int. Ed.* 52: 11158–11162.
10 (a) Burstein, C. and Glorius, F. (2004). *Angew. Chem. Int. Ed.* 43: 6205–6208. (b) Sohn, S.S., Rosen, E.L., and Bode, J.W. (2004). *J. Am. Chem. Soc.* 126: 14370–14371.
11 For reviews, see: (a) Menon, R.S., Biju, A.T., and Nair, V. (2015). *Chem. Soc. Rev.* 44: 5040–5052. (b) Nair, V., Menon, R.S., Biju, A.T. et al. (2011). *Chem. Soc. Rev.* 40: 5336–5346. (c) Nair, V., Vellalath, S., and Babu, B.P. (2008). *Chem. Soc. Rev.* 37: 2691–2698.
12 (a) Nair, V., Vellalath, S., Poonoth, M. et al. (2006). *Org. Lett.* 8: 507–509. (b) Nair, V., Vellalath, S., Poonoth, M. et al. (2007). *Synthesis* 3195–3200.
13 Dugal-Tessier, J., O'Bryan, E.A., Schroeder, T.B.H. et al. (2012). *Angew. Chem. Int. Ed.* 51: 4963–4967.
14 Sun, L.-H., Shen, L.-T., and Ye, S. (2011). *Chem. Commun.* 47: 10136–10138.
15 Li, J.-L., Sahoo, B., Daniliuc, C.-G., and Glorius, F. (2014). *Angew. Chem. Int. Ed.* 53: 10515–10519.
16 Raup, D.E.A., Cardinal-David, B., Holte, D., and Scheidt, K.A. (2010). *Nat. Chem.* 2: 766–771.
17 Lv, H., Tiwari, B., Mo, J. et al. (2012). *Org. Lett.* 14: 5412–5415.
18 Goodman, C.G., Walker, M.M., and Johnson, J.S. (2015). *J. Am. Chem. Soc.* 137: 122–125.
19 ElSohly, A.M., Wespe, D.A., Poore, T.J., and Snyder, S.A. (2013). *Angew. Chem. Int. Ed.* 52: 5789–5794.
20 Qi, J., Xie, X., Han, R. et al. (2013). *Chem. Eur. J.* 19: 4146–4150.
21 Lee, A. and Scheidt, K.A. (2014). *Angew. Chem. Int. Ed.* 53: 7594–7598.
22 (a) Wang, Z.-D., Wang, F., Li, X., and Cheng, J.-P. (2013). *Org. Biomol. Chem.* 11: 5634–5641. (b) Lakshmi, K.C.S., Paul, R.R., Suresh, E., and Nair, V. (2014). *Synlett* 853–857.
23 Jang, K.P., Hutson, G.E., Johnston, R.C. et al. (2014). *J. Am. Chem. Soc.* 136: 76–79.
24 Struble, J.R. and Bode, J.W. (2009). *Tetrahedron* 65: 4957–4967.
25 Nair, V., Poonoth, M., Vellalath, S. et al. (2006). *J. Org. Chem.* 71: 8964–8965.

26 Zhao, X., DiRocco, D.A., and Rovis, T. (2011). *J. Am. Chem. Soc.* 133: 12466–12469.
27 Cardinal-David, B., Raup, D.E.A., and Scheidt, K.A. (2010). *J. Am. Chem. Soc.* 132: 5345–5347.
28 Xu, W.-Y., Iwaki, R., Jia, Y.-M. et al. (2013)). *Org. Biomol. Chem.* 11: 4622–4639.
29 Nair, V., Vellalath, S., Poonoth, M., and Suresh, E. (2006). *J. Am. Chem. Soc.* 128: 8736–8737.
30 Chiang, P.-C., Kaeobamrung, J., and Bode, J.W. (2007). *J. Am. Chem. Soc.* 129: 3520–3521.
31 Domingo, L.R., Zaragozá, R.J., and Arnó, M. (2010). *Org. Biomol. Chem.* 8: 4884–4891.
32 He, M. and Bode, J.W. (2008). *J. Am. Chem. Soc.* 130: 418–419.
33 Lakshmi, K.C.S., Sinu, C.R., Padmaja, D.V.M. et al. (2014). *Org. Lett.* 16: 5532–5535.
34 Cohen, D.T., Cardinal-David, B., Roberts, J.M. et al. (2011). *Org. Lett.* 13: 1068–1071.
35 (a) Nair, V., Babu, B.P., Vellalath, S., and Suresh, E. (2008). *Chem. Commun.* 747–749. (b) Nair, V., Vellalath, S., Babu, B.P. et al. (2010). *Org. Biomol. Chem.* 8: 4861–4866.
36 Chen, X.-Y., Li, S., Sheng, H. et al. (2017). *Chem. Eur. J.* 23: 1–5.
37 Maki, B.E., Chan, A., and Schiedt, K.A. (2008). *Synthesis* 1306–1315.
38 Nair, V., Babu, B.P., Vellalath, S. et al. (2009). *Org. Lett.* 11: 2507–2510.
39 Yang, Y.-J., Ji, Y., Qi, L. et al. (2017). *Org. Lett.* 19: 3271–3274.
40 Nair, V., Sinu, C.R., Babu, B.P. et al. (2009). *Org. Lett.* 11: 5570–5573.
41 Maji, B., Ji, L., Wang, S. et al. (2012). *Angew. Chem. Int. Ed.* 51: 8276–8280.
42 White, N.A., DiRocco, D.A., and Rovis, T. (2013). *J. Am. Chem. Soc.* 135: 8504–8507.
43 Kaeobamrung, J. and Bode, J.W. (2009). *Org. Lett.* 11: 677–680.
44 Jiang, K., Tiwari, B., and Chi, Y.R. (2012). *Org. Lett.* 14: 2382–2385.
45 Guo, C., Schedler, M., Daniliuc, C.G., and Glorius, F. (2014). *Angew. Chem. Int. Ed.* 53: 10232–10236.
46 Wang, M., Rong, Z.-Q., and Zhao, Y. (2014). *Chem. Commun.* 50: 15309–10312.
47 Lakshmi, K.C.S., Krishnan, J., Sinu, C.R. et al. (2014). *Org. Lett.* 16: 6374–6377.
48 Izquierdo, J., Orue, A., and Scheidt, K.A. (2013). *J. Am. Chem. Soc.* 135: 10634–10637.
49 Lv, H., Jia, W.-Q., Sun, L.-H., and Ye, S. (2013). *Angew. Chem. Int. Ed.* 52: 8607–8610.
50 Li, X.-S., Zhao, L.-L., Wang, X.-K. et al. (2017). *Org. Lett.* 19: 3943–3946.
51 Chan, A. and Scheidt, K.A. (2005). *Org. Lett.* 7: 905–908.
52 Sohn, S.S. and Bode, J.W. (2005). *Org. Lett.* 7: 3873–3876.
53 Burstein, C., Tschan, S., Xie, X.L., and Glorius, F. (2006). *Synthesis* 2418–2439.
54 He, M., Struble, J.R., and Bode, J.W. (2006). *J. Am. Chem. Soc.* 128: 8418–8420.

55 Phillips, E.M., Wadamoto, M., Chan, A., and Scheidt, K.A. (2007). *Angew. Chem. Int. Ed.* 46: 3107–3110.
56 Kaeobamrung, J., Kozlowski, M.C., and Bode, J.W. (2010). *Proc. Natl. Acad. Sci. U.S.A.* 107: 20661–20665.
57 Wadamoto, M., Phillips, E.M., Reynolds, T.E., and Scheidt, K.A. (2007). *J. Am. Chem. Soc.* 129: 10098–10099.
58 Phillips, E.M., Roberts, J.M., and Scheidt, K.A. (2010). *Org. Lett.* 12: 2830–2833.
59 (a) Cohen, D.T., Eichman, C.C., Phillips, E.M. et al. (2012). *Angew. Chem. Int. Ed.* 51: 7309–7313. (b) Cohen, D.T., Johnston, R.C., Rosson, N.T. et al. (2015). *Chem. Commun.* 51: 2690–2693.
60 Nair, V., Sinu, C.R., Rejithamol, R. et al. (2011). *Org. Biomol. Chem.* 9: 5511–5514.
61 McCusker, E.O. and Scheidt, K.A. (2013). *Angew. Chem. Int. Ed.* 52: 13616–13620.
62 Liu, K., Hovey, M.T., and Scheidt, K.A. (2014). *Chem. Sci.* 5: 4026–4031.
63 Phillips, E.M., Reynolds, T.E., and Scheidt, K.A. (2008). *J. Am. Chem. Soc.* 130: 2416–2417.
64 Chan, A. and Scheidt, K.A. (2007). *J. Am. Chem. Soc.* 129: 5334–5335.
65 Chan, A. and Scheidt, K.A. (2008). *J. Am. Chem. Soc.* 130: 2740–2741.
66 Seayad, J., Patra, P.K., Zhang, Y., and Ying, J.Y. (2008). *Org. Lett.* 10: 953–956.
67 Fang, X., Jiang, K., Xing, C. et al. (2011). *Angew. Chem. Int. Ed.* 50: 1910–1913.
68 Bhunia, A., Patra, A., Puranik, V.G., and Biju, A.T. (2013). *Org. Lett.* 15: 1756–1759.
69 Sinu, C.R., Padmaja, D.V.M., Ranjini, U.P. et al. (2013). *Org. Lett.* 15: 68–71.
70 Singh, A.K., Chawla, R., Rai, A., and Yadav, L.D.S. (2012). *Chem. Commun.* 48: 3766–3768.
71 Fu, Z., Xu, J., Zhu, T. et al. (2013). *Nat. Chem.* 5: 835–839.
72 Fu, Z., Jiang, K., Zhu, T. et al. (2014). *Angew. Chem. Int. Ed.* 53: 6506–6510.
73 Jin, Z., Chen, S., Wang, Y. et al. (2014). *Angew. Chem. Int. Ed.* 53: 13506–13509.
74 Palomo, C., Oiarbide, M., García, J.M. et al. (2003). *J. Am. Chem. Soc.* 125: 13942–13943.
75 Chiang, P.-C., Rommel, M., and Bode, J.W. (2009). *J. Am. Chem. Soc.* 131: 8714–8718.
76 Chiang, P.-C., Kim, Y., and Bode, J.W. (2009). *Chem. Commun.* 4566–4568.

5

Domino Processes in NHC Catalysis

Pankaj Chauhan, Suruchi Mahajan, Xiang-Yu Chen, and Dieter Enders

Indian Institute of Technology Jammu, Department of Chemistry, Jagti, NH-44, P.O. Nagrota, Jammu 181221, India
RWTH Aachen University, Institute of Organic Chemistry, Landoltweg 1, Aachen 52074, Germany

5.1 Introduction

Because of the growing concern of environmental pollution, synthesis chemists have specified certain criteria for highly efficient and sustainable syntheses. These criteria are atom-, redox-, step-, and pot-economical synthesis by avoiding functional group protection–deprotection chemistry. This led to the tremendous progress in the development of various one-pot transformations. There are different variants of one-pot reactions including domino/cascade reactions and multicomponent consecutive reactions. One of the most efficient and interesting class of one-pot transformations is domino reactions [1]. A general and widely accepted definition of domino reactions has been given by Tietze and Beifuss, which states that "a domino reaction is a process involving two or more consecutive reactions in which subsequent reactions result as a consequence of the functionality formed by bond formation or fragmentation in the previous step" [2]. As two or more reactions are carried out in a single operation under the same reaction conditions, domino reactions have advantages over the classical synthesis and other one-pot variants.

In parallel, *N*-heterocyclic carbene (NHC) catalysis [3] is moving head to head with the other classes of catalysis for the development of new one-pot transformations including domino reactions [4]. In the early development, the NHCs showed their potential as covalently bonded nucleophilic Lewis base organocatalysts to catalyze benzoin [5] and Stetter reactions [6] through the nucleophilic Breslow intermediate **A** (Scheme 5.1a) [7]. With further progress, the NHC catalysis was applied in various cycloaddition reactions by the introduction of new types of nucleophilic enolate **B** [8], homoenolate **C** [9], dienolate **D** [10], and electrophilic α,β-unsaturated acyl azolium **E** [11] intermediates (Scheme 5.1b–e). In addition, the acyl azoliums **F** as electrophilic intermediates have been reported for direct nucleophilic substitution reactions (Scheme 5.1f) [12]. More recently, the applications of NHC catalysis have been extended in noncovalent Brønsted base catalysis [13].

N-Heterocyclic Carbenes in Organocatalysis, First Edition. Edited by Akkattu T. Biju.
© 2019 Wiley-VCH Verlag GmbH & Co. KGaA. Published 2019 by Wiley-VCH Verlag GmbH & Co. KGaA.

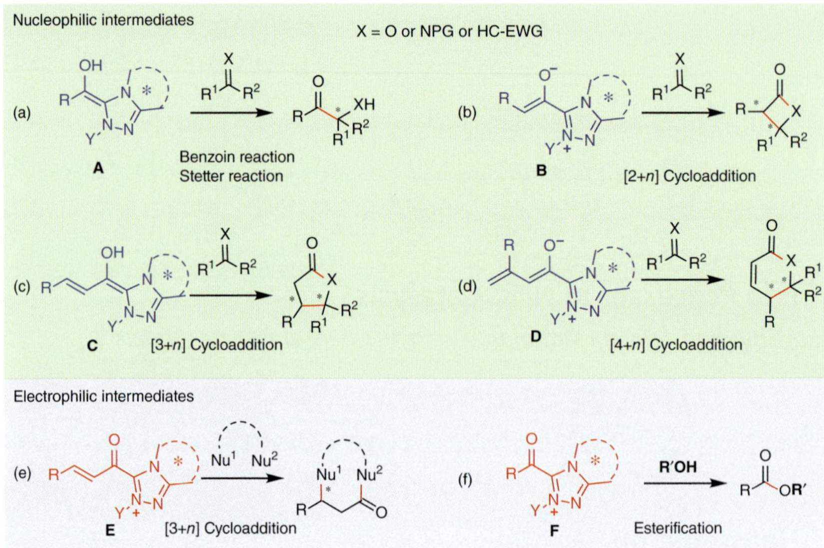

Scheme 5.1

A sequential combination of these activation modes could lead to interesting domino processes. For a better presentation and understanding, this book chapter is subdivided according to the involvement of the different types of NHC-generated intermediates.

5.2 Domino Reactions Involving Homoenolate–Enolate Intermediates

Besides the Breslow intermediates, the most commonly studied and applied intermediates in NHC catalysis are the homoenolates and enolates. The first variant of NHC-catalyzed domino reactions indeed involved the sequential formation of homoenolate–enolate intermediates.

5.2.1 Domino Reactions Involving a Michael/Aldol Reaction Sequence

The domino reactions involving Michael/aldol reaction sequence are most common and widely explored in organocatalysis including in NHC catalysis by combining homoenolate–enolate intermediates. In this context, the pioneering contribution was made by Nair's group in 2006 [14]. It was found out that in the presence of 1,3-dimesityl imidazolium chloride **C1** as NHC precatalyst, the domino reaction between different enals **1** and chalcones **2** occurs efficiently to provide cyclopentenes **3** in moderate-to-good yields. It was also observed that side reactions such as benzoin and Stetter reactions were suppressed by the bulky substituents of the catalyst (Scheme 5.2). This transformation follows a Michael/aldol/lactonization/decarboxylation domino sequence, where the first

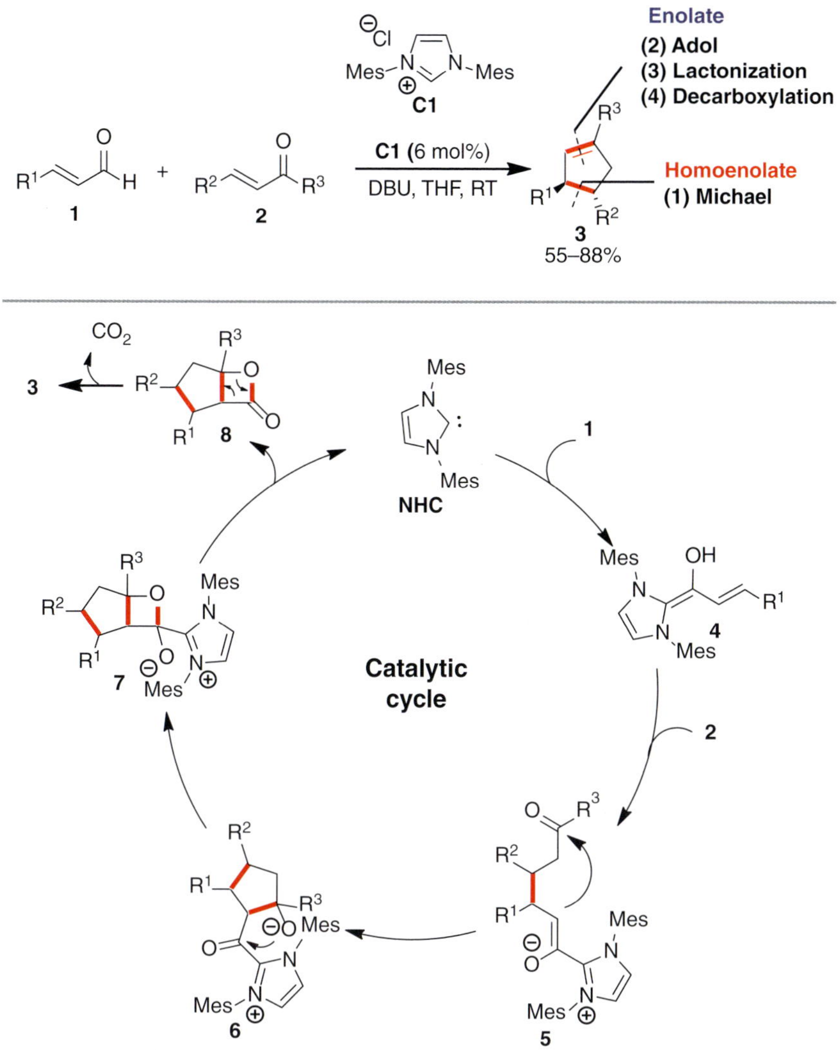

Scheme 5.2

key step involves a Michael addition of the homoenolate equivalent **4** to the Michael acceptor **2** to generate the azolium enolate **5**. The latter one undergoes an intramolecular aldol/lactonization sequence, thus giving an unstable lactone **8** via the intermediates **6** and **7** with the regeneration of the NHC catalyst. Finally, the desired cyclopentenes **3** are formed via a subsequent decarboxylation reaction.

The enantioselective version of this domino process has also been developed by using the aminoindanol-derived NHC precursor **C2** (Scheme 5.3a) [15]. However, the scope of enones was limited to 4-oxo-2-butenoates **9**, which provided cyclopentenes **10** with excellent levels of ee-values. An improved variant of this domino transformation involving **C3** as NHC precatalyst in a Lewis acid cocatalytic system provided better results in terms of substrate scope tolerance and

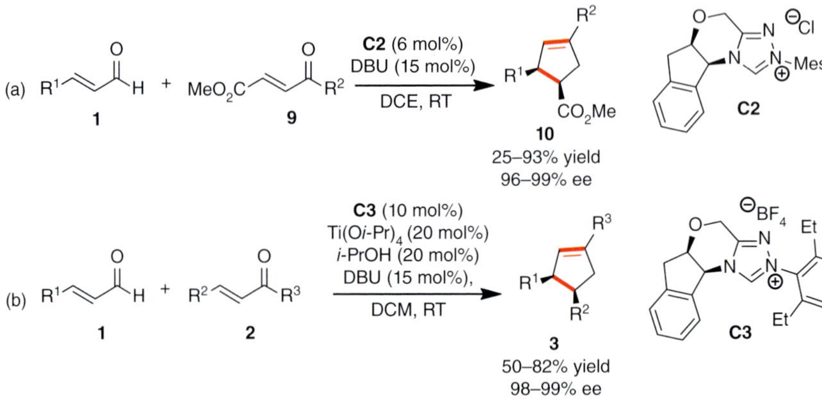

Scheme 5.3

stereoselectivities(Scheme 5.3b) [16]. It is noteworthy that the *cis*-diastereomers were exclusively obtained, which is in contrast with Nair's work [14].

With certain classes of substrates, decarboxylation does not take place, hence providing β-lactones as stable products via a Michael/aldol/lactonization reaction. The NHC precatalyst **C2** turned out to be an efficient promoter of an asymmetric domino sequence between the α,β-unsaturated aldehydes **1** and the unsaturated oxindole **11** (Scheme 5.4) [17]. This transformation leads to the formation of polycyclic oxindoles **12** in moderate-to-good yields and high stereoselectivities.

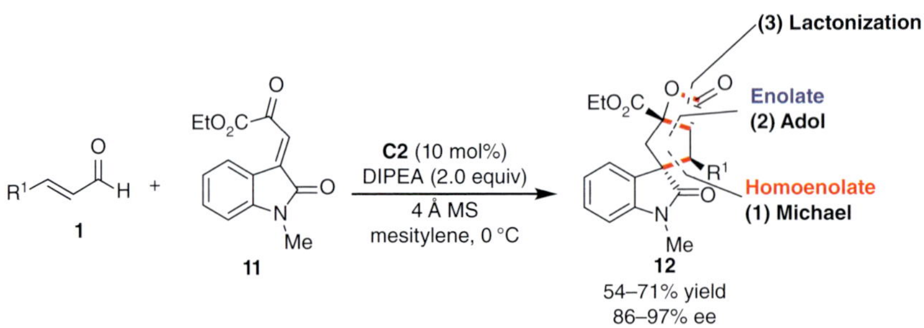

Scheme 5.4

Another important Michael/aldol/esterification domino reaction between enals **1** and unsaturated ketoesters **13** as the substrates by employing NHC generated homoenolate–enolate intermediates opens a direct entry to the highly functionalized cyclopentanes **14** with excellent stereoselectivities (Scheme 5.5) [18]. This transformation occurs in the presence of a NHC catalyst derived from the triazole salt **C3** and requires a stoichiometric amount of $Ti(Oi\text{-}Pr)_4$.

A series of heterocyclic molecules could be synthesized by introducing a nucleophilic group on the Michael acceptor instead of using an external nucleophile. Based on this approach, the *o*-hydroxy chalcones **15** reacted as a suitable substrate to undergo a diastereoselective Michael/aldol/dehydration/lactonization

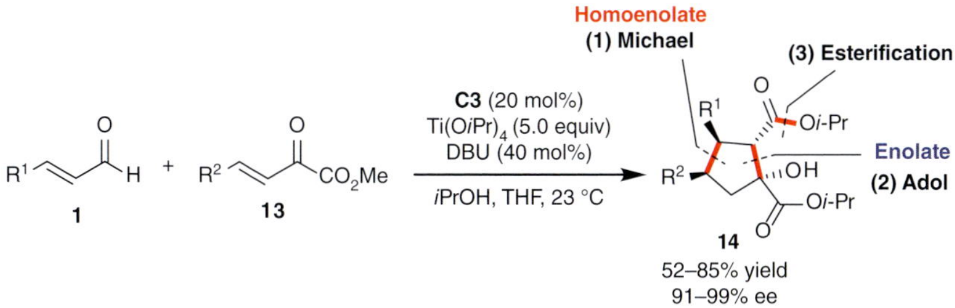

Scheme 5.5

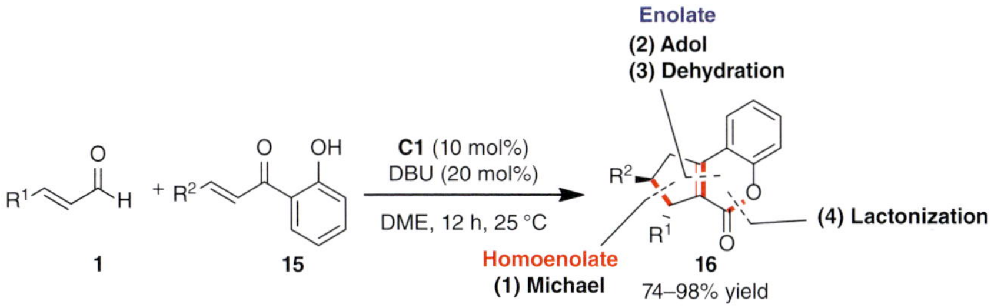

Scheme 5.6

sequence furnishing the cyclopentane-fused coumarins **16** in good yields (Scheme 5.6) [19].

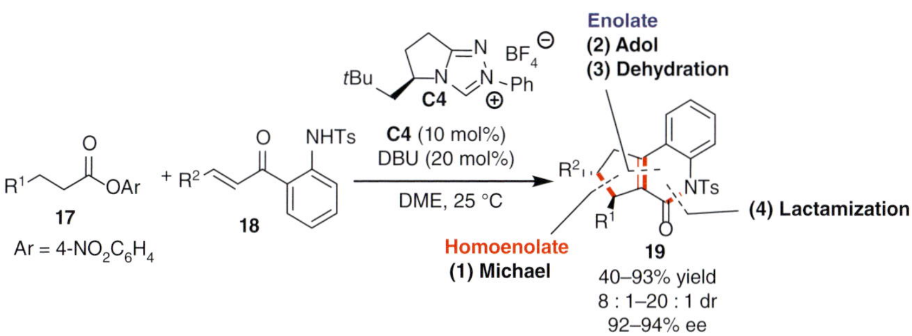

Scheme 5.7

A similar strategy was used for the asymmetric synthesis of polycyclic quinolinones **19** via a domino Michael/aldol/dehydration/lactamization sequence (Scheme 5.7) [20]. This transformation utilized saturated esters **17** and aminoenones **18** as substrates, and the reaction is catalyzed by the NHC based on the precatalyst **C4** to afford polycyclic quinolinones **19** in good yields with very good stereoselectivities.

A closely related NHC-mediated Michael/aldol/lactamization reaction between indole-derived enones **20** and α,β-unsaturated aldehydes **1** resulted in the formation of cyclopenta[3,4]pyrrolo[1,2-*a*]indol-4-ones **21** with four consecutive stereogenic centers in good yields with excellent stereoselectivities (Scheme 5.8) [21]. In this transformation, the dehydration process is

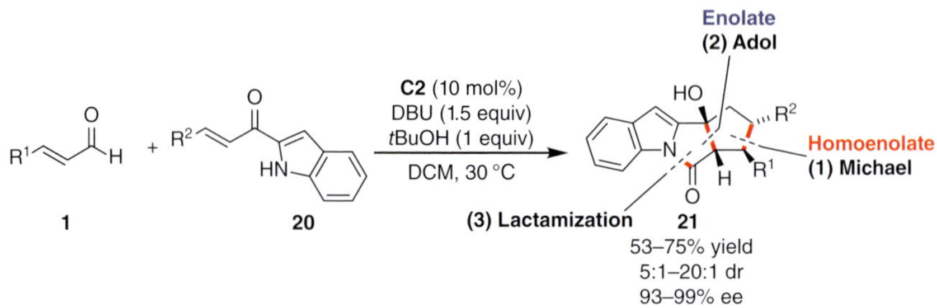

Scheme 5.8

not involved, and the other advantages include a wide substrate scope and a successful scale-up reaction.

5.2.2 Domino Reactions Involving a Michael/Michael Reaction Sequence

Another important class of domino reactions involves double Michael additions. In this regard, our group developed a Michael/Michael/esterification sequence, which involves the sequential generation of homoenolate–enolate intermediates (Scheme 5.9) [22]. Using enals **1** and nitroallylic acetates **22** as substrates, 10 mol% of the NHC precatalyst **C5** and 1.0 equiv of K_3PO_4, the asymmetric synthesis of a series of tetra-substituted cyclopentanes **23** could be achieved. In the proposed reaction pathway, the initial step involves the formation of the homoenolate **24**, which undergoes a first Michael addition to the nitroallylic acetates **22** to form the anionic intermediate **25**. This step is followed by the elimination of the acetate group from the adduct **25** to generate another Michael acceptor **26**, which then undergoes an intramolecular Michael addition to form adduct **27**. Finally, the desired cyclopentane product **23** is released through esterification with the regeneration of the NHC catalyst.

In order to achieve more complex structures, benzodienones emerged as excellent substrates in NHC catalysis to develop new variants of asymmetric domino reactions. In this context, the NHC precatalyst **C2** promoted a Michael/Michael/esterification domino reaction sequence between enals **1** and benzodienones **28** to provide the desired tricyclic products **29** with good yields and good-to-excellent diastereo- and high enantioselectivities (Scheme 5.10) [23].

The benzodienones **28** could be used to develop even more complex domino processes, which are initiated by a Michael addition reaction of the homoenolate equivalent. An NHC-catalyzed reaction between 2-aroylvinylchalcones **30** and benzodienones **28** in the presence of the achiral precatalyst **C6** provided a series of complex indane derivatives **31** via a domino Michael/Michael/Michael/lactonization sequence (Scheme 5.11) [24]. The reaction is initiated by the generation of the homoenolate equivalent **32** from the NHC and enal **30**, which undergoes a Michael addition to one of the enone moieties of the benzodienones **28**. This step generates two enolates in **33**, which subsequently lead to the adduct **34** via a double Michael addition.

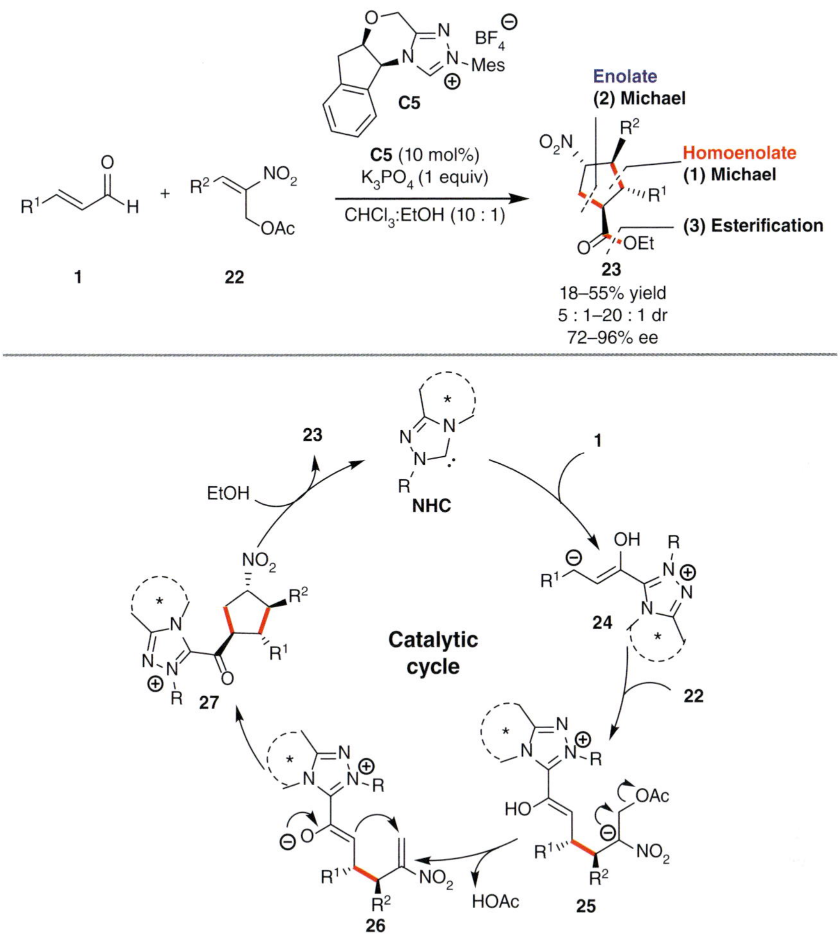

Scheme 5.9

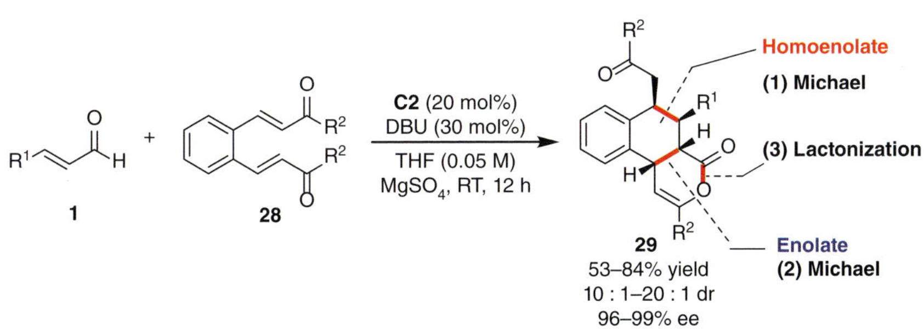

Scheme 5.10

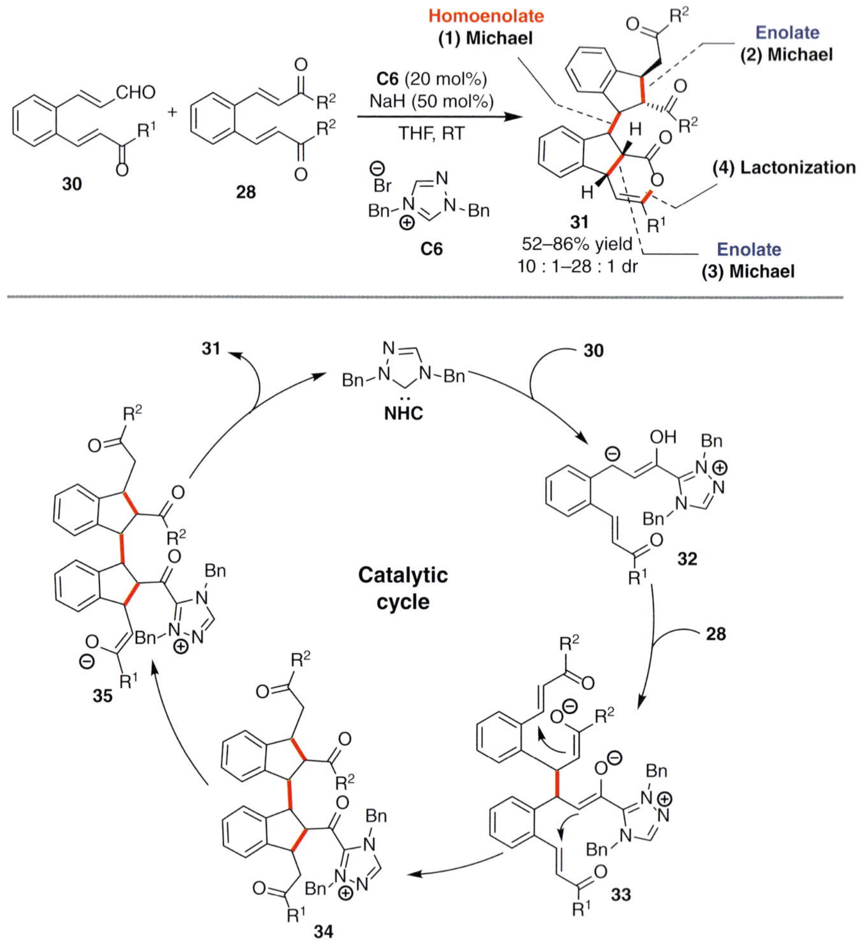

Scheme 5.11

The final lactonization through the intermediate **35** regenerates the catalyst and affords the corresponding products **31**.

An NHC-catalyzed asymmetric domino reaction of 2-aroylvinylcinnamaldehydes **30** with the unsaturated imines **36** led to the formation of indeno[2,1-*c*]pyran-1-one derivatives **37** via a Michael/Michael/lactonization sequence (Scheme 5.12) [25]. An entirely different class of products could be obtained by changing the *N*-substituent on the NHC from 4-methoxyphenyl (in **C-7**) to the mesityl group (***ent*-C2**). In the latter case, indenocyclopentan-1-ones **38** were obtained via a Michael/Michael/acylation sequence.

5.2.3 Domino Reactions Involving a Michael/Mannich Reaction Sequence

The strategy of homoenolate–enolate activation modes could be applied to develop domino reactions initiated by a Michael/Mannich sequence, where

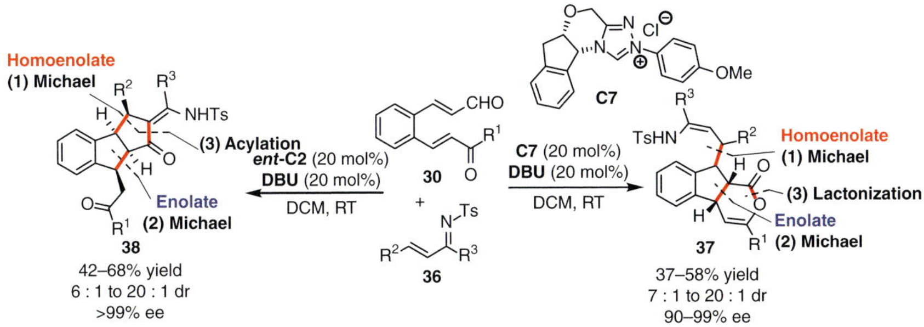

Scheme 5.12

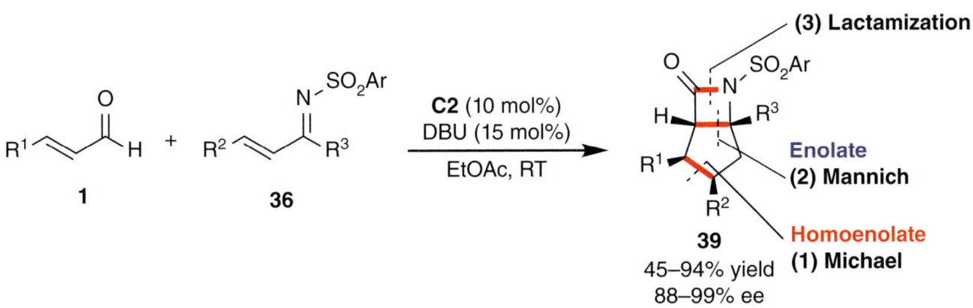

Scheme 5.13

α,β-unsaturated imines are used as one of the substrates. In this context, the domino Michael/Mannich/lactamization sequence between α,β-unsaturated imines **36** and enals **1** required **C2** as a precatalyst to provide an efficient entry to the cyclopentane-fused β-lactams **39** in moderate-to-good yields with good-to-excellent stereoselectivities (Scheme 5.13) [26]. An extended nonenantioselective version of a similar domino sequence utilized the isatin-derived imines **40** to construct cyclopentane-fused β-lactams **41** bearing an oxindole moiety in low-to-good diastereoselectivities (Scheme 5.14) [27].

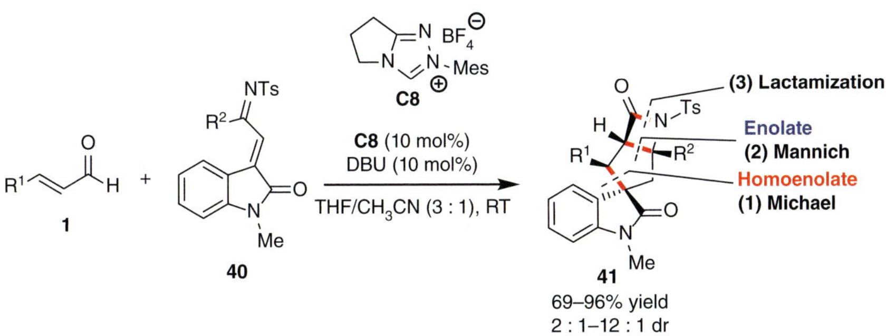

Scheme 5.14

A related strategy resulted in the formation of spirocyclopentane oxindole fused β-lactams **43** by using isatin-derived enals **42** and α,β-unsaturated imines **36** as reacting partner in the presence of the NHC precatalyst **C3** (Scheme 5.15) [28]. This Michael/Mannich/lactamization reaction provided a direct access

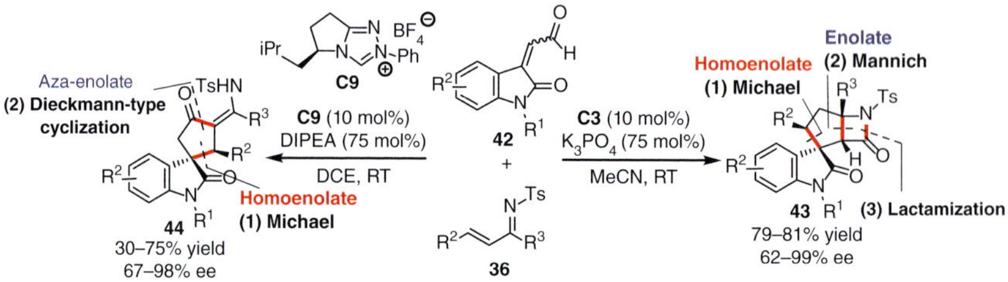

Scheme 5.15

to the spirocyclic oxindoles **43** bearing four contiguous stereocenters, including one quaternary and one tetrasubstituted stereocenter in good yields and moderate-to-excellent enantioselectivities. Utilizing the same substrates and a different NHC precursor **C9**, a different class of products, i.e. spirocyclopentane oxindoles **44**, could be synthesized via a Michael/aza-Dieckmann-type cyclization sequence.

5.2.4 Domino Reactions Involving a Homo-aldol/Michael Addition Sequence

The application of aroylvinylcinnamaldehydes **30** has been used to develop a domino sequence initiated by the homo-aldol reaction. A domino homo-aldol/Michael/lactonization reaction between 2-aroylvinylcinnamaldehydes **30** and different aromatic aldehydes **45** in the presence of a chiral triazole carbene precursor ***ent*-C2** and Ti(O*i*-Pr)$_4$ leads to the formation of the 8-(aroylmethyl)-3-arylindeno[1,2-*c*]furan-1-ones **46** bearing four stereogenic centers in moderate-to-excellent yields with excellent stereoselectivities (Scheme 5.16) [29].

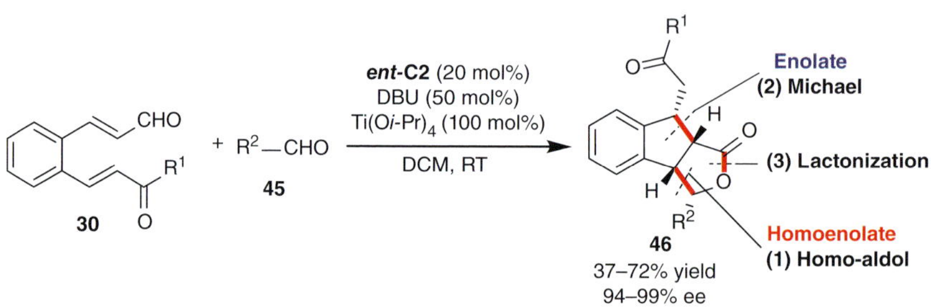

Scheme 5.16

5.3 Domino Reactions Involving Dienolate–Enolate Intermediates

NHC-generated azolium dienolates have emerged as an excellent class of intermediates for the remote γ-functionalization of various carbonyl compound

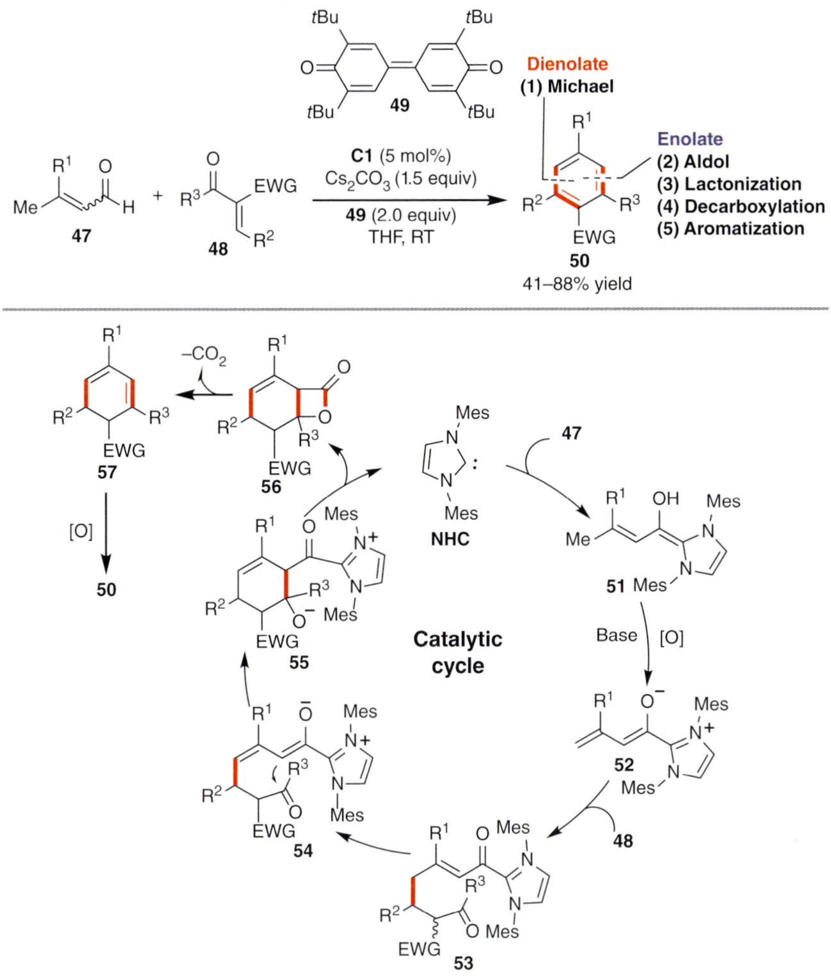

Scheme 5.17

and have been explored to develop a wide range of cycloaddition reactions [10]. These dienolate species could be combined with other intermediates such as azolium enolates to furnish successful domino reactions. In this context, dienolate–enolate intermediates have been explored for the synthesis of multisubstituted benzene derivatives (Scheme 5.17) [30]. The strategy employs **C1** as the NHC precursor and an oxidant **49** to promote the Michael/aldol/lactonization/decarboxylation/aromatization sequence between enals **47** and enones **48**. The domino sequence is initiated by the generation of the enol species **51** from the NHC and the enal **47**, which under oxidative conditions provides the azolium dienolate **52**. Then, a Michael addition of the azolium dienolate **52** to the enone **48** occurs to provide the adduct **53**, which then undergoes an intramolecular aldol reaction through an azolium enolate **54** to afford adduct **55**. The latter one subsequently regenerates the NHC catalyst and provides a β-lactone **56**, which leads to the desired products **50** via decarboxylation and oxidative aromatization through the cyclohexadiene **57**.

A similar strategy of combining dienolate–enolate intermediates was applied for the domino reaction between enals **47** with furanone **58** (Scheme 5.18) [31]. Employing **C8** as NHC precursor, a variety of enals **47** reacted well with furanone **58** to give the desired coumarin-type products **59** in good yields. The reaction is initiated by the addition of the NHC to the enal **47** to form enol **60**, followed by the further oxidation and deprotonation, leading to the unsaturated azolium dienolate intermediate **61**. The latter undergoes a vinylogous Michael addition to the electron-deficient 2,4-diene **58**, leading to the adduct **62**, which after subsequent cyclization provides the spiro products **63** and releases the NHC catalyst. The spiro adduct **63** gives the dienol **64** through tautomerization, which then undergoes an intramolecular transesterification to form the intermediate product **65**. Finally, dehydration of **65** yields the final aromatic product **59**.

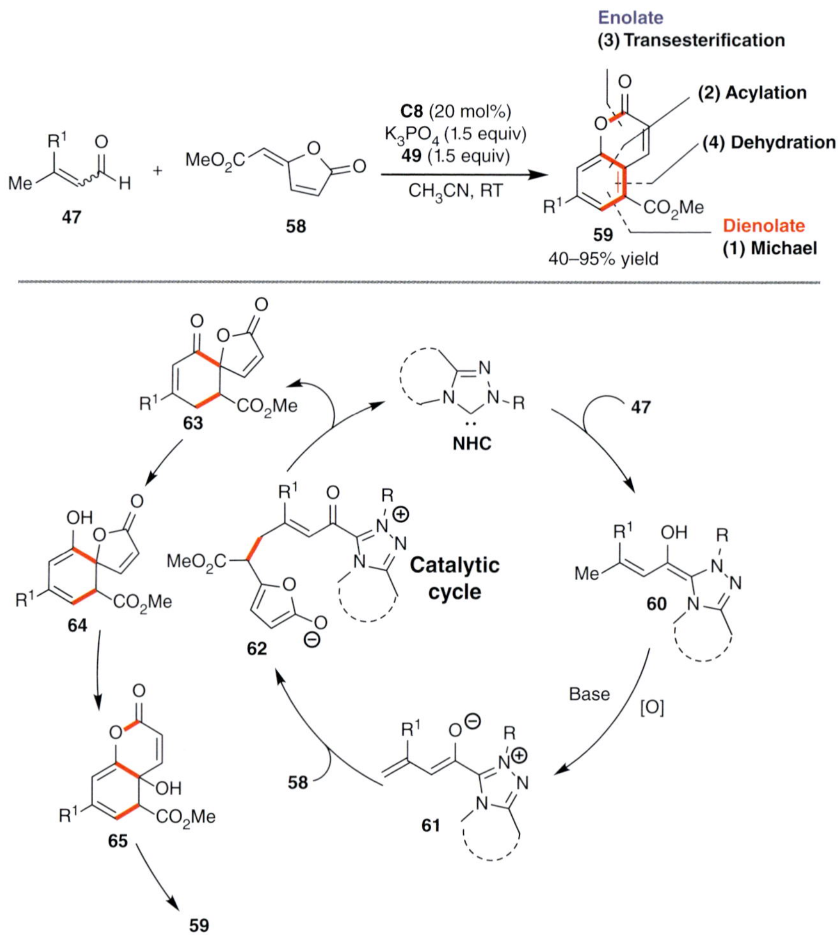

Scheme 5.18

5.4 Domino Reactions Involving Unsaturated Acyl Azolium–Enolate Intermediates

The α,β-unsaturated acyl azolium species generated by an NHC from various substrates have been widely explored to develop various cycloaddition and domino reactions by the sequential combination with enolate intermediates.

5.4.1 Domino Reactions Involving a Michael/Aldol Sequence

The α,β-unsaturated acid fluorides are identified as an efficient class of substrates to generate α,β-unsaturated acyl azolium intermediates in NHC catalysis. A domino Michael/aldol/lactonization reaction between α,β-unsaturated acid fluorides **66** with the donor–acceptor cyclopropanes **67** provided a direct access to the cyclopentane-fused β-lactones **68** (Scheme 5.19) [32, 33]. This domino transformation is initiated by the addition of the NHC (obtained from **C10** and KHMDS) to the acid fluoride **66** to generate the α,β-unsaturated acyl

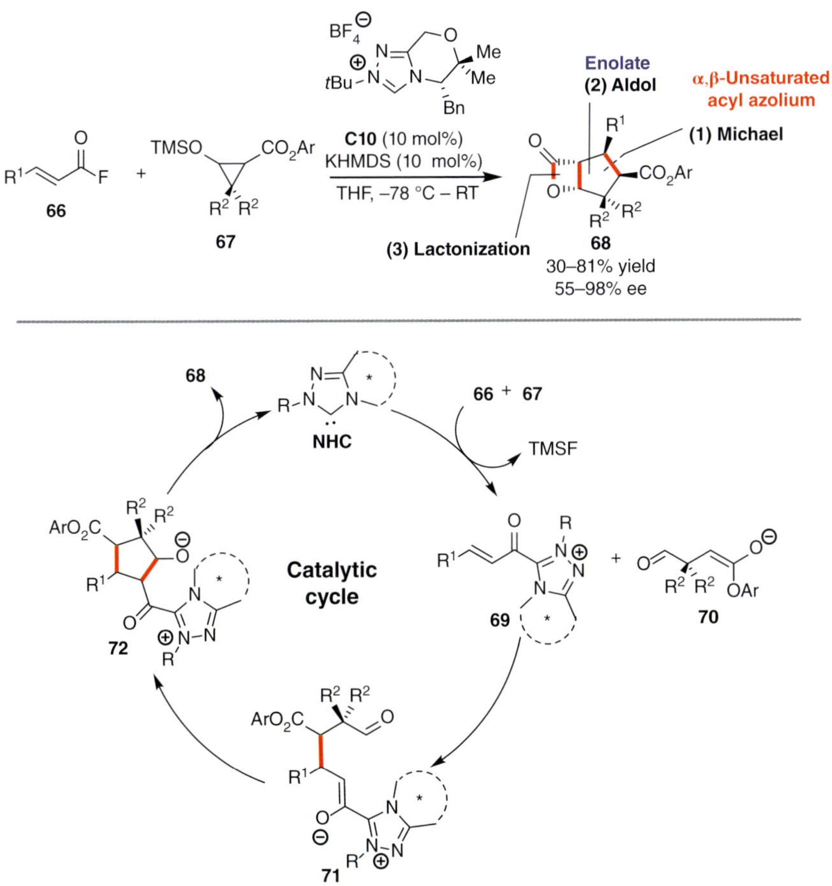

Scheme 5.19

azolium **69**. Simultaneously, an enolate **70** is generated by fluorodesilylation and retro-aldol reaction of the cyclopropanes **67**. These two intermediates combine via a Michael addition to generate a new enolate **71**, which undergoes an intramolecular aldol cyclization to afford the intermediate **72**. The latter one undergoes a lactonization reaction to afford β-lactones **68** with the libration of the NHC catalyst for the next catalytic cycle.

Closely related β-lactones **74** could be synthesized via a similar domino sequence by employing the malonates **73** and enals **1** instead of the donor–acceptor cyclopropanes and α,β-unsaturated acid fluorides as substrates under oxidative NHC catalysis (Scheme 5.20) [34]. In this transformation, **C2** acts as the NHC precursor, and lithium chloride is required to achieve high enantioselectivities.

In the presence of a triazolium salt NHC precatalyst **C11**, the α-amino ketones **75** and enals **1** react in a domino aza-Michael/aldol/lactonization sequence to produce a new series of pyrrolidine-fused β-lactones **76** in moderate-to-excellent levels of enantioselectivities (Scheme 5.21) [35]. The domino sequence works very well for differently substituted α-amino ketones **75** and a broad range of aromatic and aliphatic enals **1**.

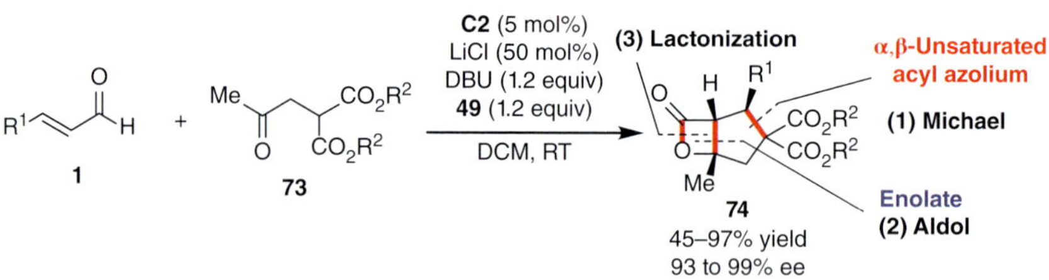

Scheme 5.20

With certain classes of substrate combinations, an additional decarboxylation reaction occurs to provide the corresponding cyclopentenes instead of a lactone

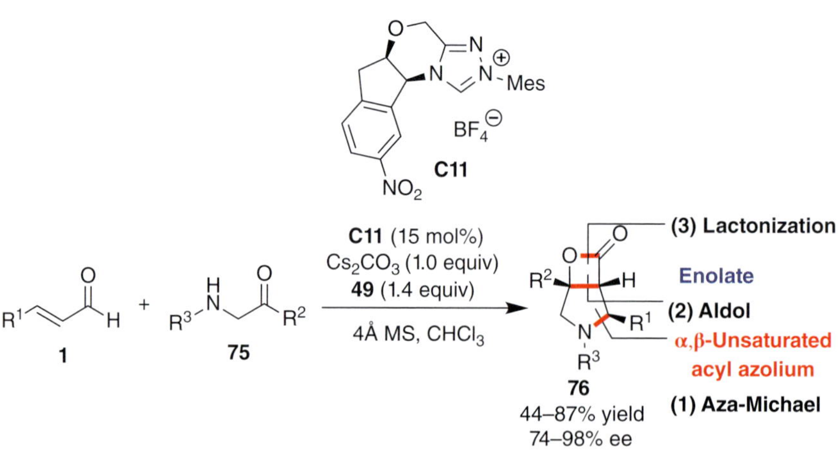

Scheme 5.21

product. In this context, by employing α-bromoenals **77** and malonates **78** bearing aryl substituents as substrates and **C2** as the NHC precatalyst, a domino Michael/aldol/lactonization/decarboxylation sequence occurs to afford the corresponding cyclopentenes **79** (Scheme 5.22) [36]. In the presence of **C5** as NHC precursor, the α-bromoenals and ketone-tethered oxindoles **80** reacted via a similar domino sequence to provide spirooxindoles **81** in moderate-to-good yields and stereoselectivities.

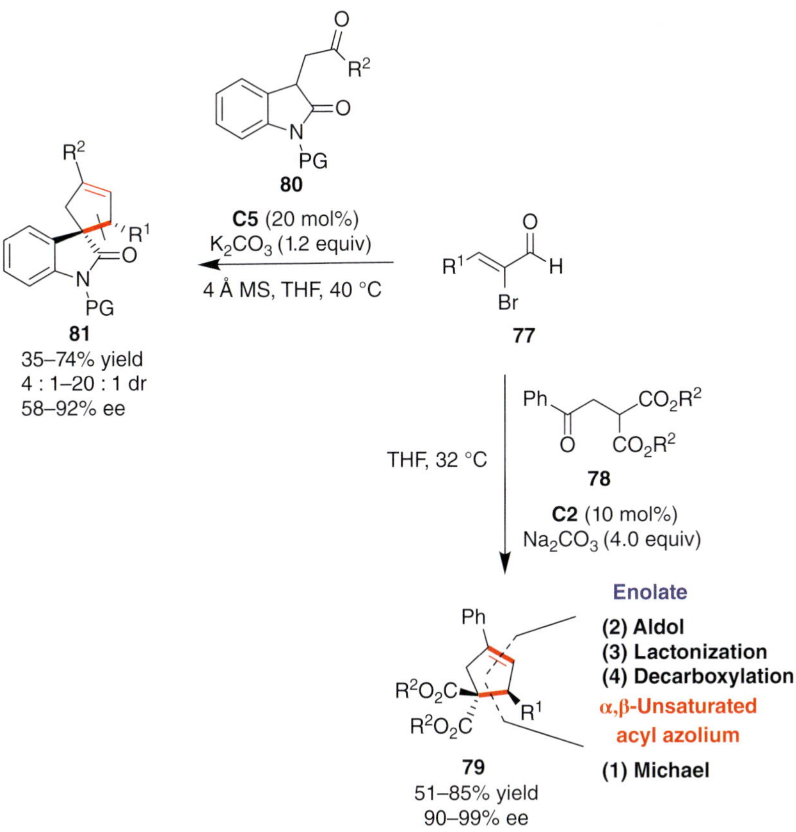

Scheme 5.22

A highly diastereoselective synthesis of cyclohexadienes **83** could be achieved via a Michael/aldol/lactonization/decarboxylation reaction between α,β-unsaturated acid fluorides **66** and silyl dienol ethers **82** (Scheme 5.23) [37]. Using the imidazolylidene NHC catalyst **C12**, the corresponding 1,3-cyclohexadienes **83** have been isolated in good yields with excellent diastereoselectivities.

Two different NHC precatalysts **C1** or **C13** under oxidative conditions provided an efficient route for the synthesis of multisubstituted benzenes **85** via a Michael/aldol/lactonization/decarboxylation/aromatization sequence (Scheme 5.24) [38, 39]. In both cases, enals **1** and dienolate precursors **84** have

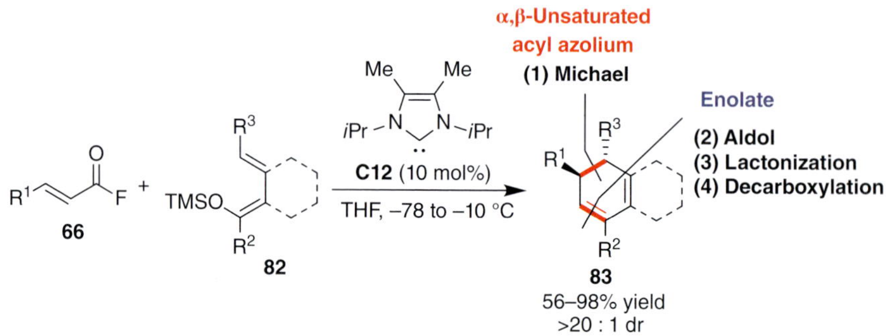

Scheme 5.23

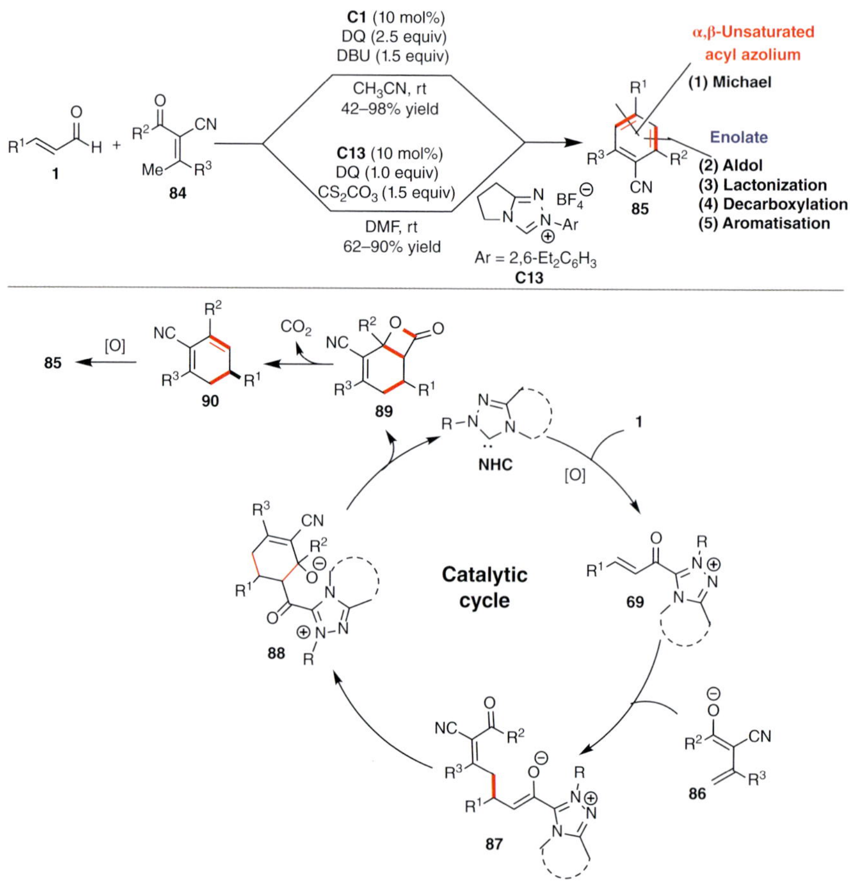

Scheme 5.24

been used as reacting species under slightly different reaction conditions. The proposed reaction pathway includes the initial Michael addition of the dienolate **86** to the α,β-unsaturated acyl azolium intermediate **69** to afford the azolium enolate **87**. The subsequent aldol/lactonization via the intermediate **88** furnishes the β-lactone-fused cyclohexene **89** and regenerates the NHC catalyst. The unstable β-lactone **89** undergoes a decarboxylation to afford the cyclohexadienes **90**, which were further oxidized to afford the final benzonitriles **85**.

Another example of a Michael/aldol/lactonization/decarboxylation reaction of diketones **91** with enals **1** allows a direct access to 1,2-dihydronaphthalenes **92** (Scheme 5.25) [40]. This process is catalyzed by the precatalyst **C5** under oxidative conditions to furnish a series of 1,2-dihydronaphthalenes **92** bearing two adjacent stereocenters with high yields and stereoselectivities.

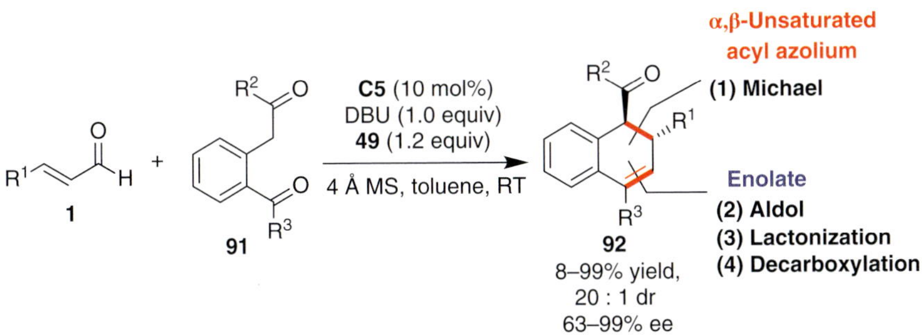

Scheme 5.25

A Michael/aldol/lactonization/decarboxylation domino reaction of α,β-unsaturated acyl azolium precursors **93** with TMS enol ethers **94** efficiently provided indene derivatives **95** in low-to-high levels of ee-values (Scheme 5.26) [41]. Both electron-rich and electron-poor acetophenone-derived TMS enol ethers reacted smoothly in the presence of the NHC catalyst derived from **C5**. It was observed that the electron-poor TMS enol ethers give better yields of the desired products. Even the TMS enol ether of acetone is well tolerable, which gives high enantioselectivity.

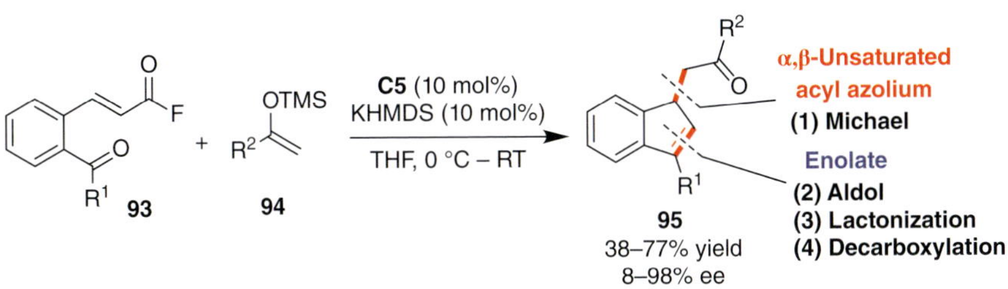

Scheme 5.26

5.4.2 Domino Reactions Involving a Michael/Michael Addition Sequence

The strategy of α,β-unsaturated acyl azolium–enolate activation modes could also be utilized to develop valuable domino sequences via Michael/Michael addition reactions. In this regard, a Michael/Michael/lactonization reaction of oxindoles

97 with the alkynyl aldehydes **96** afforded the tricyclic oxindole systems **98** with four consecutive stereogenic centers and low-to-good diastereoselectivities in the presence of precatalyst **C1** (Scheme 5.27) [42].

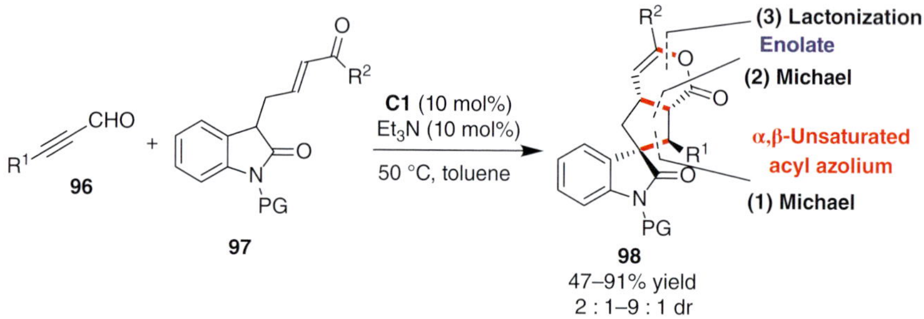

Scheme 5.27

The NHC precatalyst **C2** [43] or **C5** [44] in the presence of LiCl as a cocatalyst catalyzed a Michael/Michael/lactonization reaction of enals **1** with bifunctional malonates **99** (Scheme 5.28). Both catalysts led to the formation of cyclopentane-fused δ-lactones **100** in good-to-high yields, excellent diastereoselectivities combined with high enantioselectivities. It is worth mentioning that LiCl is necessary to achieve a high enantioselectivity. The same series of products could be synthesized by using α,β-unsaturated carboxylic esters as the α,β-unsaturated acyl azolium precursors [45].

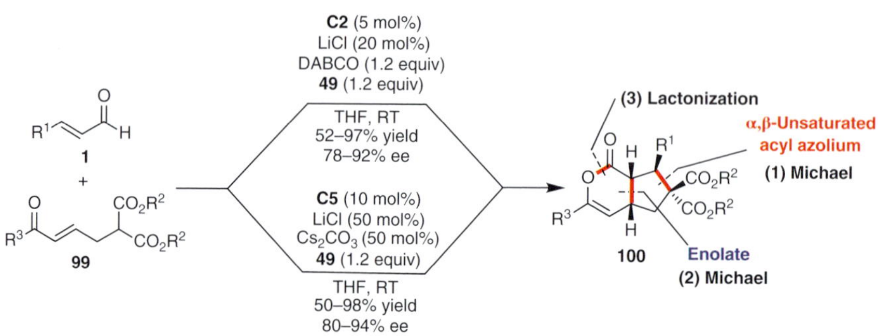

Scheme 5.28

A highly stereo-enriched tricyclic δ-lactones **102** bearing three vicinal stereocenters have been synthesized by the NHC-catalyzed functionalization of benzylic C(sp^3)—H bonds of dinitrotoluene derivatives **101** (Scheme 5.29) [46]. The NHC-derived from **C2** catalyzed a Michael/Michael/lactonization sequence between enals **1** and dinitrotoluene derivatives **101** under oxidative conditions to afford the δ-lactones **102** in good yields and excellent diastereo- and enantioselectivities for a broad range of substrates (Scheme 5.29).

A stereoselective aza-Michael/Michael/lactonization domino reaction of 2′-aminophenylenones **103** with the 2-bromoenals **77** could be catalyzed by

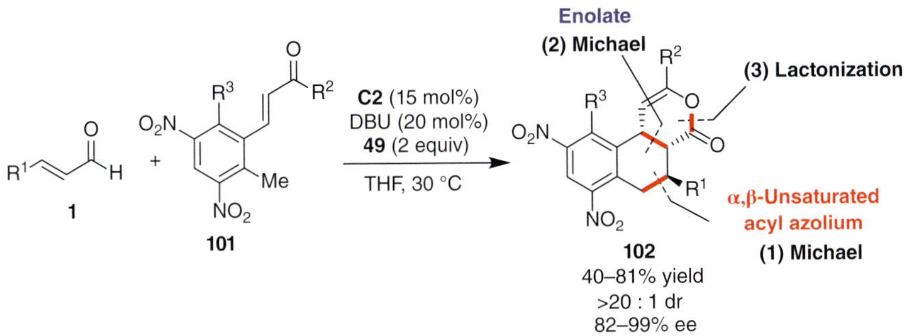

Scheme 5.29

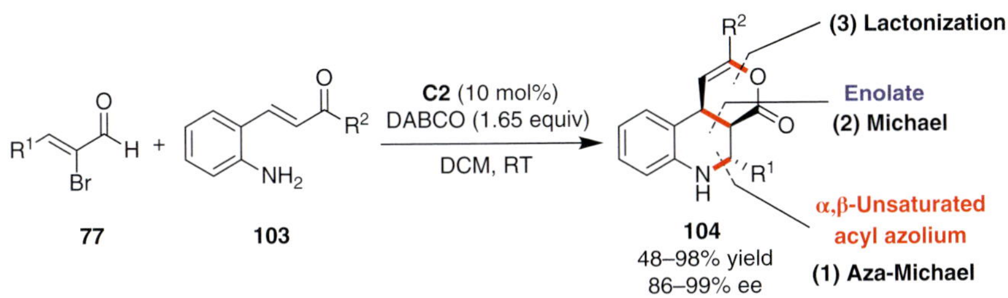

Scheme 5.30

the NHC precatalyst **C2** (Scheme 5.30) [47]. This domino process provides a series of functionalized tetrahydroquinolines **104** bearing three adjacent stereogenic centers in moderate-to-high yields and excellent enantioselectivities with remarkably high diastereoselectivities.

The thioesters **105** could be utilized as substrates to develop a similar sulfa-Michael/Michael/lactonization sequence (Scheme 5.31) [48]. In the presence of precatalyst **C14**, the highly functionalized thiochroman derivatives **106**

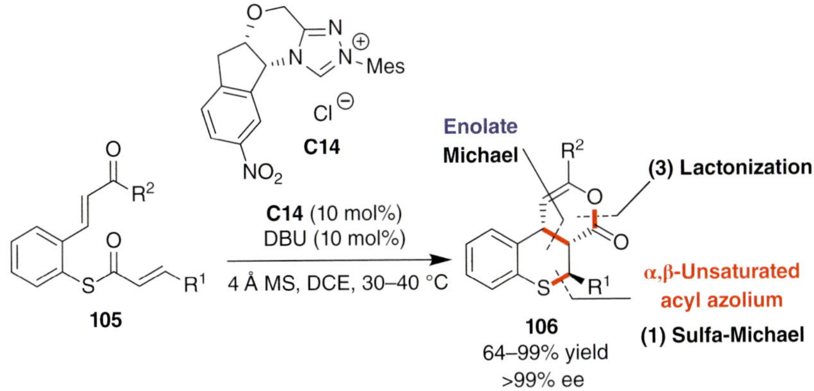

Scheme 5.31

could be obtained in good-to-high yields with excellent diastereo- and enantioselectivities with a broad range of thioesters. The transformations is initiated by the generation of a α,β-unsaturated acyl azolium intermediate and a thiolate species from the thioesters **105** in the presence of the NHC derived from **C14**.

A domino Michael/Michael/lactonization reaction between β-diketones **107** and bifunctional enals **30** initiated by the α,β-unsaturated acyl azolium species has been reported (Scheme 5.32) [49]. In the presence of the NHC precatalyst **C2**, the desired tricyclic δ-lactones **108** could be synthesized in moderate-to-good yields with excellent levels of stereoselectivities.

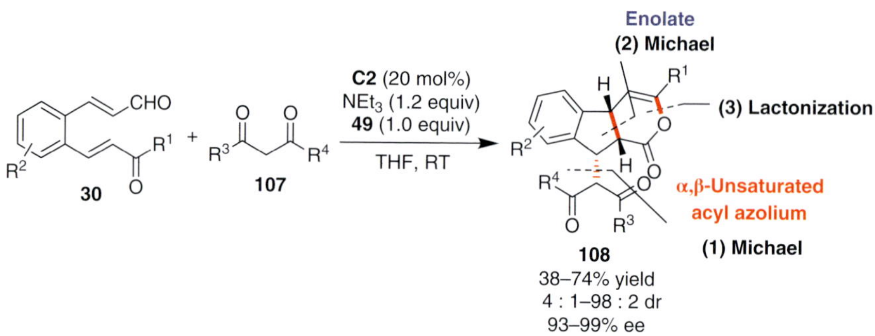

Scheme 5.32

5.4.3 Domino Reactions Involving a Michael/Mannich Reaction Sequence

An NHC-catalyzed reaction between *o*-aminoarylideneaminomalonate imines **109** and 2-bromoenals **77** initiates a Michael/Mannich/lactamization sequence leading to the corresponding pyrrolo-[3,2-*c*]quinolines **110** (Scheme 5.33) [50]. This transformation requires **C2** as the NHC precatalyst to achieve the desired products **110** bearing three consecutive stereocenters in moderate-to-good yields and excellent stereoselectivities.

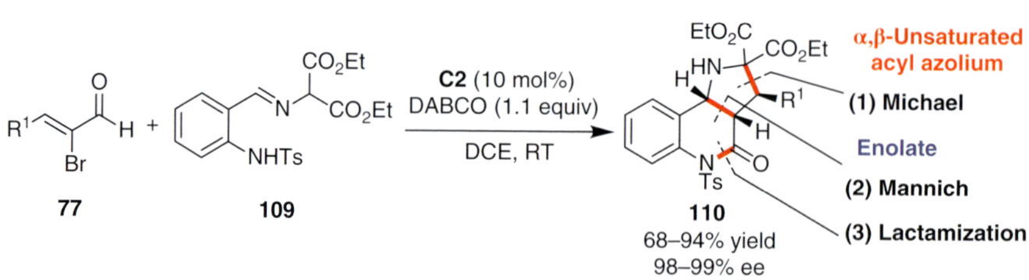

Scheme 5.33

5.4.4 Domino Reactions Involving a Michael/S_N2 Reaction Sequence

Employing a NHC-generated α,β-unsaturated acyl azolium–enolate intermediate and a sulfur ylide **111** as the substrate, highly substituted cyclopropyl carboxylic esters **112** could be synthesized in moderate-to-excellent diastereo- and enantioselectivities (Scheme 5.34) [51]. The transformation utilizes NHC precatalyst **C2** to promote the domino Michael/S_N2/esterification reaction between α,β-unsaturated aldehydes **1** and sulfur ylides **111**. The selectivities of the reaction were strongly dependent on the electronic effects of the substrates, showing better performances with electron-deficient substituents.

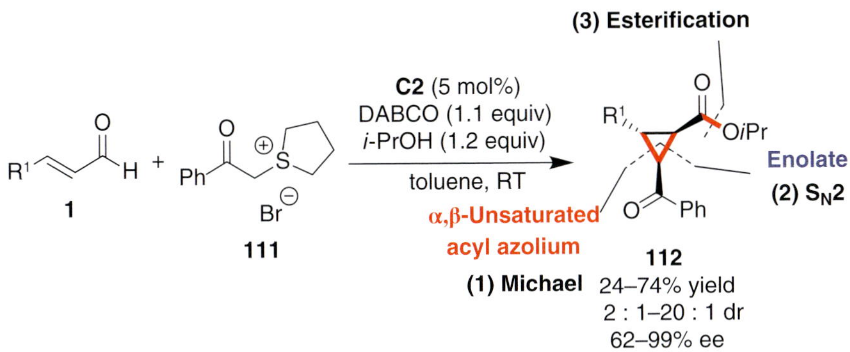

Scheme 5.34

5.5 Conclusions and Outlook

In summary, merging NHC catalysis with the concept of domino reactions turned out to be an excellent strategy for the synthesis of highly functionalized carbo- and heterocyclic molecules. The NHC-mediated domino processes not only worked well for the synthesis of achiral molecules but also facilitated various asymmetric transformations to achieve an excellent level of stereocontrol. The other salient features of NHC-catalyzed domino reactions include simple operational procedures as well as the readily available NHC catalysts and substrates. Until now, the NHC-catalyzed domino reactions involved only two activation modes via an initial intermolecular step followed by an intramolecular reaction. The further progress in the development of new and synthetically important domino processes employing more than two activation modes and the extension toward the development of multicomponent intermolecular domino sequences will be a future goal for the synthesis chemists. Moreover, research toward the direct application of these domino sequences toward specific targets, such as natural products, pharmaceuticals, and agrochemicals synthesis, will be carried out. At the current stage, we expect that this book chapter will inspire chemists to further develop some new NHC-mediated activation modes and to apply them in novel and practical domino processes.

References

1 (a) Enders, D., Grondal, C., and Hüttl, M.R. (2007). *Angew. Chem. Int. Ed.* 119: 1590–1601; (2007). *Angew. Chem. Int. Ed.* 46: 1570–1581. (b) Grondal, C., Jeanty, M., and Enders, D. (2010). *Nat. Chem.* 2: 167–178. (c) Pellissier, H. (2012). *Adv. Synth. Catal.* 354: 237–294. (d) Volla, C.M.R., Atodiresei, I., and Rueping, M. (2014). *Chem. Rev.* 114: 2390–2461. (e) Chauhan, P., Mahajan, S., Kaya, U. et al. (2015). *Adv. Synth. Catal.* 357: 253–281. (f) Chauhan, P., Mahajan, S., and Enders, D. (2017). *Acc. Chem. Res.* 50: 2809–2821.

2 Tietze, L.F. and Beifuss, U. (1993). *Angew. Chem. Int. Ed. Engl.* 105: 137–170; (1993). *Angew. Chem. Int. Ed. Engl.* 32: 131–163.

3 (a) Enders, D. and Balensiefer, T. (2004). *Acc. Chem. Res.* 37: 534–541. (b) Enders, D., Niemeier, O., and Henseler, A. (2007). *Chem. Rev.* 107: 5606–5655. (c) Ryan, S.J., Candish, L., and Lupton, D.W. (2013). *Chem. Soc. Rev.* 42: 4906–4917. (d) Wang, M.H. and Scheidt, K.A. (2016). *Angew. Chem. Int. Ed.* 128: 15134–15145; (2016). *Angew. Chem. Int. Ed.* 55: 14912–14922. (e) Flanigan, D.M., Romanov-Michailidis, F., White, N.A., and Rovis, T. (2015). *Chem. Rev.* 115: 9307–9387.

4 (a) Grossmann, A. and Enders, D. (2012). *Angew. Chem. Int. Ed.* 124: 320–332; (2012). *Angew. Chem. Int. Ed.* 51: 314–325. (b) Reyes, E., Uria, U., Carrillo, L., and Vicario, J.L. (2017). *Synthesis* 49: 451–471. (c) Chen, X.-Y., Li, S., Vetica, F. et al. (2018). *iScience* 2: 1–26.

5 Ukai, T., Tanaka, R., and Dokawa, T. (1943). *J. Pharm. Soc. Jpn.* 63: 296–300.

6 (a) Stetter, H. (1976). *Angew. Chem. Int. Ed. Engl.* 88: 695–704; (1976). *Angew. Chem. Int. Ed. Engl.* 15: 639–647. (b) Enders, D., Breuer, K., Runsink, J., and Teles, J.H. (1996). *Helv. Chim. Acta* 79: 1899–1902.

7 (a) Lapworth, A. (1903). *J. Chem. Soc. Trans.* 83: 995–1005. (b) Breslow, R. (1958). *J. Am. Chem. Soc.* 80: 3719–3726.

8 (a) Reynolds, N.T., Read de Alaniz, J., and Rovis, T. (2004). *J. Am. Chem. Soc.* 126: 9518–9519. (b) He, M., Uc, G.J., and Bode, J.W. (2006). *J. Am. Chem. Soc.* 128: 15088–15089. (c) Zhang, Y.-R., He, L., Wu, X. et al. (2008). *Org. Lett.* 10: 277–280. (d) Douglas, J., Churchill, G., and Smith, A.D. (2012). *Synthesis* 44: 2295–2309. (e) Chen, X.-Y. and Ye, S. (2013). *Synlett* 24: 1614–1622.

9 (a) Burstein, C. and Glorius, F. (2004). *Angew. Chem. Int. Ed.* 116: 6331–6334; (2004). *Angew. Chem. Int. Ed.* 43: 6205–6208. (b) Sohn, S.S., Rosen, E.L., and Bode, J.W. (2004). *J. Am. Chem. Soc.* 126: 14370–14371. (c) Nair, V., Menon, R.S., Biju, A.T. et al. (2011). *Chem. Soc. Rev.* 40: 5336–5346. (d) Menon, R.S., Biju, A.T., and Nair, V. (2015). *Chem. Soc. Rev.* 44: 5040–5052.

10 (a) Shen, L.T., Shao, P.L., and Ye, S. (2011). *Adv. Synth. Catal.* 353: 1943–1948. (b) Chen, X.-Y., Liu, Q., Chauhan, P., and Enders, D. (2018). *Angew. Chem. Int. Ed.* 130: 3924–3935; (2018). *Angew. Chem. Int. Ed.* 57: 3862–3875.

11 (a) Vora, H.U., Wheeler, P., and Rovis, T. (2012). *Adv. Synth. Catal.* 354: 1617–1639. (b) Sarkar, S.D., Biswas, A., Samanta, R.C., and Studer, A. (2013). *Chem. Eur. J.* 19: 4664–4678. (c) Mahatthananchai, J. and Bode, J.W. (2014).

Acc. Chem. Res. 47: 696–707. (d) Zhang, C., Hooper, J.F., and Lupton, D.W. (2017). *ACS Catal.* 7: 2583–2596.
12 Castells, J., Llitjos, H., and Moreno-Mañas, M. (1977). *Tetrahedron Lett.* 18: 205–206.
13 Chen, J. and Huang, Y. (2014). *Nat. Commun.* 3437: 1–8.
14 Nair, V., Vellalath, S., Poonoth, M., and Suresh, E. (2006). *J. Am. Chem. Soc.* 128: 8736–8737.
15 Chiang, P.-C., Kaeobamrung, J., and Bode, J.W. (2007). *J. Am. Chem. Soc.* 129: 3520–3521.
16 Cardinal-David, B., Raup, D.E.A., and Scheidt, K.A. (2010). *J. Am. Chem. Soc.* 132: 5345–5347.
17 Zhang, J.-Q., Li, N.-K., Yin, S.-J. et al. (2017). *Adv. Synth. Catal.* 359: 1541–1551.
18 Cohen, D.T., Cardinal-David, B., and Scheidt, K.A. (2011). *Angew. Chem. Int. Ed.* 123: 1716–1720; (2011). *Angew. Chem. Int. Ed.* 50: 1678–1682.
19 Bhunia, A., Patra, A., Puranik, V.G., and Biju, A.T. (2013). *Org. Lett.* 15: 1756–1759.
20 Fu, Z., Jiang, K., Zhu, T. et al. (2014). *Angew. Chem. Int. Ed.* 126: 6624–6628; (2014). *Angew. Chem. Int. Ed.* 53: 6506–6510.
21 Yang, Y.-J., Ji, Y., Qi, L. et al. (2017). *Org. Lett.* 19: 3271–3274.
22 Shu, T., Ni, Q., Song, X. et al. (2016). *Chem. Commun.* 52: 2609–2611.
23 Fang, X., Jiang, K., Xing, C. et al. (2011). *Angew. Chem. Int. Ed.* 123: 1950–1953; (2011). *Angew. Chem. Int. Ed.* 50: 1910–1913.
24 Fan, X.-W. and Cheng, Y. (2014). *Org. Biomol. Chem.* 12: 123–131.
25 Wang, Z.-T., Zhao, Y., Wang, Z.-Y., and Cheng, Y. (2015). *J. Org. Chem.* 80: 1727–1734.
26 He, M. and Bode, J.W. (2008). *J. Am. Chem. Soc.* 130: 418–419.
27 Jiang, K., Tiwari, B., and Chi, Y.R. (2012). *Org. Lett.* 14: 2382–2385.
28 Wang, L., Li, S., Blümel, M. et al. (2017). *Angew. Chem. Int. Ed.* 129: 8636–8641; (2017). *Angew. Chem. Int. Ed.* 56: 8516–8521.
29 Wang, G., Wang, Z.-Y., Niu, S.-S. et al. (2016). *J. Org. Chem.* 81: 8276–8286.
30 Zhu, T., Zheng, P., Mou, C. et al. (2014). *Nat. Commun.* 5: 5027–5032.
31 Huang, X., Zhu, T., Huang, Z. et al. (2017). *Org. Lett.* 19: 6188–6191.
32 Candish, L. and Lupton, D.W. (2013). *J. Am. Chem. Soc.* 135: 58–61.
33 Candish, L., Forsyth, C.M., and Lupton, D.W. (2013). *Angew. Chem. Int. Ed.* 125: 9319–9322; (2013). *Angew. Chem. Int. Ed.* 52: 9149–9152.
34 Bera, S., Samanta, R.C., Daniliuc, C.G., and Studer, A. (2014). *Angew. Chem. Int. Ed.* 126: 9776–9780; (2014). *Angew. Chem. Int. Ed.* 53: 9622–9626.
35 Wu, X., Hao, L., Zhang, Y. et al. (2017). *Angew. Chem. Int. Ed.* 129: 4265–4269; (2017). *Angew. Chem. Int. Ed.* 56: 4201–4205.
36 Mondal, S., Yetra, S.R., Patra, A. et al. (2014). *Chem. Commun.* 50: 14539–14542.
37 Ryan, S.J., Candish, L., and Lupton, D.W. (2011). *J. Am. Chem. Soc.* 133: 4694–4697.
38 Zhang, C.-L., Gao, Z.-H., Liang, Z.-Q., and Ye, S. (2016). *Adv. Synth. Catal.* 358: 2862–2866.
39 Jia, Q. and Wang, J. (2016). *Org. Lett.* 18: 2212–2215.

40 Perveen, S., Zhao, Z., Zhang, G. et al. (2017). *Org. Lett.* 19: 2470–2473.
41 Zhang, C. and Lupton, D.W. (2017). *Org. Lett.* 19: 4456–4459.
42 Zhou, B., Luo, Z., and Li, Y. (2013). *Chem. Eur. J.* 19: 4428–4431.
43 Bera, S., Daniliuc, C.G., and Studer, A. (2015). *Org. Lett.* 17: 4940–4943.
44 Liang, Z.Q., Wang, D.L., Zhang, H.M., and Ye, S. (2015). *Org. Lett.* 17: 5140–5143.
45 Fu, Z., Wu, X., and Chi, Y.R. (2016). *Org. Chem. Front.* 3: 145–149.
46 Mukherjee, S., Ghosh, A., Marelli, U.K., and Biju, A.T. (2018). *Org. Lett.* 20: 2952–2955. https://doi.org/10.1021/acs.orglett.8b00998.
47 Zhang, H.-R., Dong, Z.-W., Yang, Y.-J. et al. (2013). *Org. Lett.* 15: 4750–4753.
48 Lu, H., Zhang, J.-L., Liu, J.-Y. et al. (2017). *ACS Catal.* 7: 7797–7802.
49 Biswas, A., Sarkar, S.D., Fröhlich, R., and Studer, A. (2011)). *Org. Lett.* 13: 4966–4969.
50 Yang, Y.-J., Zhang, H.-R., Zhu, S.-Y. et al. (2014). *Org. Lett.* 16: 5048–5051.
51 Biswas, A., Sarkar, S.D., Tebben, L., and Studer, A. (2012). *Chem. Commun.* 48: 5190–5192.

6

N-Heterocyclic Carbene Catalysis via the α,β-Unsaturated Acyl Azolium

Changhe Zhang and David Lupton

Monash University, School of Chemistry, Faculty of Science, RM 238, Bldg. 23(S) Clayton, Melbourne, VIC 3800, Australia

6.1 Introduction

In 2006, Zeitler reported the formation of the α,β-unsaturated acyl azolium (**1**) from ynals and its esterification to give a series of cinnamates [1]. This report, while introducing a simple esterification, has provided a framework for the discovery of a host of reactivity patterns that allow bond-forming events to be orchestrated at the α-, β-, γ-, and acyl groups of the α,β-unsaturated acyl azolium (**1**).

Over the past decade, the α,β-unsaturated acyl azolium (**1**) has been extensively examined in modern *N*-heterocyclic carbene (NHC) organocatalysis. In general, reactions of this species exploit the reactivity patterns outlined in Scheme 6.1. Guided by these patterns, this chapter reviews key discoveries in this rapidly evolving field of research.

6.2 Generation of the α,β-Unsaturated Acyl Azolium

There are five general methods currently known to access the α,β-unsaturated acyl azolium intermediates. In four of these methods, a redox approach is exploited that involves initial formation of an acyl anion intermediates (i.e. **11**, Scheme 6.2) from an appropriate aldehyde. At this point, either reducible functionality within the aldehyde substrate or an external oxidant allow access to the α,β-unsaturated acyl azolium. An example of the former strategy can be seen in Zeitler's 2006 report, with the conversion of ynals **7** to the corresponding α,β-unsaturated acyl azolium. The process commences with the addition of the NHC to the aldehyde carbonyl, thereby forming the Breslow intermediate, which is protonated to give allenol **13** that tautomerizes to the α,β-unsaturated acyl azolium **1** [1]. A further example of this type of approach was reported by Ye and coworkers in 2011 who accessed α,β-unsaturated acyl azoliums by exploiting α-bromoenals **8**. The process begins with the formation of the Breslow intermediate that gives α-bromo acyl azolium **14** and is followed by the

N-Heterocyclic Carbenes in Organocatalysis, First Edition. Edited by Akkattu T. Biju.
© 2019 Wiley-VCH Verlag GmbH & Co. KGaA. Published 2019 by Wiley-VCH Verlag GmbH & Co. KGaA.

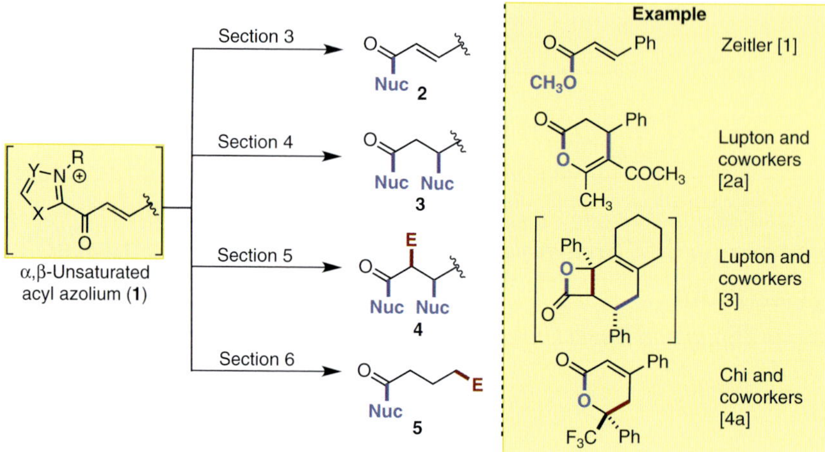

Scheme 6.1 Summary of reactions of the α,β-unsaturated acyl azolium.

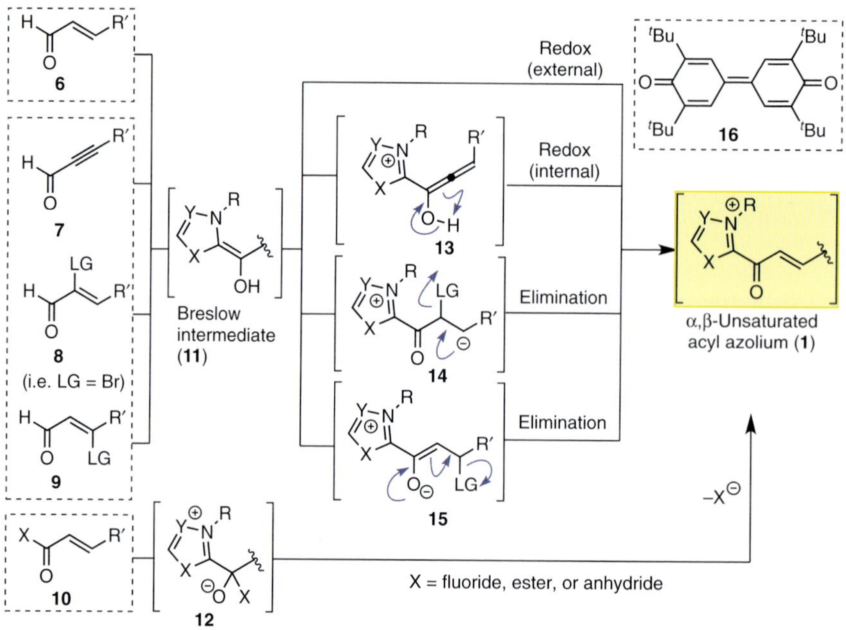

Scheme 6.2 Common approaches to the generation of the α,β-unsaturated acyl azolium.

elimination of the bromide [5]. Finally, Ma and coworkers described the use of β-bromoenals (i.e. **9**) as a method to prepare α,β-unsaturated acyl azolium **1** [6].

Although internal redox approaches are common, the use of external oxidants to access the acyl azolium is most often exploited. Specifically, the Kharasch oxidant **16** has been extensively used to generate the acyl azolium from the acyl anion in an approach pioneered by Studer and coworkers [7].

In contrast to the above methods that exploit redox processes to access the α,β-unsaturated acyl azoliums, it is also possible to exploit substrates in the ester oxidation state (i.e. **10**), such as esters, acyl fluorides, or anhydrides with substitution chemistry leading to the key species [2].

6.3 Esterification of the α,β-Unsaturated Acyl Azolium

The simplest reactions of the α,β-unsaturated acyl azolium involve bond formation at the acyl carbon to provide ester products. In 2006, Zeitler reported redox isomerization of alkynal aldehydes (i.e. **7a**) to α,β-unsaturated esters (i.e. **17a**) in the presence of IMes NHC (**A1**) and an appropriate alcohol (Eq. (1)) [1].

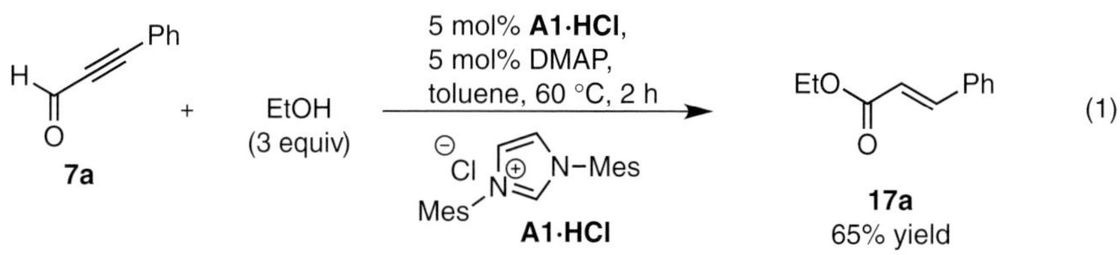

In 2007, Scheidt and coworkers reported the use of MnO_2 as the terminal oxidant for the conversion of both allylic alcohols and α,β-unsaturated aldehydes (i.e. **6a**) to the α,β-unsaturated acyl azolium (i.e. **21**) [8]. Beyond simple esterifications in this study, the use of chiral NHCs (i.e. **B1**) was examined allowing the desymmetrization of mesodiol **18** to provide enantioenriched ester **19** (Eq. (2)) (Scheme 6.3).

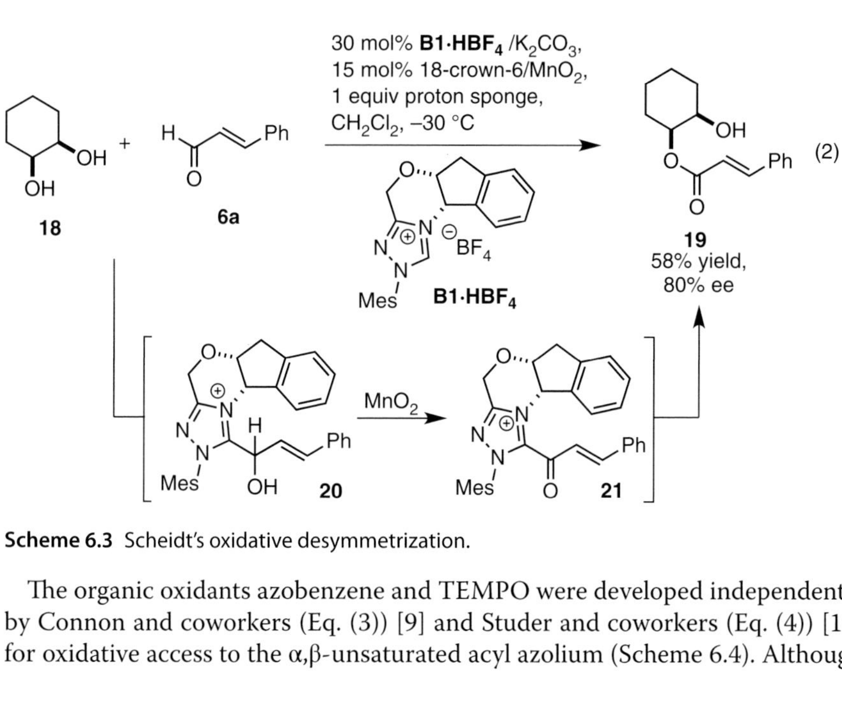

Scheme 6.3 Scheidt's oxidative desymmetrization.

The organic oxidants azobenzene and TEMPO were developed independently by Connon and coworkers (Eq. (3)) [9] and Studer and coworkers (Eq. (4)) [10] for oxidative access to the α,β-unsaturated acyl azolium (Scheme 6.4). Although

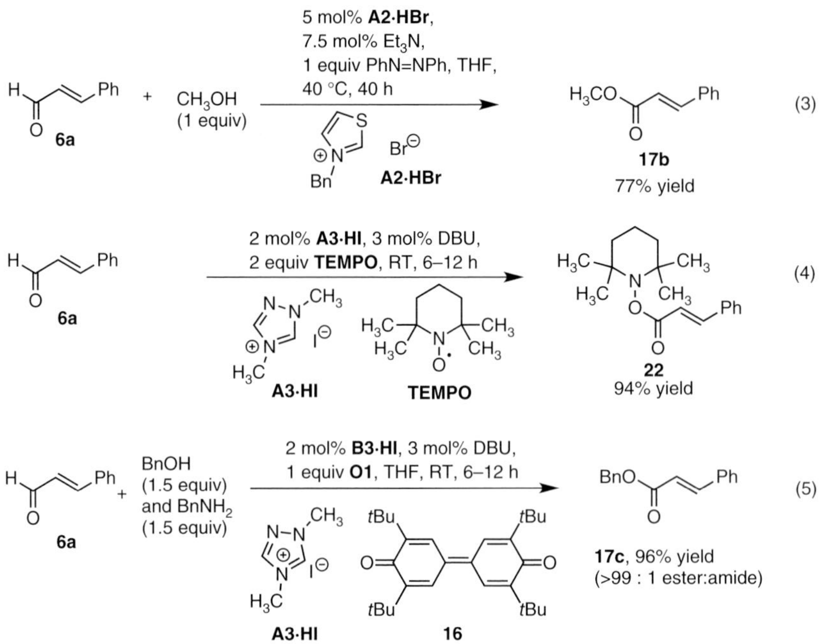

Scheme 6.4 Various oxidative esterifications.

both gave acceptable yields of the expected cinnamate ester, the use of TEMPO was limited to the formation of TEMPO cinnamate **22** with rapid consumption of azolium by TEMPO$^-$ (Eq. (4)). This limitation was addressed by the use of the Kharasch oxidant **16**, which Studer found to be compatible with a variety of alcohols (Eq. (5)) [11]. The Kharasch oxidant **16** has become the most common oxidant used in acyl azolium catalysis. This study also identified the acyl azoliums preference for esterification over amidation (Eq. (5)). Specifically, when equimolar quantities of benzylalcohol and benzylamine were subjected to the reaction conditions, only ester **17c** formed. This feature of acyl azolium catalysis is notable in subsequent studies where amides and lactams have generally proved more challenging to access than esters and lactones (*vide infra*).

6.4 [3+*n*] Annulations of the α,β-Unsaturated Acyl Azolium

The α,β-unsaturated acyl azolium can be considered a LUMO-lowered species, thereby allowing bond formation at the unsaturation and the acyl carbon. If a bis-nucleophile is exploited, this defines a [3+*n*] annulation. Implementation of such a strategy was first reported in 2009 by Lupton, with independent studies by Studer and Bode following closely. Subsequent studies by Mayr have indicated that the electrophilicity of these species is modest and is similar to a simple Knoevenagel adduct. In this section, annulations of the α,β-unsaturated acyl azolium with bond formation at the acyl and β-position are discussed and have been separated based on the type of bis-nucleophile implemented.

6.4.1 Annulation with Enolates

Lupton and coworkers described the first annulation of the α,β-unsaturated acyl azolium in 2009 (Eq. (6), Scheme 6.5) [2a]. The α,β-unsaturated acyl azolium was generated from the corresponding acid fluoride (e.g. **23a**) and reacted with unmasked nucleophiles (e.g. **27**) at the β-position, either by direct 1,4-addition or by 1,2-addition to the carbonyl followed by Claisen rearrangement. The resultant azolium enolate **29** tautomerized and cyclized to give the dihydropyranone **25** with regeneration of the catalyst. The reaction was also achieved starting from various enol esters (i.e. **30**). In this case, a moderately enantioselectivity variant was also reported with chiral catalyst **B2** (Eq. (7)). Various mechanistic studies have been performed on this reaction, and a fragmentation/addition pathway from the enol ester was deemed unlikely, with a Claisen rearrangement or tight ion pair mechanism favored [12].

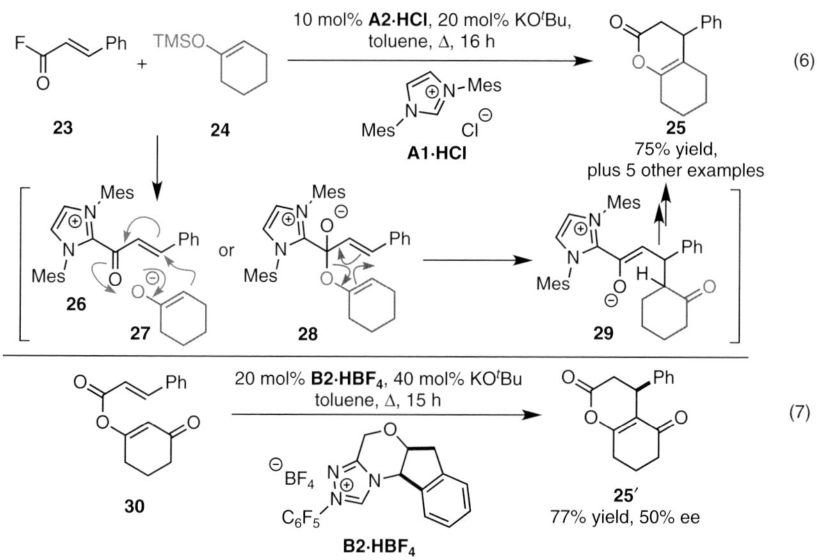

Scheme 6.5 Lupton's [3+3] annulation with enolates.

The Lupton group later applied this NHC-catalyzed transformation in the synthesis of (−)-7-deoxyloganin **33** (Scheme 6.6) [13]. Starting with enantiopure

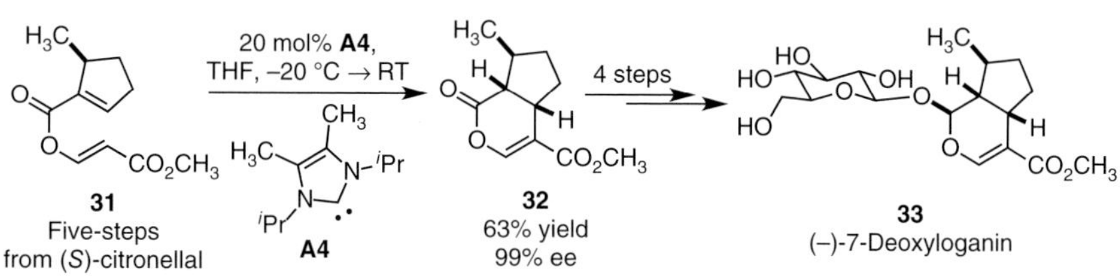

Scheme 6.6 Lupton's total synthesis of (−)-7-deoxyloganin.

enol ester **31**, prepared in five steps from (*S*)-citronellal, rearrangement with tetraalkyl imidazole catalyst **A4** gave dihydropyranone **32** in 63% yield and with high enantiopurity. Chemoselective reduction, glycosylation, and deprotection then completed the total synthesis.

Numerous annulations of α,β-unsaturated acyl azoliums (e.g. **35**) with the enolates of 1,3-dicarbonyls (e.g. **34′**) have subsequently been reported. These studies have generally focused on either the introduction of a new method for acyl azolium formation or they have allowed new chiral NHC (pre)catalysts to be reported (Scheme 6.7). Contributions of this type have included the synthesis of dihydropyranone **36** via the α,β-unsaturated acyl azolium using external oxidation from precursors including cinnamaldehyde **6a** [14], dihydrocinnaldehyde **37** [15], or potassium 2-oxo-3-enoate 38 [11]. In addition, internal redox isomerization of ynals **7a** [16], α-[5, 17], and β-[6] bromocinnaldehydes (**8a** and **9a)** also allow its formation. Finally, the novel NHC catalyst **B4** was found to be effective in the synthesis of chiral dihydropyranones [18].

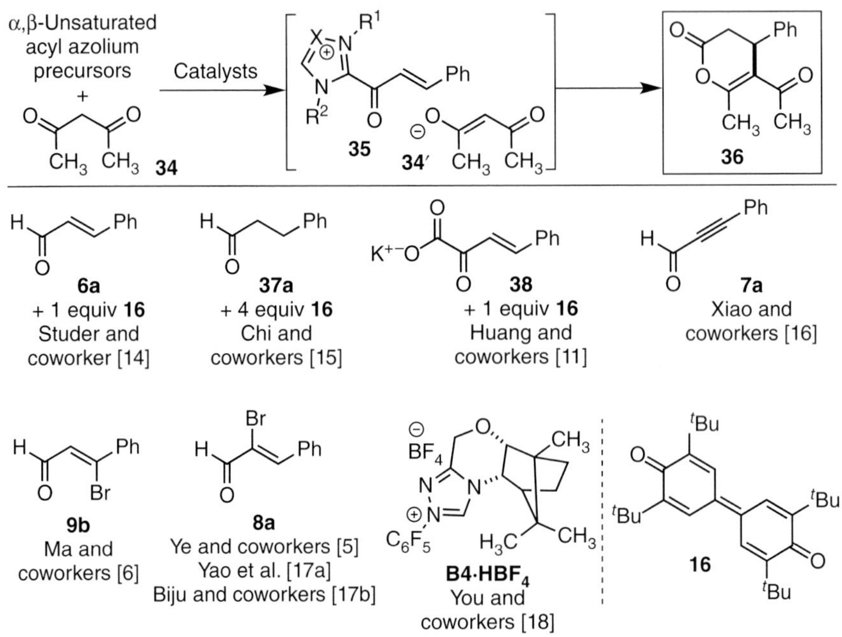

Scheme 6.7 Various oxidative [3+3] annulations.

In addition to simple 1,3-diketones, alternate enolates have been exploited in related [3+*n*] annulations, selected examples are discussed below. In a very early report in the area, Bode and coworkers reported the enantioselective synthesis of pyranones by the reaction of kojic acid derivatives with the α,β-unsaturated acyl azolium intermediate formed by redox isomerization of ynals 7 (Eq. (8), Scheme 6.8) [19]. In addition to kojic acid adducts **40**, isolated as the methyl ester derivatives, dihydropyranone **41** from a pyruvate, and the tricyclic naphthol-derived product **42** could be prepared with enantioselectivity. A 1,4-addition mechanism was eliminated based on the kinetic analysis

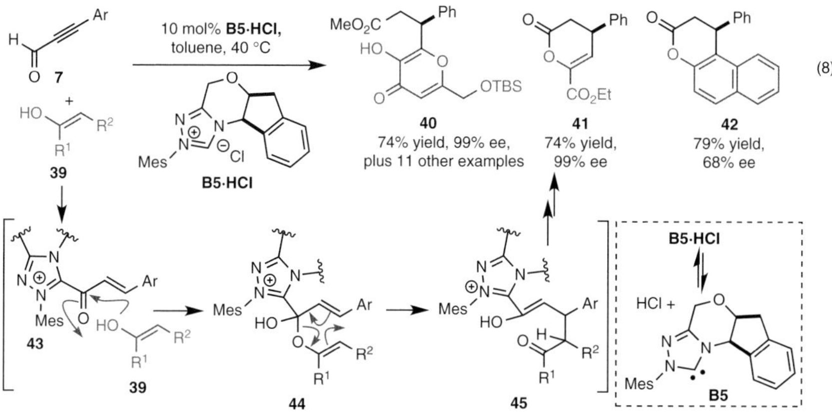

Scheme 6.8 Bode's enantioselective [3+3]-annulations

with a Claisen rearrangement from intermediate **44** to give **45** the preferred mechanism.

You and coworkers have exploited naphthols (e.g. **46a**) [20] and electron-rich phenols (e.g. **46b**) [21] as bis-nucleophiles for the enantioselective synthesis of dihydropyranones using a novel chiral NHC precatalysts **B6·HBF$_4$** prepared from L-phenylalanine (Eq. (9)).

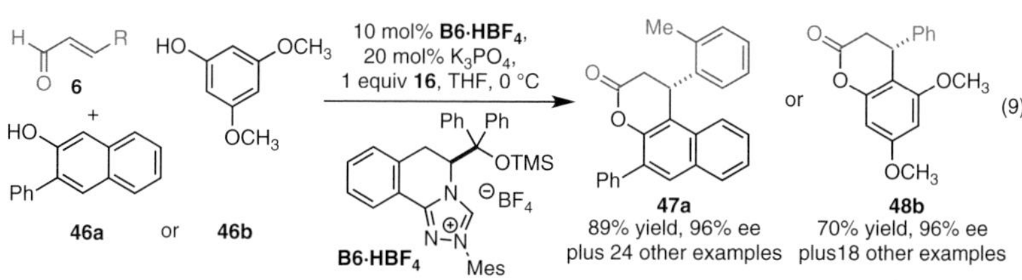

Less common indolin-3-one **49** has also been examined as a bis-nucleophile for coupling to α,β-unsaturated acyl azoliums, in this case to provide dihydropyranindolone **50**. In 2013, Lu and coworkers applied the external oxidation strategy to access the α,β-unsaturated acyl azoliums en route to racemic dihydropyranindolone **50** (Eq. (10), Scheme 6.9) [22]. Subsequently, Enders and coworkers reported the enantioselective variant of this reaction exploiting α-bromocinnamaldehyde derivatives as the precursors to the α,β-unsaturated acyl azoliums (Eq. (11)) [23].

In 2012, Alexakis and coworkers described the application of diketone **51** as a novel bis-nucleophile in the reaction with α,β-unsaturated acyl azoliums from ynals (e.g. **7a**), to give bicyclic lactones **52** (Eq. (12), Scheme 6.10) [24]. This reaction is catalyzed by achiral catalyst **A5** in the presence of 4 Å molecular sieves. The conjugate addition of the bis-enol tautomer of diketone **51** to the α,β-unsaturated acyl azolium followed by tautomerization of the resulting azolium enolate gives acyl azolium **53**. Rather than direct formation of an enol ester from **53**, lactolization gives hemiketal **54**, which provides bicyclic lactone product **52**.

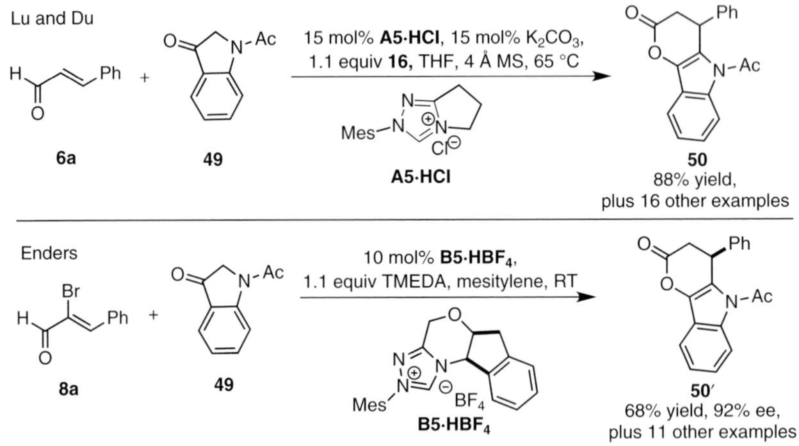

Scheme 6.9 [3+3] annulations with indolinone.

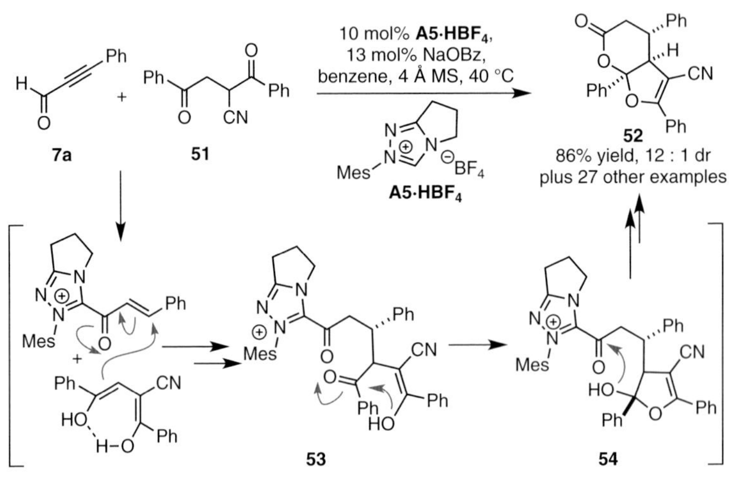

Scheme 6.10 Alexakis' intercepted [3+3] annulation.

Beyond the use of novel bis-nucleophiles, Lu and coworkers have examined novel α,β-unsaturated acyl azoliums to allow the synthesis of spirocyclic quaternary carbon-containing dihydropyranones **57** from the α,β-unsaturated acyl azolium precursor **55** (Eq. (13)) [25]. Unfortunately, the enantioselective version of this reaction was unsuccessful.

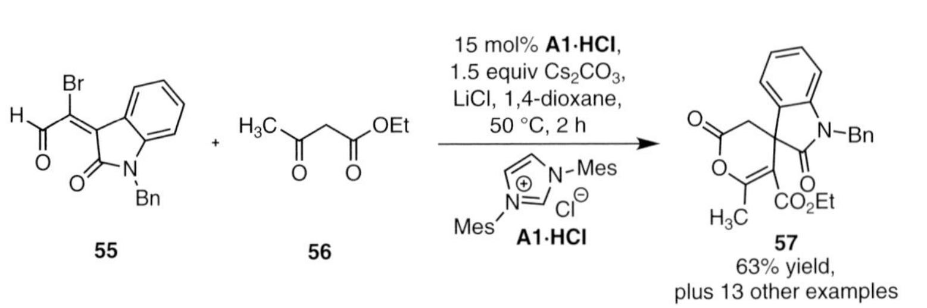

(13)

6.4.2 Annulation with Enamines

The annulation of the α,β-unsaturated acyl azolium with enamines to produce dihydropyridinone was reported first by Bode and coworkers (Eq. (14), Scheme 6.11) [26]. The α,β-unsaturated acyl azolium could be derived from either enal **6** or enal equivalent **58** and was reacted with a variety of enamines **59** to give dihydropyridinones **60** containing various functional groups including the ester (**60a**), nitro (**60b**), and nitrile (**60c** and **60d**). It was proposed that the reaction proceeds via the hemiaminal intermediate **61**, which underwent sigmatropic rearrangement to give azolium enol **62**, followed by tautomerization and lactamization to complete the reaction.

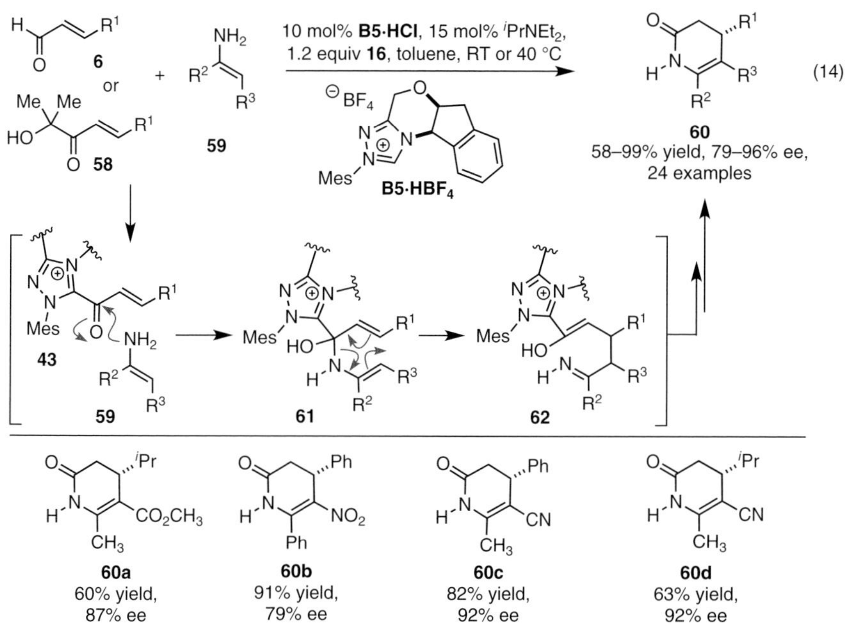

Scheme 6.11 Bode's [3+3] annulation with enamines.

More recently, Qi and coworkers utilized *N*-Ts-2-aminoacrylate **63** in an NHC-catalyzed annulation reaction with the α,β-unsaturated acyl azolium (Eq. (15)) [27]. Interestingly, either the 5,6-dihydropyridinone **64** or 3,4-dihydropyridinone **65** could be formed depending on the base utilized. Although most examples are nonenantioselective, high enantioselectivity was also achieved in a limited number of cases.

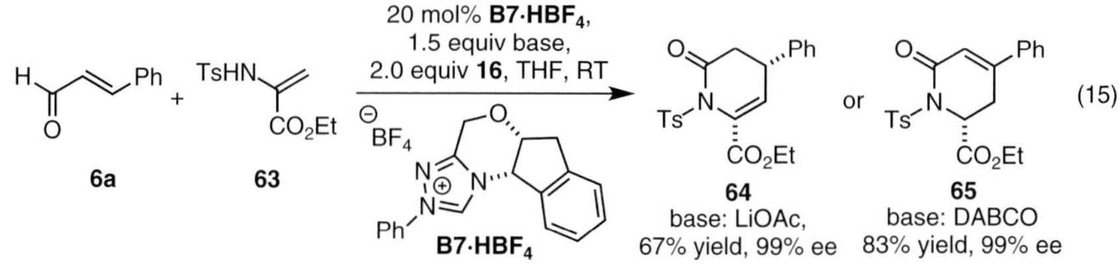

Cyclic *N*-sulfonylimines **66** have also been exploited as enamine precursors allowing the synthesis of dihydropyridinones **67** as reported by Bode and coworkers (Eq. (16), Scheme 6.12) [28]. Enals **6** bearing up to three substituents (R^1, R^2, R^3) were applied as the acyl azolium precursors producing tricyclic *N*-sulfonyl dihydropyridinones **67** with high enantioselectivity (up to 99% ee), including examples containing a quaternary center (i.e. **67b**) and with α-substitution (i.e. **67c**, from the six-membered sulfonylimine **66b**).

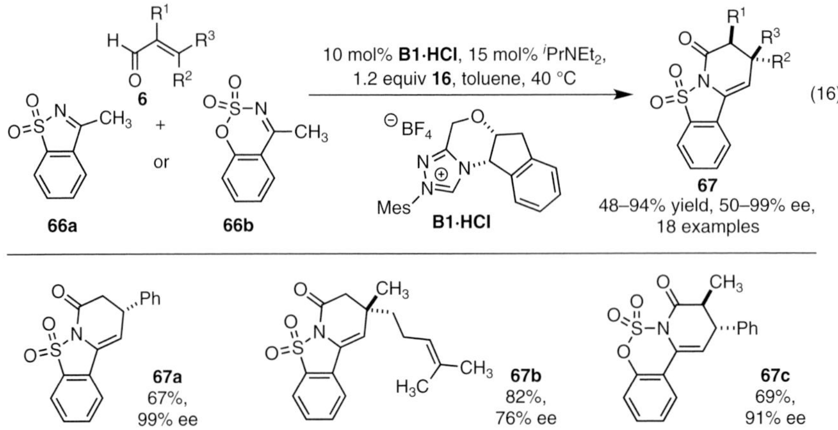

Scheme 6.12 Bode's [3+3] annulations with enamines from sulfonylimines.

Variants of this reaction have been developed using a number of precursors to the α,β-unsaturated acyl azolium (Scheme 6.13). In 2013, the Chi group exploited 4-nitrophenol ester **68** as the precursors to the α,β-unsaturated acyl azolium, which coupled with acylic sulfonylimines **72** [29]. In 2014, Ye and coworkers developed a similar reaction with mixed anhydride **69** or α-bromoenal **8** as precursors to the α,β-unsaturated acyl azoliums coupling with sulfonyl imines, including cyclic **71** or the acyclic variants **72** derived from acetophenones [2b, 30]. In 2015, Zhong and coworkers accomplished a similar reaction with *N*-hydroxyphthalimide ester **70** as the precursor to the acyl azolium [31]. All reactions were catalyzed by **B1·HBF$_4$** or its enantiomer **B5·HBF$_4$** and gave products with very high enantioselectivity (typically >95% ee).

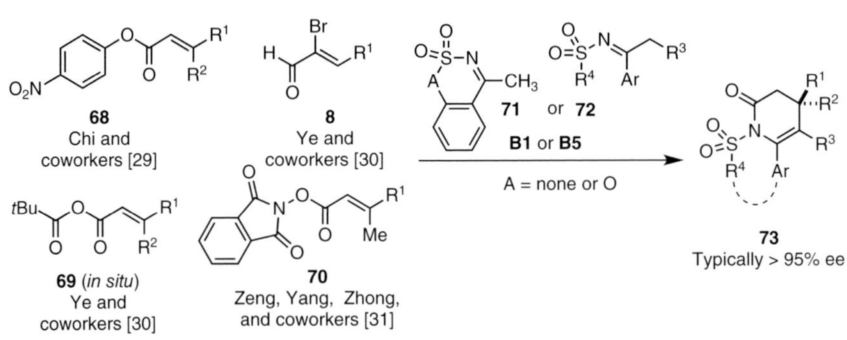

Scheme 6.13 Alternate [3+3] annulations with enamines from sulfonylimines.

Enders and coworkers reported the use of benzoxazolyl, benzothiozolyl acetates, and acetonitriles **74** serving as enamine precursors for annulation reactions with the α,β-unsaturated acyl azoliums derived from α-bromoenals **8** (Eq. (17), Scheme 6.14) [32]. A series of dihydrobenzoxazolyl- or benzothiozolyl-fused dihydropyranones **75** were prepared using catalyst **B5**; however, low-to-moderate enantioselectivity (32–73% ee) was observed.

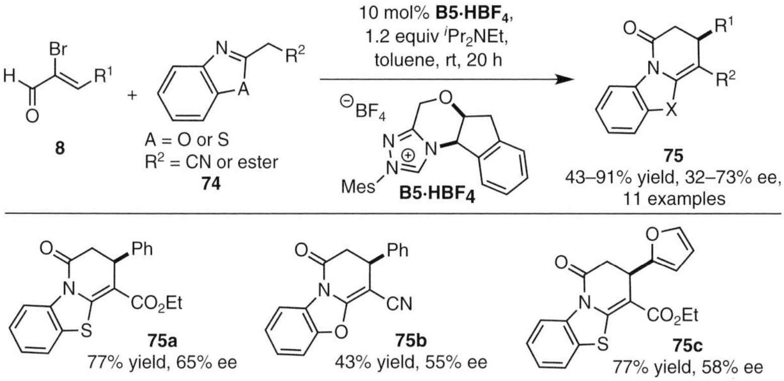

Scheme 6.14 Ender's [3+3] annulations with enamines from sulfonylimines.

In 2017, Ye and coworkers utilized indolin-2-imines (e.g. **76**) as the enamine precursor reacting with α,β-unsaturated acyl azoliums to produce amidine **77** (Eq. (18), Scheme 6.15) [33]. The reaction presumably proceeds by conjugate addition of the enamine anion **79** with α,β-unsaturated acyl azolium **78**.

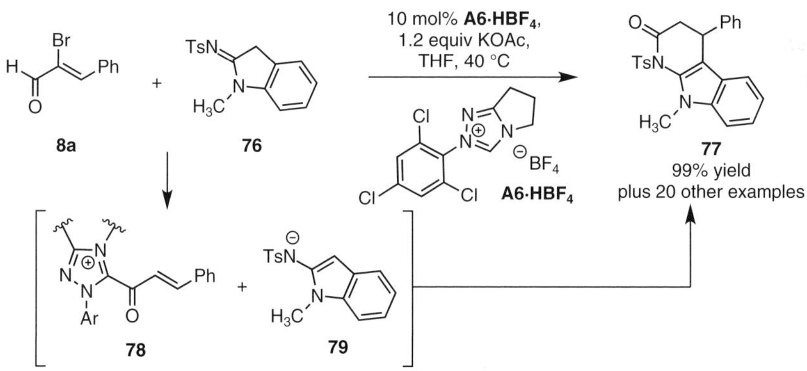

Scheme 6.15 Ye's [3+3] annulation with enamines from amidine **76**.

In 2017, Hui and coworkers discovered the NHC-catalyzed [3+4] annulation of hydrazone **80** with the α,β-unsaturated acyl azolium from α-bromoenals (e.g. **8a**) (Eq. (19)) [34]. Presumably, the enamine of hydrazone **80** undergoes conjugate addition to the α,β-unsaturated acyl azolium, which then cyclizes to give the seven-membered tetrahydro-1*H*-1,2-diazepines **81** with high enantiomeric integrity.

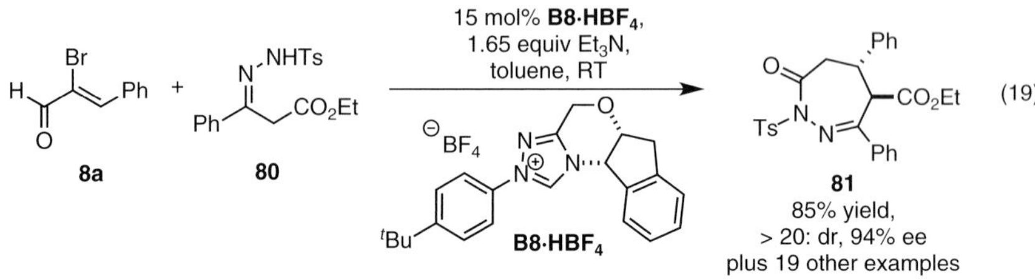

6.4.3 Annulation with Other Nucleophiles

In addition to the use of enamines in the annulations of α,β-unsaturated acyl azolium intermediates, alternate bis-nucleophiles including α-aminoketones **83** and 3-aminooxindoles **85** have been exploited (Scheme 6.16). In 2014, Ye and coworkers demonstrated that the α,β-unsaturated acyl azoliums could be generated from the *in situ* formed mixed anhydride **69** and reacted with *N*-sulfonyl-α-aminoketones **83** to afford lactams such as **84** (Eq. (20), Scheme 6.16). A similar strategy was reported by Ye in 2016 with 3-aminooxindoles **85** acting as the bis-nucleophile, which gave spirocyclicoxindoles **86** in a reaction catalyzed by **B1** (Eq. (21)) [35].

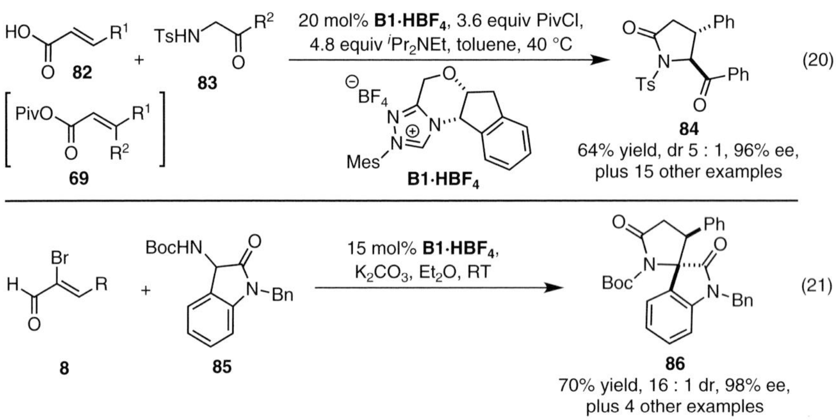

Scheme 6.16 Ye's [3+2] annulations with aminoketone **83**.

Benzils can also serve as the bis-nucleophiles allowing the synthesis of γ-lactones via the α,β-unsaturated acyl azolium (Eq. (22), Scheme 6.17) [36]. In 2015, Huang and coworkers reported a novel one-pot two-step coupling of α-bromoenal **8a** and benzyl **89**; the latter is formed from benzaldehyde **87** via a benzoin reaction mediated by triazolylidene NHC **A7**. Addition of α-bromoenal **8a** and imidazolidene catalyst **A1** then gives the α,β-unsaturated acyl azolium intermediate **26**, which subsequently reacts with the enol form of benzyl **89**. Cyclization of the resultant acyl azolium **90** completes the catalytic cycle, giving γ-lactones **88** with good levels of diastereoselectivity.

Although all bis-nucleophiles discussed above react with the unsaturation of the α,β-unsaturated acyl azolium first to make a new C—C bond, recently, Chi and coworkers reported that protected *N*,*N*′-tosyl-protected hydrazine **91** react

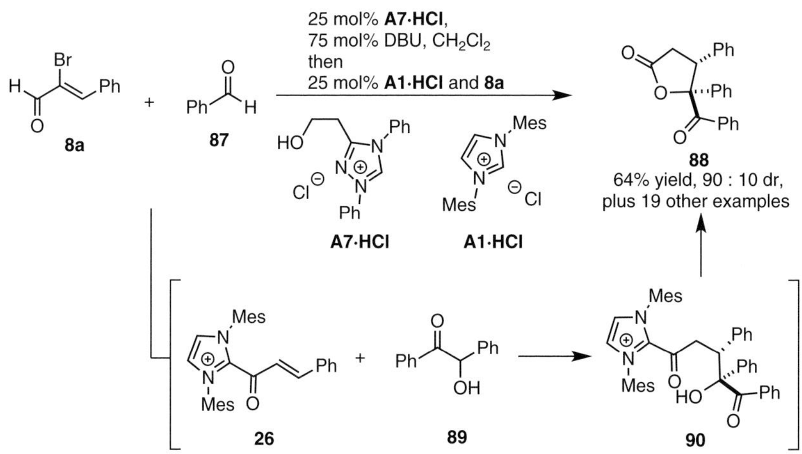

(22)

Scheme 6.17 Huang's [3+2] annulations with *in situ* formed benzil.

to make a C—N bond (Eq. (23)) [37]. Pyrazolidones **92** was formed with high enantioselectivity catalyzed by morpholinone-based catalyst **B9**.

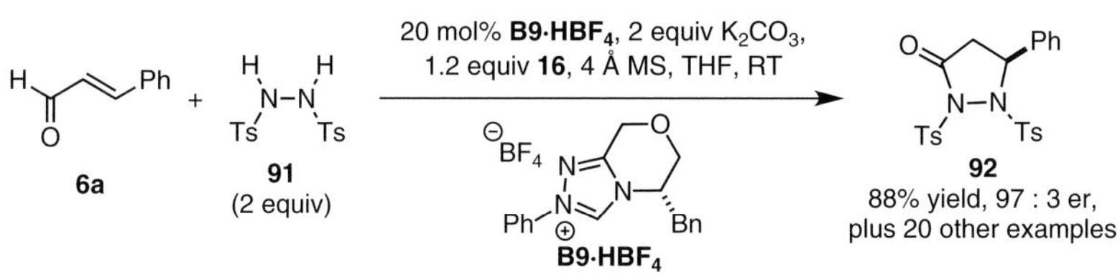

(23)

Pyrazolinone **93** can serve as a novel bis-nucleophile allowing the formation of two new C—C bonds in a [3+3] annulation reaction catalyzed by NHC **B5**, reported by Biju and coworkers (Eq. (24), Scheme 6.18) [38]. Conjugate addition of the deprotonated pyrazolinone **95** to the acyl azolium **43** afforded the

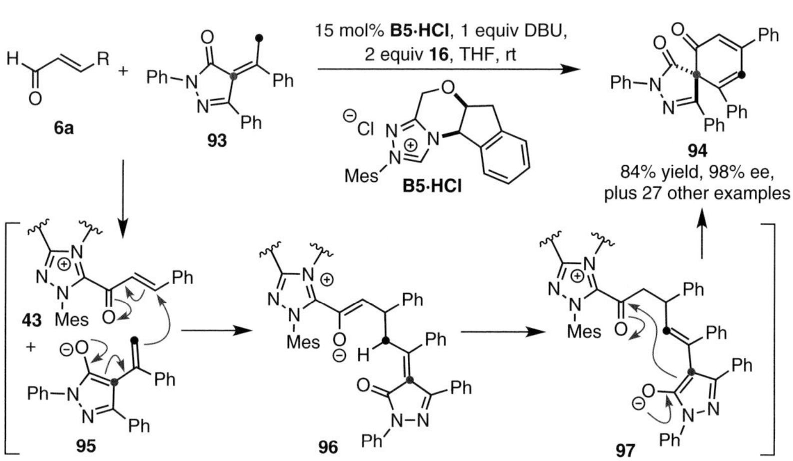

(24)

Scheme 6.18 Biju's [3+3] annulation with dienolate **95**.

azolium enolate **96**. Tautomerization, forming **97**, and cyclization then gave **94** as the product.

Thiolates are known to undergo conjugate addition with electron-poor olefins. However, only very recently has the α,β-unsaturated acyl azolium been examined with thiolate anions. Du and coworkers discovered the enantioselective [3+4] annulation of α,β-unsaturated acyl azoliums and aminobenzenethiols in the synthesis of 1,5-benzothiazepines (Eq. (25)) [39]. α-Bromoenal **8a** serves as the precursor to the α,β-unsaturated acyl azolium intermediate, while the aminobenzethiol **98** is the bis-nucleophile.

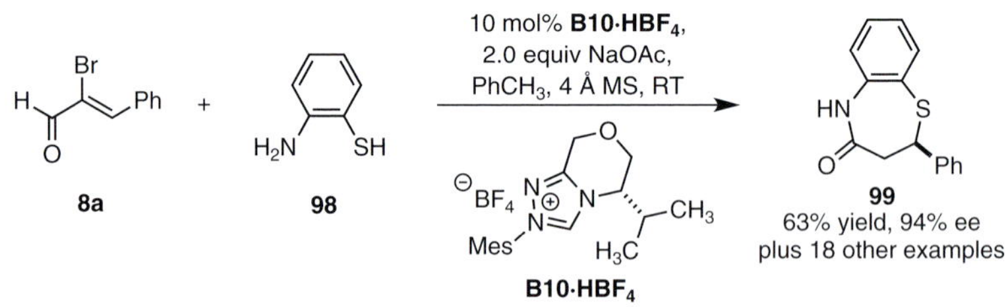

(25)

6.5 [2+*n*] Annulations of the α,β-Unsaturated Acyl Azolium

In the previous section following nucleophilic addition to the α,β-unsaturated acyl azolium, the resultant azolium enolate was generally protonated and acylated. However, the azolium enolate **100** can be exploited as a nucleophile to access more functionalized products. When an appropriate substrate is employed, multiple-ring systems can be accessed via bond formations at the α-, β-, and acyl carbons in [2+*n*] annulation reactions (Scheme 6.19).

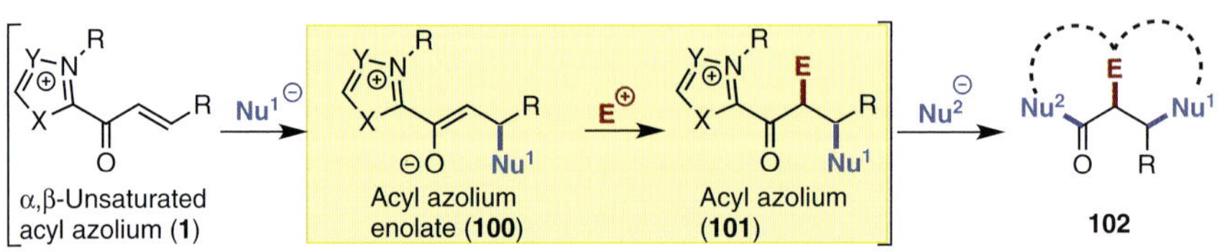

Scheme 6.19 Overview of [2+*n*] annulation reactions.

6.5.1 [2+4] Annulations Terminating in β-Lactonization

In 2011, the Lupton group reported the nonenantioselective NHC-catalyzed all-carbon [2+4] annulation of α,β-unsaturated acyl fluorides (e.g. **23**) and silyl dienol ethers (e.g. **103**) to give cyclohexadienes (e.g. **104**) (Eq. (26), Scheme 6.20) [40]. α,β-Unsaturated acyl azolium **105** was generated by nucleophilic substitution of imidazolium NHC **A4** on the carbonyl of **23**, with defluorination and desilylation of **103** giving acyl azolium **105** and dienolate **106**. Secondary kinetic isotope effect analysis and computational studies implicate a mechanism involving stepwise annulation, via vinylogous Michael addition of **106** to **105** giving

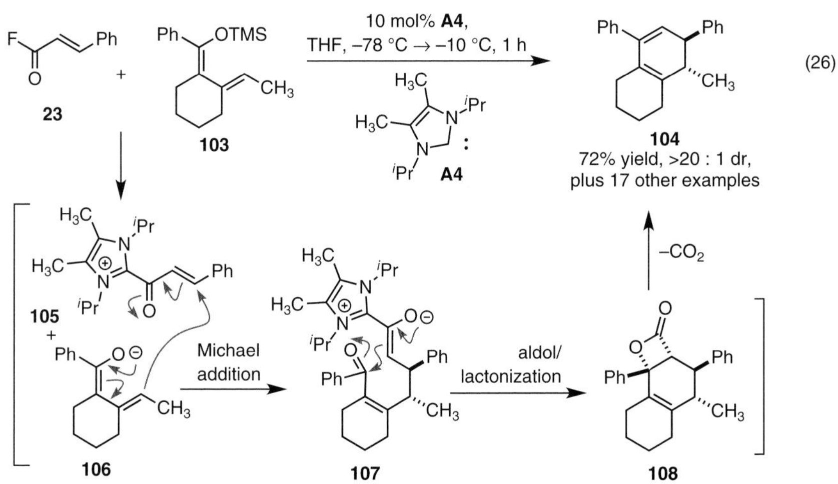

Scheme 6.20 Lupton's [2+4] annulations with dienolates.

107 and then aldol addition, followed by lactonization and decarboxylation, to provide the cyclohexadienes (e.g. **104**) as single diastereoisomers.

Unfortunately, this [2+4] annulation has a number of drawbacks including poor chemoselectivity with undesired enol ester formation, limited scope, and incompatibility with homochiral NHCs. To address these problems, dienol esters (e.g. **109**) were examined and found to provide the corresponding β-lactones (e.g. **112**) (Eq. (27), Scheme 6.21) [41]. The optimal precatalyst for this transformation was **B11** with a morpholinone-based triazolium bearing *tert*-butyl *N*-substituent [42]. Crossover studies suggested the reaction commenced with the fragmentation of the dienyl ester **109** to give acyl azolium **110** and dienolate **111**. Vinylogous Michael addition of **111** to **110** followed by isomerization,

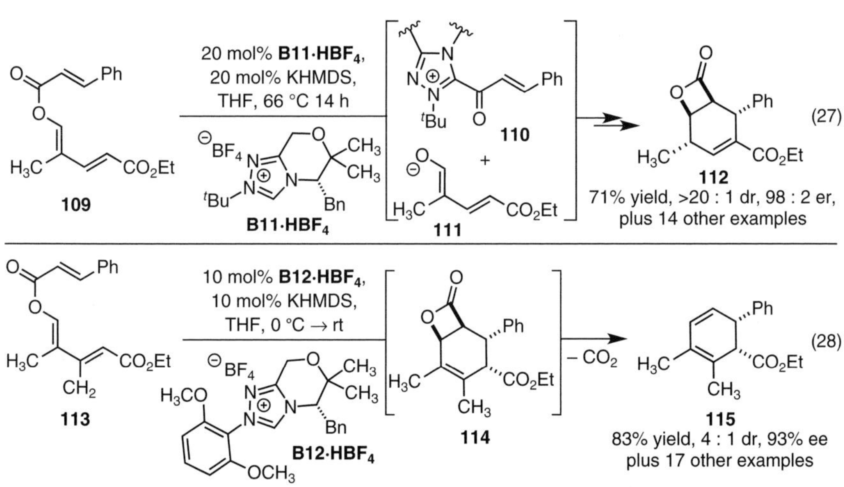

Scheme 6.21 Lupton's enantioselective [2+4] annulations from dienol esters.

proton transfer, and lactonization provided the β-lactones (**112**). Dienol ester **113**, with one additional substituent compared to **109**, gave a different product, namely cyclohexadiene **115**. In this case, olefin isomerization does not occur, and hence, decarboxylation becomes facile leading to the diene products (Eq. (28), Scheme 6.2) [43].

Very recently, Fang and coworkers described a similar all-carbon [2+4] annulation with the union of α,β-unsaturated acyl azolium **119** and dienolate **116′**, from α-bromoenal **8a** and α-benzoyloxy ketone **116**, respectively (Scheme 6.22) [44]. This reaction represents the first enantioselective intermolecular [2+4] annulation.

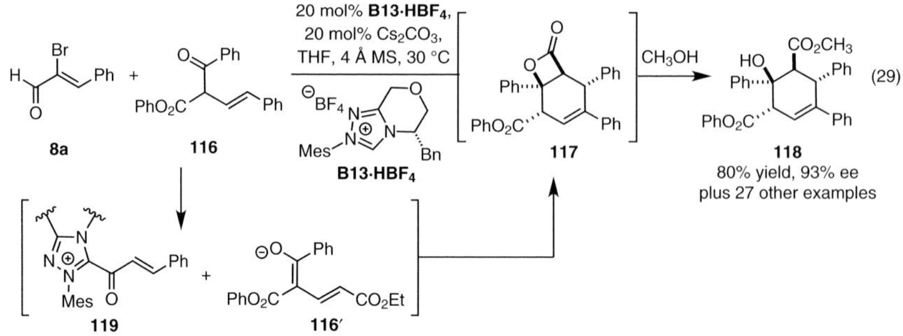

Scheme 6.22 Fang's enantioselective intermolecular [2+4] annulations.

Concurrently, the same group demonstrated that 1,5-diketones (e.g. **120**) could act as the bis-nucleophiles in the synthesis of chiral 1,2-dihydronaphthalenes (e.g. **121**) (Eq. (30), Scheme 6.23) [45]. Similarly, Michael addition of the enolate **120′**, from deprotonation of diketone **120**, to α,β-unsaturated acyl azolium **43a**, affords the azolium enolate **122**. Aldol cyclization followed by lactonization and decarboxylation of lactone **123** gives product **121** in high yield and with high enantiomeric integrity.

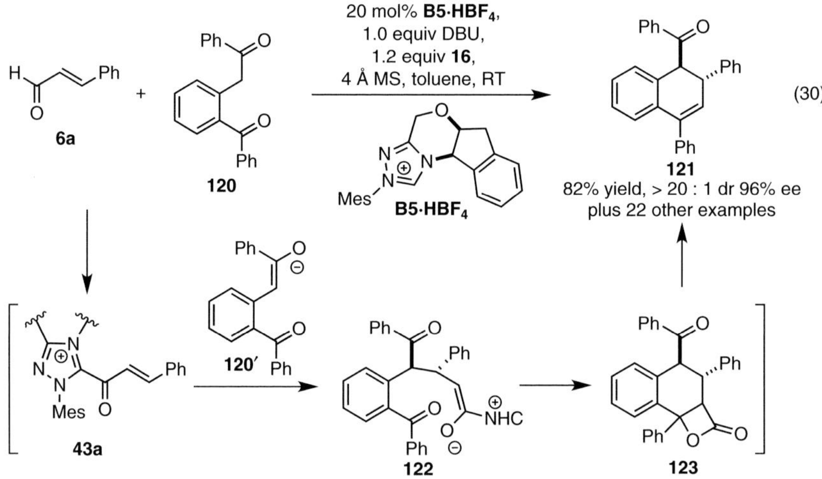

Scheme 6.23 Fang's enantioselective [2+4] annulations with diketone**120**.

In 2016, an alternate approach to enantioenriched cyclohexyl β-lactones by NHC organocatalysis was reported by Lupton and coworkers involving the ring-opening of donor–acceptor cyclobutanes (e.g. **125**) (Eq. (31), Scheme 6.24) with acid fluorides (e.g. **124**) to provide cyclohexyl β-lactones (e.g. **126**) [46]. Mechanistic studies provide evidences that the reaction is likely to be via [3,3]-sigmatropic rearrangement of adduct **129**, which is formed from the α,β-unsaturated acyl azolium **127** and enolate **128** by NHC-mediated defluorination–desilylation cascade.

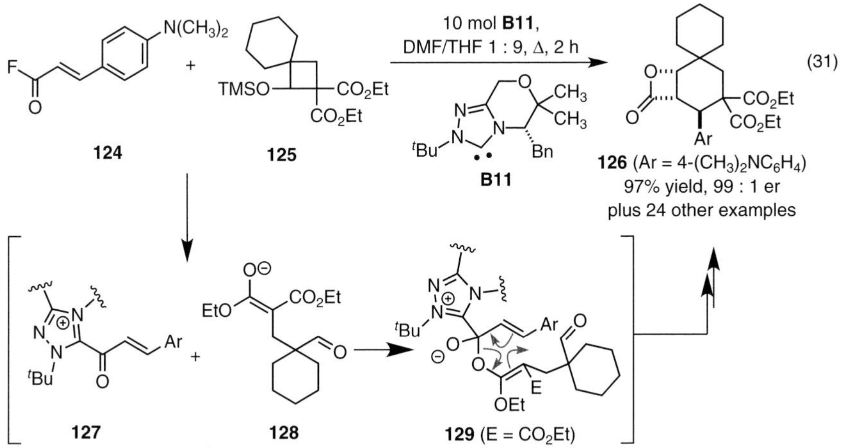

Scheme 6.24 Lupton's enantioselective [2+4] annulations with DA-cyclobutanes**125**.

The groups of Wang and Ye independently demonstrated the NHC-catalyzed synthesis of polysubstituted benzenes from α-cyanoenones involving the α,β-unsaturated acyl azolium (Eq. (32), Scheme 6.25, Ye's example) [47]. Mechanistically, vinylogous Michael addition of deprotonated dienolate **132** with acyl azolium **26**, or Diels–Alder cycloaddition of the two components, suggested by Wang and coworker [47a] and Ye and coworkers [47b], respectively,

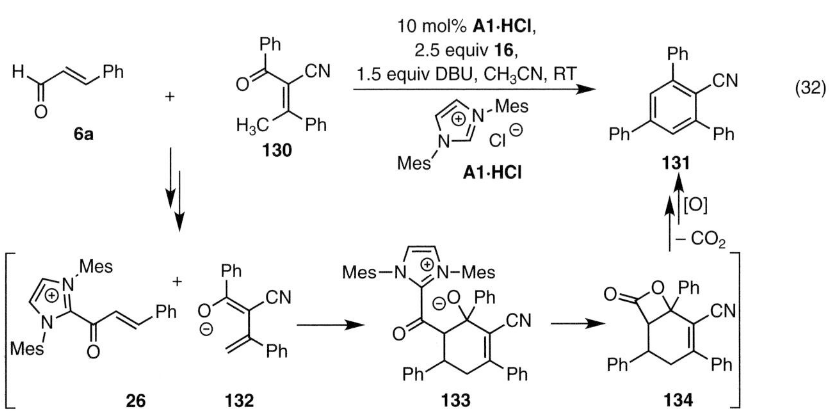

Scheme 6.25 Ye's [2+4] annulations to prepare substituted benzonitrile**131**.

affords adduct **133** that lactonizes to give **134**. Decarboxylation of **134** gives a cyclohexadiene, which is oxidized to give the benzonitrile product **131**.

6.5.2 [2+4] Annulations Terminating in δ-Lactonization

In 2015, the groups of Studer [48] and Ye [49] both reported the NHC-catalyzed construction of cyclopentane and cyclohexane-fused δ-lactones such as **136** and **137** by annulations of α,β-unsaturated acyl azoliums from enals (i.e. **6a**) and malonates **135** bearing Michael acceptors (Eq. (33), Scheme 6.26). LiCl served as a Lewis acidic cocatalyst in both studies, and a transition state that includes lithium is proposed, e.g. **138**. Specifically, oxidative activation of enals (e.g. **6a**) with an NHC leads to acyl azolium that reacts with deprotonated malonate from **135** by Michael addition, giving azolium enolate **139**, which then undergoes a second Michael addition to give acyl azolium **140**. Lactonization of **140** gives **136** or **137**. In general, cyclohexanes **137** were isolated with higher enantioselectivity than the cyclopentane variants, whereas yields were not compromised.

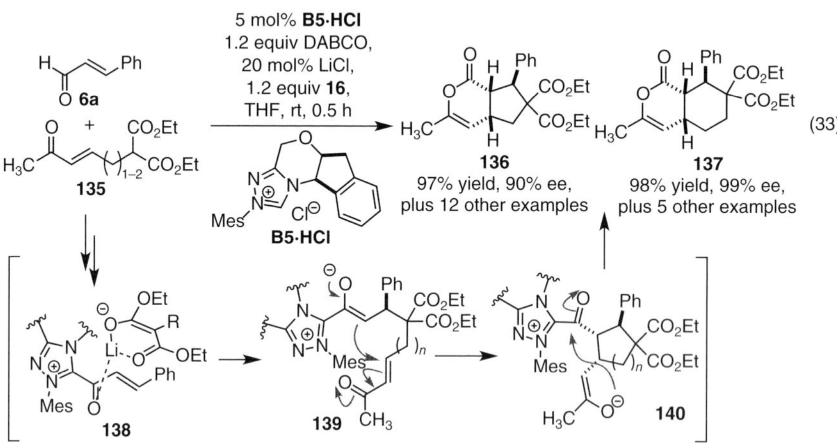

Scheme 6.26 Studer's [2+4] and [2+3] annulations with malonate **135**.

Xu and coworkers discovered that thioesters could be the precursors to α,β-unsaturated acyl azoliums [50]. Thioester **141** with an *ortho*-enone substituent reacts with nitro-substituted NHC **B14**, giving thiolate **143** and α,β-unsaturated acyl azolium **144** (Eq. (34), Scheme 6.27). Conjugate addition of **143** with acyl azolium **144** gives azolium enolate **145** in which a second conjugate addition occurs. After lactonization of **146** and regeneration of the catalyst, the thiochroman product **142** is obtained.

6.5.3 [2+3] Annulations Terminating in β-Lactonization

In 2012, Lupton and coworkers reported an enantioselective synthesis of cyclopentyl-fused β-lactones (e.g. **148**) via the [2+3] annulation of α,β-unsaturated acid fluorides (e.g. **23**) with donor–acceptor cyclopropanes (e.g. **147**) catalyzed by NHCs (Eq. (35), Scheme 6.28). Although achiral catalyst

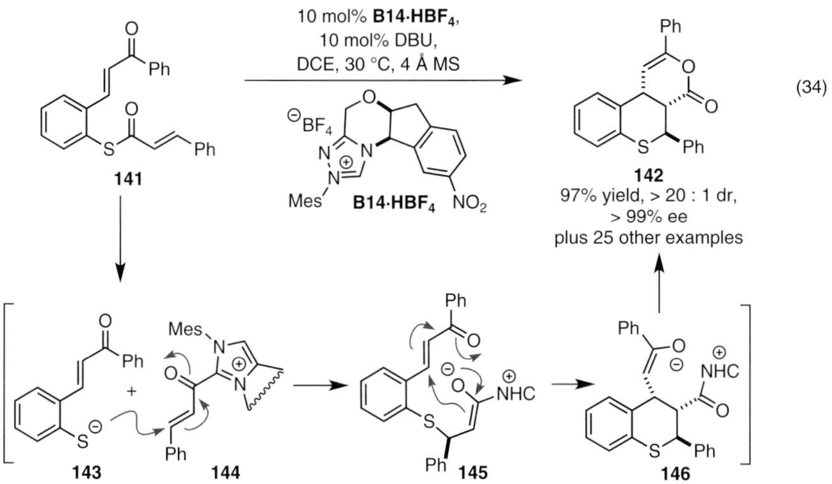

Scheme 6.27 Xu's [2+4] annulations exploiting thioesters.

IMes **A1** gives racemic **148** with good yields and high diastereoselectivity (all dr > 20 : 1) [51], an examination of new chiral catalysts designed for this reaction revealed that morpholinone-derived **B11** with a *tert*-butyl *N*-substituent provided good-to-excellent enantioselectivity without erosion of diastereoselectivity [42]. Mechanistically, it is believed that substitution of the carbene to acid fluoride **23** affords acyl azolium **149** and fluorodesilylation of **147** gives enolate **150**. Their union provides **151**, and finally aldol cyclization of **151** provides **152**, which then transforms to β-lactone **148** by lactonization.

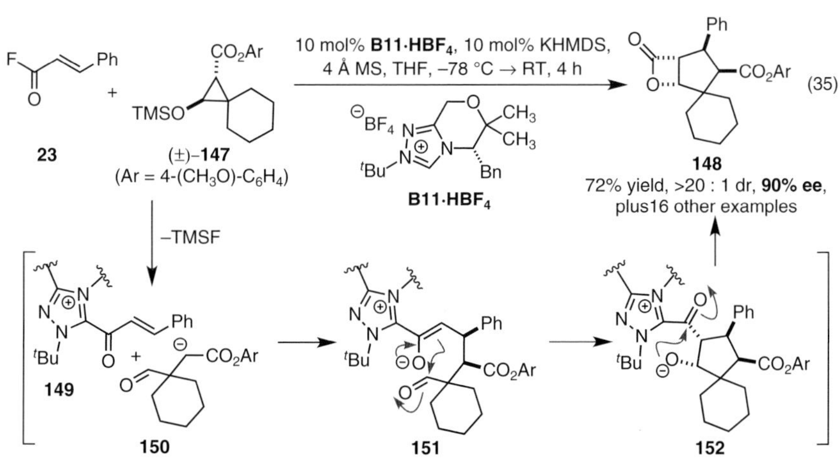

Scheme 6.28 Lupton's enantioselective [2+3] annulation with DA-cyclopropanes **147**.

In 2014, Studer and coworkers reported a related NHC-catalyzed [2+3] annulation between enals (e.g. **6a**) and ketone–malonates (e.g. **153**), allowing the synthesis of cyclopentyl-fused β-lactones (e.g. **154a**, Scheme 6.29) [52]. When

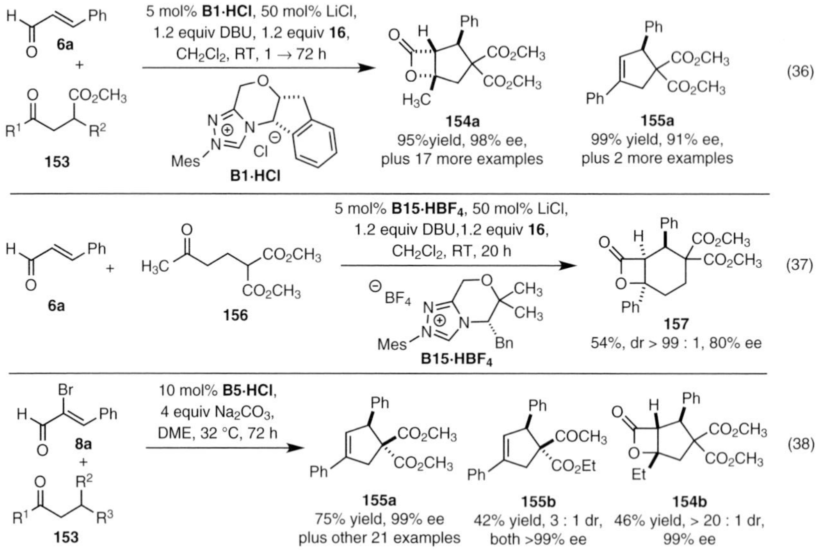

Scheme 6.29 Studer and Biju's [2+3] annulations with malonates and derivatives.

LiCl was used as the cooperative Lewis acid catalyst, reaction with malonates **153a** (R^2 = ester) gave β-lactones with excellent diastereoselectivity (all dr > 99 : 1) and enantioselectivity (94–99% ee) (Eq. (36)). In contrast, β-ketoester **154b** ($R^2 = COCH_3$) under the same conditions reacted with low-to-moderate diastereoselectivity (1.1 : 1 to 7 : 1). When the R^1 substituent is a phenyl group, decarboxylation of the β-lactones occurred and the corresponding cyclopentenes (e.g. **155a**) were obtained.

Extension of the tether between the malonate and ketone moiety also gave cyclohexyl-fused β-lactone **157** with excellent diastereoselectivity (>99 : 1 dr), albeit with diminished enantioselectivity (80% ee) using catalyst **B15** (Eq. (37)). Biju and coworkers subsequently reported a related reaction exploiting α-bromoenals (e.g. **8a**) as the acyl azolium precursors to give the [3+2] annulation products including cyclopentenes (such as **155a** and **155b**) and β-lactones (e.g. **154b**) (Eq. (38)) [53].

6.5.4 [2+1] Annulations

Annulation of the α,β-unsaturated acyl azoliums can also produce cyclopropanes via (2+1) annulation. In 2012, Studer and coworkers reported the NHC-catalyzed synthesis of cyclopropyl carboxylic acid esters (e.g. **159**) from enals (e.g. **6a**), sulfur ylides (e.g. **158**), and alcohols (e.g. ^{i}PrOH, Eq. (39), Scheme 6.30) [54]. Activation of enals in the presence of the NHC catalyst and oxidant **16** affords acyl azolium **43a**, which reacts with ylide **160**, generated from the corresponding salt precursor **158**, to give acyl azolium **161**, which undergoes esterification with alcohols to produce the final product ester **159**.

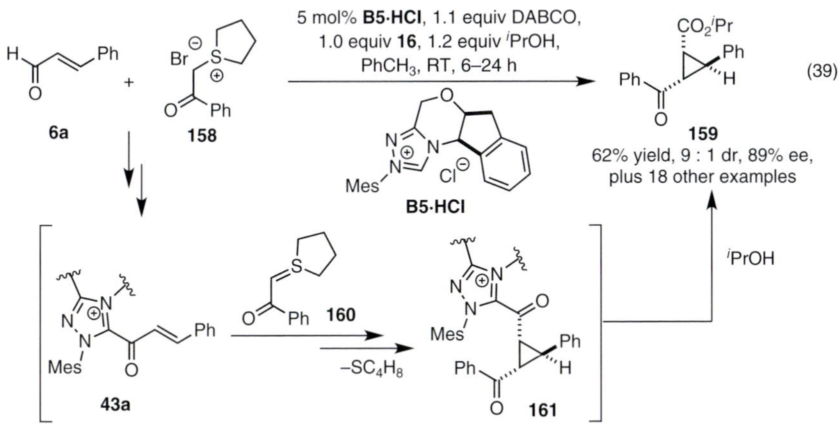

Scheme 6.30 Studer's [2+1] annulations with sulfur ylides.

6.6 Cascades Involving Bond Formation at the γ-Carbon and Acyl Carbon

When the α,β-unsaturated acyl azolium bears a methyl group at the β-position (i.e. **162**), the sp^3 γ-C–H is relatively acidic and deprotonation results in the azolium dienolate **163** (Scheme 6.31). Although related intermediates have previously been accessed from the corresponding ketene, with notable work from Ye and Smith [55], access from the α,β-unsaturated acyl azolium has delivered new opportunities for reaction design and discovery. In this section, reactions of the γ- and acyl carbons are examined.

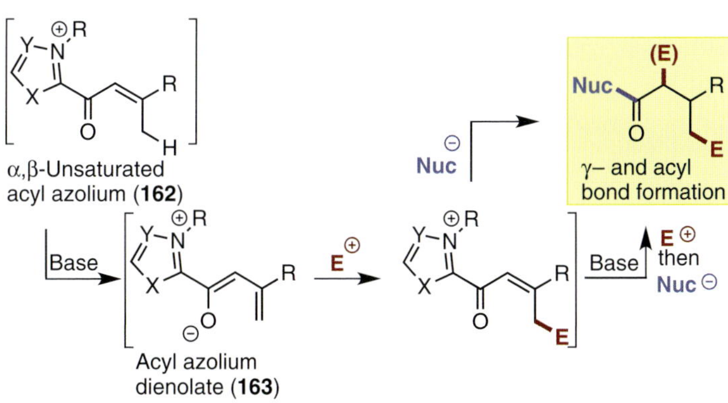

Scheme 6.31 Overview of [4+*n*] annulation reactions of the acyl azolium dienolate.

6.6.1 Annulations with Ketones and Imines

The pioneering work in this area was achieved by the group of Chi [4]. In 2012, Chi and coworkers reported the first example of this type of reactivity with the NHC-catalyzed oxidative γ-functionalization of enals to deliver unsaturated δ-lactones **166** (Scheme 6.32) [4a]. It was postulated that α,β-unsaturated acyl azolium **167** forms and is deprotonated to give a dienolate, which cyclizes with trifluoromethyl ketones (e.g. **165a**) via the chelated intermediate **168** (Eq. (40)). The reaction proceeded with moderate yield and enantioselectivity in the absence of Lewis acids, whose role was assigned as a coordinating reagent bringing ketone electrophile and the acyl azolium dienolate proximal. Chi and coworkers subsequently expanded this transformation to 2-methyl indole, benzofuran, and benzothiophene derivatives (**169**) [4b]. Because of the increased acidity of the benzylic hydrogen, the reaction does not require Lewis acid cocatalysts allowing ketone electrophiles including trifluoromethyl ketones and isatins **165** to give a series of dihydropyranones **170** (Eq. (41)).

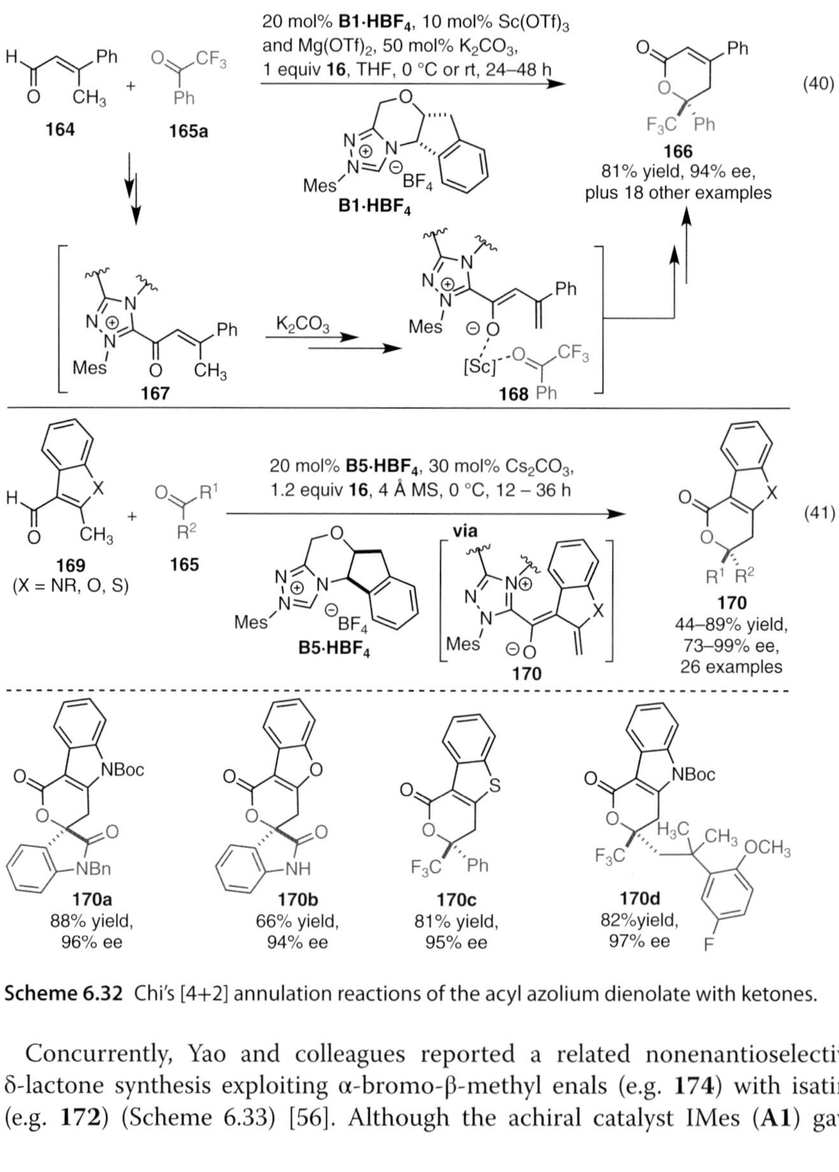

Scheme 6.32 Chi's [4+2] annulation reactions of the acyl azolium dienolate with ketones.

Concurrently, Yao and colleagues reported a related nonenantioselective δ-lactone synthesis exploiting α-bromo-β-methyl enals (e.g. **174**) with isatins (e.g. **172**) (Scheme 6.33) [56]. Although the achiral catalyst IMes (**A1**) gave

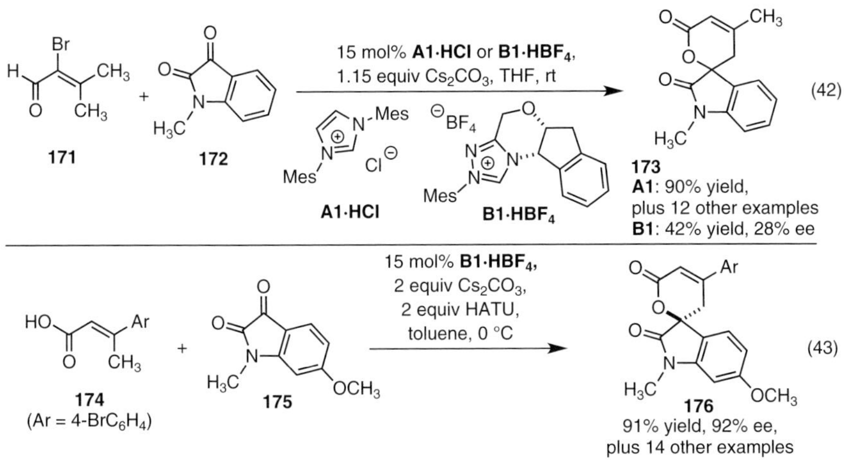

Scheme 6.33 Yao's [4+*n*] annulation reactions of the acyl azolium dienolate with ketones.

spirocylic lactones **173** with high yields, chiral catalyst **B1** gave **173** with only modest yields and enantioinduction (Eq. (42)). In 2016, the Yao group reported that acyl azolium dienolates can also be generated from α,β-unsaturated carboxylic acid (e.g. **174**) with HATU as the activating reagent [57]. In this work, they were able to access chiral spirocyclic dihydropyranones **176** with far improved enantioselectivity (Eq. (43)).

The azolium dienolate intermediates can also be generated from other α,β-unsaturated acyl azolium precursors (Scheme 6.34) [58]. In 2013, Chi

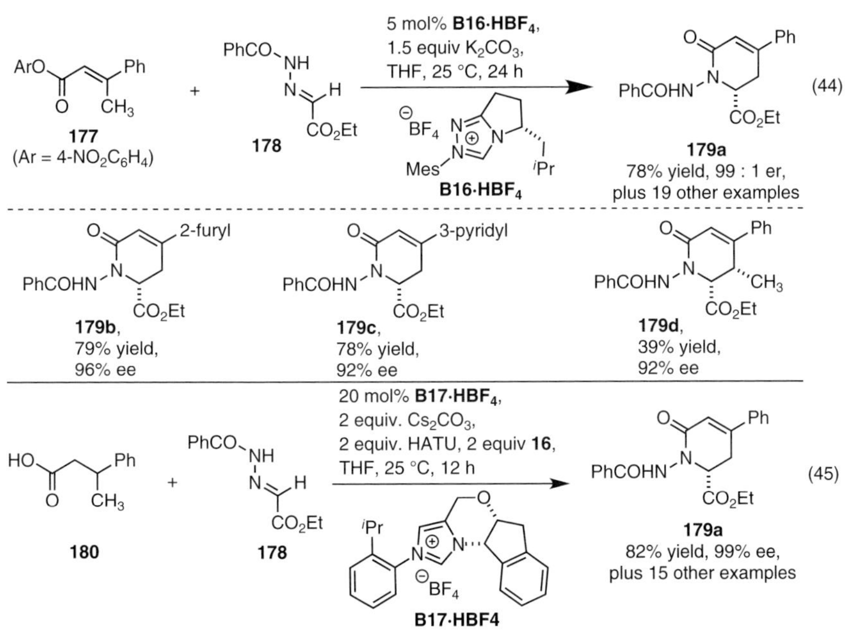

Scheme 6.34 Other methods to achieve reactions via the acyl azolium dienolate.

and coworkers reported the NHC-catalyzed annulation of α,β-unsaturated esters (**177**) with hydrazones (**178**) (Eq. (44)) [58a], whereas in 2015, Yao and coworkers reported that saturated β-methyl carboxylic acids **180** with HATU as the acid-activating reagent and quinones as oxidant allowed a related annulation (Eq. (45)) [58b]. Both reactions gave δ-lactams **179** with high enantiopurity via related reactive intermediates.

In 2014, Chi and coworkers reported a [4+3] annulation of the acyl azolium dienolate with racemic azomethine imine **181** (Eq. (46), Scheme 6.35) [59]. Oxidative formation of acyl azolium **183** as previously discussed and deprotonation gives azolium dienolates **184**, which undergoes a resolving nucleophilic addition to imine **181** to form acyl azolium **185**. Cyclization and release of the catalyst provides the dinitrogen-fused seven-membered heterocyclic products **182** in high enantiopurity while leaving unreacted, and enantioenriched, imine **181**. By changing the stoichiometry of the reaction, a kinetic resolution of azomethine imines such as **181** could be achieved with *S* factor up to 339 (Scheme 6.35).

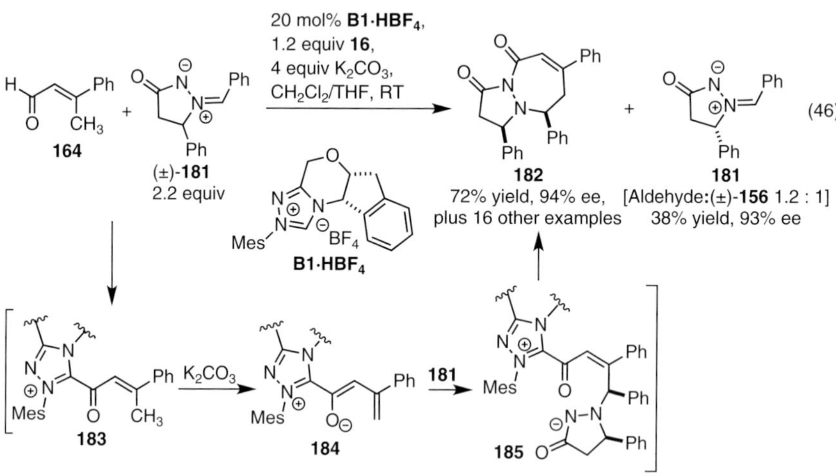

Scheme 6.35 Chi's [4+3] annulation reaction of the acyl azolium dienolate.

6.6.2 [4+2] Annulations with Electron-Poor Olefins

In addition to annulations across C=O and C=N bonds (Section 6.6.1), the acyl azolium dienolate can participate in the annulations of electron-poor C=C bonds (Scheme 6.36). In 2014, the Chi group reported the NHC-catalyzed synthesis of benzene derivatives by [4+2] cycloaddition of enal **164** with activated enone **186** (Eq. (47)) [60]. The reaction proceeds via azolium dienolate **188**, which reacts with enone **186** by Michael addition to form azolium intermediate **189**. Subsequent tautomerization gives azolium dienolate **190**, which cyclizes to provide β-lactone **191**. Finally, decarboxylation and oxidation gives the aromatized products **187**. This reaction provides tetrasubstituted benzenes with moderate-to-high yields.

In 2015, a mechanistically related [4+2] annulation was reported by the Yao group (Eq. (48), Scheme 6.35). In this reaction, the azolium dienolate derived

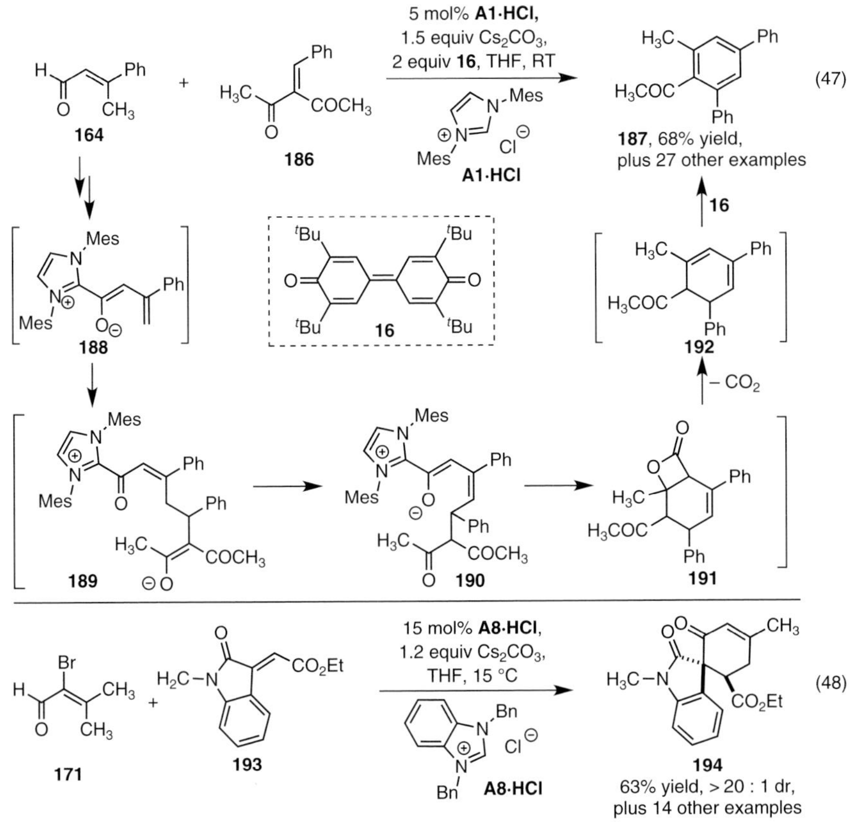

Scheme 6.36 Chi and Yao's [4+2] annulations of the acyl azolium dienolate with electron poor olefins.

from enal **171** reacts with isatin-derived olefin **193** to give spirocylic product **194** via a vinylogous Michael/Claisen condensation cascade with generally excellent diastereoselectivity (most >20 : 1) (Eq. (48)) [59].

6.7 Other Reactions of the α,β-Unsaturated Acyl Azolium

In the previous sections, a series of reaction cascades have been developed, which, in general, involve the use of bifunctional partners that couple to simple α,β-unsaturated acyl azoliums. An alternate, although less thoroughly examined, approach to reaction discovery involves bifunctional α,β-unsaturated acyl azolium partners.

In 2011, Studer and coworkers reported an oxidative NHC-catalyzed cascade for the synthesis of δ-lactone-fused indanes (e.g. **196**) from *ortho*-functionalized enals (e.g. **195**) and 1,3-dicarbonyls (e.g. **134**, Eq. (49), Scheme 6.37) [61]. Reaction of enal **195** with an NHC in the presence of oxidant **16** gave acyl azolium

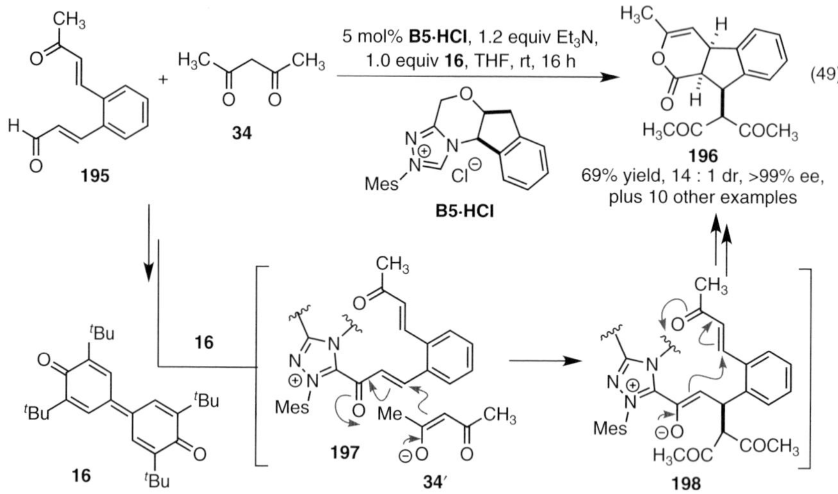

Scheme 6.37 Studer's use of bifunctional α,β-unsaturated acyl azoliums.

197. Conjugate addition of the deprotonated 1,3-dicarbonyl **34′**, then provided azolium enolate **198**, which cyclized by conjugate addition into the enone moiety, and lactonization affords the indane **196**.

In 2017, the Lupton group presented a related cascade reaction of bifunctional fluorides with TMS enol ethers as a method for the synthesis of enantioenriched indenes (Eq. (50), Scheme 6.38) [62]. The reaction occurs via a cascade Michael addition into the bifunctional acyl azolium **202** with enolate **203**, then aldol reaction of enolate **204** and β-lactonization. Finally, decarboxylation gives the indene products **201**. The reaction has a broad generality and moderate-to-high enantioselectivity.

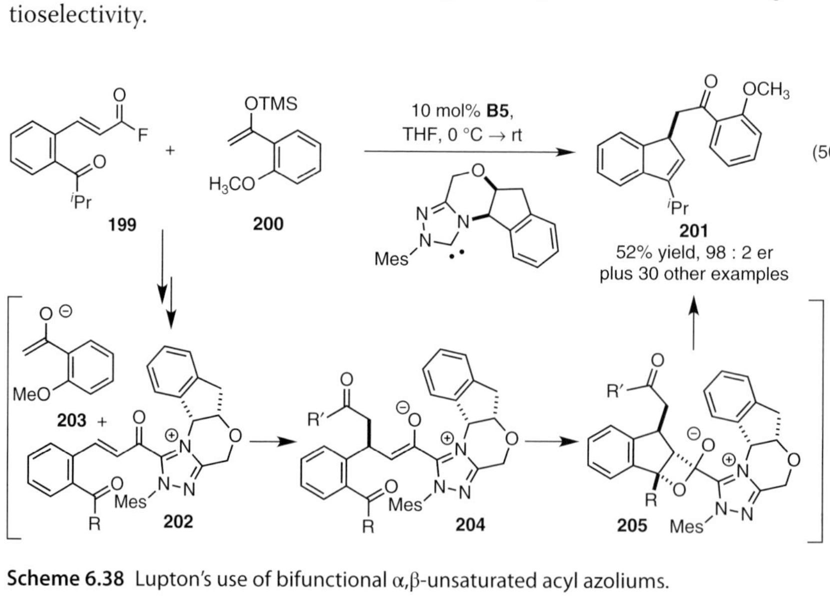

Scheme 6.38 Lupton's use of bifunctional α,β-unsaturated acyl azoliums.

6.8 Conclusions and Outlook

In the past decade, the α,β-unsaturated acyl azolium has played an important role in a number of reaction discoveries. In general, it has been demonstrated that it can serves as a two, three, or four-atom donor in annulative reactions with bifunctional coupling partners. Within these limitations, a range of reactions have been discovered with enantioselectivity generally achievable using a range of catalysts based around the chiral morpholinone or indanol scaffolds. In contrast to the chemistry of acyl anions, the use of the mesityl (and related) *N*-substituents has been found to be most advantageous.

Although the chemistry of this important intermediate is becoming well developed, there are still a number of frontiers that remain to be examined. Firstly, the electrophilicity of the α,β-unsaturated acyl azolium is modest, and as a consequence, the range of nucleophilic partners is somewhat limited. Presumably, catalyst design approaches may well be able to address this limitation.

Other frontiers being investigated and likely to continue to bear fruit involve studies on analogs of the acyl azolium, examination of extended vinylogy and interception of radical intermediates formed *en route* to the α,β-unsaturated acyl azolium. Significant opportunities exist in these areas such that much work remains to be done with this novel reactive intermediate.

References

1 Zeitler, K. (2006). *Org. Lett.* 8: 637.
2 (a) Ryan, S.J., Candish, L., and Lupton, D.W. (2009). *J. Am. Chem. Soc.* 131: 14176. (b) Chen, X.-Y., Gao, Z.-H., Song, C.-Y. Zhang, C.L., Wang, Z.X., and Ye, S. (2014). *Angew. Chem. Int. Ed.* 53: 11611.
3 Ryan, S.J., Candish, L., Lupton, D.W. (2011). *J. Am. Chem. Soc.* 133: 1634.
4 (a) Mo, J.M., Chen, X.K., and Chi, Y.R. (2012). *J. Am. Chem. Soc.* 134: 8810. (b) Chen, X.K., Yang, S., Song, B.A., and Chi, Y.R. (2013). *Angew. Chem. Int. Ed.* 52: 11134.
5 Sun, F.G., Sun, L.H., and Ye, S. (2011). *Adv. Synth. Catal.* 353: 3134.
6 Wang, G., Chen, X., Miao, G., Yao, W., and Ma, C. (2013). *J. Org. Chem.* 78: 6223.
7 De Sarkar, S., Grimme, S., and Studer, A. (2010). *J. Am. Chem. Soc.* 132: 1190.
8 Maki, B.E., Chan, A., Phillips, E.M., and Scheidt, K.A. (2007). *Org. Lett.* 9: 371.
9 Noonan, C., Baragwanath, L., and Connon, S.J. (2008). *Tetrahedron Lett.* 49: 4003.
10 Guin, J., De Sarkar, S., Grimme, S., and Studer, A. (2008). *Angew. Chem. Int. Ed.* 47: 8727.
11 Gao, Y., Ma, Y., Xu, C., Li, L., Yang, T., Sima, G., Fu, Z., and Huang, W. (2017). *Adv. Synth. Catal.* 360: 479–484.
12 Candish, L. and Lupton, D.W. (2011). *Org. Biomol. Chem.* 9: 8182.
13 Candish, L. and Lupton, D.W. (2010). *Org. Lett.* 12: 4836.
14 De Sarkar, S. and Studer, A. (2010). *Angew. Chem. Int. Ed.* 49: 9266.

15 Mo, J., Shen, L., and Chi, Y.R. (2013). *Angew. Chem. Int. Ed.* 52: 8588.
16 Zhu, Z.-Q., Zheng, X.-L., Jiang, N.-F., Wan, X., and Xiao, J.C. (2011). *Chem. Commun.* 47: 8670.
17 (a) Yao, C., Wang, D., Lu, J., Li, T., Jiao, W., and Yu, C. (2012). *Chem. Eur. J.* 18: 1914. (b) Yetra, S.R., Bhunia, A., Patra, A., Mane, M.V., Vanka, K., and Biju, A.T. (2013). *Adv. Synth. Catal.* 355: 1089.
18 Rong, Z.-Q., Jia, M.-Q., and You, S.-L. (2011). *Org. Lett.* 13: 4080.
19 Kaeobamrung, J., Mahatthananchai, J., Zheng, P., and Bode, J.W. (2010). *J. Am. Chem. Soc.* 132: 8810.
20 Li, G.-T., Gu, Q., and You, S.-L. (2015). *Chem. Sci.* 6: 4273.
21 Li, G.T., Li, Z.K., Gu, Q., and You, S.L. (2017). *Org. Lett.* 19: 1318.
22 Lu, Y., Tang, W., Zhang, Y., Du, D., and Lu, T. (2013). *Adv. Synth. Catal.* 355: 321.
23 Ni, Q., Song, X., Raabe, G., and Enders, D. (2014). *Chem. Asian J.* 9: 1535.
24 Romanov-Michailidis, F., Besnard, C., and Alexakis, A. (2012). *Org. Lett.* 14: 4906.
25 Xu, J., Zhang, W., Liu, Y., Zhu, S., Liu, M., Hua, X., Chen, S., Lu, T., and Du, D. (2016). *RSC Adv.* 6: 18601.
26 Wanner, B., Mahatthananchai, J., and Bode, J.W. (2011). *Org. Lett.* 13: 5378.
27 Zhao, L.L., Li, X.S., Cao, L.L., Zhang, R., Shi, X.Q., and Qi, J. (2017). *Chem. Commun.* 53: 5985.
28 Kravina, A.G., Mahatthananchai, J., and Bode, J.W. (2012). *Angew. Chem. Int. Ed.* 51: 9433.
29 Cheng, J.J., Huang, Z.J., and Chi, Y.R. (2013). *Angew. Chem. Int. Ed.* 52: 8592.
30 Zhang, H.-M., Jia, W.-Q., Liang, Z.-Q., and Ye, S. (2014). *Asian J. Org. Chem.* 3: 462.
31 Zhang, Z., Zeng, X., Xie, D., Chen, D., Ding, L., Wang, A., Yang, L., and Zhong, G. (2015). *Org. Lett.* 17: 5052.
32 Ni, Q., Xiong, J., Song, X., Raabe, G., and Enders, D. (2015). *Synlett* 26: 1465.
33 Yi, L., Zhang, Y., Zhang, Z.F., Sun, D., and Ye, S. (2017). *Org. Lett.* 19: 2286.
34 Zhu, S.Y., Zhang, Y., Wang, W., and Hui, X.P. (2017). *Org. Lett.* 19: 5380.
35 Chen, K.Q., Li, Y., Zhang, C.L., Sun, D.Q., and Ye, S. (2016). *Org. Biomol. Chem.* 14: 2007.
36 Zhao, Q., Han, B., Wang, B., Leng, H.-J., Peng, C., and Huang, W. (2015). *RSC Adv.* 5: 26972.
37 Wu, X., Liu, B., Zhang, Y., Jeret, M., Wang, H., Zheng, P., Yang, S., Song, B.-A., and Chi, Y.R. (2016). *Angew. Chem. Int. Ed.* 55: 12280.
38 Yetra, S.R., Mondal, S., Mukherjee, S., Gonnade, R.G., and Biju, A.T. (2016). *Angew. Chem. Int. Ed.* 55: 268.
39 Fang, C., Lu, T., Zhu, J., Sun, K., and Du, D. (2017). *Org. Lett.* 19: 3470.
40 Ryan, S.J., Candish, L., and Lupton, D.W. (2011). *J. Am. Chem. Soc.* 133: 4694.
41 Candish, L., Levens, A., and Lupton, D.W. (2014). *J. Am. Chem. Soc.* 136: 14397.
42 Candish, L., Forsyth, C.M., and Lupton, D.W. (2013). *Angew. Chem. Int. Ed.* 52: 9149.
43 Levens, A., Zhang, C., Candish, L., Forsyth, C.M., and Lupton, D.W. (2015). *Org. Lett.* 17: 5332.

44 Zhang, G., Xu, W., Liu, J., Das, D.K., Yang, S., Perveen, S., Zhang, H., Li, X., and Fang, X. (2017). *Chem. Commun.* 53: 13336.

45 Perveen, S., Zhao, Z., Zhang, G., Liu, J., Anwar, M., and Fang, X. (2017). *Org. Lett.* 19: 2470.

46 Levens, A., Ametovski, A., and Lupton, D.W. (2016). *Angew. Chem. Int. Ed.* 55: 16136.

47 (a) Jia, Q. and Wang, J. (2016). *Org. Lett.* 18: 2212. (b) Zhang, C.-L., Gao, Z.-H., Liang, Z.-Q., and Ye, S. (2016). *Adv. Synth. Catal.* 358: 2862. (c) Zhang, C.L. and Ye, S. (2016). *Org. Lett.* 18: 6408.

48 Bera, S., Daniliuc, C.G., and Studer, A. (2015). *Org. Lett.* 17: 4940.

49 Liang, Z.Q., Wang, D.L., Zhang, H.M., and Ye, S. (2015). *Org. Lett.* 17: 5140.

50 Lu, H., Zhang, J.-L., Liu, J.-Y., Li, H.-Y., and Xu, P.-F. (2017). *ACS Catal.* 7: 7797.

51 Candish, L. and Lupton, D.W. (2012). *J. Am. Chem. Soc.* 135: 58.

52 Bera, S., Samanta, R.C., Daniliuc, C.G., and Studer, A. (2014). *Angew. Chem. Int. Ed.* 53: 9622.

53 Mondal, S., Yetra, S.R., Patra, A., Kunte, S.S., Gonnade, R.G., and Biju, A.T. (2014). *Chem. Commun.* 50: 14539.

54 Biswas, A., De Sarkar, S., Tebben, L., and Studer, A. (2012). *Chem. Commun.* 48: 5190.

55 (a) Shen, L.T., Sun, L.H., and Ye, S. (2011). *J. Am. Chem. Soc.* 133: 15894. (b) Chen, X.Y., Xia, F., Cheng, J.T., and Ye, S. (2013). *Angew. Chem. Int. Ed.* 52: 10644. (c) Douglas, J., Churchill, G., and Smith, A.D. *Synthesis* 44: 2295.

56 Yao, C., Xiao, Z., Liu, R., Li, T., Jiao, W., and Yu, C. (2013). *Chem. Eur. J.* 19: 456.

57 Zhu, L., Yu, C., Li, T., Wang, Y., Lu, Y., Wang, W., and Yao, C. (2016). *Org. Biomol. Chem.* 14: 1485.

58 (a) Xu, J.F., Jin, Z.C., and Chi, Y.R. (2013). *Org. Lett.* 15: 5028. (b) Que, Y., Xie, Y., Li, T., Yu, C., Tu, S., and Yao, C. (2015). *Org. Lett.* 17: 6234.

59 Wang, M., Huang, Z., Xu, J., Chi, R.C. (2014). *J. Am. Chem. Soc.* 136: 1214.

60 Zhu, T.S., Zheng, P.C., Mou, C.L., Yang, S., Song, B.A., and Chi, Y.R. (2014). *Nat. Commun.* 5: 5027.

61 Biswas, A., De Sarkar, S., Frohlich, R., and Studer, A. (2011). *Org. Lett.* 13: 4966.

62 Zhang, C. and Lupton, D.W. (2017). *Org. Lett.* 19: 4456.

7

Recent Activation Modes in NHC Organocatalysis

Zhichao Jin [1], Xingkuan Chen [2], and Yonggui R. Chi [1,2]

[1] *Laboratory Breeding Base of Green Pesticide and Agricultural Bioengineering, Key Laboratory of Green Pesticide and Agricultural Bioengineering, Ministry of Education, Guizhou University, 2708 Huaxi Road, Huaxi District, Guiyang 550025, China*

[2] *Division of Chemistry & Biological Chemistry, School of Physical & Mathematical Sciences, Nanyang Technological University, 50 Nanyang Avenue, Singapore 637371, Singapore*

7.1 Introduction

Since the first application of thiazolium salts as the precatalysts in the benzoin condensation in 1943 [1], *N*-heterocyclic carbene (abbreviated as NHC or carbene) molecules have been found as efficient organic catalysts for certain chemical reactions involving various aldehydes. After 15 years, Breslow proposed a possible reaction pathway for the NHC-catalyzed benzoin condensation [2], which has been extensively applied in the future development of novel activation modes enabled by NHC catalysts. However, it was not until the past two decades has NHC organocatalysis seen extraordinary development in both chiral and nonchiral synthetic transformations [3]. Numerous activation modes enabled by NHCs have been established. The classical reaction modes such as benzoin reactions, Stetter reactions, HOMO-enolate, and enolate activation reactions using various aldehydes as the reaction substrates have been extensively introduced in the previous chapters. In this chapter, we will focus on some of the recently developed reaction modes mediated by NHC organic catalysts. One of the selected recent activation modes that we are going to introduce is the ester activation reactions catalyzed by NHCs, which contains the activation of the α-, β-, and γ-carbons of various carboxylic acid derivatives. The other one is the NHC-catalyzed reactions involving single-electron transferring (SET) processes (Figure 7.1).

7.2 Activation of Carboxylic Acid Derivatives

The activation of various aldehydes, enals, and ketenes has gained great success in the first decade of this century [3]. In the typical NHC catalytic processes starting from enals, ketenes, or functionalized aldehydes, various carboxylic

N-Heterocyclic Carbenes in Organocatalysis, First Edition. Edited by Akkattu T. Biju.
© 2019 Wiley-VCH Verlag GmbH & Co. KGaA. Published 2019 by Wiley-VCH Verlag GmbH & Co. KGaA.

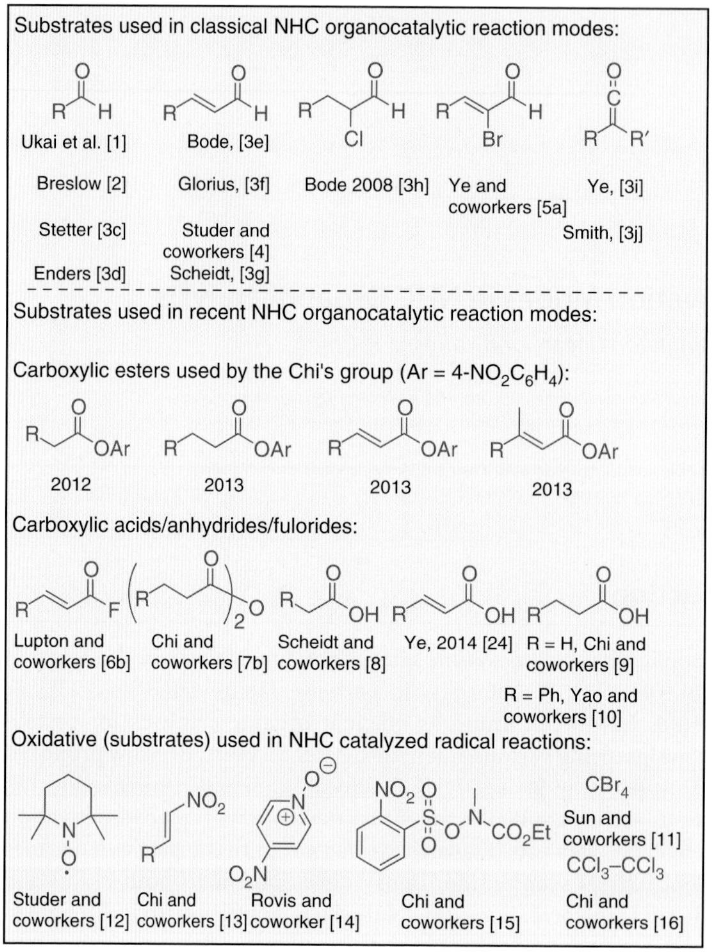

Figure 7.1 Typical substrates and oxidants used in NHC catalysis.

acid derivatives such as esters and amides are efficiently formed via certain key reaction intermediates, as discussed in the previous few chapters. In order to further explore and understand the rich chemistry in NHC catalytic reactions, the "backward" pathways starting from carboxylic esters enabled by NHC organic catalysts have been extensively studied since 2012 (Figure 7.2) [17].

7.2.1 α-Carbon Activation of Saturated Carboxylic Esters

The asymmetric organocatalytic reactions of enolate equivalents are a class of the most powerful transformations in organic synthesis [18]. Acyl azoliums are efficient reaction intermediates that can generate enolate equivalents [8]. Therefore, the generation of enolate equivalents from a stable carboxylic ester through acyl azolium intermediate enabled by NHC catalysts was initially explored [17]. The α-carbon of the carboxylic ester was activated as a nucleophilic reactive site

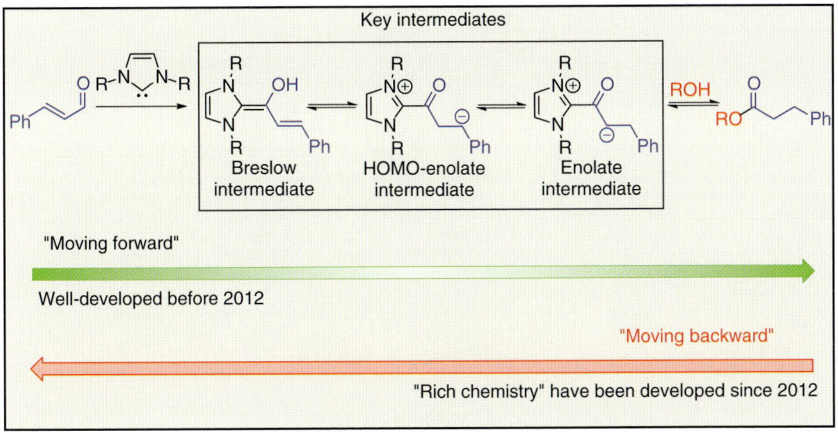

Figure 7.2 NHC-catalyzed reactions starting from aldehydes and esters.

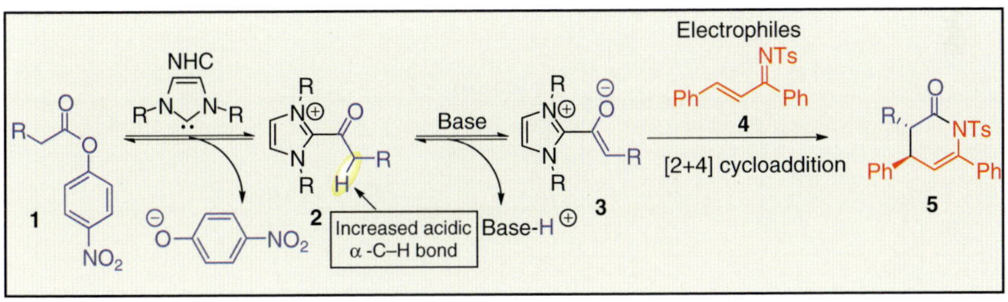

Figure 7.3 NHC-catalyzed generation of enolate intermediate from carboxylic esters.

in this transformation. As illustrated in Figure 7.3, a stable ester **1** with a good leaving group (OAr, Ar = 4-$NO_2C_6H_4$) was first attacked by the NHC organic catalyst, and a more reactive acyl azolium intermediate **2** with an increased acidic α-C—H bond was formed. Then, the α-C—H was deprotonated by an additional base to form the enolate intermediate **3**, which could then react with various electrophiles (Figure 7.3).

The activation of the α-C—H bond of the saturated carboxylic esters was first introduced by Chi and coworkers [17]. Both carboxylic esters with an aromatic (e.g. **6**) and an aliphatic (e.g. **9**) α-substituent worked well in the [2+4] cycloaddition reactions with α,β-unsaturated imine substrates (Figure 7.4) [17, 18]. The lactam products could be obtained in good yields with excellent diastereo- and enantioselectivities using the triazolium salts **7** and **10** as the NHC precatalysts.

As the critical synthetic starting materials for ester preparation, the carboxylic acids processed certain advantages over corresponding enals, functionalized aldehydes, and even the esters. Most of the acids are stable and readily available. Carboxylic acids are less prone to undergo enolate/aldol reactions than activated esters. Therefore, the direct use of carboxylic acids in NHC catalytic reactions is of great interest. The *in situ* activation of carboxylic acids using appropriate reagents is one of the most popular strategies used in NHC catalytic reactions. In 2014, Scheidt and coworkers reported the first NHC catalytic generation of acyl

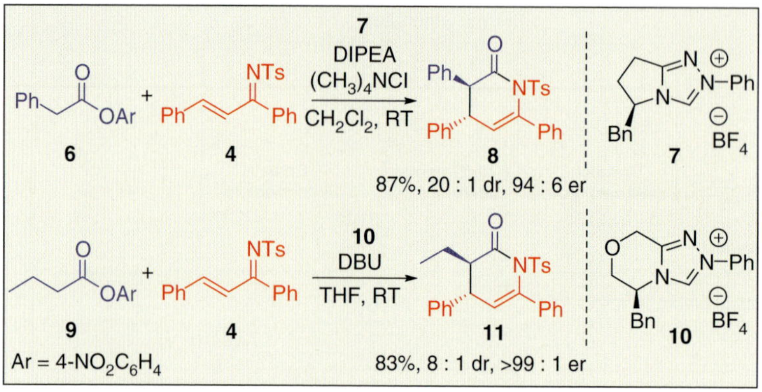

Figure 7.4 NHC-catalyzed α-activation of saturated esters.

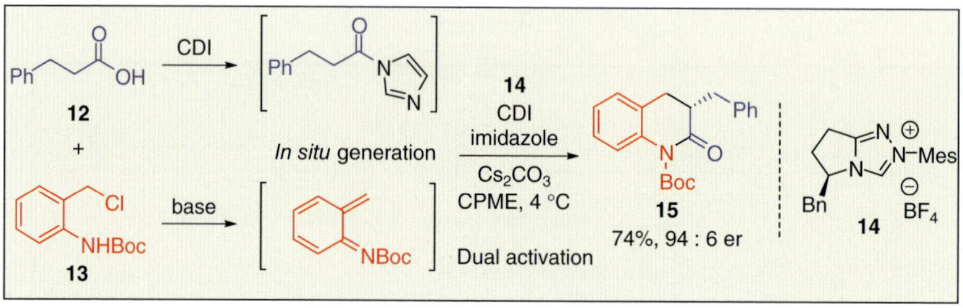

Figure 7.5 NHC-catalyzed α-activation of carboxylic acid.

azolium-derived enolate intermediates directly from saturated carboxylic acids [8]. CDI was used as the acid-activating reagent, and the saturated carboxylic acid **12** could be efficiently used in the α-carbon (enolate) activation reactions with aza-*o*-quinone methide (*in situ* formed from **13**) as the electrophilic reagent (Figure 7.5). Dihydroquinolone **15** was obtained as the product with good yield and er value.

Carboxylic acids and the related carbonyl compounds bearing two or more substituents at the α-carbons are important functional molecules. Derivatives of propionic acids bearing a stereogenic α-carbon center, such as ibuprofen and ketoprofen, are widely used as nonsteroidal anti-inflammatory drugs [19]. NHC-catalyzed activation of the carboxylic esters provide the chance to make this kind of molecules, especially in enantiomerically pure forms. In 2016, Chi and coworkers reported the first carbene-catalyzed dynamic kinetic resolution of α,α-disubstituted carboxylic esters with up to 99 : 1 er and 99% yield (Figure 7.6) [20]. The reaction starts with the addition of chiral carbene catalyst to the ester substrate **16a** (in racemic form) to give a mixture of two diastereomeric acyl azolium intermediates (**19** and **19′**) that exhibit different reactivities with an alcohol substrate **17** to achieve a chiral resolution. The carbene catalyst also played a dual role in catalyzing the isomerization (and racemization) of the ester substrate **16a** (via the acyl azolium **19′** and enolate intermediate **20**) and thus to achieve dynamic kinetic resolution of the ester substrate.

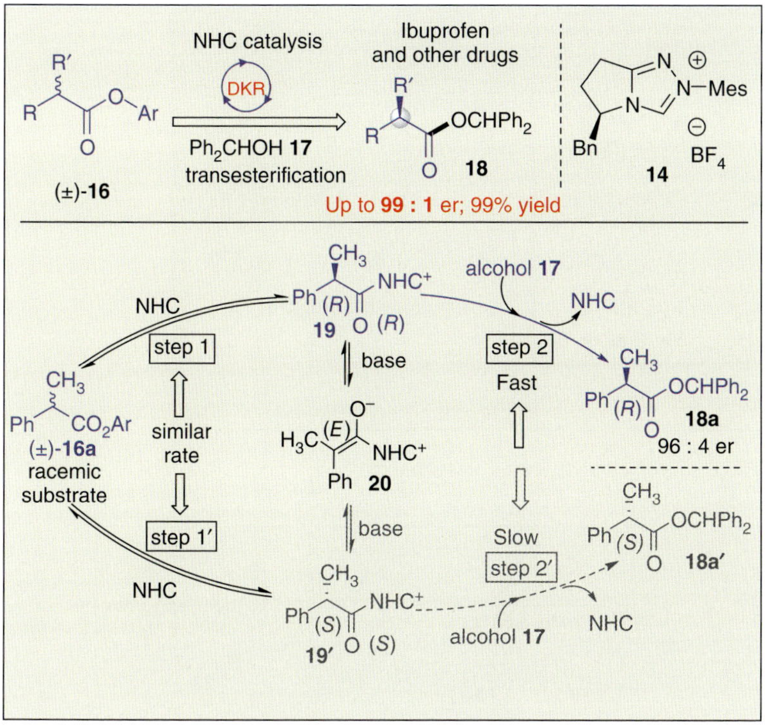

Figure 7.6 NHC-catalyzed DKR via α-activation of ester.

7.2.2 β-Carbon Activation of α,β-Unsaturated Carboxylic Compounds

α,β-Unsaturated acyl azoliums are among the most well-studied intermediates in NHC organocatalysis. This catalytic strategy provides a competent approach for the LUMO activation of α,β-unsaturated carbonyl compounds. To date, there have been several commonly used methods to generate NHC-bound α,β-unsaturated acyl azolium intermediates (Figure 7.7) [21]. For example, the NHC-catalyzed oxidative LUMO activation of α,β-unsaturated aldehydes with external oxidants have been pioneered by Studer and coworkers [4]. This strategy has been further developed by the Chi's group in 2013 with saturated aldehydes as the preoxidized reaction substrates [22]. The use of rationally designed unsaturated aldehydes with internal redox functionalities (e.g. ynals, α-bromo enals) has also been developed to deliver α,β-unsaturated acyl azoliums as the key reaction intermediates [21]. These methodologies with aldehydes as the synthetic starting materials have been introduced in previous chapters. Here, we mainly focus on the generation of α,β-unsaturated acyl azolium intermediates from various carboxylic acid derivatives.

The pioneering study on the generation of α,β-unsaturated acyl azolium intermediates using carboxylic acid derivatives is from Lupton's group in 2009 [6]. They successfully used α,β-unsaturated acyl fluorides **22** in the [3+3] cycloaddition reactions with enol ethers under NHC catalytic conditions (Figure 7.8). They proposed that the reaction was initiated by the addition of the carbene catalyst

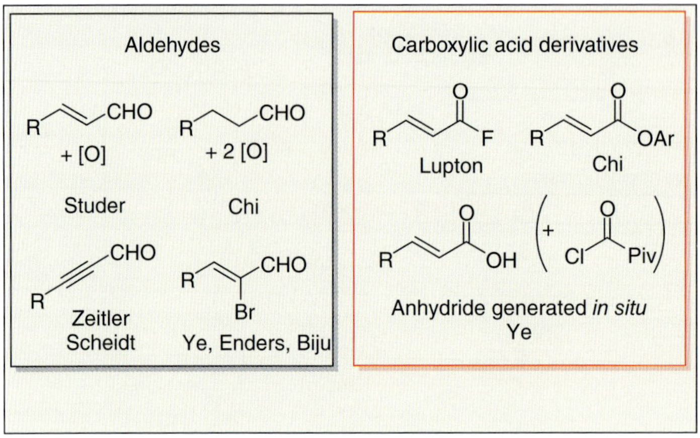

Figure 7.7 Precursors of α,β-unsaturated acyl azolium.

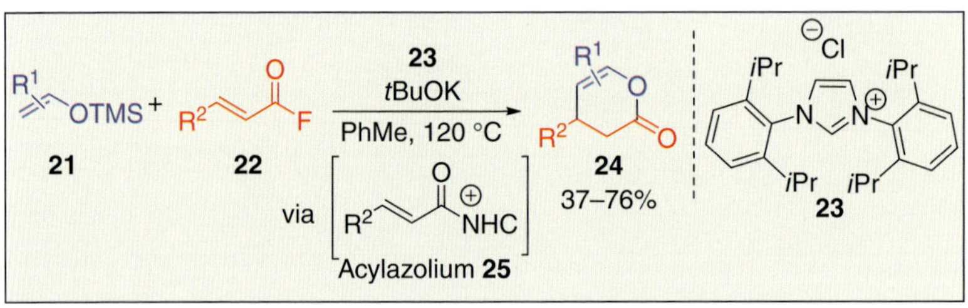

Figure 7.8 NHC-catalyzed β-activation of α,β-unsaturated acyl fluoride.

23 to the carbonyl group and followed by fluoride elimination to produce the α,β-unsaturated acyl azolium intermediate **25**. After the 1,4-addition reaction with the enol ether **21**, the afforded adduct went through an enolization and intramolecular lactone formation process to deliver dihydropyranone **24** as the final products in up to 76% yield.

In 2013, Chi and coworkers reported the [3+3] annulation reactions of α,β-unsaturated carboxylic esters with *N*-tosyl imines derived from acetophenones (Figure 7.9) [23]. The ester substrates **26** used in this methodology are easily prepared and are not sensitive to air and moisture. The reaction showed good tolerance to various substituents on both substrates, and the optically active lactam products were generally afforded in good-to-excellent yields and enantioselectivities. An α,β-unsaturated acyl azolium **30** is generated as the key intermediate from the reaction of the ester compound **26a** with the NHC catalyst. As proposed by Bode and coworkers [21b–d, f], an initial 1,2-addition reaction is likely to happen between enamide **31** that isomerized from the imine substrate **27a** and the α,β-unsaturated acyl azolium **30** to afford intermediate **32**. The key C—C bond between the two substrates is then formed through Claisen rearrangement, and the 1,4-addition intermediate **33** is afforded, which leads to the final product **29a** through lactam formation process and liberates the free NHC catalyst.

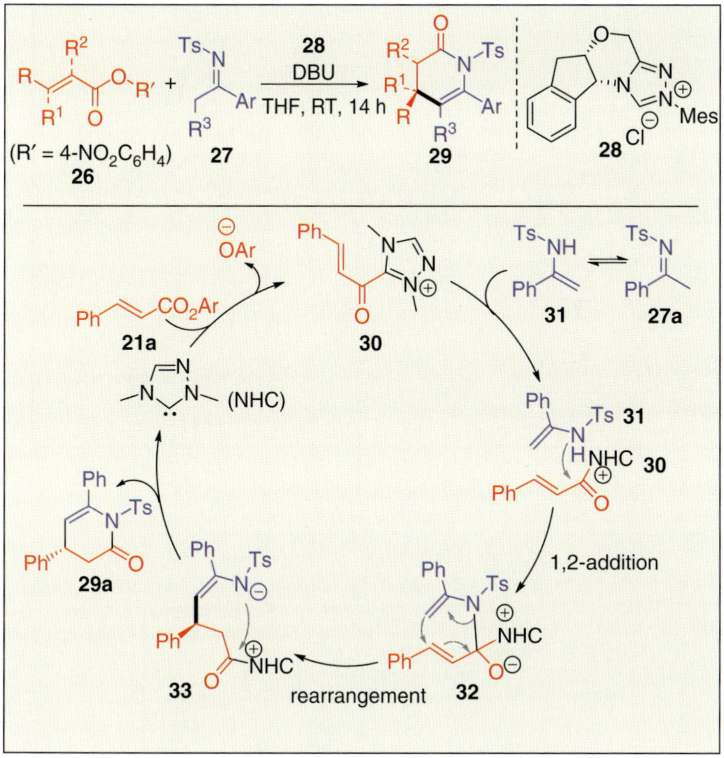

Figure 7.9 NHC-catalyzed β-activation of α,β-unsaturated esters.

Recently, Ye and coworkers further developed this strategy using the stable and inexpensive carboxylic acids as the reaction substrates to generate the α,β-unsaturated acyl azolium intermediates (Figure 7.10) [24]. This strategy was successfully developed as a general approach in the cycloaddition reactions between α,β-unsaturated carboxylic acids and various nucleophilic substrates (Figure 7.10, left part). For example, chiral pyrrolidinone **37** could be afforded in moderate yield with good diastereo- and enantioselectivities through this strategy with α-amino ketone **36** used as the synthetic starting material. [3+3] Cyclocondensations with different sulfamate-derived cyclic imine substrates **38** and **40** give dihydropyridinones **39** and **41** in good yields with excellent enantioselectivities [5]. The carboxylic acid **34** was first activated by PivCl to form mixed anhydride **35**, and the *in situ* formed mixed anhydride **35** could then be activated by NHC catalyst for further transformations. The authors proposed a Michael addition/lactam formation sequence for this reaction pathway. For example, α-amino ketone **36** could react with the α,β-unsaturated acylazolium intermediate **30** through Michael addition to give the 1,4-adduct **42**. Intermediate **42** could then go through proton transfer to afford intermediate **43** bearing a sulfonamide anion motif, which then delivers the final product **37** through intramolecular lactam formation process and liberates the free NHC catalyst to complete the catalytic cycle (Figure 7.10, right part).

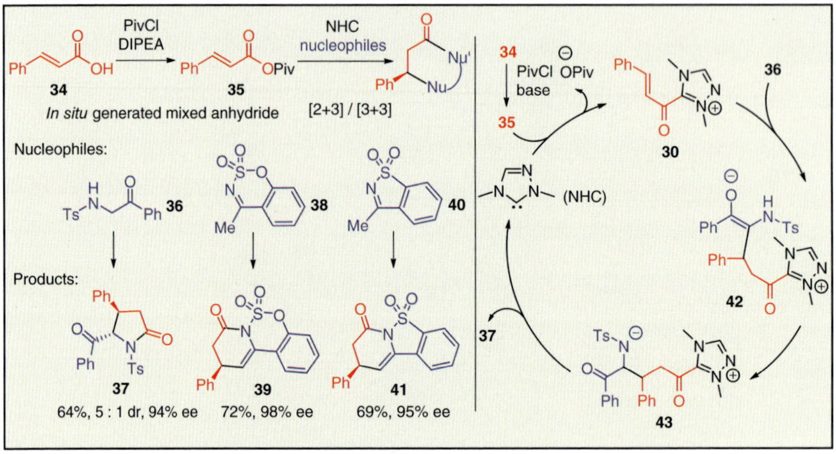

Figure 7.10 NHC-catalyzed β-activation of α,β-unsaturated carboxylic acids.

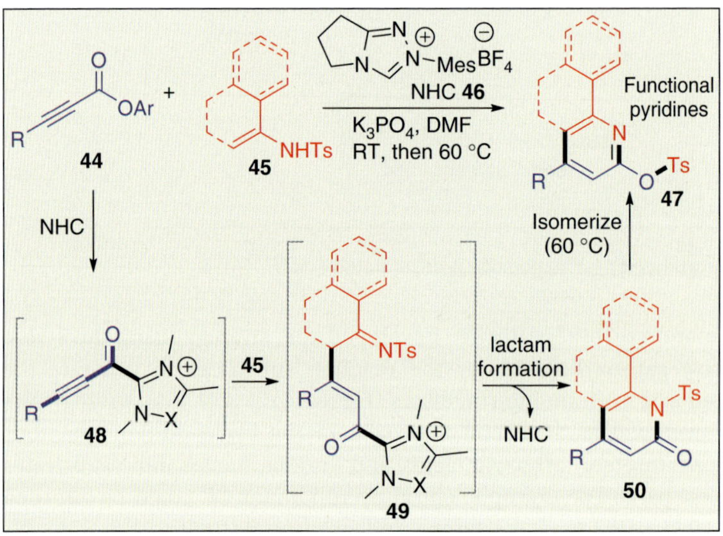

Figure 7.11 LUMO activation of alkyne esters by NHC.

Very recently, Chi and coworkers reported the NHC-catalyzed LUMO activation of α,β-unsaturated alkyne esters **44** (Figure 7.11) [25]. Functional pyridines with valuable synthetic applications could be effectively synthesized through this methodology. A previously unexplored acyl azolium intermediate **48** bearing a C—C triple bond was involved in the key step of this reaction. The products of hydroxyl pyridines **47** could be isolated in good yields and transferred to pyridine-based ligands that are frequently used in transition metal catalysis through simple protocols.

The NHC-catalyzed electrophilic β-carbon activation of saturated carboxylic esters **51** could also be achieved through oxidative strategies [26]. α,β-Unsaturated acyl azolium **30** could be afforded as the key reaction

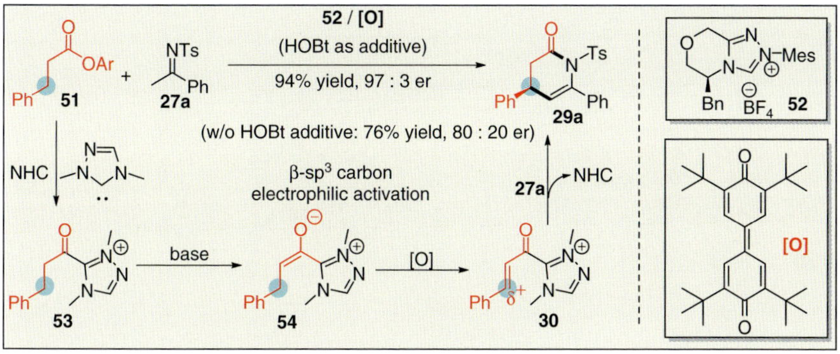

Figure 7.12 Oxidative LUMO activation of saturated esters.

intermediate from oxidation of saturated acyl azolium intermediate **53**, which react with commonly used nucleophiles (e.g. **27a**) to give lactam and lactone products in good-to-excellent yields and enantioselectivities (Figure 7.12).

7.2.3 Nucleophilic β-Carbon Activation of Saturated Carboxylic Esters

Functionalization of the α-carbons of carbonyl compounds has been one of the most basic activation modes in organic synthesis. The α-carbons are routinely activated as nucleophiles for a wide range of applications [27]. To obtain β-carbon functionalization of carbonyl compounds, α,β-unsaturated substrates are typically used. In this case, the β-carbons in α,β-unsaturated compounds can either behave as electrophiles via LUMO activation [28] or as nucleophiles via HOMO-enolate pathway [3]. Oxidative approaches have also been developed to turn saturated aldehydes into the corresponding α,β-unsaturated carbonyl derivatives as intermediates [22, 26]. However, strategies for directly activating the β-carbons of simple saturated carbonyl compounds as nucleophiles have long been elusive until the seminal reports from the Chi's group in 2013 [7]. A typical reaction pathway is depicted in Figure 7.13. NHC catalyst first attacks the ester substrate **51** to form the acyl azolium intermediate **53**, which is then deprotonated by additional bases (e.g. DBU) to afford the enolate zwitterion **54**. Intermediate **54** could then go through an intramolecular proton transfer and give the β-deprotonated intermediate **55**. Compound **55** could react with electrophiles such as imine substrates through nucleophilic addition/cyclization process (via intermediates **56** and **57**) to yield the final products such as lactams **58**.

Multiple electrophilic substrates could be used in this strategy (Figure 7.14). For example, enantio-enriched lactam **61** and lactone **63** products could be obtained through [3+2] cycloaddition reactions between the saturated carboxylic ester **59** and corresponding activated imine **60** and ketone **62** substrates. Chiral cyclopentene products **65** and **67** could be afforded from reactions with α,β-unsaturated ketones **64** and **66** through cascade annulation procedures [7].

As aryl substituents are generally required at the β-carbons of the ester substrates to enable effective reactions and acceptable yields, Chi and coworkers

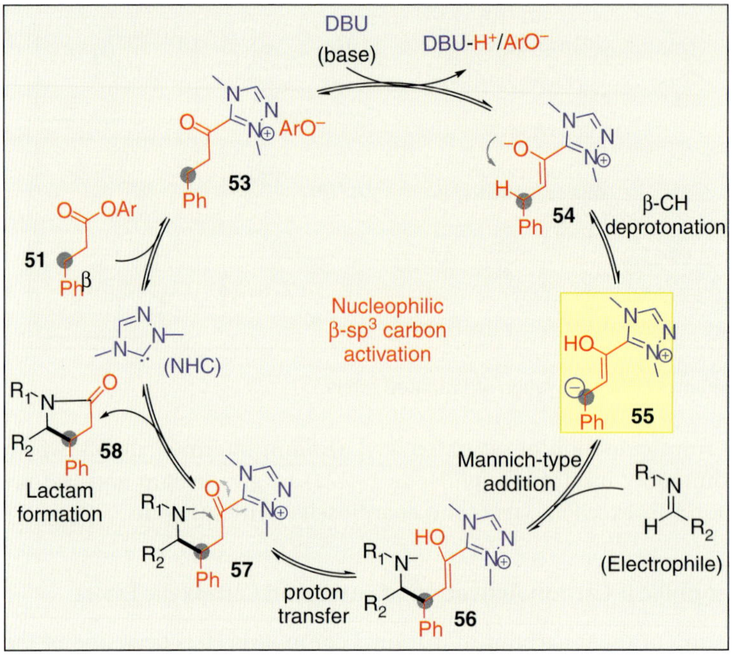

Figure 7.13 NHC-catalyzed nucleophilic β-activation of esters.

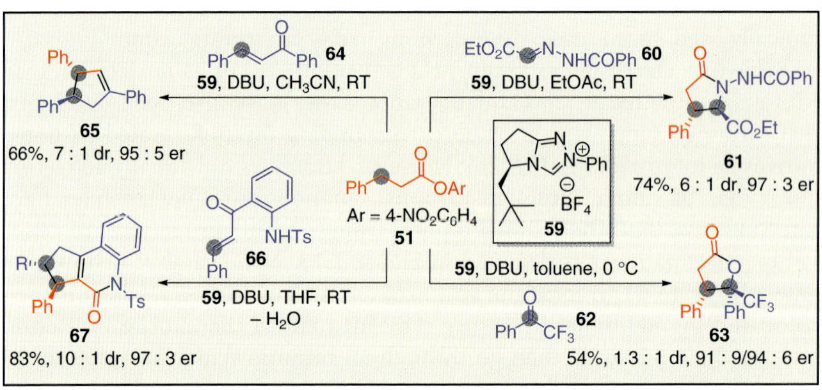

Figure 7.14 Examples of electrophiles used in NHC-catalyzed ester β-activation.

further developed this methodology and reported one solution to this challenging problem in 2014. The alternative approach involves an NHC-catalyzed β-activation of carboxylic anhydrides bearing β-alkyl substituents (e.g. butyric anhydride **67**) [13]. A variety of electrophiles (e.g. **68**, **69**, **70**, etc.) could be used as substrates to react with the β-carbon of the structurally simple and inexpensive aliphatic anhydrides with the corresponding products afforded in good yields and er values (Figure 7.15).

Having established the nucleophilic β-activation of both carboxylic esters and anhydrides, the direct use of carboxylic acid in this strategy has attracted

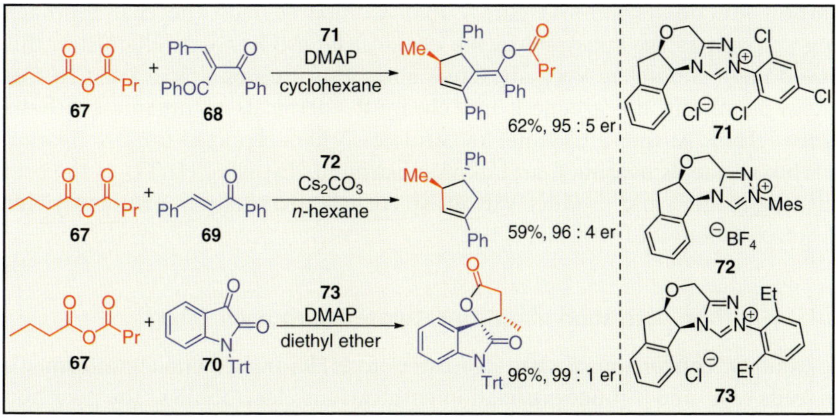

Figure 7.15 Nucleophilic β-carbon activation of simple anhydrides.

much interest. At this time, Chi and coworkers focused on the development of a stable, green, and inexpensive reagent as the smallest 3-C synthon in HOMO enolate chemistry (Figure 7.16) [9]. Therefore, propionic acid **74** has been extensively studied in the one-pot duel activation strategy by using EDCI as the dehydrating reagent for acid activation and NHC as the catalyst for the terminal

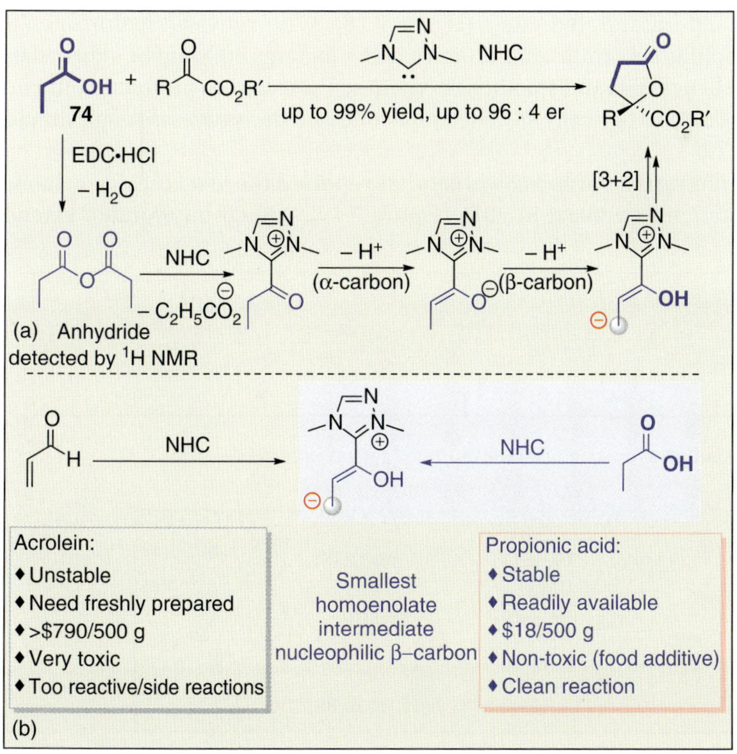

Figure 7.16 NHC-catalyzed β-activation of propionic acid.

β-carbon functionalization (Figure 7.16a). Compared with the enal approach using acrolein as the substrate in the classical HOMO enolate chemistry, this methodology should be very attractive since the propionic acid used in this strategy is among the greenest and cheapest chemicals in organic synthesis (Figure 7.16b). At the same time, Yao and coworkers reported the β-activation of carboxylic acids bearing β-aryl/alkyl substituents by using HATU as the acid activating reagent and produce various spirocyclic oxindoles in excellent yields and stereoselectivities [10].

7.2.4 γ-Carbon Activation of α,β-Unsaturated Carboxylic Esters

The catalytic conversion of saturated esters to NHC-bound enolate intermediates leads to α- and β-functionalization with various electrophiles. The β-carbon of α,β-unsaturated esters could also be activated through NHC-catalyzed LUMO activation pathway. Therefore, the direct NHC-enabled γ-carbon activation of α,β-unsaturated esters is reasonable. The key steps involved in this strategy are illustrated in Figure 7.17. Addition of the NHC catalyst to the α,β-unsaturated ester substrate **75** forms acyl azolium intermediate **76**. Deprotonation of **76** leads to the azolium dienolate **77** bearing a γ-nucleophilic reactive site (see resonance structures). Cycloaddition reactions of intermediate **77** with a variety of electrophiles produce corresponding heterocyclic compounds as the final products.

In 2013, Chi and coworkers reported the first direct organocatalytic γ-carbon activation of α,β-unsaturated esters (Figure 7.18). With activated hydrazone **78** as the electrophilic substrate, enantio-enriched δ-lactams **80** could be afforded in good-to-excellent yields and enantioselectivities. These lactam products could be easily transformed to optically enriched pipecolic acid derivatives through simple procedures [29].

It is worth noting that Ye and coworkers has reported the first access to catalytically generated vinyl enolate in 2011 (Figure 7.19). An α,β-unsaturated ketene

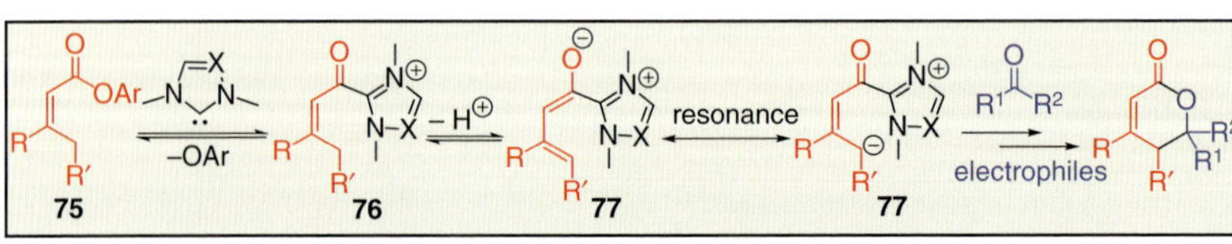

Figure 7.17 γ-Carbon activation of α,β-unsaturated carboxylic esters.

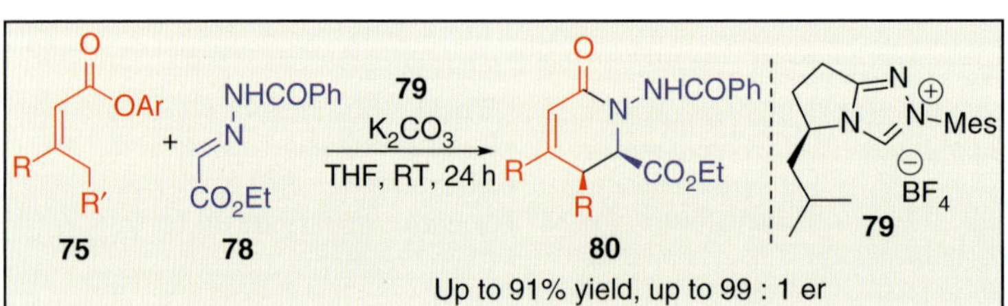

Figure 7.18 NHC-catalyzed γ-activation of esters.

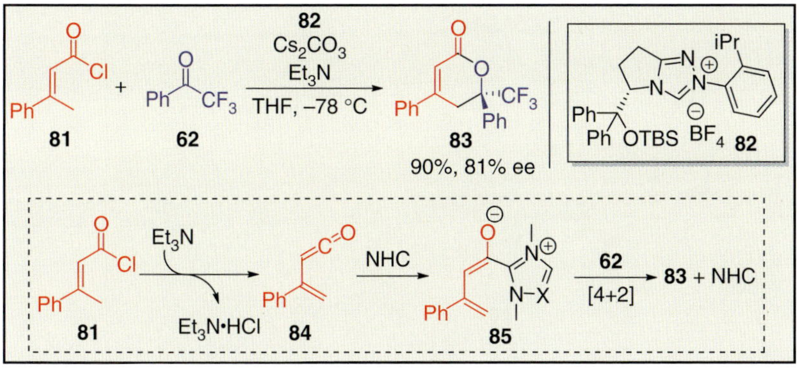

Figure 7.19 γ-Activation of acyl chloride by NHC.

intermediate **84** is generated from α,β-unsaturated acyl chloride **81** under basic conditions. The addition of NHC catalyst **82** to the *in situ* formed ketene intermediate **84** gives the key azolium dienolate intermediate **85**. Intermediate **85** then reacts with the activated electrophilic ketone **62** and affords [4+2] cycloaddition lactone product **83** with 90% yield and 81% ee value [5a].

7.3 Radical Reactions Catalyzed by NHC Organic Catalysts

As described in previous sections regarding both of the classical and recent catalytic reaction modes, NHC molecules have been developed as one class of the most robust organic catalysts for both chiral and nonchiral synthetic transformations. However, most of the NHC catalytic reactions proceed through electron-pair transferring procedures. SET processes have been much less developed in NHC organocatalytic reactions. Nevertheless, novel and unique transformations could be realized through NHC catalytic SET reactions. Therefore, enormous efforts are being devoted in this field, and new achievements are emerging continuously.

7.3.1 Lessons from Nature

Nature has always been the best teacher for organic chemists. Numerous catalytic reactions are processing continuously in natural environments [30]. One of the basic chemical reactions in living systems is the catalytic acylation of CoASH. Thiamine pyrophosphate **86** (vitamin B1, abbreviated as TPP) is the critical cofactor of the enzyme that catalyzes the oxidative decarboxylation reaction of pyruvate anion and transforms it to the acylation reagent to react with CoASH [31]. In 2001, Fontecilla-Camps and coworkers reported the crystal structure of the free radical intermediate of pyruvate ferredoxin oxidoreductase (abbreviated as $[Fe_4S_4]^{2+}$ or PFOR) and demonstrated that the CoAS-Ac synthesis by PFOR proceeds via SET processes (Figure 7.20) [32].

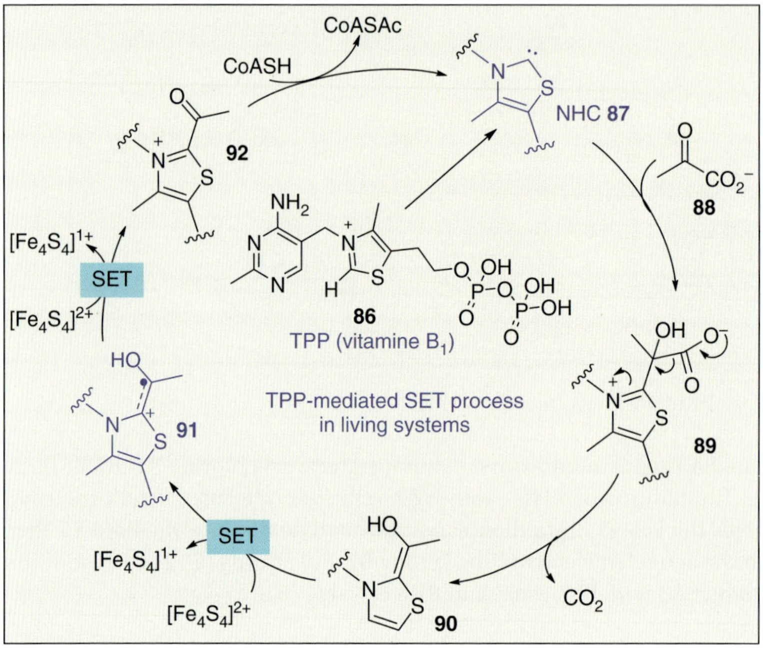

Figure 7.20 Acylation of CoASH catalyzed by TPP and PFOR.

The catalytic reactivity of the TPP molecule **86** is provided by the thiazolium motif in its chemical structure. The thiazolium motif could be deprotonated to generate the carbene catalyst **87**, which could react with pyruvate **88** to form intermediate **89**. Decarboxylation of intermediate **89** affords the Breslow intermediate **90**, which could be then oxidized by PFOR/$[Fe_4S_4]^{2+}$ through two SET procedures (via radical intermediate **91**) to afford the acyl azolium intermediate **92**. Compound **92** could react as the efficient acylation reagent with CoASH to yield CoAS-Ac, which is then used in further chemical and biological transformations.

7.3.2 Pioneering SET Reactions in NHC Organocatalysis

Inspired by the natural oxidative decarboxylation process of pyruvates enabled by TPP/PFOR system (Figure 7.20), the biomimetic oxidation of the Breslow intermediates through radical pathways was then intensively studied using various oxidants as the radical generators. Pioneering report in this area came from Studer's group in 2008 [12]. They described an NHC-catalyzed oxidative esterification of aldehydes using TEMPO as the single electron oxidant to mimic the PFOR in living systems. Both aromatic and aliphatic aldehydes worked well in this catalytic approach. The Breslow intermediate **97** generated from the reaction of the aldehyde **93** and NHC catalyst could be oxidized by TEMPO radical to afford the radical cation **98** and the anion of $TEMPO^-$. Deprotonation of the radical cation **98** by $TEMPO^-$ generated the electroneutral radical intermediate **99** and TEMPOH. Radical intermediate **99** could be further oxidized

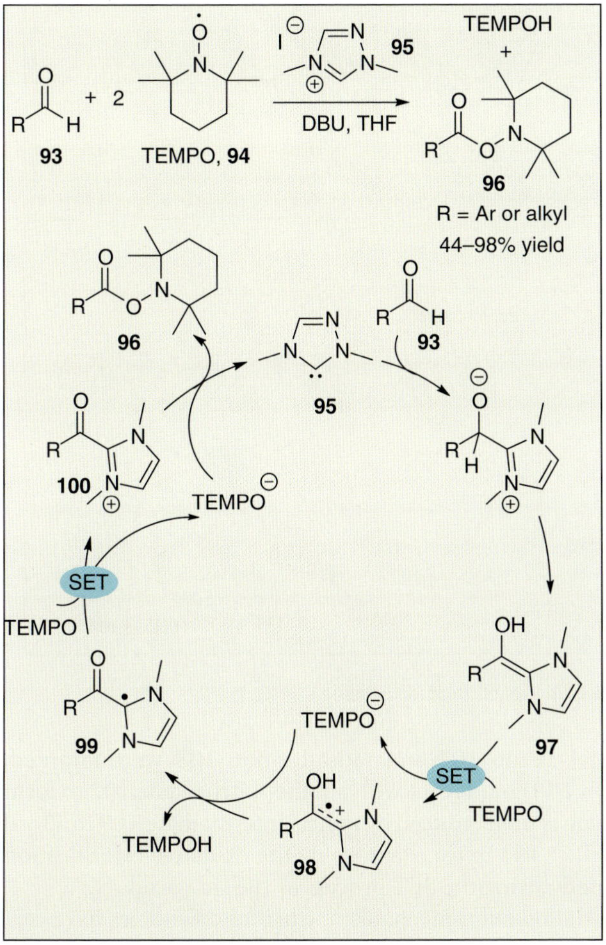

Figure 7.21 Biomimetic oxidations of aldehydes.

by another TEMPO radical to form the acyl azolium intermediate **100** and TEMPO⁻. Acylation of the afforded TEMPO⁻ by acyl azolium **100** generated the ester product **96** and releases the free NHC catalyst **95** for another catalytic cycle (Figure 7.21).

7.3.3 NHC-Catalyzed Reductive β,β-couplings of Nitroalkenes

One of the most challenging tasks in NHC-catalyzed SET reactions is to identify suitable oxidants that can selectively oxidize the Breslow intermediates to generate the reactive radical intermediates. Nitroalkenes were proved to be one class of efficient oxidants for this purpose. Reductive β,β-coupling reactions of nitroalkenes mediated by NHCs through SET pathways were reported by the Chi's group in 2014 (Figure 7.22) [15]. The Breslow intermediate **106** generated from aldehyde **101** with NHC catalyst **103** was oxidized by nitroalkene **102**,

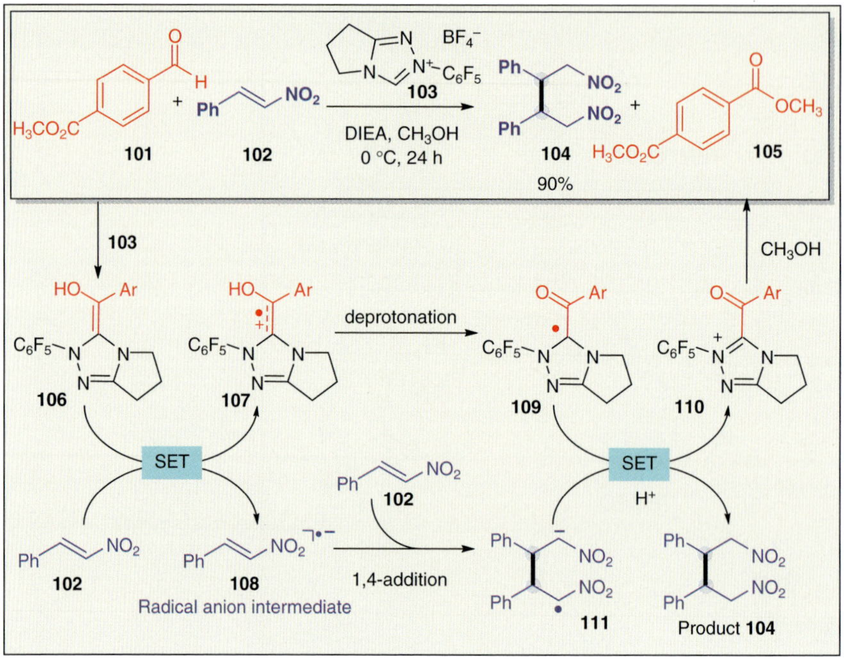

Figure 7.22 Reduction β,β-coupling reaction of nitroalkene.

and corresponding radical cation **107** and radical anion **108** were afforded, respectively. Radical anion **108** could react with another nitroalkene **102** to form the radical anion **111**, which is then reduced by radical intermediate **109** to afford the β,β-dimerization product **104** in excellent yield. The oxidative esterification product **105** is also afforded during the SET process as the side product.

To further understand the reaction mechanism, Chi and coworkers have carried out the electron paramagnetic resonance (abbreviated as EPR) experiment to detect and analyze the radical intermediates proposed in this process. As shown in Figure 7.23, signals from EPR spectra clearly expressed the formation of the radical anions **108a** and **108b**, which strongly supported the postulated reaction mechanism.

7.3.4 NHC-Catalyzed Benzylation of Electrophiles

4-Nitrobenzyl compounds have been well explored in radical chemistry. They could serve as efficient electron acceptors to selectively oxidize nucleophiles and generate radical intermediates for SET reactions [33]. Breslow intermediates are electron-rich specious and could serve as the reducing agent in the radical reactions involving 4-nitrobenzyl compounds. In 2016, Chi and coworkers reported the benzylation reaction of various electrophiles with 4-nitrobenzyl bromide via SET process using Breslow intermediate **106** as the critical electron donor (Figure 7.24) [34]. Nitro groups on the benzyl bromide **112** are crucial for the SET reaction process. The Breslow intermediate **106** could be oxidized by the 4-nitrobenzyl bromide **112** and yield the radical cation **107** and anion

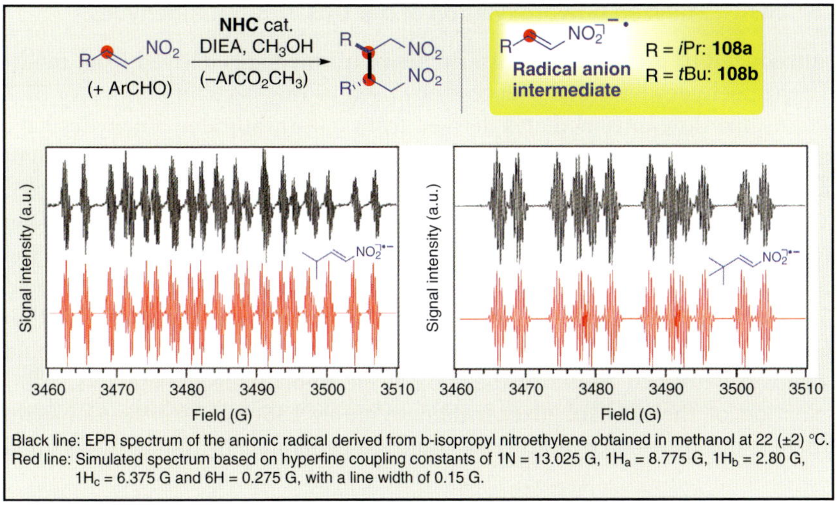

Figure 7.23 EPR experiments for evidence of radical intermediates.

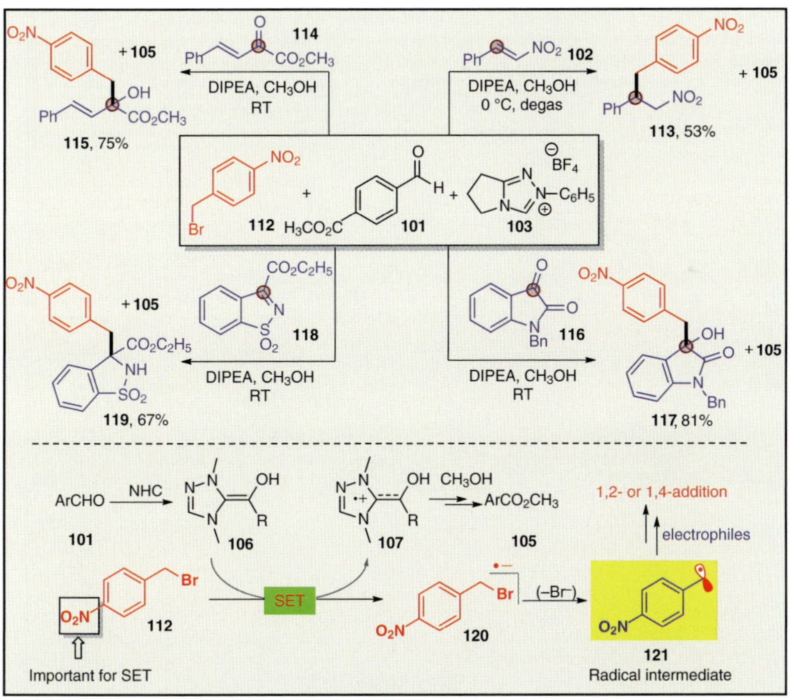

Figure 7.24 NHC-catalyzed benzylation of electrophiles.

120. Bromide anion could be released from radical anion intermediate **120** and electron neutral radical intermediate **121** is generated. Then, radical **121** could react with various electrophiles (e.g. **102, 114, 116, 118**), and the afforded intermediates could be further reduced by Breslow intermediate **106** to afford the corresponding 1,4- (e.g. **113**) and 1,2-addition (e.g. **115, 117, 119**) products in good yields.

7.3.5 NHC-Catalyzed β-hydroxylation of α,β-Unsaturated Aldehydes

As introduced in previous sections, Breslow intermediates derived from modified benzaldehydes have been successfully used as the electron donors for SET reactions. Breslow intermediates derived from α,β-unsaturated aldehydes have also played critical roles in the NHC-catalyzed SET reactions. In 2014, Rovis' group [14] and Chi's group [35] independently reported the NHC-catalyzed asymmetric β-hydroxylation reaction of α,β-unsaturated aldehydes using nitroarene derivatives as the oxidants via SET processes (Figure 7.25). The Breslow intermediate **125** generated from α,β-unsaturated aldehyde **122** could be oxidized by the nitroarene derivatives **123** and affords the radical cation **126** and radical anion **131**. The oxygen radical on the nitro group of **131** could be coupled with **126** to afford the β-oxidized anion intermediate **134**. After elimination of the nitrosobenzenesulfonate anion **135** and esterification with methanol, the β-hydroxyl ester **124** could be generated in moderate-to-good

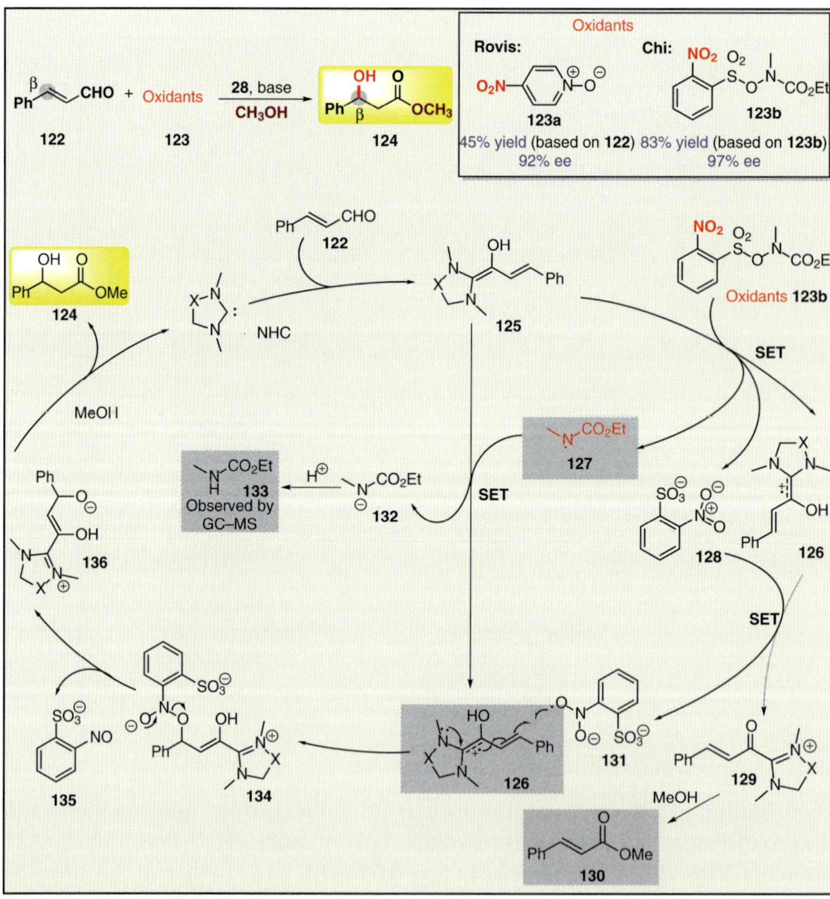

Figure 7.25 NHC-catalyzed β-hydroxylation of α,β-unsaturated aldehydes.

yield with excellent enantioselectivity. The α,β-unsaturated carboxylic ester **130** was isolated as the major side product in this reaction. Ethyl methylcarbamate **133** generated from radical intermediate **127** could be observed by GC analysis, which supported the proposed reaction pathway depicted in Figure 7.25.

7.3.6 Synthesis of Chiral 3,4-diaryl Cyclopentanones Through SET Process

Based on the mechanism of the SET process for the β-hydroxylation of the α,β-unsaturated aldehydes, Rovis' group reported the NHC-catalyzed enantioselective synthesis of 3,4-diaryl-substituted cyclopentanones from cinnamaldehydes through SET pathway (Figure 7.26) [36]. Once the radical cation intermediate **126** is generated from oxidation of the Breslow intermediate **125** by 4-nitropyridine N-oxide **123a**, it is then attacked by another Breslow intermediate **125** to form a new radical cation **140**. Intermediate **140** then undergoes a second SET process to the radical anion **139** and gives acyl azolium **143**. One equivalent of the NHC catalyst could be released through the intramolecular attack by the enolate azolium motif in **143** and the cyclopentanone intermediate **144** is formed. This cyclopentanone **144** is then attacked by water to liberate the second equivalent of the NHC catalyst and produce the β-ketoacid **145**. Decarboxylation of the β-ketoacid **145** affords the 3,4-diphenylcyclopentanone **138** as the final product.

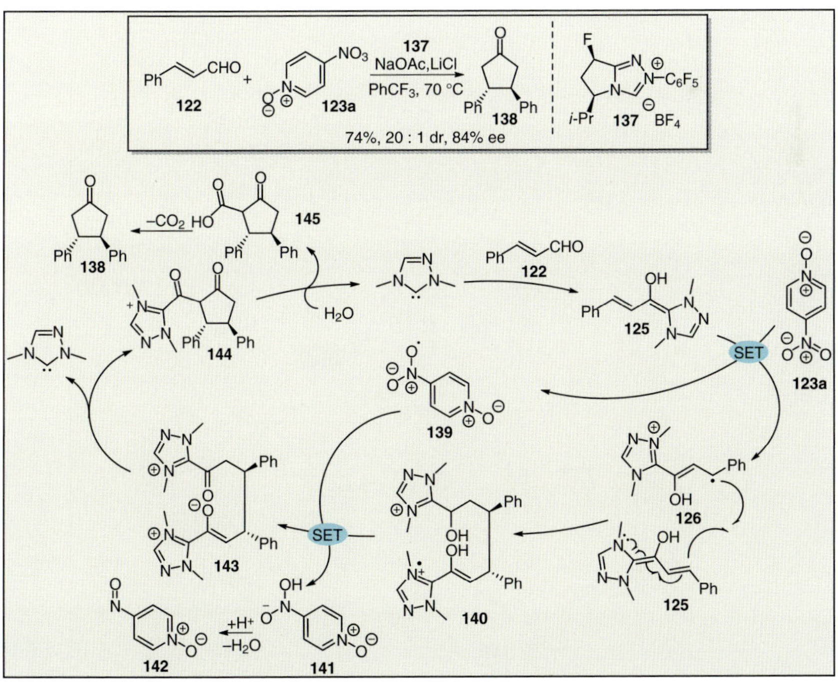

Figure 7.26 Synthesis of 3,4-diaryl cyclopentanones through NHC-catalyzed SET reactions.

7.3.7 Polyhalides as Oxidants for NHC-Catalyzed Radical Reactions

Having successfully established the SET reaction modes with NHCs as the critical organic catalysts, scientists began to make efforts to identify structurally simple and inexpensive oxidants for this kind of transformations. Polyhalides are one type of the successful examples that have been used in a variety of NHC-catalyzed SET reactions. They are introduced and extensively studied by the groups of Sun and Chi independently in the years of 2016 and 2017. Meanwhile, they are also the pioneering examples that use carbon-centered radicals as the external SET partners in NHC-catalyzed SET reactions.

In 2016, Sun and coworkers reported an efficient γ-dihalomethylenation of enals by an intermolecular C—C bond formation through NHC-catalyzed SET process using CBr_4 as the sole oxidative reactant (Figure 7.27) [11]. A plausible

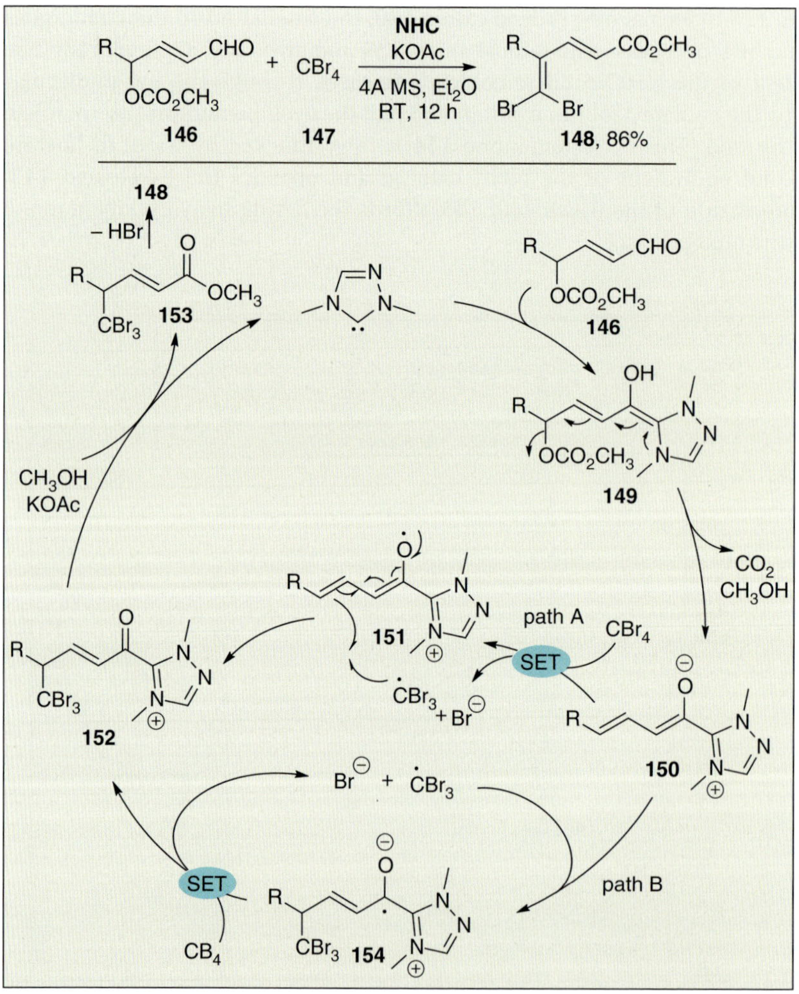

Figure 7.27 NHC catalyzed γ-dihalomethylenation of enals.

reaction mechanism was proposed by the authors as depicted in Figure 7.26. Breslow intermediate **149** that generated through the addition of the NHC catalyst to enal substrate **146** could undergo elimination and deprotonation to form the vinylogous enolate intermediate **150**. Intermediate **150** may react with CBr_4 to form the acyl azolium species **152** through two possible pathways. In path A, the enolate zwitterion **150** could be oxidized by CBr_4 through SET process and form the radical cation intermediate **151** with concomitant formation of the radical $Br_3C^•$ and bromide anion. The activated acyl azolium **152** could be then afforded through radical–radical recombination of **151** and $Br_3C^•$. Alternatively, as illustrated in path B, the vinylogous enolate **150** could be attacked by the electrophilic radical $Br_3C^•$ at the γ-position to form a new C—C bond and gives the radical zwitterion **154**. Further oxidation of **154** by CBr_4 through SET process gives the same acyl azolium intermediate **152**. Acyl substitution of **152** by methanol under basic condition regenerates the free NHC catalyst and delivers the γ-tribromomethalated intermediate **153**, which finally leads to the observed γ-dibromomethylenated product **148** by the elimination of 1 equiv of HBr.

Chi and coworkers applied C_2Cl_6 as the efficient oxidant in the NHC-catalyzed oxidative reactions of various aldehydes (Figure 7.28) [16]. The reactions

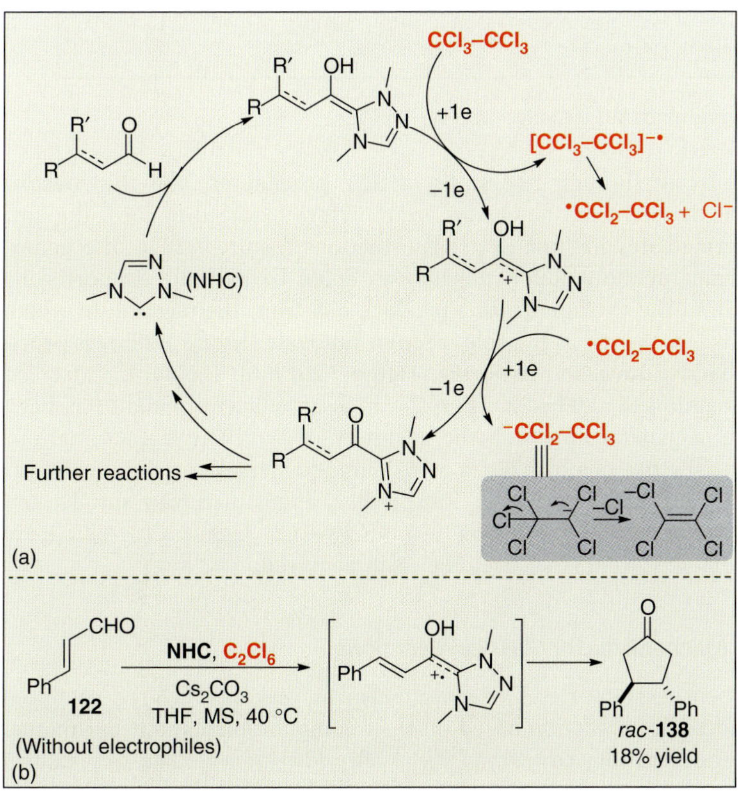

Figure 7.28 C_2Cl_6 as oxidant in NHC organocatalysis.

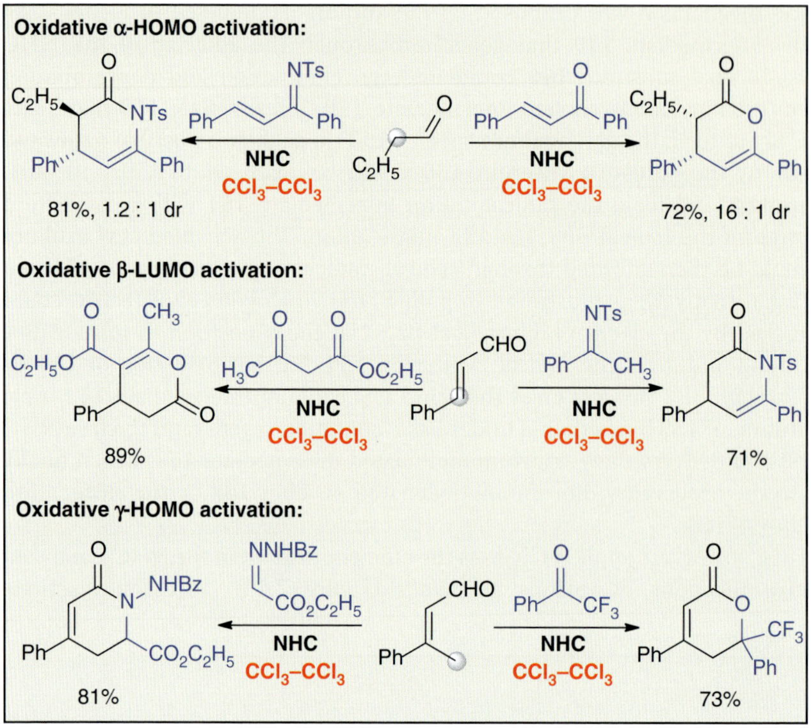

Figure 7.29 Scope of the SET oxidative process with C_2Cl_6.

are believed to go through two steps of SET procedures, and the Breslow intermediates derived from aldehydes are therefore oxidized to form the acyl azolium intermediates for further transformations (Figure 7.28a). The generation of radicals in the reaction mechanism could be partially supported by the formation of the cyclopentanone product **138** from the reaction in which cinnamaldehyde was used as the sole reaction substrate (Figure 7.28b), as previously described by Rovis and coworker (Figure 7.26) [36]. This NHC-catalyzed oxidative activation of aldehydes using C_2Cl_6 as an efficient oxidant has been developed as a general approach for a variety of synthetic transformations (Figure 7.29). The use of polyhalides as oxidants possesses certain advantages as polyhalides are readily available, inexpensive, and structurally simple, and the corresponding reduced oxidants (e.g. $CCl_2{=}CCl_2$ from $CCl_3{-}CCl_3$) are essentially volatile solvents that could be easily removed.

7.3.8 New Mechanisms for Classical Reactions

In 2015, Rehbein et al. reported a new mechanism for the classical NHC-catalyzed benzoin condensation reaction based on their computational and experimental studies on the reaction pathway [37]. A Breslow-intermediate-like radical **158** was observed in the reaction process through EPR studies. Results from multiaspects finally led to a revised catalytic cycle that incorporates the radical pair (**157** and **158**) as a second key intermediate in the NHC-catalyzed benzoin

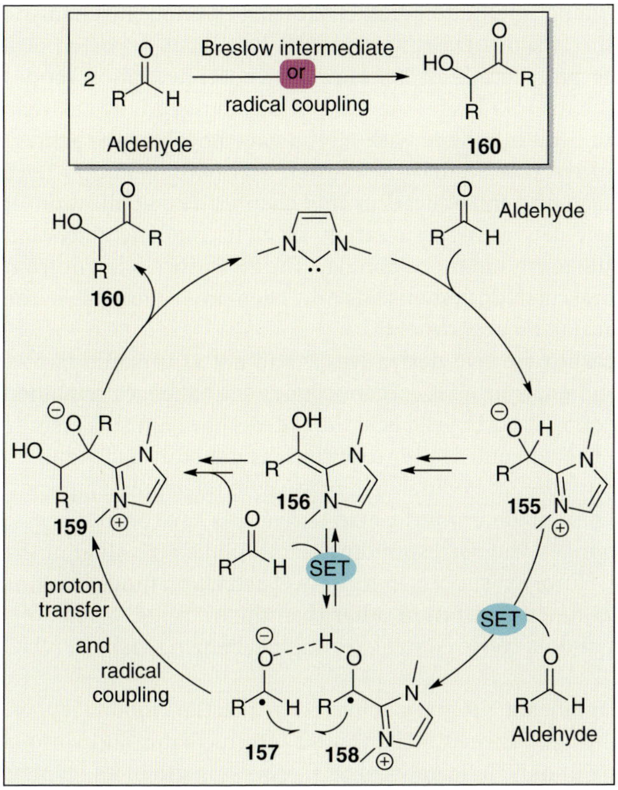

Figure 7.30 SET process for benzoin condensation.

condensation reaction (Figure 7.30). The zwitterion intermediate **155** generated from the addition of the aldehyde and the NHC catalyst could be partially oxidized by another equivalent of the aldehyde through SET process and gives the radical pair of **157** and **158**. Radical coupling reaction between intermediates **157** and **158** leads to the azolium zwitterion **159**, which then delivers the final benzoin product **160** and liberates the NHC catalyst for additional catalytic cycles. It is worth noting that the radical pair intermediates of **157** and **158** could also be generated from the Breslow intermediate **156**, which has been proved by the authors through control experiments, and therefore, the reaction mechanisms involving both paired electron transfer steps and SET radical recombination steps could be united to formulate a new catalytic cycle as depicted in Figure 7.30.

7.4 Summary and Outlook into the Future NHC Organocatalysis

NHC organocatalysis has seen tremendous developments in the last two decades. The unique ability of NHCs of reversing the polarity of certain functional groups such as aldehydes and enals has made NHC catalysis one of the most popular research fields in synthetic methodologies. Following limited mechanistic principles, reactions using various starting materials have been reported at a breathtaking pace in the past decade or so. However, novel activation modes in NHC organocatalysis have also been emerging continuously. Many catalytic transformations that had been infeasible with NHCs have now become achievable. Two examples of them that we have witnessed and participated in have been typically summarized and introduced in this chapter. The application of carboxylic acids and their derivatives in a variety of NHC catalytic activation modes has been becoming more and more popular. The SET process learnt from the natural existed enzyme catalytic systems has now been proved feasible in the laboratory NHC organocatalytic experiments.

However, the use of carboxylic acid derivatives is still not as versatile as corresponding aldehydes and enals in NHC organocatalysis. Challenges remained in several aspects. For example, carboxylic acid derivatives have not successfully been used in the simple benzoin and Stetter reactions, as it is still challenging to activate the carbonyl groups of carboxylic acid derivatives as nucleophilic species. As the β-carbons of the saturated carboxylic acid derivatives have been successfully activated as nucleophiles, the corresponding electrophilic activation of the α-carbons is reasonable. However, success in this novel activation mode remains uncovered. Remote C–H functionalization other than the α-, β-, or γ-positions of carbonyl compounds and its steric control are still challenging tasks in NHC organocatalysis.

Compared with the significant progress achieved in reaction modes involving paired electron pathways between NHC-bound intermediates and various substrates, the development of NHC-catalyzed SET process is still in its infancy. Limited examples in this direction have been reported until now. Typically, the Breslow intermediates generated from NHCs and aldehydes are electron-rich species and could be selectively oxidized by certain oxidants through SET process and give radical ions for further transformations. Developing suitable methods to trap and harness these reactive radical species in the reaction pathways for synthetic purpose represents one of the most challenging but interesting tasks in NHC organocatalysis. Efforts in this research field are being made, and fruitful results are believed to spring up in the near future.

References

1 Ukai, T., Tanaka, R., and Dokawa, T. (1943). *J. Pharm. Soc. Jpn.* 63: 296.
2 Breslow, R. (1958). *J. Am. Chem. Soc.* 80: 3719.
3 For selected reviews, see: (a) Flanigan, D.M., Romanov-Michailidis, F., White, N.A., and Rovis, T. (2015). *Chem. Rev.* 115: 9307. (b) Hopkinson, M.N., Richter, C., Schedler, M., and Glorius, F. (2014). *Nature* 510: 485; for selected examples, see: (c) Stetter, H., and Schreckenberg, M. (1973). *Angew. Chem. Int. Ed.* 12: 81. (d) Enders, D. *In Stereoselective Synthesis* Springer-Verlag: Heidelberg, Germany, (1993). p 63. (e) Sohn, S. S., Rosen, E. L., and Bode, J. W. (2004). *J. Am. Chem. Soc.* 126: 14370. (f) Burstein, C. Glorius, F. (2004). *Angew. Chem. Int. Ed.* 43: 6205. (g) Samanta, R. C., Maji, B., De Sarkar, S.,

Bergander, K., Fröhlich, R., Mück-Lichtenfeld, C., Mayr, H., and Studer, A. (2012). *Angew. Chem. Int. Ed.* 51: 5234. (h) He, M., Beahm, B. J., and Bode, J. W. (2008). *Org. Lett.* 10: 3817. (i) Zhang, Y.-R., He, L., Wu, X., Shao, P.-L., and Ye, S. (2007). *Org. Lett.* 10: 277. (j) Duguet, N., Campbell, C. D., Slawin, A. M. Z., and Smith, A. D. (2008). *Org. Biomol. Chem.* 6: 1108.

4 De Sarkar, S., Grimme, S., and Studer, A. (2010). *J. Am. Chem. Soc.* 132: 1190.

5 (a) Shen, L.-T., Shao, P.-L., and Ye, S. (2011). *Adv. Synth. Catal.* 353: 1943. (b) Wang, X.-N., Zhang, Y.-Y., and Ye, S. (2010). *Adv. Synth. Catal.* 352: 1892.

6 (a) Ryan, S.J., Candish, L., and Lupton, D.W. (2011). *J. Am. Chem. Soc.* 133: 4694. (b) Ryan, S.J., Candish, L., and Lupton, D.W. (2009). *J. Am. Chem. Soc.* 131: 14176.

7 (a) Fu, Z., Xu, J., Zhu, T. et al. (2013). *Nat. Chem.* 5: 835. (b) Fu, Z., Jiang, K., Zhu, T.S. et al. (2014). *Angew. Chem. Int. Ed.* 53: 6506.

8 Lee, A., Younai, A., Price, C.K. et al. (2014). *J. Am. Chem. Soc.* 136: 10589.

9 Jin, Z., Fu, Z., Jiang, K. et al. (2015). *Chem. A. Eur. J.* 21: 9360.

10 Xie, Y., Yu, C., Li, T. et al. (2015). *Chem. Eur. J.* 21: 5355.

11 Yang, W., Hu, W., Dong, X. et al. (2016). *Angew. Chem. Int. Ed.* 128: 16015.

12 Guin, J., De Sarkar, S., Grimme, S., and Studer, A. (2008). *Angew. Chem. Int. Ed.* 47: 8727.

13 Jin, Z., Chen, S., Wang, Y. et al. (2014). *Angew. Chem. Int. Ed.* 53: 13506.

14 White, N.A. and Rovis, T. (2014). *J. Am. Chem. Soc.* 136: 14674.

15 Du, Y., Wang, Y., Li, X. et al. (2014). *Org. Lett.* 16: 5678.

16 Wu, X., Zhang, Y., Wang, Y. et al. (2017). *Angew. Chem. Int. Ed.* 56: 2942.

17 Hao, L., Du, Y., Lv, H. et al. (2012). *Org. Lett.* 14: 2154.

18 (a) Douglas, J., Churchill, G., and Smith, A.D. (2012). *Synthesis* 44: 2295. (b) Hao, L., Chen, S., Xu, J. et al. (2013). *Org. Lett.* 15: 4956. (c) Chen, S., Hao, L., Zhang, Y. et al. (2013). *Org. Lett.* 15: 5822.

19 (a) Harrison, I.T., Lewis, B., Nelson, P. et al. (1970). *J. Med. Chem.* 13: 203. (b) Shen, T.Y. (1972). *Angew. Chem. Int. Ed. Engl.* 11: 460. (c) Lednicer, D. and Mitscher, L.A. (1977). *The Organic Chemistry of Drug Synthesis*, vol. 1. New York: Wiley.

20 Chen, X., Fong, J.Z.M., Xu, J. et al. (2016). *J. Am. Chem. Soc.* 138: 7212.

21 (a) Zeitler, K. (2006). *Org. Lett.* 8: 637. (b) Kaeobamrung, J., Mahatthananchai, J., Zheng, P., and Bode, J.W. (2009). *J. Am. Chem. Soc.* 132: 8810. (c) Daeobamrung, J., Mahatthananchai, J., Zhang, P., and Bode, J.W. (2010). *J. Am. Chem. Soc.* 132: 8810. (d) Mahatthananchai, J., Zheng, P., and Bode, J.W. (2011). *Angew. Chem. Int. Ed.* 50: 1673. (e) Zhu, Z.-Q. and Xiao, J.-C. (2010). *Adv. Synth. Catal.* 352: 2455. (f) Lyngvi, E., Bode, J.W., and Schoenebeck, F. (2012). *Chem. Sci.* 3: 2346.

22 Mo, J., Shen, L., and Chi, Y.R. (2013). *Angew. Chem. Int. Ed.* 52: 8588.

23 Cheng, J., Huang, Z., and Chi, Y.R. (2013). *Angew. Chem. Int. Ed.* 52: 8592.

24 Chen, X.-Y., Gao, Z.-H., Song, C.-Y. et al. (2014). *Angew. Chem. Int. Ed.* 53: 11611.

25 Mou, C., Wu, J., Huang, Z. et al. (2017). *Chem. Commun.* 53: 13359.

26 Liu, B., Wang, W., Huang, R. et al. (2013). *Org. Lett.* 20: 260.

27 List, B., Lerner, R.A., and Barbas, C.F. III, (2000). *J. Am. Chem. Soc.* 122: 2395.

28 Ahrendt, K.A., Borths, C.J., and MacMillan, D.W.C. (2000). *J. Am. Chem. Soc.* 122: 4243.
29 Xu, J., Jin, Z., and Chi, Y.R. (2013). *Org. Lett.* 15: 5028.
30 Ragsdale, S.W. (2003). *Chem. Rev.* 103: 2333.
31 Mansoorabadi, S.O., Seravalli, J., Furdui, C. et al. (2006). *Biochemistry* 45: 7122.
32 Chabriere, E., Vernède, X., Guigliarelli, B. et al. (2001). *Science* 294: 2559.
33 Wang, Y., Wu, X., and Chi, Y.R. (2017). *Chem. Commun.* 53: 11952.
34 Li, B.-S., Wang, Y., Proctor, R.S.J. et al. (2016). *Nat. Commun.* 7: 12933.
35 Zhang, Y., Du, Y., Huang, Z. et al. (2015). *J. Am. Chem. Soc.* 137: 2416.
36 White, N.A. and Rovis, T. (2015). *J. Am. Chem. Soc.* 137: 10112.
37 Rehbein, J., Ruser, S.-M., and Phan, J. (2015). *Chem. Sci.* 6: 6013.

8

N-Heterocyclic Carbene-Catalyzed Reactions via Azolium Enolates and Dienolates

Zhao-Fei Zhang, Chun-Lin Zhang, and Song Ye

Institute of Chemistry Chinese Academy of Sciences, Laboratory of Molecular Recognition and Function, No. 2, 1st North Street, Zhongguancun, Beijing 100190, P.R. China

8.1 Introduction

The catalytic generation of chiral enolates and their equivalents is a vital strategy in organic synthesis. Chiral metal complexes [1], which display basic and/or acidic properties, have been successfully employed to generate the chiral enolate from carbonyl compounds. In the area of organocatalysis, chiral secondary amines [2] have been used as the catalysts to generate enamines from aldehydes and ketones. In addition, cinchona alkaloids [3] and planar chiral 4-*N*,*N*-dimethylaminopyridine (DMAP) derivatives [4] featured as nucleophilic catalysts furnishing excellent enantioselectivities for a range of reactions.

The *N*-heterocyclic carbene (NHCs)-catalyzed generation of azolium enolates has also been well developed from various organic compounds (Scheme 8.1). α-Functionalized aldehydes, including α-chloroaldehydes, α-epoxyaldehydes, and enals, are the most applied starting materials for the NHC-catalyzed transformation. In addition, ketenes have also been demonstrated as suitable reagents for the transformation. Recently, several carboxylic derivatives have also been successfully employed for the reaction.

The generated azolium enolates can be protonated and then work as acylation reagents to give carbonyl derivatives, such as esters and amides. More importantly, the azolium enolates can undergo a range of formal [2+*n*] annulation reactions to give a wide variety of heterocycles.

8.2 Azolium Enolates from α-Functionalized Aldehydes

8.2.1 Synthesis of Carboxylic Compounds

The first reported NHC-catalyzed reaction involving azolium enolate intermediate was the redox reaction α-functionalized aldehydes. In 2004, Bode and coworkers [5] reported a facile approach to β-hydroxyesters **1** from

N-Heterocyclic Carbenes in Organocatalysis, First Edition. Edited by Akkattu T. Biju.
© 2019 Wiley-VCH Verlag GmbH & Co. KGaA. Published 2019 by Wiley-VCH Verlag GmbH & Co. KGaA.

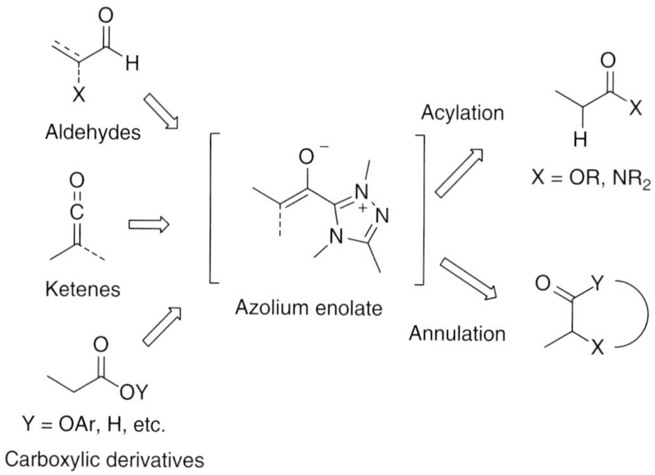

Scheme 8.1 NHC-catalyzed generation and transformation of azolium enolate intermediate.

epoxyaldehydes catalyzed by a thiazolium catalyst **2**. A variety of epoxyaldehydes and alcohols worked well for this process, giving the corresponding *anti*-β-hydroxyesters **3** in good yield and diastereoselectivity. *N*-Tosyl-β-aminoester can also be obtained from α,β-aziridinylaldehyde using the same methodology (Scheme 8.2). Simultaneously, Rovis and coworkers [6] independently reported the NHC-catalyzed synthesis of esters from α-haloaldehydes via azolium enolate intermediate.

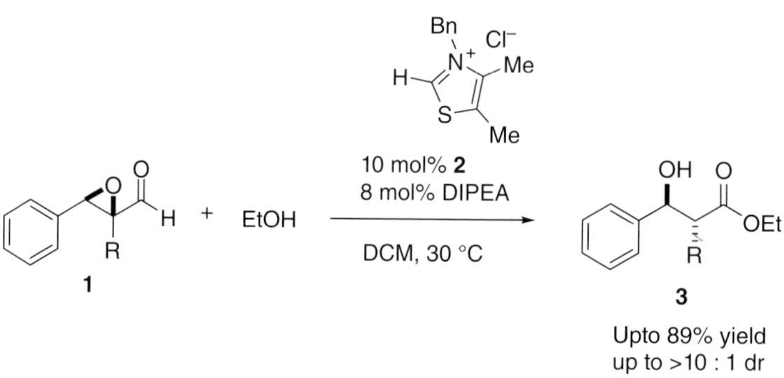

Scheme 8.2

The catalytic cycle of the reaction is initiated by the addition of NHC to epoxyaldehydes, followed by epoxide opening to give the thiazolium homoenolate intermediate. Subsequent proton transfer gives the azolium β-hydroxyenolate, which is protonated to form an acyl thiazolium intermediate. It is then trapped by alcohol to give β-hydroxyester as the final product and regenerated the NHC catalyst (Scheme 8.3).

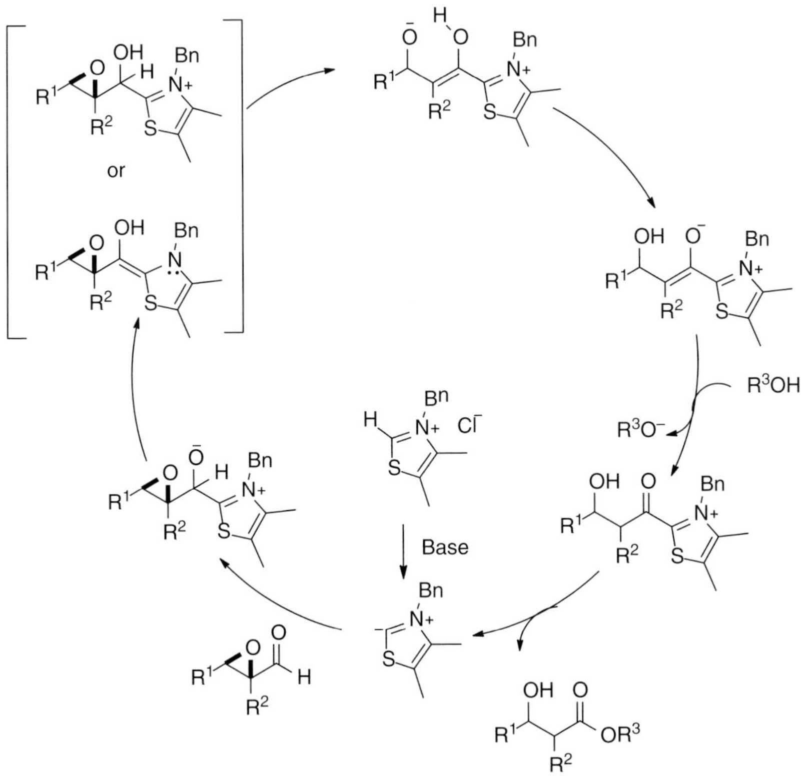

Scheme 8.3

Chiral formylcyclopropanes can also be employed in the NHC-catalyzed redox esterifications for the synthesis of enantioenriched β-substituted esters and thioesters (Scheme 8.4) [7].

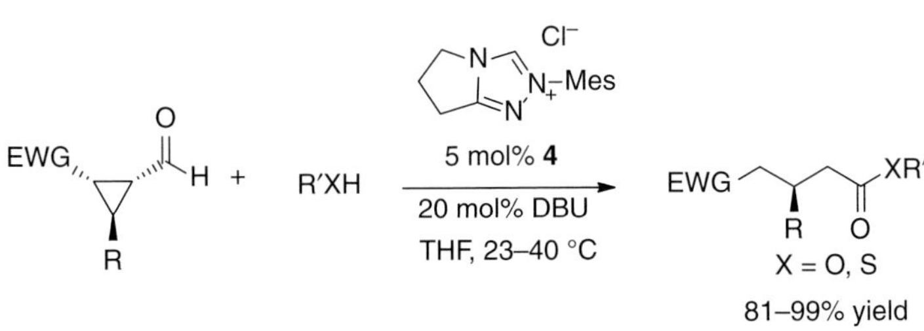

Scheme 8.4

In 2005, Rovis and coworker [8] developed a novel method to access chiral α-chloroesters based on an enantioselective protonation of *in situ* generated chiral α-haloenolates. A bulky, acidic phenol **5** was used to form the base reservoir to avoid the background epimerization. Using the precatalyst **6** derived from aminoindan-2-ol, a wide range of phenolic esters was obtained in good yields (typically 64–79%) and excellent enantioselectivity (84–93% ee) (Scheme 8.5).

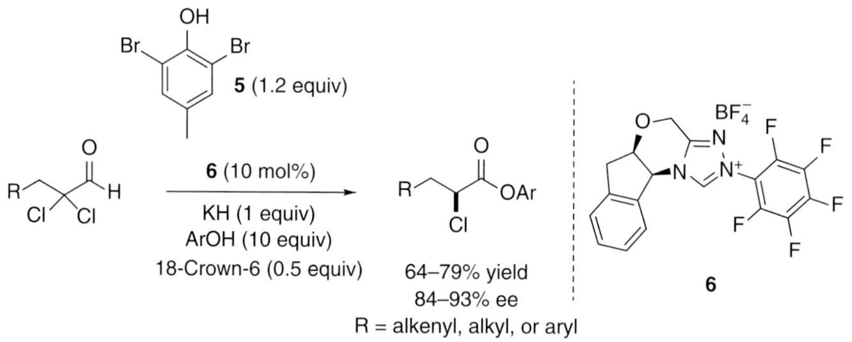

Scheme 8.5

Other kinds of carboxylates can also be accessed using this methodology. Bode and Rovis and coworkers [9] independently reported NHC-catalyzed amide bond-forming reaction using α-functionalized aldehydes. The introduction of 1-hydroxy-7-azabenzotriazole (HOAt) or imidazole as the cocatalysts is necessary for these transformations (Scheme 8.6).

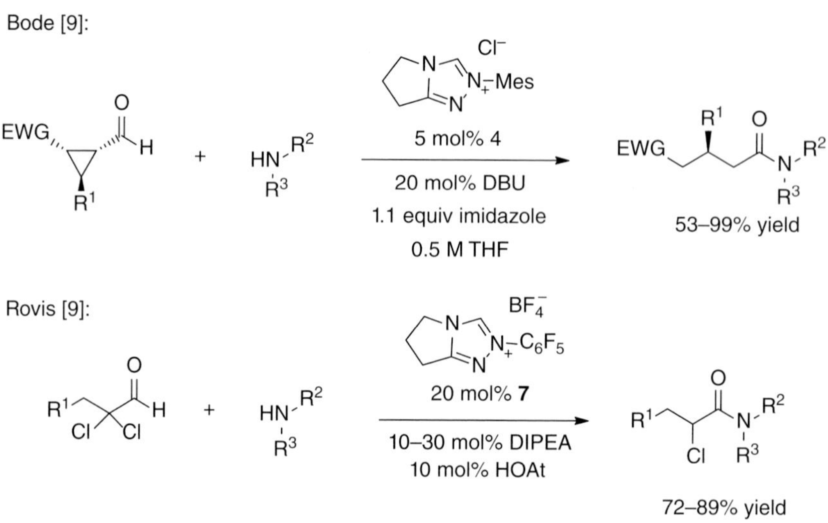

Scheme 8.6

Rovis and coworkers [10] extended this methodology to access chiral α-fluoro carboxylic acids via the use of water as the nucleophile. One molar aqueous potassium carbonate was introduced as the base and the nucleophile. Using the precatalyst **8**, the pharmaceutically important α-fluoro carboxylic acids were synthesized in excellent yield and enantioselectivity (Scheme 8.7).

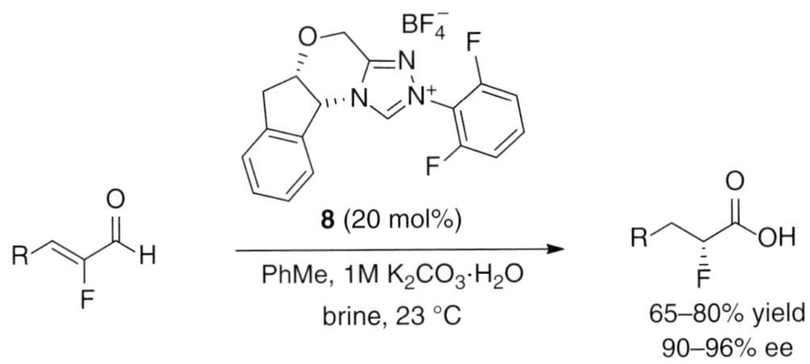

Scheme 8.7

In 2009, Scheidt and coworkers [11] reported that α-aryloxyacetaldehyde **9** could also serve as the enolate equivalent in the NHC-catalyzed Mannich reaction with imines, generating β-amino acid derivatives with excellent enantioselectivity. Other carboxylic compounds such as acid, ester, and amide can also be accessed using the correct trapping agents (Scheme 8.8).

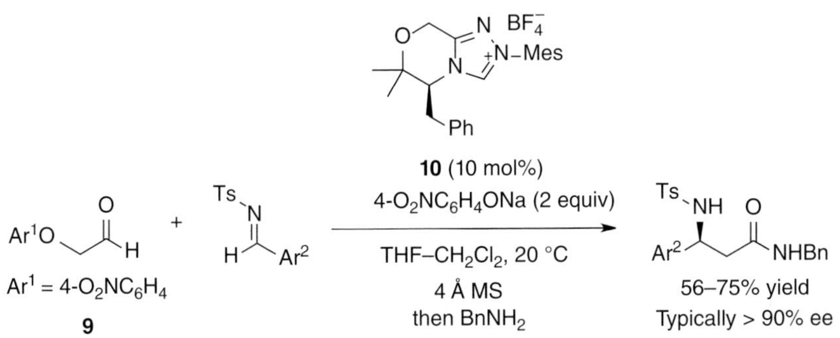

Scheme 8.8

The proposed catalytic cycle is initiated by the addition of a carbene to α-aryloxyacetaldehyde, followed by the elimination of an aryloxy anion to give the enolate. The Mannich reaction of the enolate with activated imines affords β-amino acyl azolium intermediates. The aryloxy anion can "rebound" by re-entering the catalytic cycle, regenerating the catalyst, and delivering a useful activated intermediate, which can be trapped by amine to afford the products (Scheme 8.9).

8.2.2 Formal [2+4] Cycloaddition

The use of α-functionalized aldehydes as azolium enolate precursors can also be employed in various formal cycloadditions. In 2006, Bode and coworkers [12] disclosed a [2+4] cycloaddition of α-chloroaldehydes and enones using aminoindanol-derived catalyst **11**. This process provides a range of corresponding 3,4,6-trisubstituted dihydropyran-2-ones in excellent yield (up to 98%), and high diastereo- and enantioselectivity (up to >20 : 1 dr and typically 99% ee) under mild conditions with low catalyst loading of 0.5 mol% (Scheme 8.10).

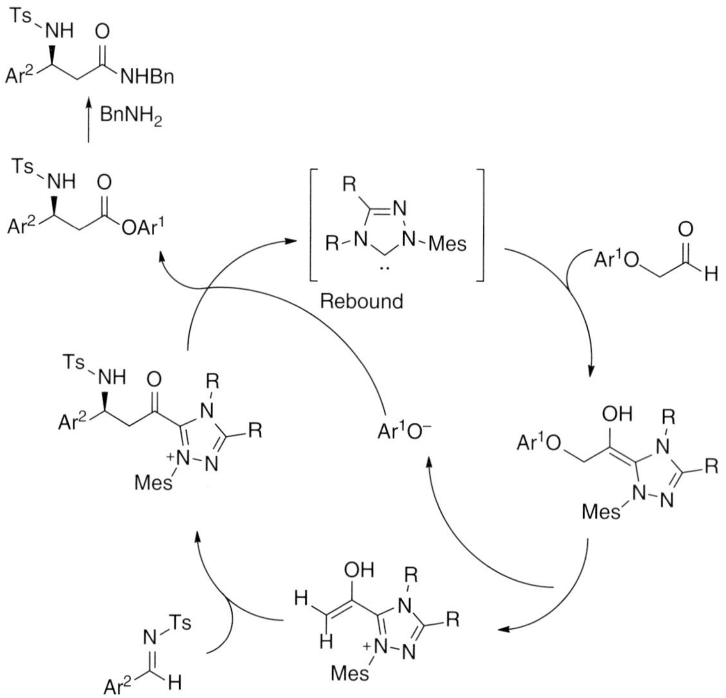

Scheme 8.9

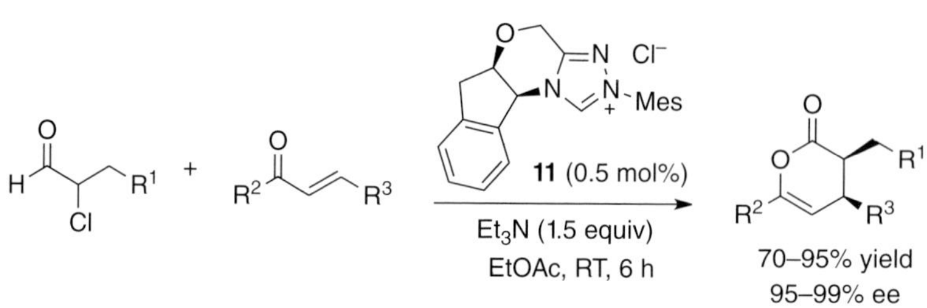

Scheme 8.10

The bench-stable α-chloroaldehyde bisulfite salts can also be employed as the enolate precursors [13]. Similar high yields and excellent enantioselectivity (up to 98% yield and 99% ee) were obtained in the biphasic system (Scheme 8.11).

The formal [2+4] cycloaddition of α,β-disubstituted enones **12** and 2-chloropropanal was developed by Kobayashi et al. [14]. The corresponding nitrogen-substituted dihydropyran-2-ones were obtained in high enantioselectivity (up to 98% ee) (Scheme 8.12).

The [2+4] annulation of α-chloroaldehydes and azadienes was developed by Ye and coworkers [15], using a NHC catalyst **13** (Scheme 8.13). The reaction was applicable with a range of substrate scope to provide the *cis*-dihydropyridinones in good yield with good-to-excellent enantioselectivity.

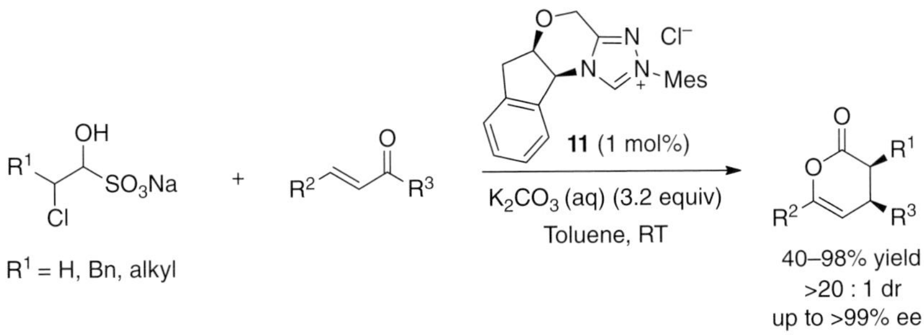

Scheme 8.11

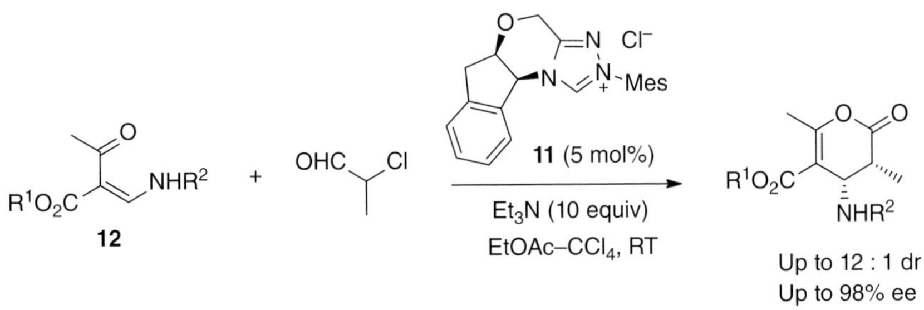

Scheme 8.12

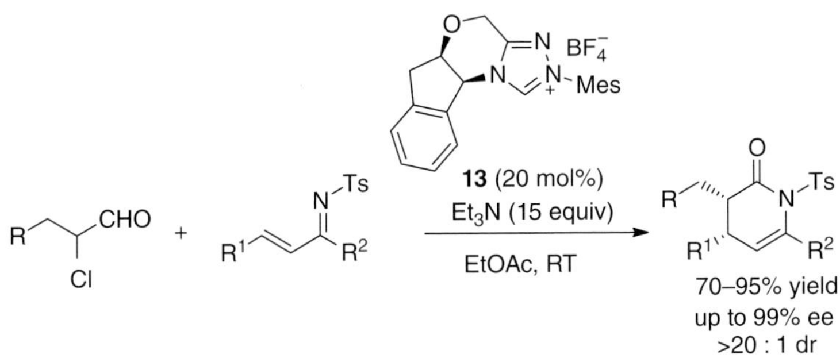

Scheme 8.13

Using 4-arylidenylpyrazolones [16] as the acceptor, the [2+4] annulation could give dihydropyrano[2,3-c]pyrazol-6-(1H)-ones in high yield with good diastereoselectivity and excellent enantioselectivity (Scheme 8.14). Various chloroaldehydes are well tolerated for the reaction.

Bioactive indenopyrones were synthesized via the NHC-catalyzed [2+4] cyclization of α-chloroaldehyde and arylidene indanediones [17]. High yields with exclusive *cis*-selectivities and excellent enantioselectivities were obtained with a broad range of scope (Scheme 8.15).

Trifluoromethyl *N*-Boc azadienes has been demonstrated as an effective coupling partner in the [2+4] cyclization with α-chloroaldehyde [18]. The corresponding 3,4-disubstituted-6-trifluoromethyldihydropyridin-2(1H)-ones

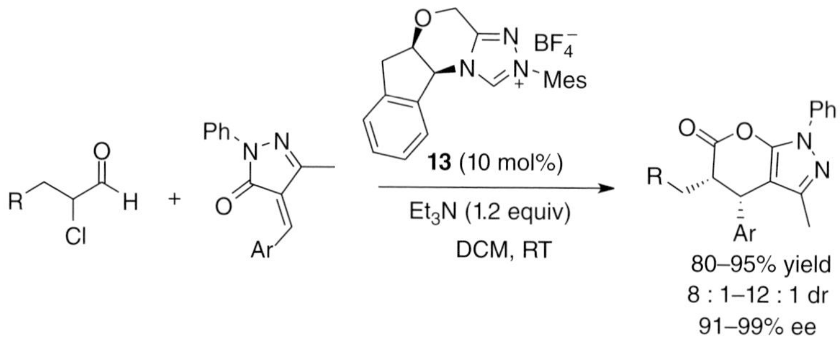

Scheme 8.14

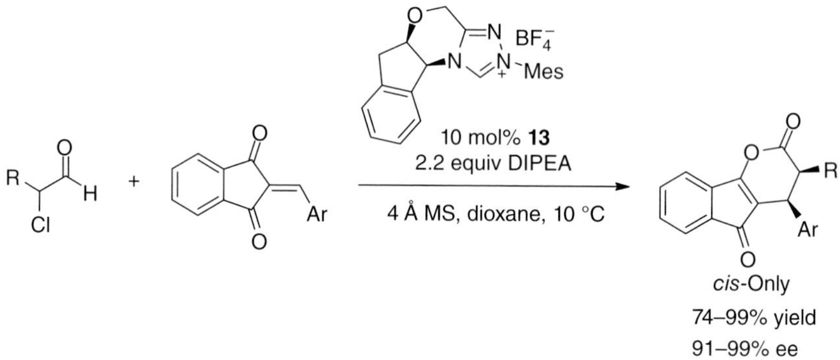

Scheme 8.15

were synthesized in good yields with exclusive *cis*-selectivities and excellent enantioselectivities (Scheme 8.16).

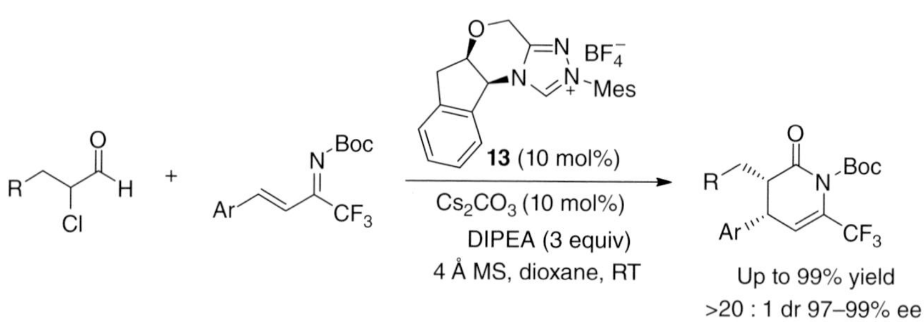

Scheme 8.16

High levels of stereocontrol in the NHC-catalyzed [2+4] cyclocondensation of α-chloroaldehydes and saccharine-derived 1-azadiene (Scheme 8.17) were realized by Ye and coworkers [19]. The corresponding dihydropyridinones were obtained in good yield with exclusive *cis*-selectivity and excellent enantioselectivity.

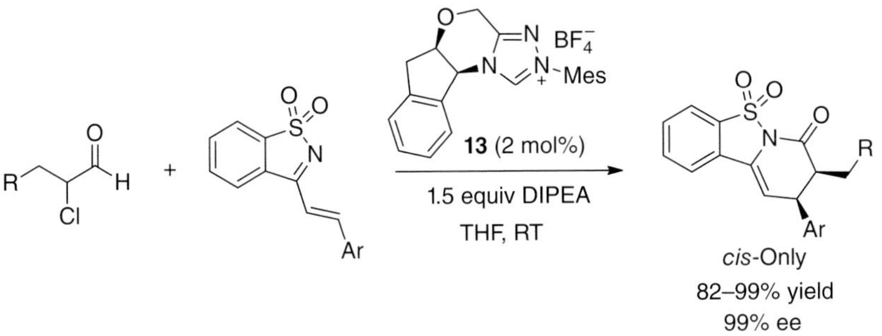

Scheme 8.17

NHC-catalyzed formal hetero-Diels–Alder reactions of formylcyclopropanes with chalcone were developed by Chi and coworkers [20]. NHC-bounded enolate intermediates generated from formylcyclopropanes were captured for new C—C bond-forming reactions. Various chalcones were tolerated in good yields with high diastereo- and enantioselectivities (Scheme 8.18).

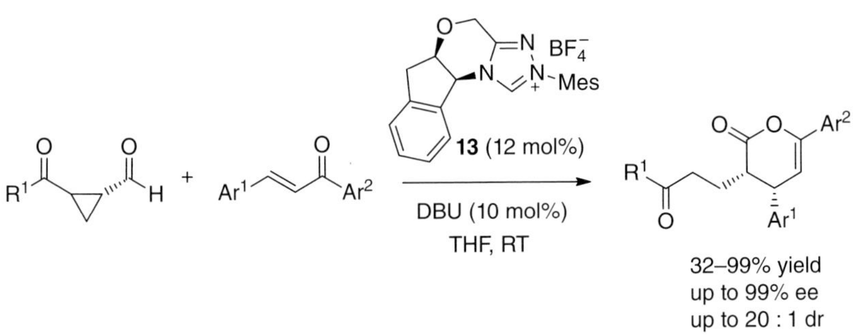

Scheme 8.18

α-Aroyloxyaldehydes have also been demonstrated as azolium enolate precursors by Smith and coworkers [21]. Using α,β-unsaturated trichloromethyl ketones or trifuloromethyl ketones as the coupling partners, the [2+4] cycloaddition could occur and provide the dihydropyranones with a broad range of scope (Scheme 8.19).

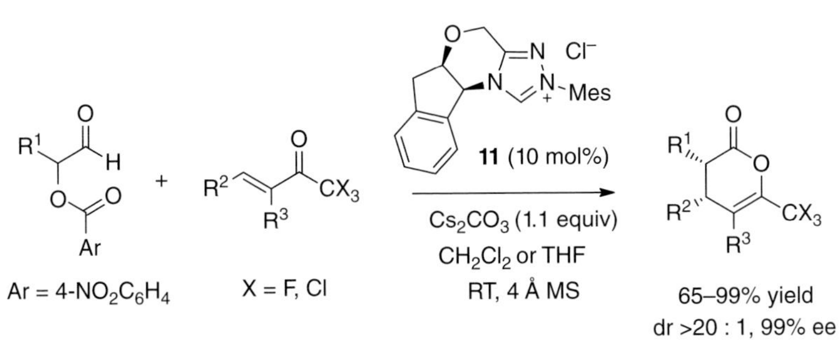

Scheme 8.19

Bioactive α-hydrazino esters [22] can be obtained from NHC-catalyzed annulation of α-aroyloxyaldehydes with *N*-aryl-*N*-aroyldiazenes and subsequent ring-opening reaction with alcohol. Further derivatization gives the corresponding N-aryl amino esters through samarium(II) iodide-mediated N—N bond cleavage (Scheme 8.20).

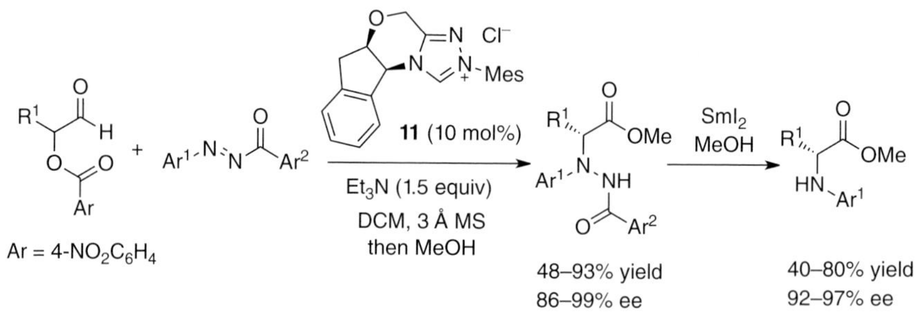

Scheme 8.20

8.2.3 Formal [2+2] Cycloaddition

NHC-catalyzed formal [2+2] cycloaddition from α-aroyloxyaldehydes with perfluorinated ketones has been developed [23]. Ring opening of the cycloaddition product by various amine nucleophiles give β-hydroxy carbonyl compounds in excellent yield, diastereo- and enantioselectivity. A series of fluorinated oxetanes can be obtained through a reductive ring opening and subsequent cyclization in good yield as a single diastereomer (Scheme 8.21).

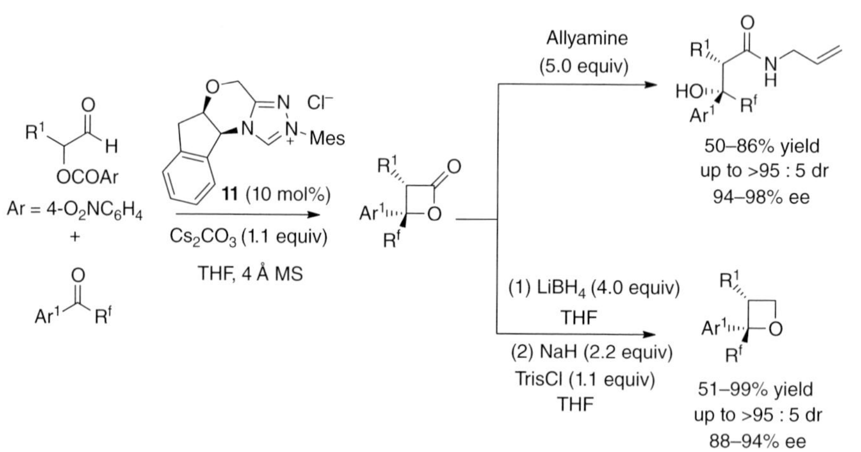

Scheme 8.21

8.2.4 Formal [2+3] Cycloaddition

In 2013, Enders and coworkers [24] reported a NHC-catalyzed [2+3] annulation with nitrovinylindoles for the synthesis of *trans*-disubstituted pyrroloindolones (Scheme 8.22). Screening of the enolate precursor revealed that α-chloroaldehydes worked best for the reaction, giving the desired pyrroloindolones in moderate to good yields with excellent diastereo- and enantioselectivities with a broad range of scope.

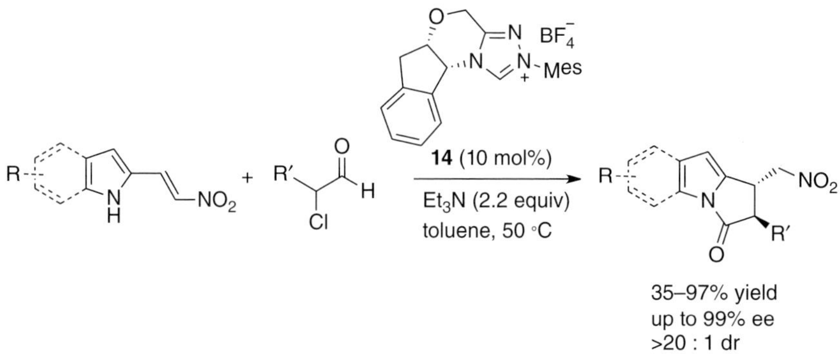

Scheme 8.22

Another example of NHC-catalyzed [2+3] cyclocondensation (Scheme 8.23) is reported by Ye and coworkers [25]. Various α-chloroaldehydes with azomethine imines were well tolerated giving the desired pyrazolidinones in good-to-high yields, with moderate-to-good diastereoselectivities and excellent enantioselectivities.

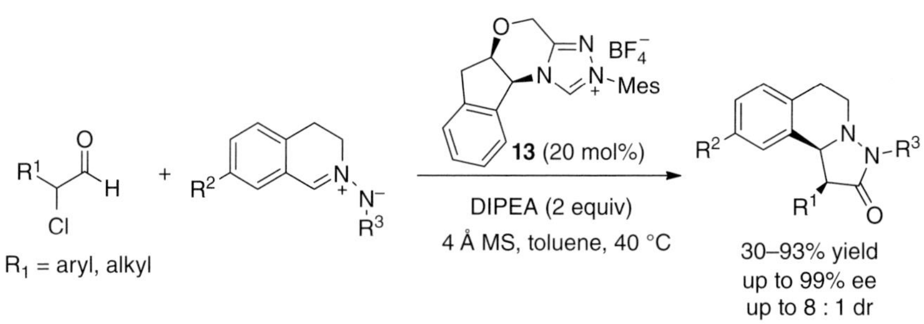

Scheme 8.23

8.3 Azolium Enolate from Ketenes

Catalytic enantioselective cycloaddition reactions of ketenes have been intensively explored over the past several decades. Various chiral catalysts have been developed for these cycloadditions, such as cinchona alkaloids, planar chiral DMAP-derived catalysts, and Lewis acid-based catalysts [4, 26]. In recent years, NHCs have also been demonstrated as efficient catalysts for the cycloaddition of ketenes to give various heterocycles.

8.3.1 Formal [2+2] Cycloaddition

In 2008, Ye and coworkers found that IMes could catalyze the Staudinger [2+2] cycloaddition reaction of phenyl(ethyl)ketene with *N*-tosylphenylmethanimine to give the corresponding β-lactam in good yield. After screening a series of chiral NHCs, NHC precursor **15**, derived from L-pyroglutamic acid, was found optimum. Good yield, moderate diastereoselectivity, and excellent enantioselectivity were realized for various ketenes and *N*-*tert*-butoxycarbonyl-substituted (*N*-Boc) imines [27]. An advantage of this work was that the resulting *N*-Boc-substituted β-lactams could be easily deprotected to afford the free β-lactams in high yield without erosion of the enantiopurity (Scheme 8.24). A similar excellent reaction was reported by Smith group using imidazolinium precatalyst.

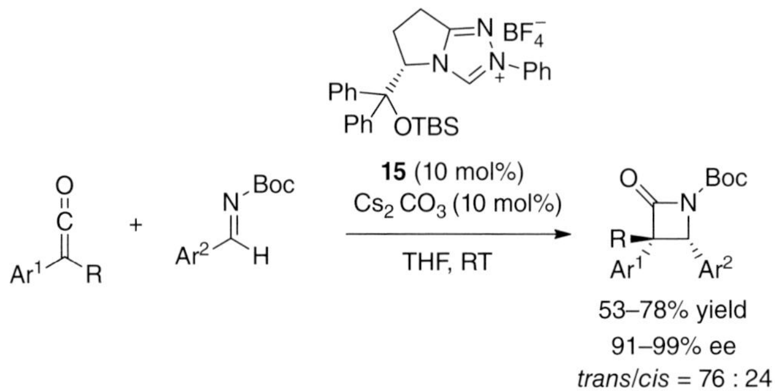

Scheme 8.24

A common mechanistic rationale for formal [2+2] cycloadditions is shown in Scheme 8.25. Attack of NHC to ketene yields (*Z*)-enolate, followed by addition to imine-producing zwitterionic intermediate. Collapse of this species generates β-lactam and releases NHC for the next catalytic cycle.

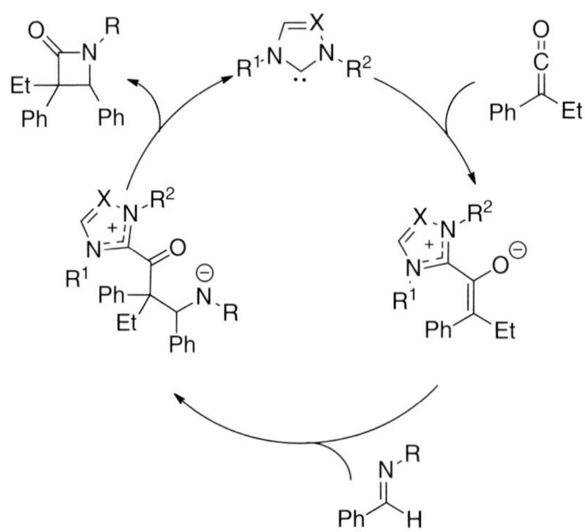

Scheme 8.25

Isatin-derived ketimines were also employed as the coupling partner in the Staudinger reaction with ketenes [28]. The bifunctional NHCs **16** with a free hydroxyl group played an important role for the enantioselective control. Various ketenes and ketimines were well tolerated, providing corresponding spirocyclic oxindolo-β-lactams in high yields with excellent diastereo- and enantioselectivities (Scheme 8.26).

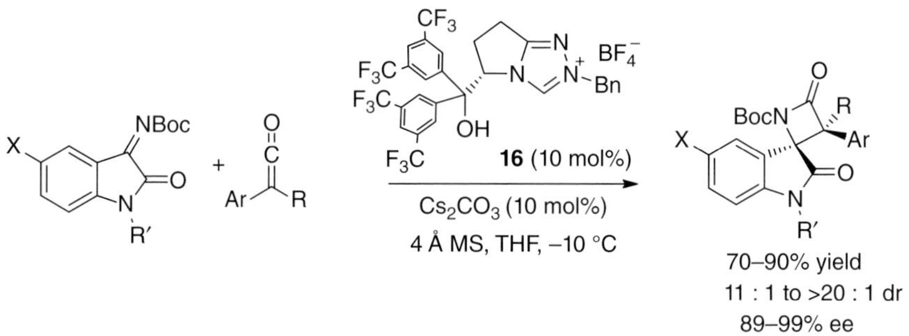

Scheme 8.26

The NHC-catalyzed enantioselective [2+2] cycloaddition between ketenes and aryl aldehydes was disclosed by Smith group [29]. The scope of aldehyde is limited to 2-nitrobenzaldehyde and 2- and 4-pyridinecarboxaldehydes, and the reaction with a range of disubstituted ketenes proceeds with generally excellent levels of enantio- and diastereoselectivity, allowing access to highly functionalized β-lactones (Scheme 8.27).

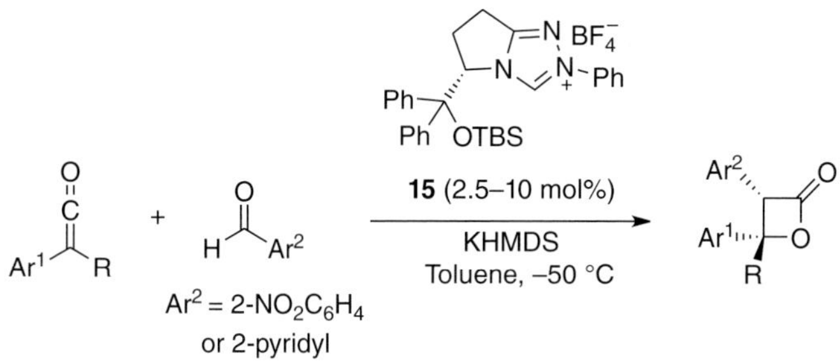

Scheme 8.27

In 2008, Ye and coworkers [30] found that 2-oxoaldehydes were suitable electrophiles for the NHC-catalyzed [2+2] cycloaddition reaction of ketenes. The corresponding acyl-substituted β-lactones were isolated in good-to-excellent yield with high diastereo- and enantioselectivity (Scheme 8.28).

Trifluoromethyl lactones [31] were generated using the same strategy by using trifluoromethyl ketones as the coupling partner. The reaction proceeded smoothly to give the desired products in impressive yields and stereoselectivities (typically >80% yield and up to >20 : 1 dr and 99% ee) using precatalyst **17** with an *N*-(2-isopropyl)phenyl substituent (Scheme 8.29).

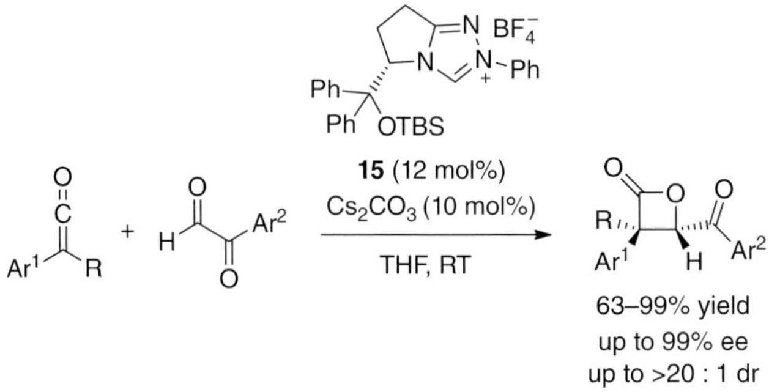

Scheme 8.28

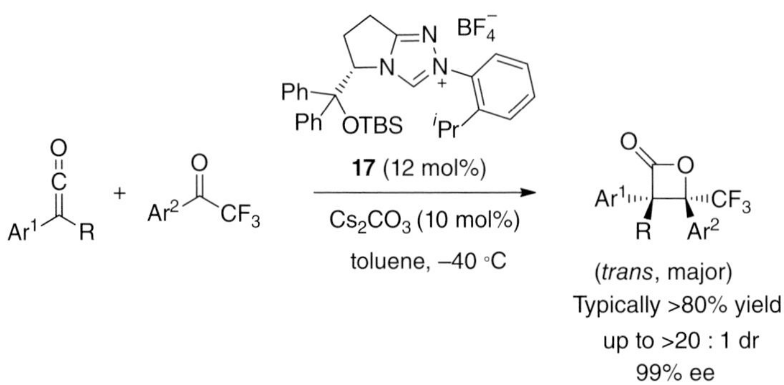

Scheme 8.29

Isatins were also found to be suitable electrophile for the [2+2] cycloaddition of ketenes using a NHC precatalyst **17** [32]. High enantioselectivities (13 examples of 99% ee) were achieved, giving the corresponding biologically important spirocyclic oxindolo-β-lactones. Ring opening with Grignard reagents gave the corresponding 3-hydroxy-oxindoles in good yield without erosion of enantiointergrity (Scheme 8.30).

The enantioselective [2+2] cycloaddition of ketenes with azodicarboxylates catalyzed by NHC precatalyst **18** was realized [33]. High enantioselectivities were obtained, giving the aza-β-Lactams in excellent yields (typically 80–90%) and good-to-excellent enantioselectivities (typically 80–90% ee). The N-substituent (carboxylate vs. benzoyl) of the diazenes plays an important role to control the reaction modes (formal [2+2] vs. [4+2] cycloadditions) (Scheme 8.31).

The [2+2] cycloaddition between ketenes and nitroso compound was reported by Ye and coworkers [34]. The precatalyst **15** were found to be the efficient catalyst of the reaction, giving the corresponding 1,2-oxazetidin-3-ones in moderate-to-good yields with high enantioselectivities (Scheme 8.32). The synthetically useful 1,2-oxazetidin-3-ones were shown to undergo facile zinc-mediated reductive cleavage to render the corresponding α-hydroxy amides without erosion of enantiopurity.

Asymmetric dimerization of disubstituted ketenes catalyzed by *N*-heterocyclic carbene was disclosed by Ye and coworkers [35]. Using **19** bearing a free

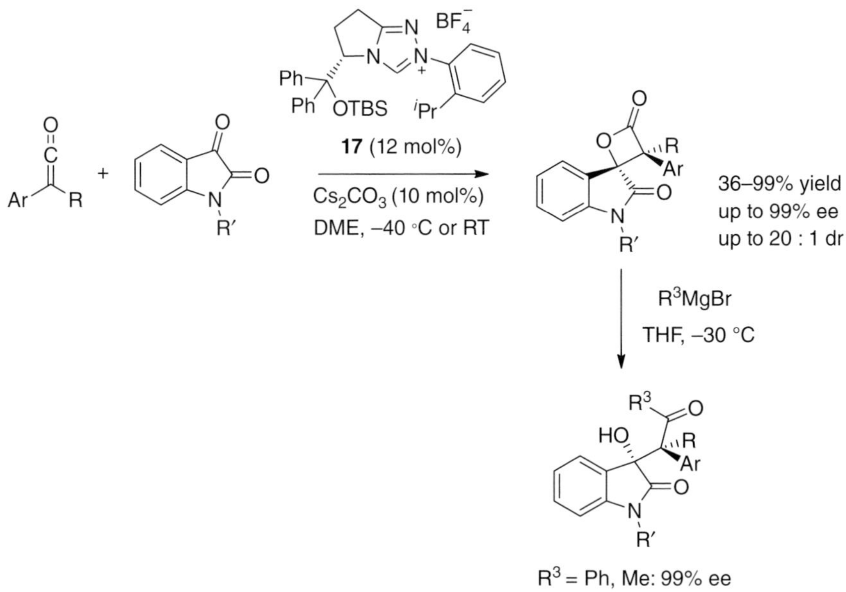

Scheme 8.30

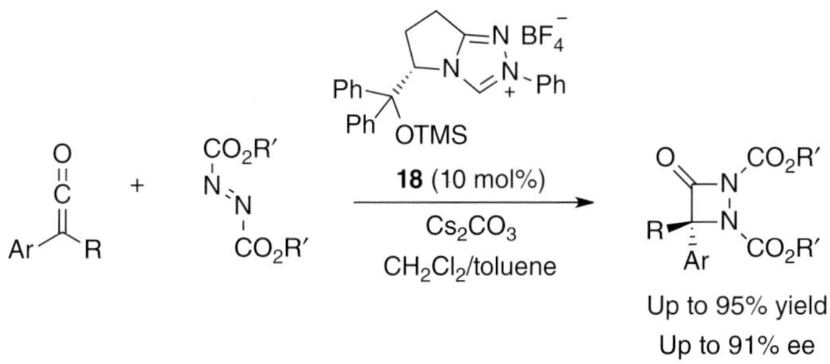

Scheme 8.31

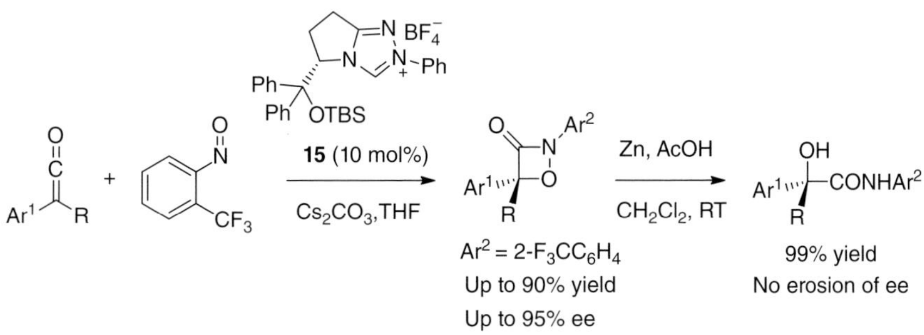

Scheme 8.32

hydroxyl group as the precatalyst, highly functionalized α-quaternary β-alkylidenyl-β-lactones were achieved in good yield (62–85%) and with excellent enantioselectivity (typically >90% ee) (Scheme 8.33). The synthetic application of the products has been demonstrated by ring-opening reaction and alcoholysis reaction. One-pot ketene generation and dimerization were also conducted in 57% yield with 96% ee. The author also confirmed that the catalytic cycle initiated by the addition of the NHC to the ketene because there are no other electrophiles than ketenes in this reaction.

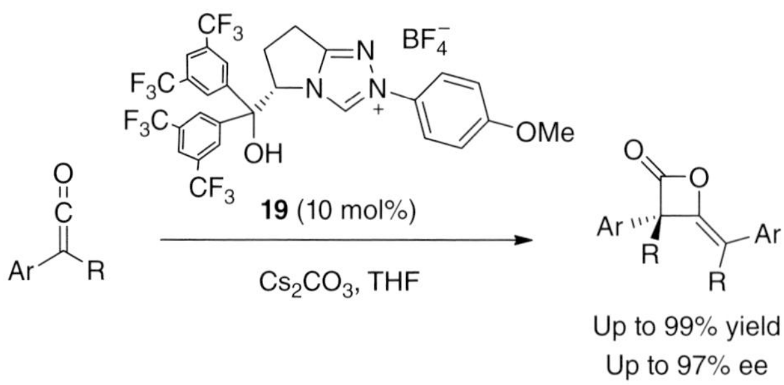

Scheme 8.33

β-Sultams can be synthesized via the formal [2+2] cycloaddition between ketenes and *N*-sulfinylanilines [36]. Both enantiomers of the products could be obtained by choosing either NHC precursor **15** or **20** derived from L-pyroglutamic acid or aminoindan-2-ol, respectively, as the catalyst. The catalyst loading can be reduced to 1 mol% with 4 Å MS as the additive (Scheme 8.34).

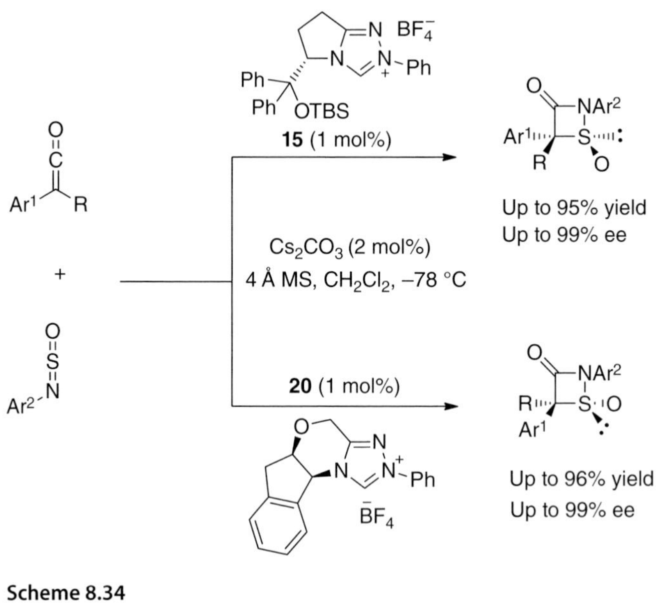

Scheme 8.34

In addition, the cycloaddition product can undergo a range of transformations (Scheme 8.35). The oxidation of **21** with 3-chloroperoxybenzoic acid gave the corresponding 3-oxo-β-sultam. Reduction at −78 °C with diisobutylaluminum hydride resulted in ring opening to afford α-sulfanyl-substituted amide in good yield and with excellent enantioselectivity, whereas β-sulfanyl-substituted amine could be formed in good yield when the reduction was carried out at room temperature. Aminolysis of **21** with pyrrolidine gave the sulfonamide in good yield and high enantioselectivity.

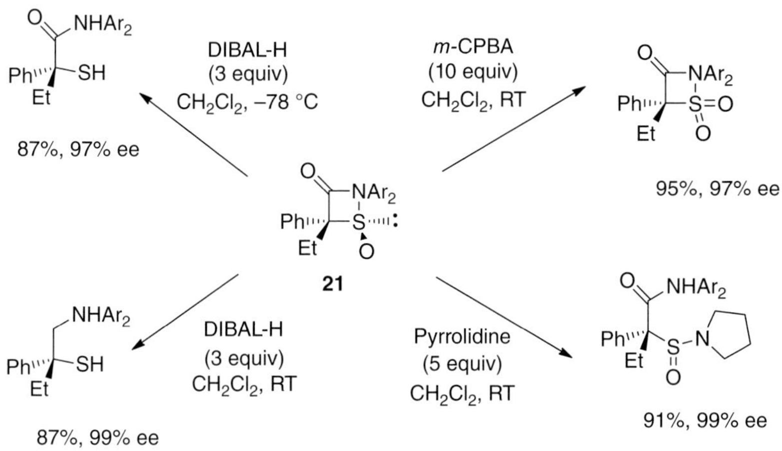

Scheme 8.35

4-Nitrophenyl isothiocyanate [37] was also employed as a coupling partner in the [2+2] cycloaddition with ketenes using a NHC catalyst **16**, yielding biologically important 4-thioxo-2-azetidinones (thioxo-β-lactams) in good yield (62–85%) and with excellent enantioselectivity (typically >90% ee). It is interesting to note that [2+2+2] cycloadduct was obtained when benzoyl isothiocyanate was used instead using precatalyst **17** (Scheme 8.36).

Another example of [2+2+2] cycloaddition of ketenes with carbon disulfide using an NHC catalyst was reported by Ye group [38]. The corresponding 1,3-oxathian-6-ones were obtained in good yield and with excellent enantioselectivity (92–97% ee). Mechanistic studies revealed that the catalytic reaction is initiated by the addition of NHC to ketenes rather than carbon disulfide (Scheme 8.37).

Ketenimine generated by Cu-catalyzed reaction of alkyne with TsN_3 could also be activated by the NHC catalyst to form azolium enamide intermediate [39]. Each of the two catalysts, copper metal catalyst and NHC organocatalyst, could work independently without quenching each other (Scheme 8.38).

The activated azolium enamide could facilitate effective formal [2+2] cycloaddition with isatin-derived imines to give the spiro-azetidine oxindoles as a single diastereomer (>99 : 1 dr) with good yields. Using isatins as the electrophile, bioactive 3-alkenyl-oxindoles were obtained in excellent yield (Scheme 8.39).

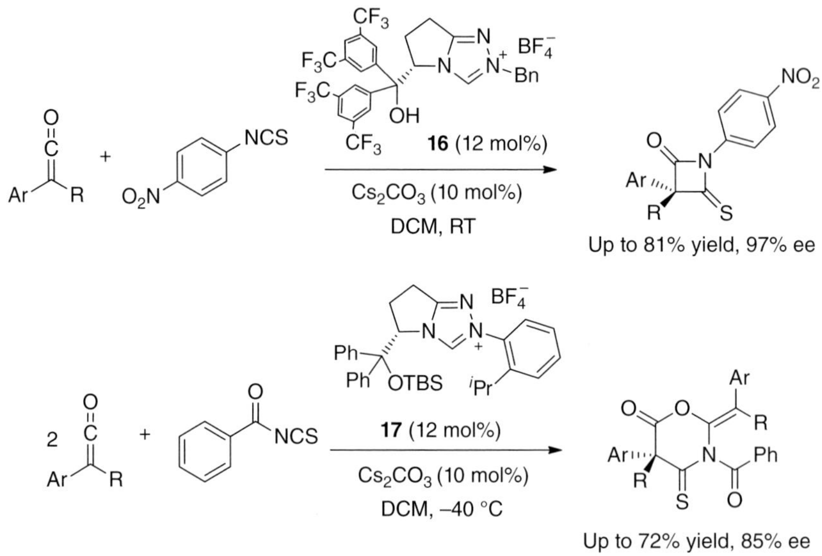

Scheme 8.36

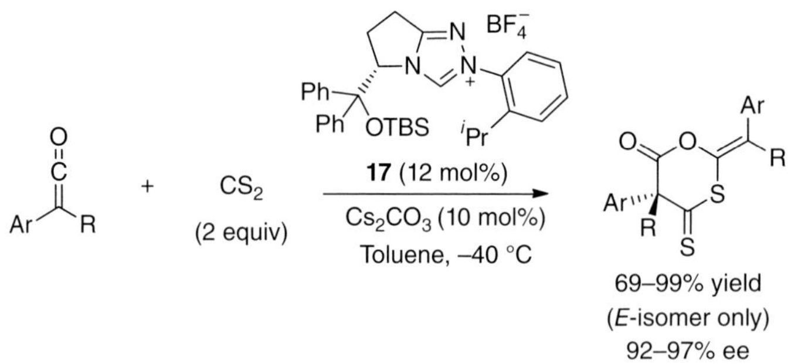

Scheme 8.37

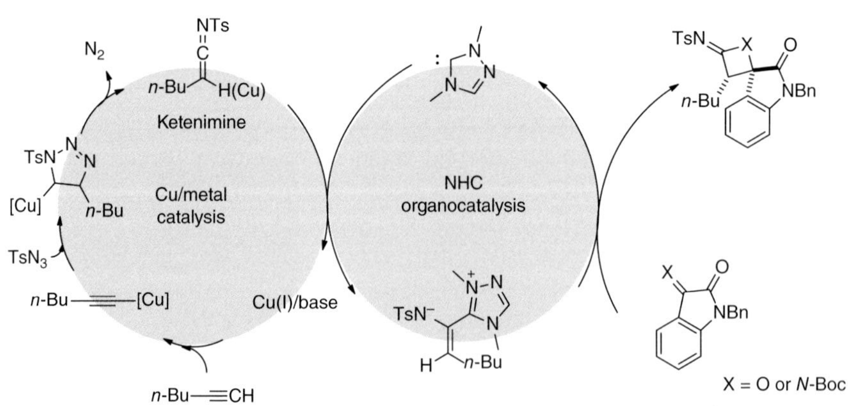

Scheme 8.38

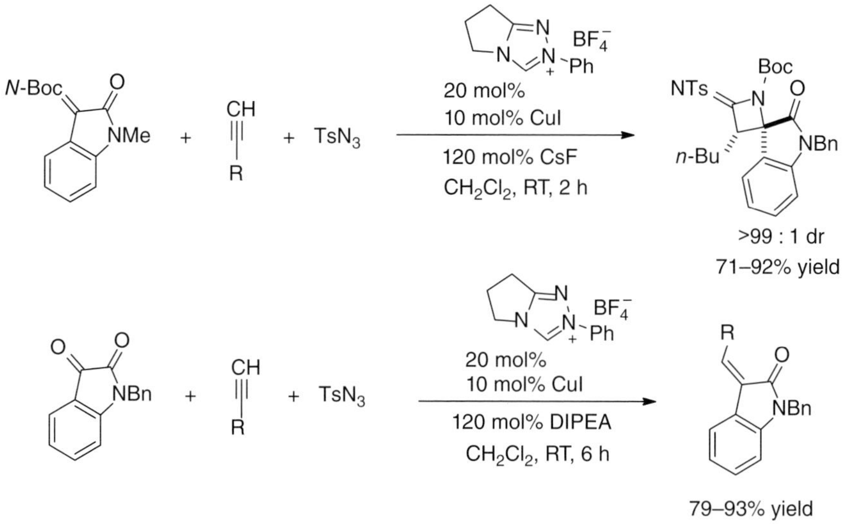

Scheme 8.39

8.3.2 Asymmetric Formal [2+3] Cycloadditions

In 2010, Ye and coworkers [40] developed a novel [2+3] cycloaddition between ketenes and oxaziridine (Scheme 8.40). In the presence of 10 mol% of NHC precursor **22**, the reaction of ketenes with oxaziridine gave [2+3] cycloadduct in good yield and with good diastereoselectivity and high enantioselectivity. Interestingly, the opposite isomer can also be accessed when NHC precursor **23**, bearing an N-benzyl rather than an N-aryl substituent, and an *O*-trimethylsilyl ether rather than a free hydroxy substituent was used.

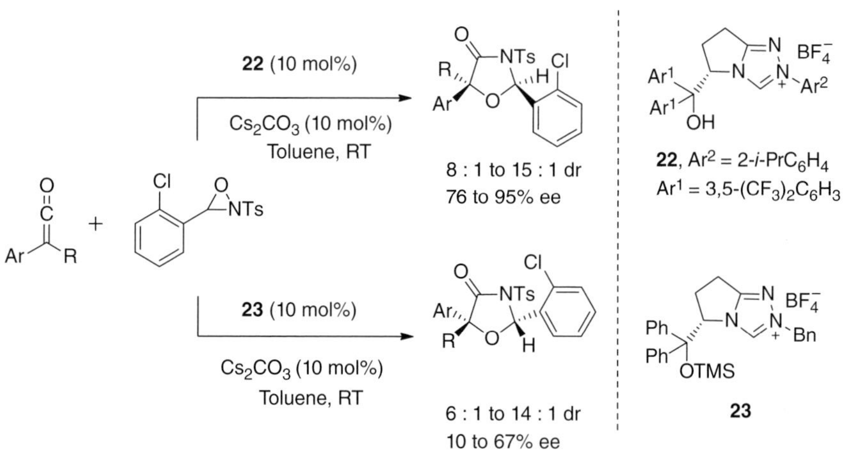

Scheme 8.40

Derivation of the oxazolidin-4-one by saponification and reduction could generate the corresponding α-hydroxy acid and 1,2-diol in good yield without erosion of enantiopurity (Scheme 8.41).

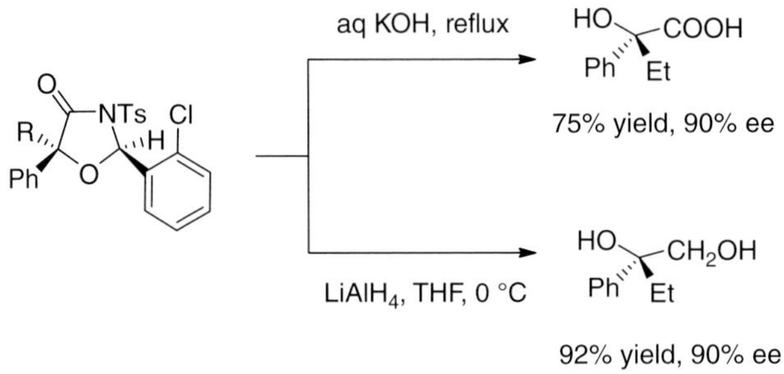

Scheme 8.41

8.3.3 Asymmetric Formal [2+4] Cycloadditions

The [2+4] cycloaddition reaction of ketenes is an efficient approach to six-membered heterocycles. In 2008, Ye and coworkers [41] demonstrated that the NHC precursor **15** was an efficient precatalyst for the [2+4] of ketenes with enones, giving the corresponding δ-lactones, containing α-quaternary and β-tertiary stereocenters, in good yield with good diastereoselectivity and high enantioselectivity. Ketenes generated *in situ* from the corresponding acyl chloride in the presence of excess base also worked well for this reaction (Scheme 8.42).

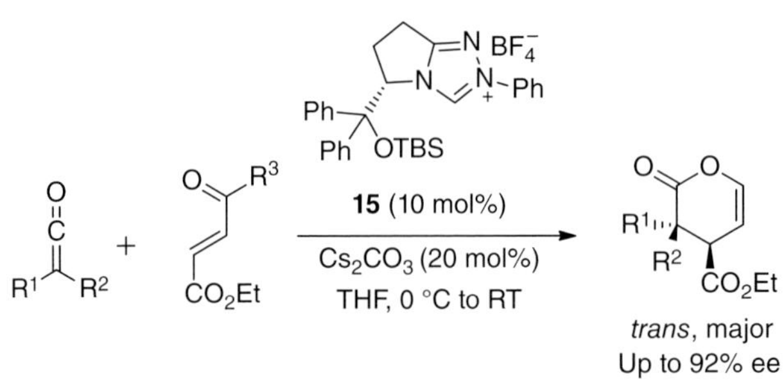

Scheme 8.42

Isatin-derivated unsaturated ester has also been employed in the NHC-catalyzed reaction with ketenes [42]. Precatalyst **24** turns out to be the superior catalyst, providing the indole-fused dihydropyranones in excellent yield (88–99%), good diastereoselectivity (up to 10 : 1), and good enantioselectivity (typically >80% ee) with a broad range of scope (Scheme 8.43).

Optically active, highly functionalized 3,4-dihydropyridin-2-ones were synthesized via NHC-catalyzed [2+4] enantioselective cycloaddition of ketenes and 1-azadienes [43]. A broad range of ketenes were tolerated, giving the corresponding dihydropyridinones with good levels of enantiocontrol (83–93% ee) in excellent yield (typically >80%) (Scheme 8.44).

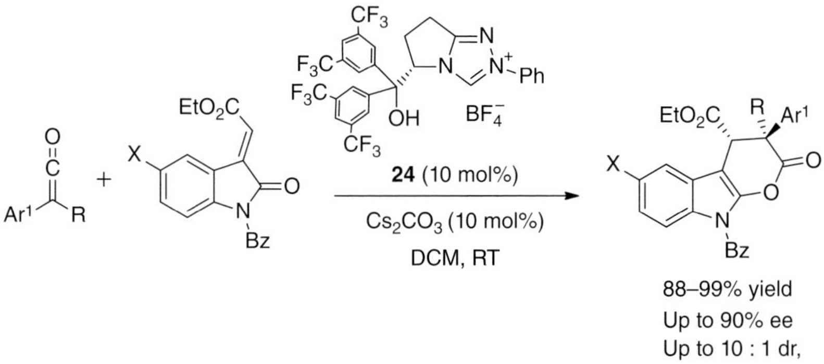

Scheme 8.43

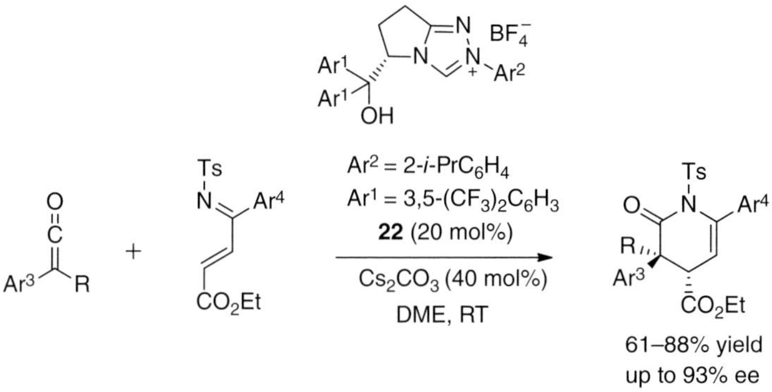

Scheme 8.44

Dihydrocoumarin derivatives are present as active ingredients in many traditional Chinese herbal medicines. A series of chiral 3,3,4-trisubstituted dihydrocoumarins were synthesized via the NHC-catalyzed [2+4] cycloaddition of ketenes to *o*-quinone methide [44]. It was unexpected that methanol was required as an additive to achieve stable results for the reaction (Scheme 8.45).

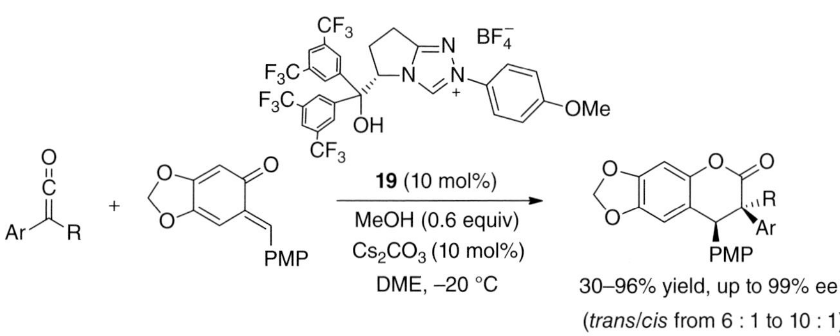

Scheme 8.45

The NHC-catalyzed [2+4] cycloaddition reaction of ketenes with 3-(arylcarbonyl)coumarins has also been developed [45]. The NHC precatalyst **25** derived from L-pyroglutamic acid bearing a free hydroxyl group was used as the catalyst, giving the desired dihydrocoumarin-fused dihydropyranones in high yield with good-to-high diastereo- and enantioselectivity (Scheme 8.46).

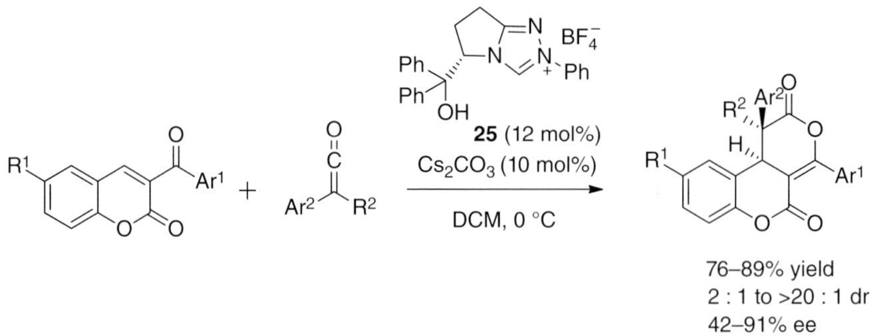

Scheme 8.46

The NHC-catalyzed [2+4] cycloaddition of disubstituted ketenes to 9,10-phenanthrenequinone was also developed [46]. Cycloaddition products were typically obtained in high yield (up to 98%) and with excellent levels of enantiocontrol (>80% ee at either RT or 0 °C) (Scheme 8.47).

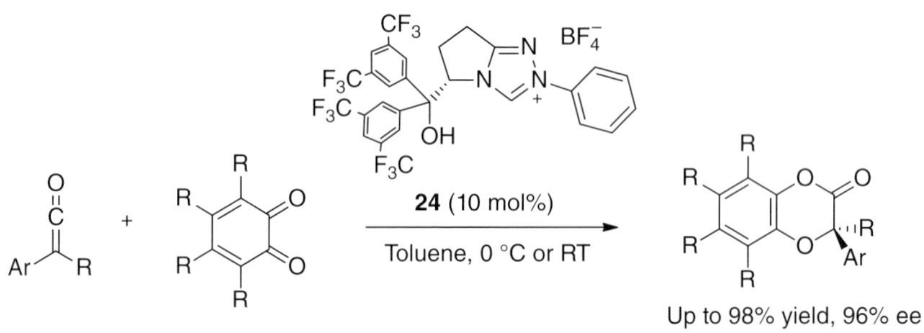

Scheme 8.47

Before the reported formal [2+2] cycloaddition of ketenes and diazene carboxylates, the formal [4+2] cycloaddition was realized via the use of *N*-benzoyldiazenes [47]. Opposite enantioselectivities were observed for the reactions catalyzed by NHC precursors **15** and **26**. The bulky mesityl substituent and the free hydroxy group in catalyst **26** may play a role in switching the enantioselectivity (Scheme 8.48).

The resulting 1,3,4-oxadiazin-6-one could be transformed into the α-amino ester derivative (Scheme 8.49), by one-pot alcoholysis and reductive nitrogen–nitrogen bond cleavage using samarium(II) iodide.

The formal [2+4] cycloadditions of alkylarylketenes and β,γ-unsaturated α-ketocarboxylic esters were developed by Smith group [48]. Divergent

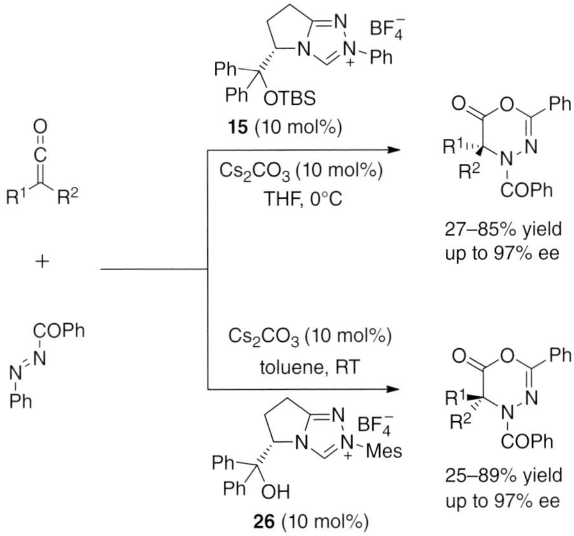

Scheme 8.48

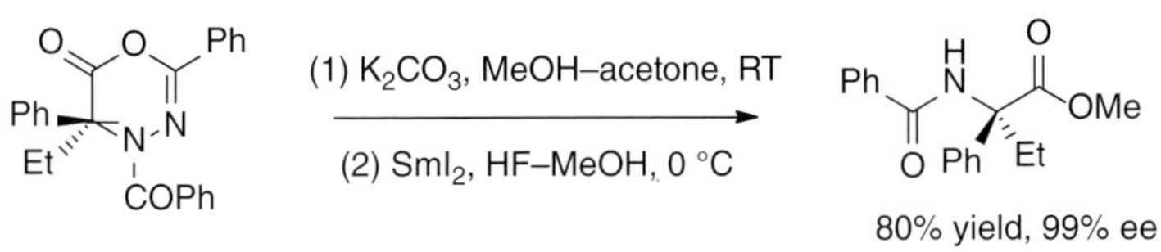

Scheme 8.49

diastereoselectivity is observed in this process, with γ-aryl-β,γ-unsaturated α-ketocarboxylic esters and amides giving preferentially *syn*-dihydropyranones (up to 68 : 32 dr, *syn:anti*, up to 98% ee), whereas γ-alkyl-substrates generate *anti*-dihydropyranones (up to 18 : 82 dr, *syn:anti*, up to 75% ee) (Scheme 8.50).

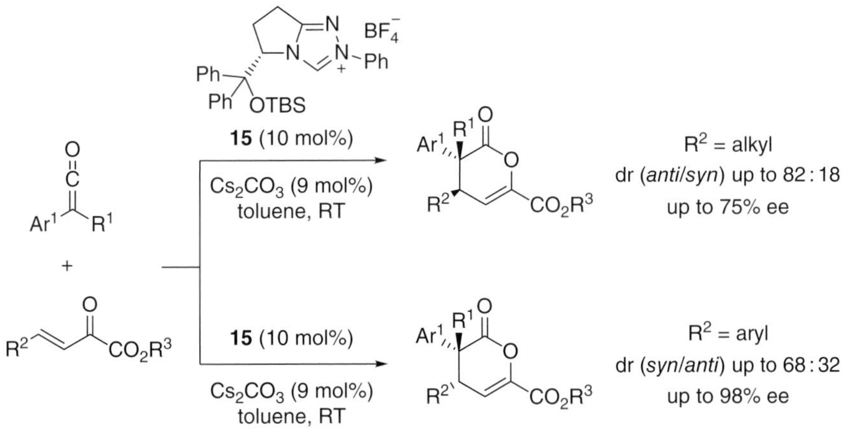

Scheme 8.50

8.3.4 Asymmetric Protonation and Halogenation

The asymmetric esterification of ketenes catalyzed by an NHC is developed by Ye group [49]. The bulk nucleophilic reagent benzhydrol was crucial for the enantioselective control, and the pyroglutamic-acid-derived precatalyst **15** was employed to promote the reaction. Various ketenes were well tolerated, giving the corresponding esters in good yields (typically 60–75%) and enantioselectivity (up to 95% ee) (Scheme 8.51).

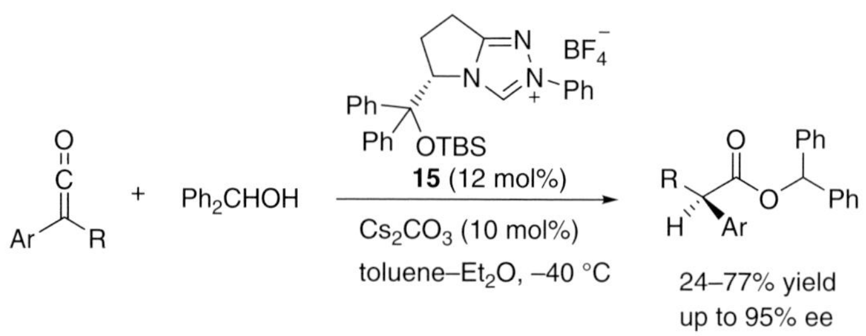

Scheme 8.51

The mechanism (Scheme 8.52) is postulated to proceed via typical nucleophilic addition of the NHC to the ketene to generate enolate followed by stereoselective protonation of benzhydrol. Displacement of the NHC catalyst from intermediate with the alkoxide provides the required ester.

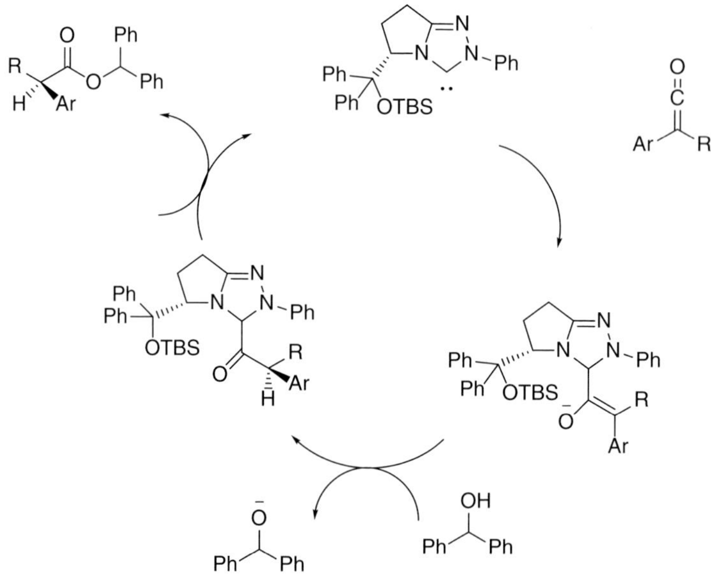

Scheme 8.52

A complementary methodology was developed by Smith group [50]. Under the NHC catalyst **27**, using potassium hexamethyldisilazide (KHMDS) as the base, the chiral variant of the esterification can be realized when 2-phenylphenol was employed. The desired α-alkyl-α-arylacetic acid derivatives were obtained in similar yields (65–91%) and enantioselectivities (up to 84% ee) to that using benzhydrol as the nucleophilic reagent (Scheme 8.53).

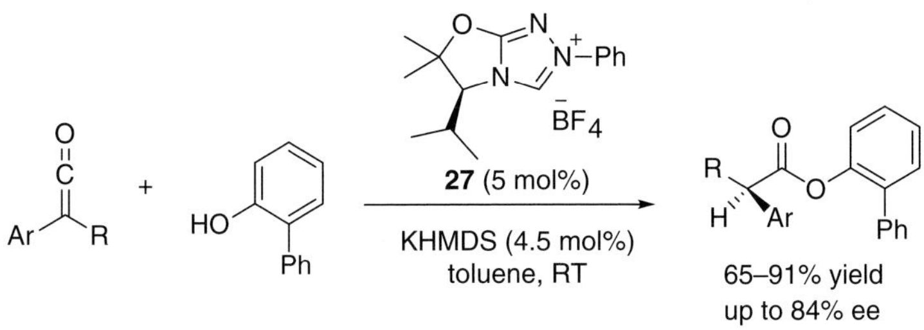

Scheme 8.53

The NHC-catalyzed asymmetric halogenation of alkylarylketenes has been disclosed by Smith and coworkers [51]. After screening a series of chlorinating agents, the 2,3,4,5,6,6-hexachlorocyclohexa-2,4-dienone was used, giving the α-chloro esters in good yield (up to 97%) but in only moderate enantioselectivities (26–61% ee) (Scheme 8.54). Employing the 2,4,4,6-tetrabromocyclohexa-2,5-dienone as the brominating agent, the corresponding α-bromo ester could be obtained in 92% yield but with a moderate 45% ee.

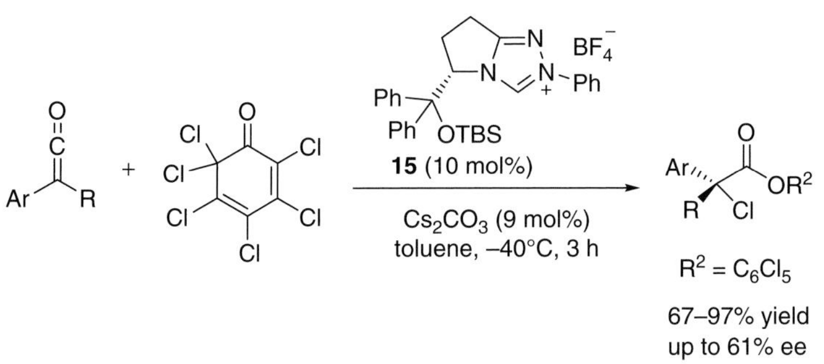

Scheme 8.54

8.4 Azolium Enolate from Enals

The NHC-bound homoenolate intermediate from enal could be converted to the enolate by the β-protonation under the appropriate reaction conditions. In 2006, Bode and coworkers [52] first demonstrated the ability of enal and precatalyst **11** to generate the azolium enolate, and [2+4] cycloaddition could occur in the presence of cinnamaldehyde-derived imines. The biologically active dihydropyridinone products were obtained in good yield and with remarkable enantioselectivities (Scheme 8.55). The reaction is postulated to

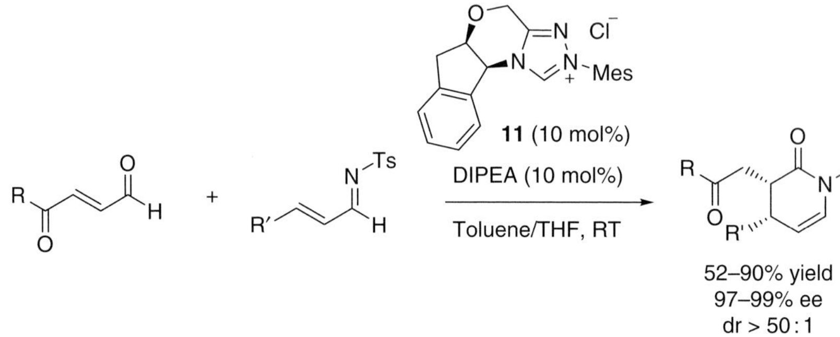

Scheme 8.55

proceed via a highly favored *endo* inverse-electron-demanded Diels–Alder reaction from the (*Z*)-enolate, thus accounting for the exceptionally high level of stereocontrol.

In 2010, Bode and coworkers [53] have identified conditions for the selective generation of NHC-bound enolates from enals and their use in highly enantioselective hetero-Diels–Alder reactions with electron-deficient enones. The use of a weak amine base, such as *N*-methylmorpholine (NMM) or 4-dimethylaminopyridine (DMAP), is crucial for selectivity of enolate instead of homoenolate pathway. It facilitates the protonation of the homoenolate from the conjugate acid of the amine base at a faster rate than direct homoenolate carbon–carbon bond formation, thus minimizing this competitive reaction process. A range of substituted dihydropyran-2-ones were synthesized with outstanding levels of diastereo- and enantioselectivity (typically >20 : 1 dr and 99% ee) (Scheme 8.56).

A complementary method to generate enolate from enals is fulfilled by the addition of catalytic amount acetic acid as the cocatalyst, allowing for individual access to diverse products from the same substrates [54]. Chi and coworkers have used this strategy to synthesize the lactone derivatives in good yield, excellent diastereoselectivity and enantioselectivity with a broad range of enals (Scheme 8.57).

In 2013, Chi and coworkers [55] found that by introducing an electron-withdrawing group (EWG) to the α-positions of chalcone substrates, the typical reactivity modes of NHC-activated enals can be alternated (Scheme 8.58). The generated enolate could facilitate effective Diels–Alder reactions with alkylidene diketones to give fully substituted cyclic enol esters. β-Arylenal is well tolerated, giving the Diels–Alder products in essentially 100% optical purities with good yields. With β-alkyl substrates, the use of a bulky imidazolium-based NHC catalyst IMes was necessary to obtain the Diels–Alder product.

In 2007, Scheidt and coworkers [56] developed a highly selective route to α,α-disubstituted cyclopentenes catalyzed by NHCs. Using the precatalyst **28** and the diisopropylethylamine (DIPEA) as the base, the selective generation of

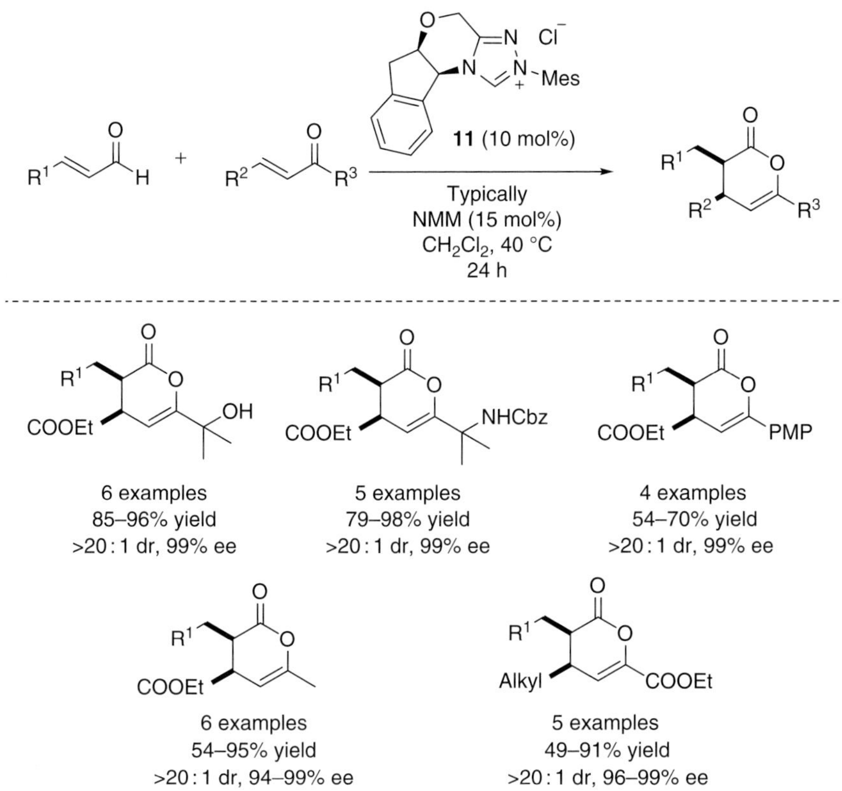

Scheme 8.56

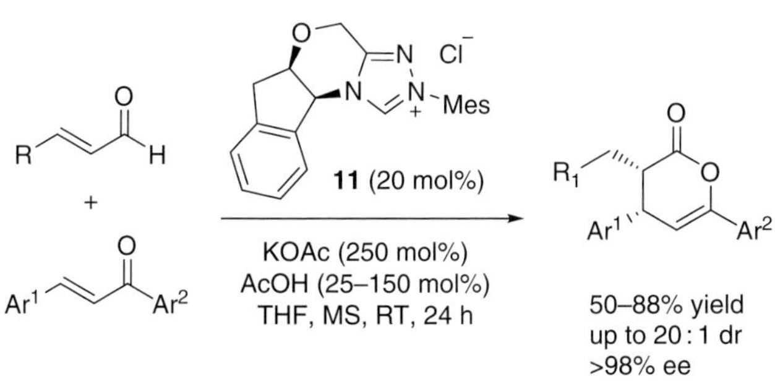

Scheme 8.57

enolate was realized to facilitate the desymmetrizing aldol addition and subsequent decarboxylation. Diastereoisomeric purity and excellent enantioselectivity (up to 96% ee) were obtained (Scheme 8.59).

This methodology was successfully applied for the synthesis of the bakkenolides [57]. The key intermediate was obtained via NHC-catalyzed desymmetrization of a 1,3-diketone with excellent enantio- and diastereoselectivity. Elaboration of the key intermediate can lead to the *bakkenolide* family (Scheme 8.60).

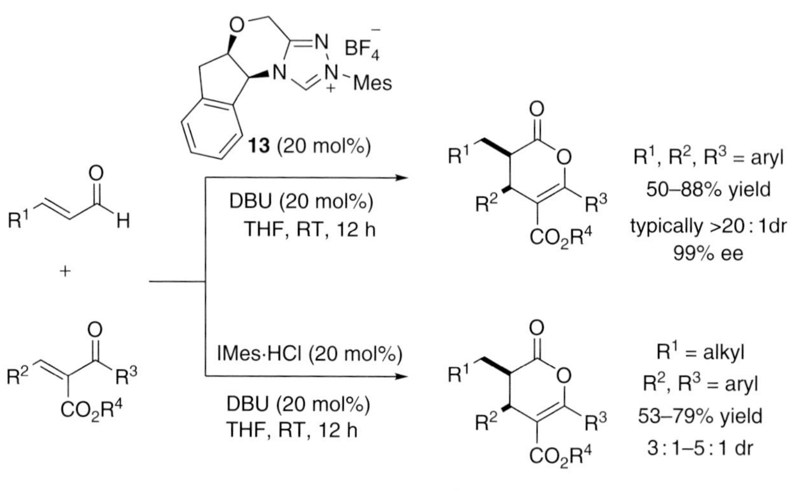

Scheme 8.58

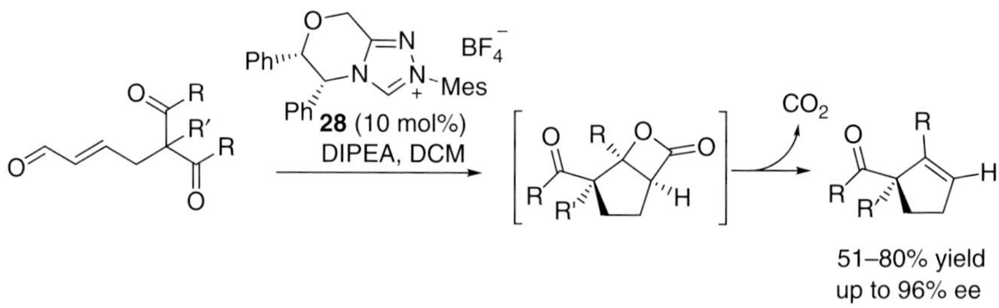

Scheme 8.59

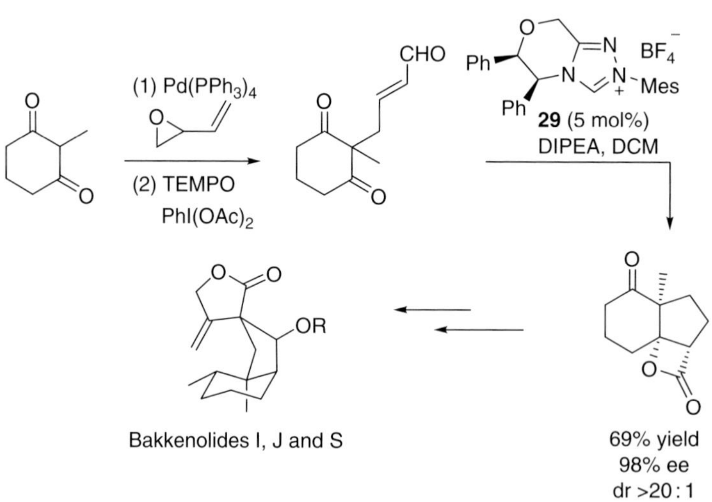

Scheme 8.60

Scheidt group has extended this methodology by employing the enones as the acceptor in the intermolecular Michael–lactonization process [58]. Aryl and alkyl substituents are suitable for the reaction, and high enantioselectivity is achieved when chiral triazolium salt **13** is used (Scheme 8.61). Similar work was reported by You and coworkers for the corresponding reactions catalyzed by a camphor-derived NHC catalyst [59].

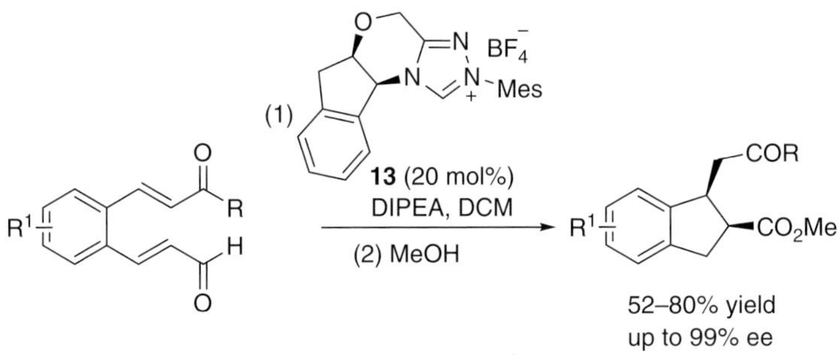

Scheme 8.61

NHC-catalyzed cascade reaction of enals and nitroalkenes to prepare enantioenriched 3,4-dihydrocoumarins is disclosed by Wang and coworkers [60]. The reaction proceeds through Michael addition via an enolate intermediate from the internal β-protonation and subsequent lactonization step. Excellent enantio- and diastereoselectivities were obtained with a broad range of scope (Scheme 8.62).

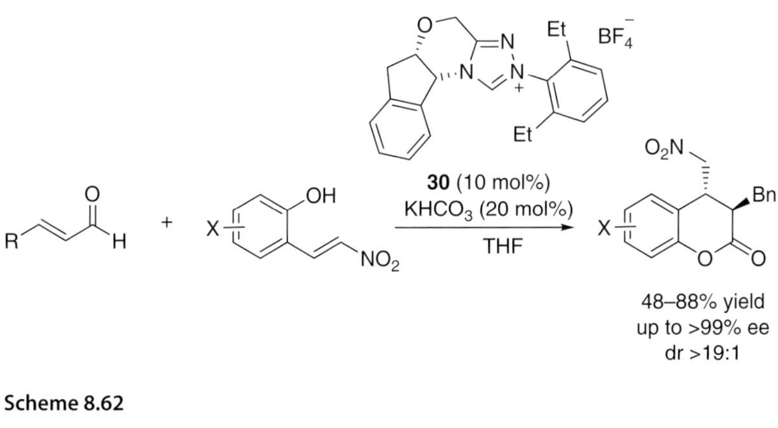

Scheme 8.62

Recently, regiodivergent annulation reaction of enals with aromatic azomethine imines has been developed by Glorius and coworkers [61]. The reaction mode is determined by the N-substituent of the NHCs. The precatalyst **30** with 1,8-diazabicyclo[5.4.0]undec-7-ene (DBU) as the base provided the [3+3] cycloaddition product, whereas the precatalyst **31** with K_2CO_3 as the base supplied the [2+3] cycloaddition product (Scheme 8.63).

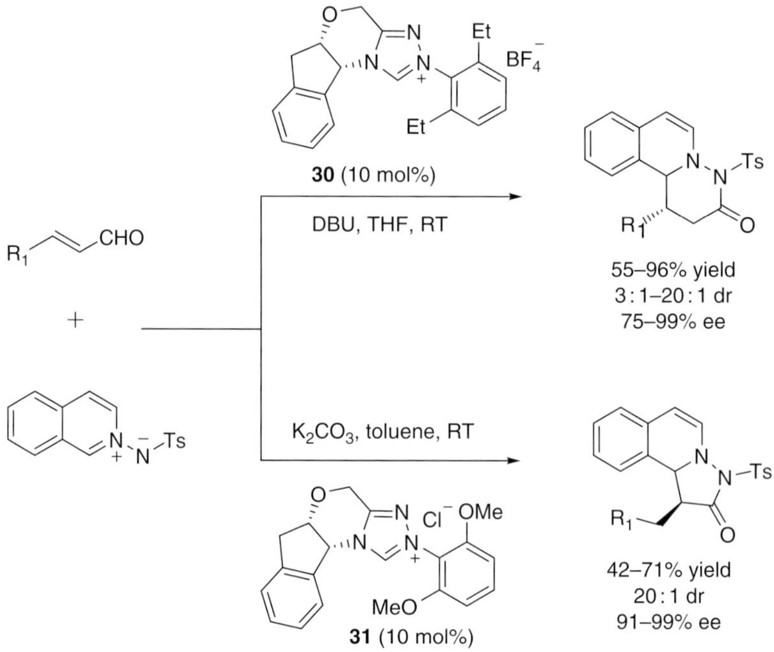

Scheme 8.63

8.5 Azolium Enolate from Aldehydes with Oxidant

In 2012, Rovis and coworkers [62] developed a novel method to generate azolium enolate from simple aliphatic aldehydes. Addition of the aliphatic aldehyde to an NHC would give the Breslow intermediate, and subsequent oxidation and deprotonation of the intermediate would yield an enolate (Scheme 8.64). In the presence of oxidant **32**, the [2+4] annulation of aldehydes with α,β-unsaturated ketamine provided the *trans*-lactams in high yields with high enantioselectivities (Scheme 8.65).

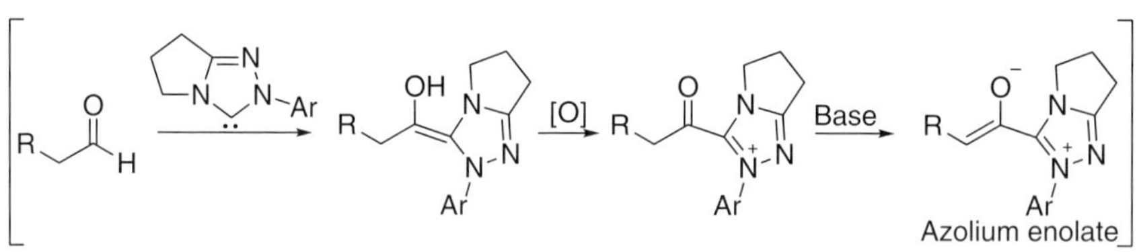

Scheme 8.64

Simultaneously, Chi and coworkers [63] reported an analogous method by employing **33** as the oxidant (Scheme 8.66). The generated ester enolate intermediates can readily undergo stereoselective reactions with enones and trifluoromethyl ketones using **13** as the precatalyst.

N-fluorobenzenesulfonimide (NFSI) can also be used as the oxidant in the reaction of oxidative generation of enolates, which is reported by Wang group

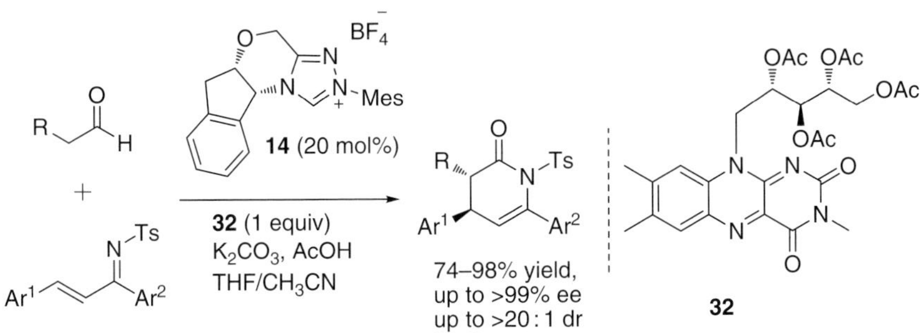

Scheme 8.65

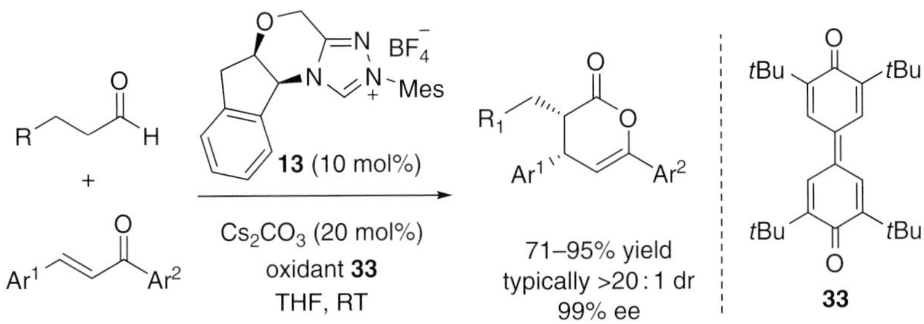

Scheme 8.66

[64a] and Sun group [64b] independently. After initial addition reaction of aldehydes with the NHC catalyst forms the Breslow intermediate **I**. Nucleophilic addition of Breslow intermediate to NFSI generates the NHC-bound fluorinated intermediate **II**. Then, elimination of hydrofluoride (HF) forms the azolium enolate in the presence of base (Scheme 8.67). Meanwhile, NFSI can also serve as an electrophilic fluorinating resource. Different α-fluorocarboxylic compounds were synthesized with high enantioselectivities (Scheme 8.68).

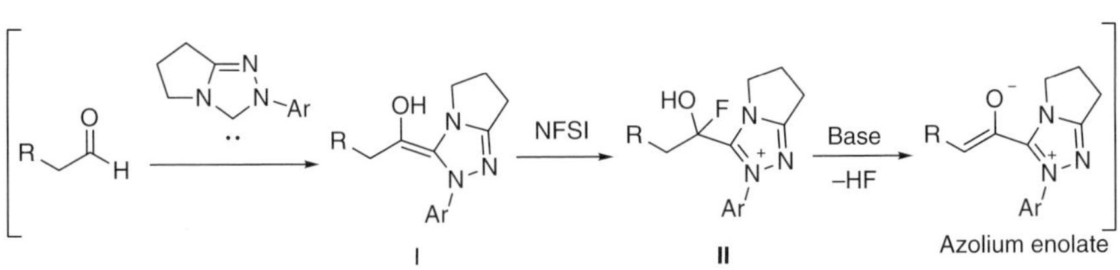

Scheme 8.67

The oxidative-generated enolate could undergo additional oxidative process in the presence of excess oxidant to form an NHC bound α,β-unsaturated ester intermediate, which can react as formal Michael acceptors [65]. This conversion could be processed via single-electron transfer or electron-pair transfer mechanism. The reaction with 1,3-dicarbonyl compounds could provide β-C(sp^3) functionalization products (Scheme 8.69).

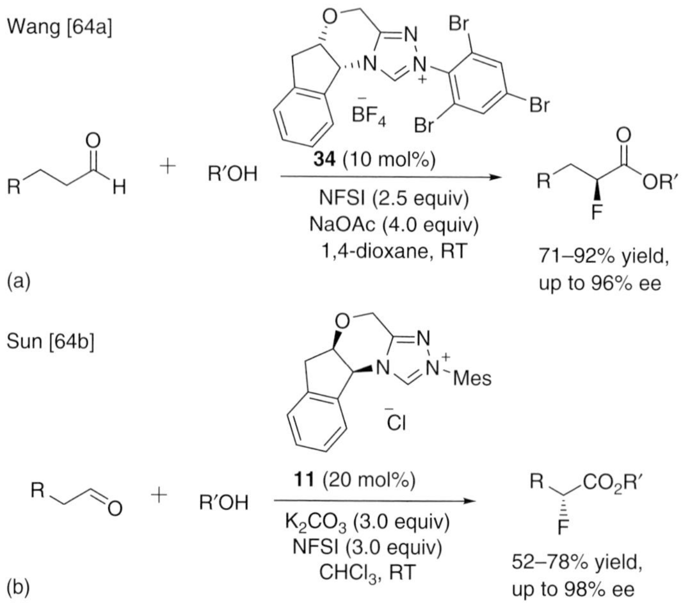

Scheme 8.68

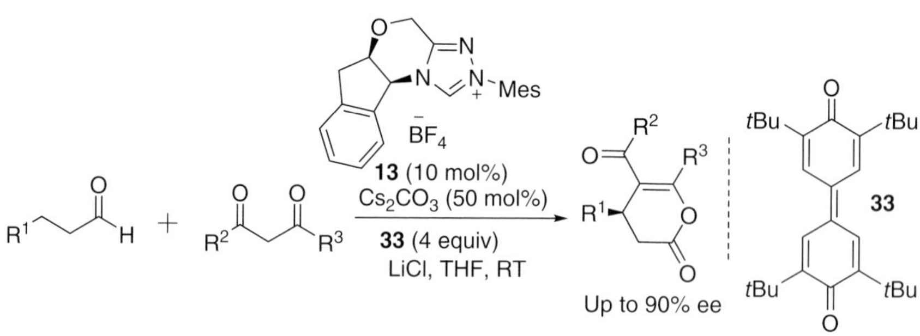

Scheme 8.69

8.6 Azolium Enolates from Activated Esters

In 2012, Chi and coworkers [66] showed that activated arylacetic acid aryl esters can act as the precursors of azolium enolates. A stable ester (**I**) bearing a good leaving group may react with an NHC to form a more reactive intermediate (**II**) with increased acidities of the α-hydrogen. The ester intermediate (**II**) subsequently undergoes deprotonation to generate enolate (**III**) as a key intermediate that can react with electrophiles (Scheme 8.70). 4-Nitro-phenol is the optimum leaving group and various β-aryl esters can behave as effective substrates for the [2+4] cycloaddition with α,β-unsaturated imine (Scheme 8.71).

Alkylacetic esters were also used as the precursors of azolium enolate using triazolium **36** as the precatalyst [67]. Excellent diastereoselectivity and enantioselectivity were obtained with a broad range of scope (Scheme 8.72).

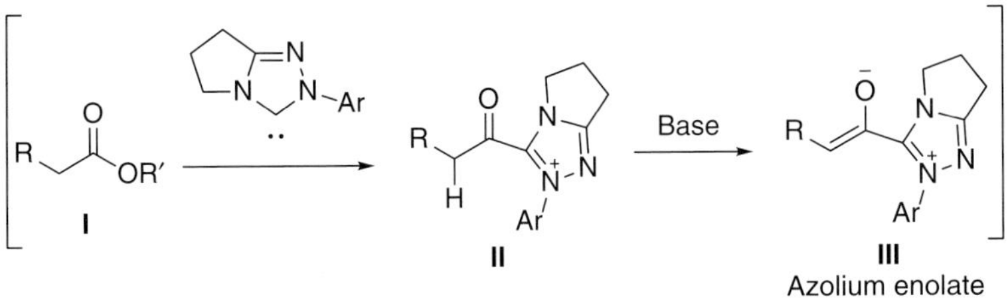

Scheme 8.70

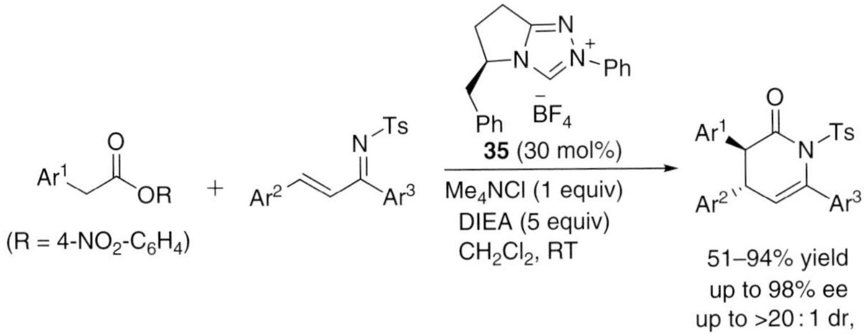

Scheme 8.71

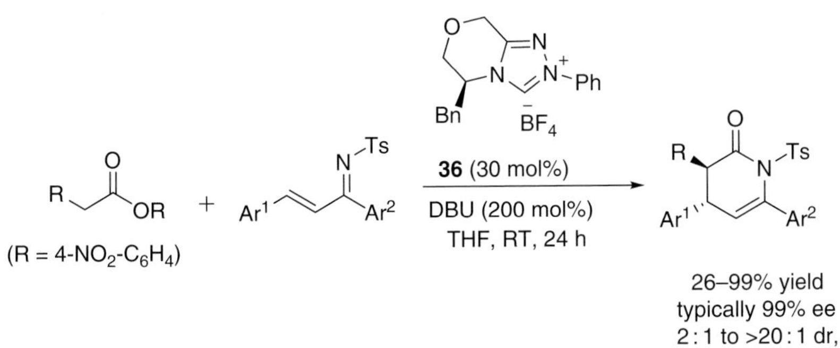

Scheme 8.72

The simple acetic ester substrates [68] could also be employed as the azolium enolate precursor. The generated two-carbon enolate intermediates served as two-carbon building blocks could readily undergo highly enantioselective reactions with α,β-unsaturated ketones and imines to give α-unsubstituted δ-lactones and lactams, respectively (Scheme 8.73).

In 2013, Chi and coworkers [69] expanded the use of esters in the inert β-sp^3 carbon of saturated esters as nucleophiles. The initial azolium enolate was generated from esters and NHC. The electron-withdrawing nature of the azolium substituent in this compound renders the β-CH_2 protons relatively acidic and can be deprotonated in the presence of excess DBU to afford the homoenolate equivalent, which is analogous to that obtained under conventional conditions with

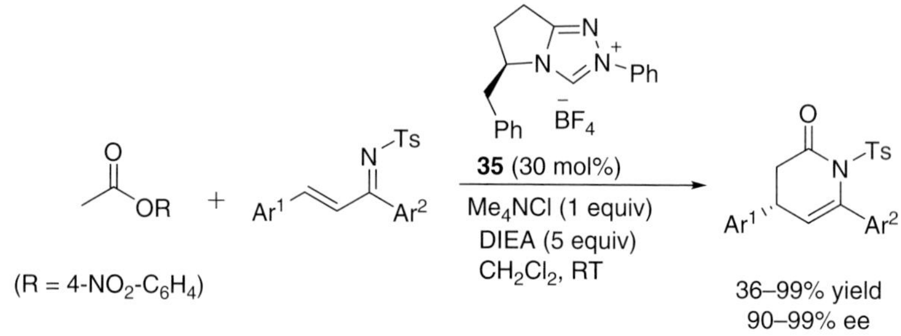

Scheme 8.73

enal. It can then readily react as a nucleophile to afford β-functionalized products (Scheme 8.74). Using the NHC precatalyst **35**, high enantioselectivities were obtained during the annulation reactions with five-membered carbo- and heterocycles, using enone, trifluoroketone, or hydrazone as electrophiles (Scheme 8.75). The β-activation of carboxylic anhydrides was also disclosed by the same group.

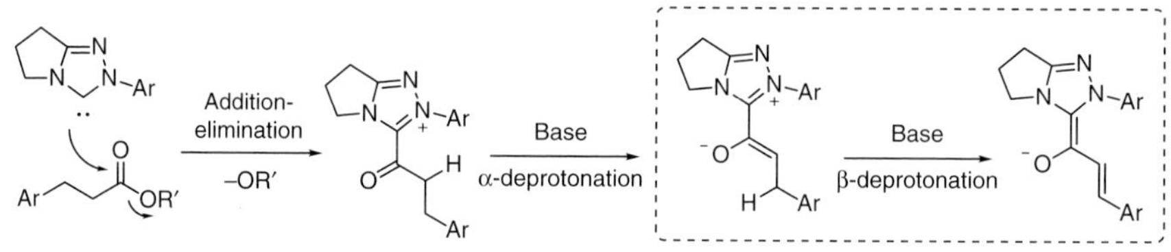

Scheme 8.74

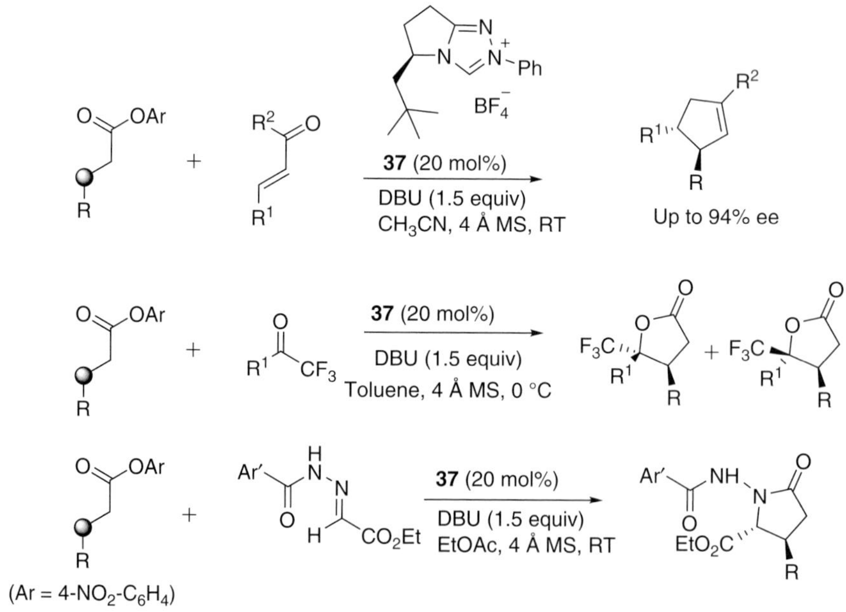

Scheme 8.75

8.7 Azolium Enolates from Acids

Carboxylic acids served as precursors to NHC enolates were disclosed by Scheidt group [70]. In the presence of *N,N′*-carbonyldiimidazole (CDI) as the acid-activating reagent, acyl imidazole was generated from acid and transferred to azolium enolate under NHC catalyst (Scheme 8.76). Simultaneously, a reactive aza-*o*-quinone methide was generated under the basic conditions. The reaction between enolate and aza-*o*-quinone methide could proceed through [2+4] cycloaddition, providing dihydroquinolones in good yield and excellent enantioselectivity (Scheme 8.77).

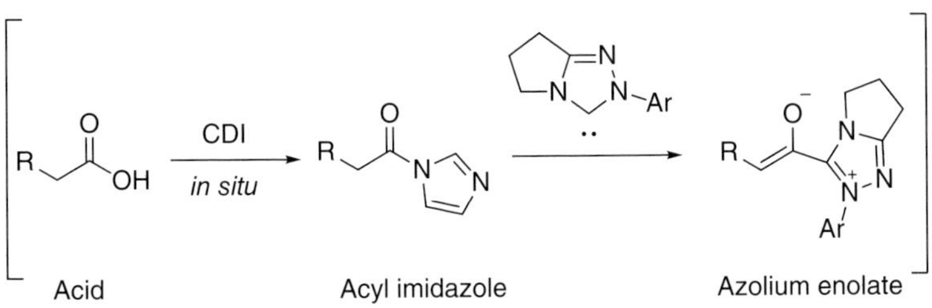

Scheme 8.76

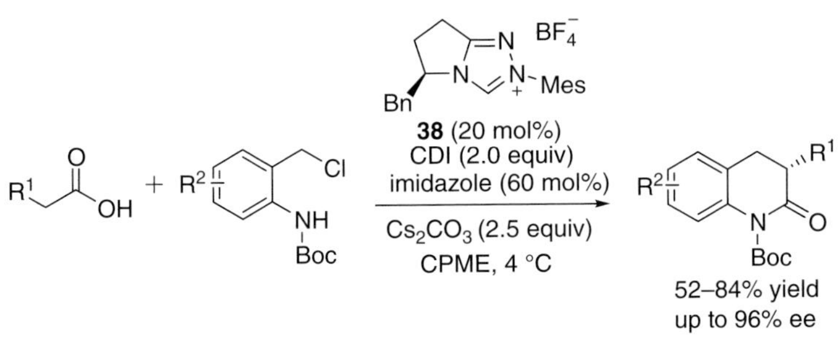

Scheme 8.77

Lately, Scheidt group [71] developed a more convenient method to generate the NHC-enolate by employing the acyl imidazole as the starting material. The reaction of azolium enolate with the *in situ* generated activated *o*-quinone methide could proceed to provide dihydrocoumarin derivatives in good yield with good-to-excellent enantioselectivity (Scheme 8.78).

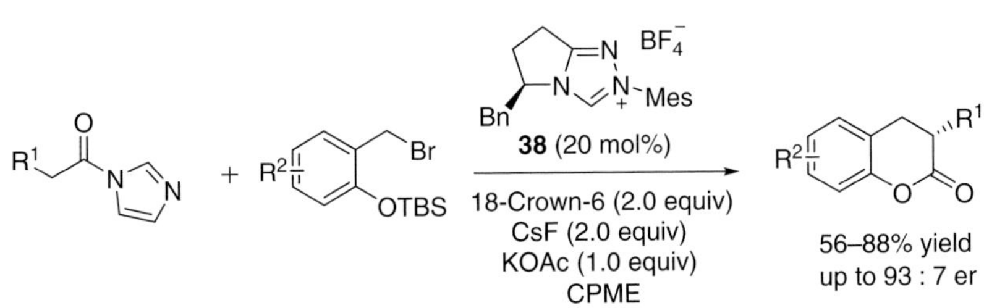

Scheme 8.78

A complementary method to generate the azolium enolate from acid is developed by Ye and coworkers [72] using pivaloyl chloride as the activating reagent (Scheme 8.79). The mixed anhydride was generated *in situ*, followed by the addition of NHC to give azolium enolate. The [2+4] cycloaddition with enones afforded the corresponding δ-lactones in good yields with good diastereo- and excellent enantioselectivities (Scheme 8.80).

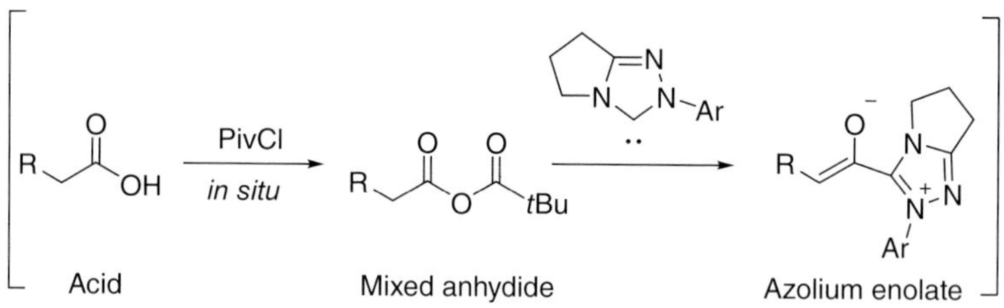

Scheme 8.79

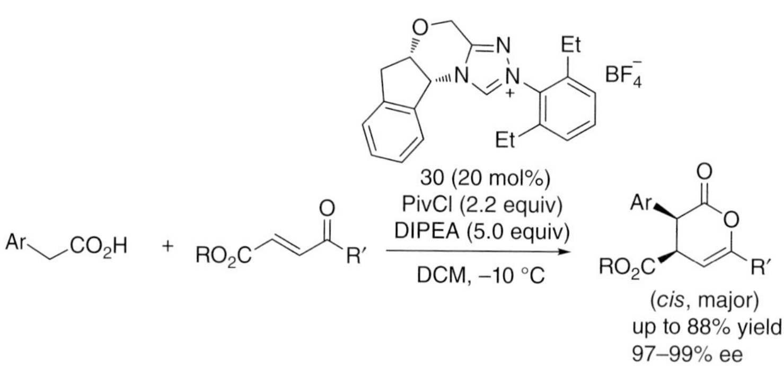

Scheme 8.80

The azolium enolate generated from acid can also be converted to homoenolate intermediate by proton transfer in the presence of adequate base, a similar route as the β-activation of esters. In 2015, Yao and coworkers [73] have developed an NHC-catalyzed β-functionalization of saturated carboxylic acid by employing 2-(7-aza-1H-benzotriazole-1-yl)-1,1,3,3-tetramethyluronium hexafluorophosphates (HATU) as the acid-activating reagent (Scheme 8.81). The

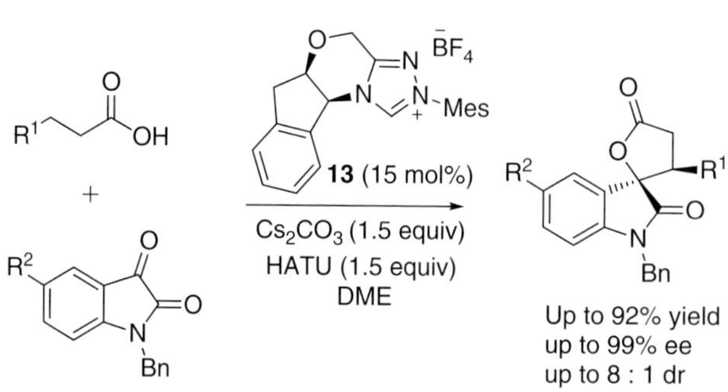

Scheme 8.81

desired spirocyclic oxindolo-γ-butyrolactones were obtained in good-to-high yields with excellent enantioselectivity.

Direct β-carbon activation of propionic acid($C_2H_5CO_2H$) by carbene organocatalysis has been developed by Chi and coworkers [74]. Compared to azolium homoenolate from acrolein, this process operates from cheap, stable, and safe starting material. The generated smallest azolium homoenolate intermediate served as a three-carbon nucleophile for enantioselective formal [3+2] cycloadditions with activated ketones (Scheme 8.82).

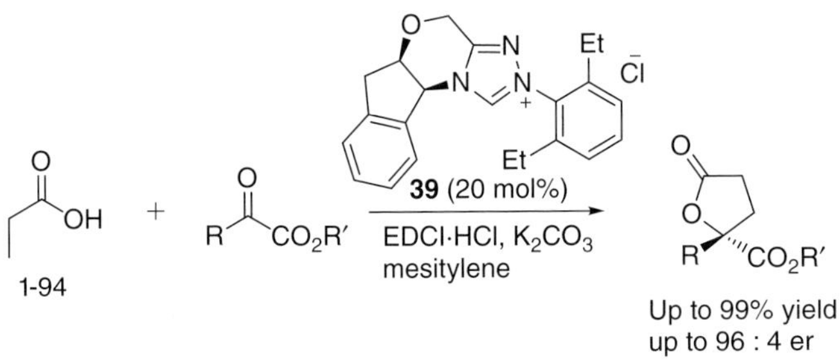

Scheme 8.82

8.8 Azolium Dienolate

NHC-enolate can be extended to dienolate, which can facilitate γ-activation of the substrate and served as four-carbon building blocks in various cycloadditions. In 2011, Ye and coworkers [75] have demonstrated that unsaturated acyl chlorides can be used as the precursor in the presence of a base. The activated ketones, such as isatins and trifluoromethyl ketones, have been employed as the effective [2+4] coupling partners with NHC precursor **17** as the optimum catalyst (Scheme 8.83).

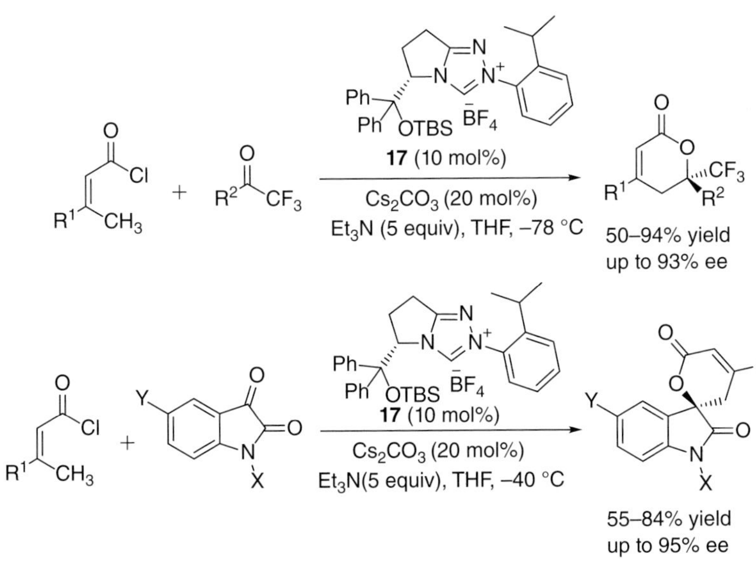

Scheme 8.83

The NHCs were also found to be efficient catalysts for the [4+2] annulation of α,β-unsaturated acyl chlorides with azodicarboxylates or 3-alkylideneoxindoles. However, very low enantioselectivities were observed for the investigated reactions. Fortunately, high enantioselectivities were realized by the corresponding synthesis catalyzed by cinchona alkaloids [76].

The first NHC-catalyzed asymmetric fluorination reaction is disclosed by Sun and coworkers [77]. The key dienolate intermediate can be generated through addition of an NHC to enal and subsequent elimination and deprotonation. In the presence of NFSI, the C—F bond formation occurs efficiently at the a position of enals, providing β,γ-unsaturated a-fluoroesters with excellent enantioselectivities (Scheme 8.84).

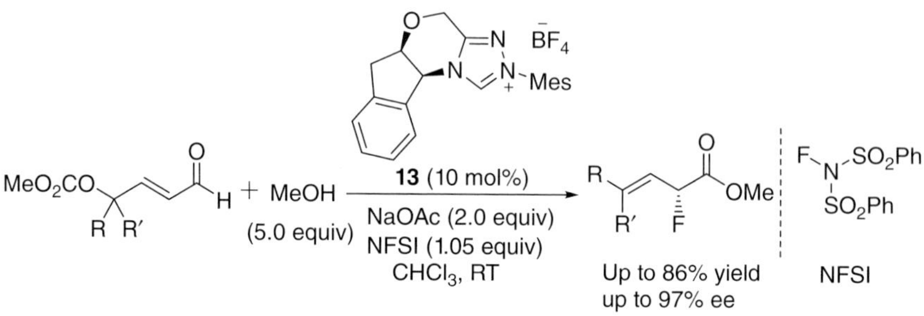

Scheme 8.84

This methodology can be extended in the [2+4] annulation with azodicarboxylates allowing access to corresponding dihydropyridazinones [78]. Various enals were well tolerated with good yield and excellent enantioselectivities. The synthetic application of the cycloaddition product has been demonstrated for the synthesis of γ-amino acids and their derivatives (Scheme 8.85).

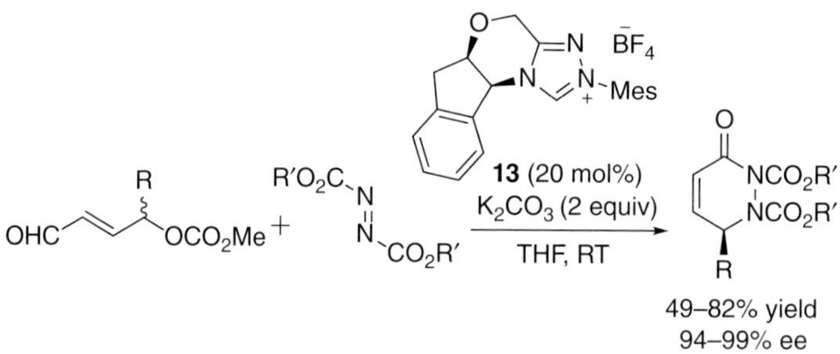

Scheme 8.85

NHC-catalyzed [4+2] cycloaddition of unsubstituted dienolate from enals with isatins was developed by Ye and coworkers [79]. The corresponding spirocyclic oxindolo-dihydropyranones were obtained in moderate-to-good yields with moderate-to-good enantioselectivities (Scheme 8.86).

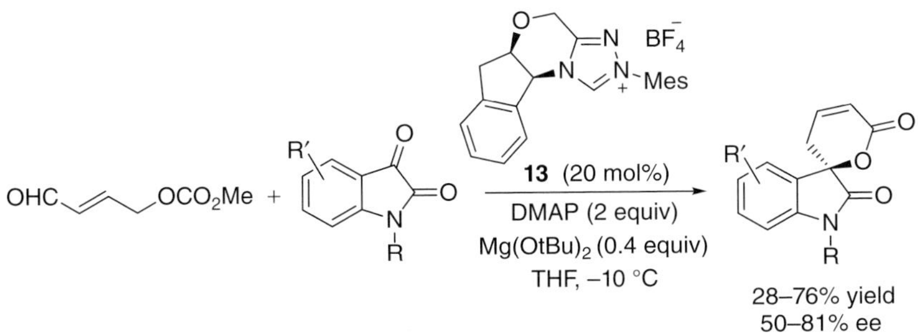

Scheme 8.86

Recently, Sun and coworkers [80] developed an efficient NHC-catalyzed dihalomethylenation process, giving unsaturated δ,δ-dihalo esters in good yields. The merge of NHC catalysis with single-electron chemistry was realized for the first time. The versatile carbon-centered trihalomethyl radicals have been demonstrated, for the first time, to be compatible with an NHC-bound intermediate, thus leading to efficient and regioselective intermolecular C—C bond formation (Scheme 8.87).

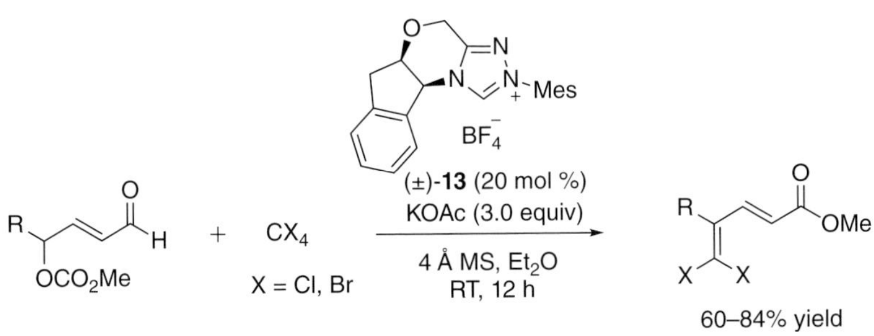

Scheme 8.87

The proposed mechanism (Scheme 8.88) initiates with the generation of dienolate intermediate, which may react in two possible ways. In pathway a, **II** undergoes single-electron oxidation by CBr_4 to form the radical cation, and subsequent radical–radical coupling gives the activated acyl species **IV**. Alternatively, as shown in pathway b, **II** could form a C—C bond with the electrophilic radical Br_3CC to generate the radical zwitterion **V**, which then undergoes SET oxidation by CBr_4 to generate the same intermediate **IV**. Finally, acyl substitution by methanol regenerates the NHC catalyst and delivers the δ-tribromomethylated intermediate, which leads to the observed product by HBr elimination.

NHC-catalyzed oxidative generation of vinyl enolates from enals is disclosed by Chi and coworkers [81]. In the presence of oxidant **33**, unsaturated acyl azolium was formed from enal, and subsequent γ-deprotonation generates the vinyl enolate intermediate (Scheme 8.89), which could lead to [4+2] cycloaddition with a range of trifluoroacetophenone. Lewis acid cocatalysts, $Mg(OTf)_2$ and $Sc(OTf)_3$,

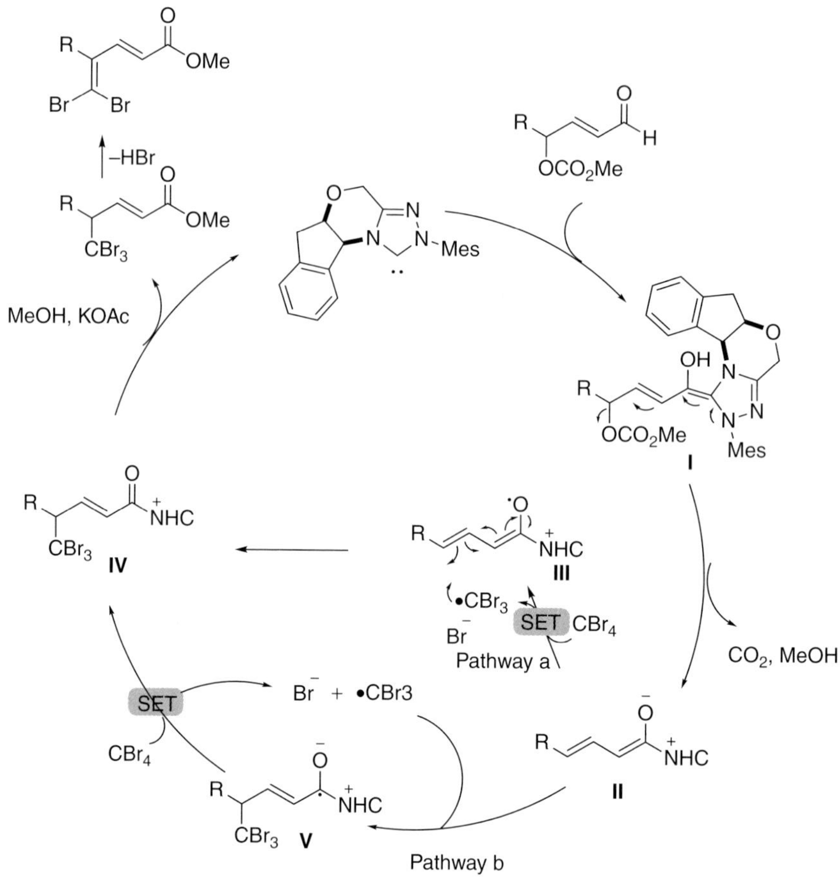

Scheme 8.88

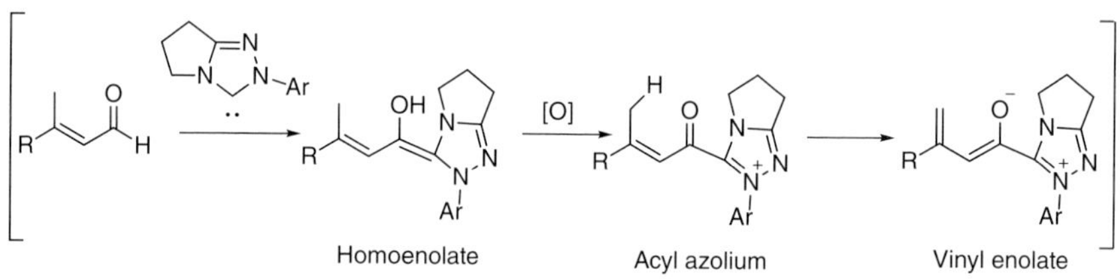

Scheme 8.89

are crucial for the enantioselective control of the γ-functionalization of enals. A series of unsaturated δ-lactones were obtained in good yield and excellent enantioselectivities (Scheme 8.90).

The NHC-catalyzed activation of the aromatic carbon atoms and the branched carbon atoms is realized [82]. NHC-bound heterocyclic ortho-quinodimethane from 2-methylindole-3-carboxaldehyde could be generated, which can readily undergo [4+2] cycloaddition with trifluoromethyl ketones and isatins to give multicyclic and spirocyclic lactones, respectively. High enantioselectivities were obtained with a broad range of scope (Scheme 8.91).

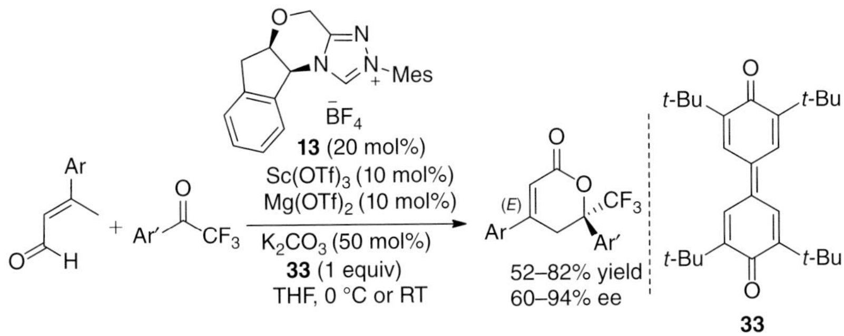

Scheme 8.90

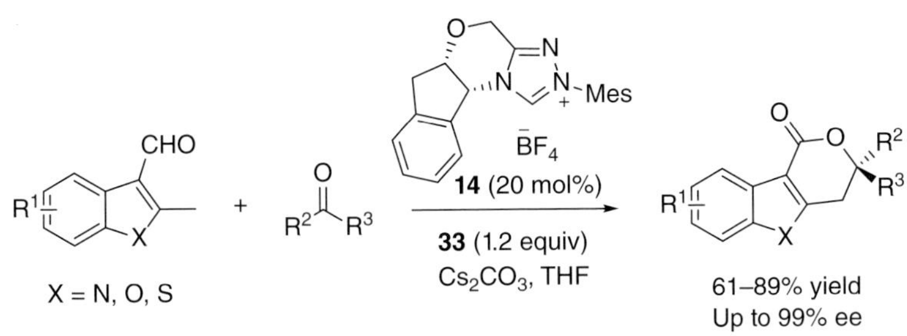

Scheme 8.91

This methodology was extended by employing the oxidative vinyl enolate from enals in the [4+3] reaction with azomethine imines to give the desired dinitrogen-fused seven-membered heterocyclic compounds [83]. In addition, the reaction enables highly effective kinetic resolution of azomethine imines (*S*-factor up to 339). Various azomethine imines and enals are well tolerated with high diastereoselectivities (>20 : 1 dr) and enantioselectivities (84–99% ee) (Scheme 8.92).

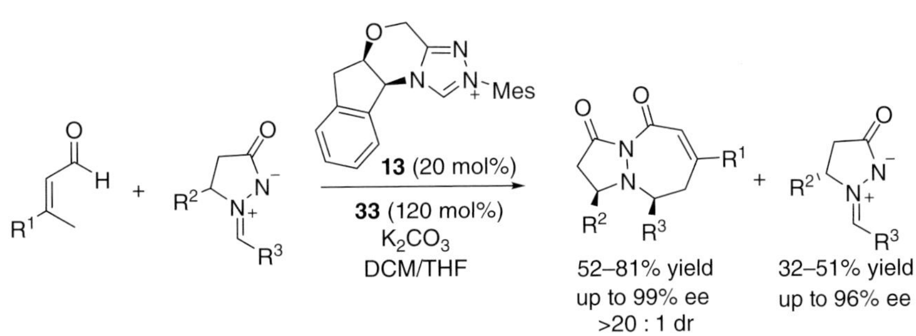

Scheme 8.92

Bioactive multisubstituted benzenes were synthesized by using NHC-bound vinyl enolate as the 1,4-diploarelectrophiles in the formal [3+3] annulation with enones [84]. Excess oxidants were added to generate the key vinyl enolate intermediate and to generate the final product. The reaction operates efficiently within one step in good-to-excellent yields (Scheme 8.93).

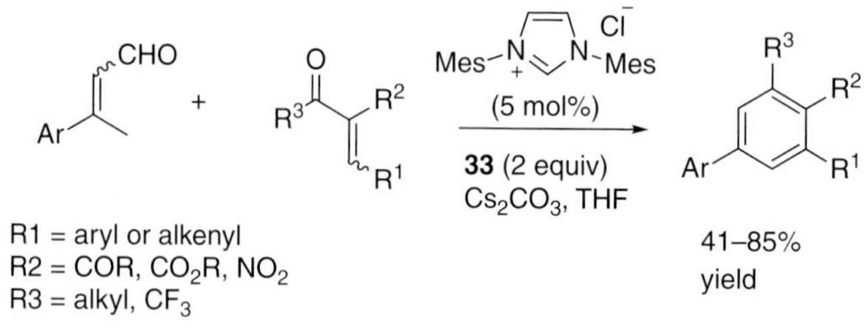

Scheme 8.93

NHC-catalyzed [4+2] annulation of α-bromo-α,β-unsaturated aldehydes with isatin has been developed by Yao et al. [85]. The vinyl enolate was generated by the γ-deprotonation of α,β-unsaturated acyl azolium, which was formed by the addition of NHC to bromoenal (Scheme 8.94). The enantioselective variant of the annulation was also realized with NHC and Lewis acid as the cocatalyst [86].

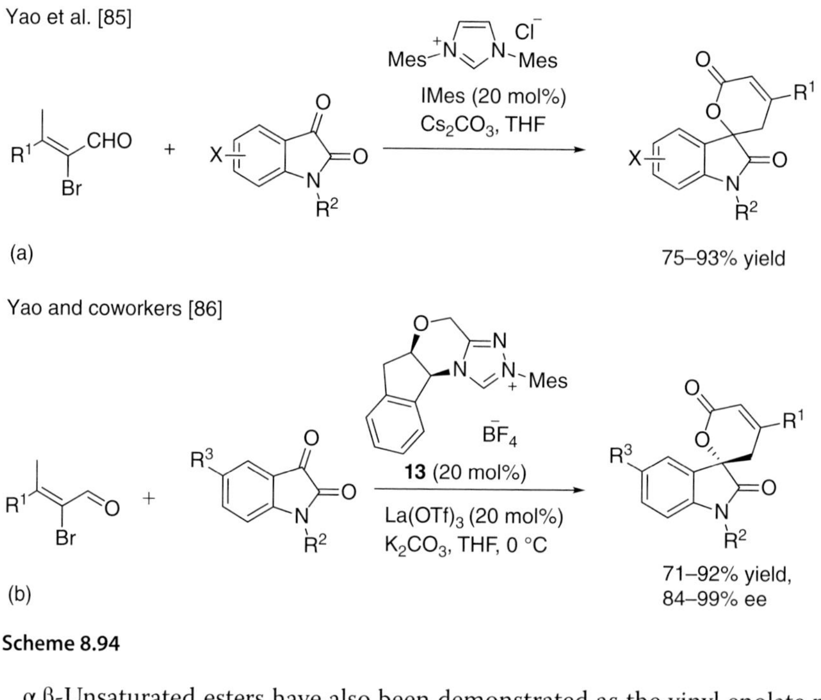

Scheme 8.94

α,β-Unsaturated esters have also been demonstrated as the vinyl enolate precursor for the synthesis of δ-lactam [87]. γ-Aminoalkylation of unsaturated ester was realized with high levels of enantioselective control (Scheme 8.95).

N-Hydroxybenzotriazole esters have also been used as the vinyl enolate precursor in the [2+4] annulation to prepare spirocyclic oxindole dihydropyranones [88]. The reaction operated in good yield with high enantioselectivity (Scheme 8.96).

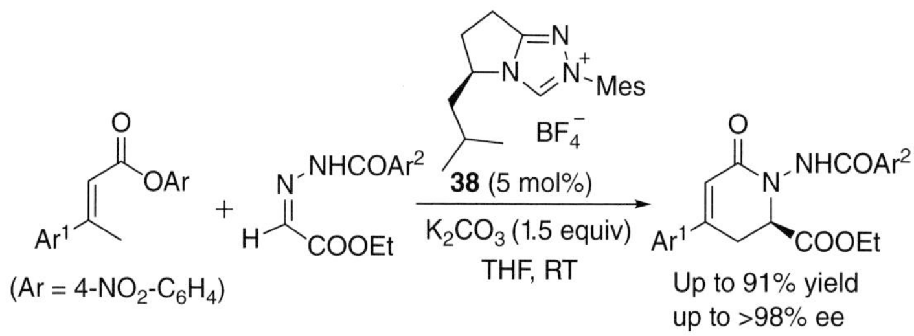

Scheme 8.95

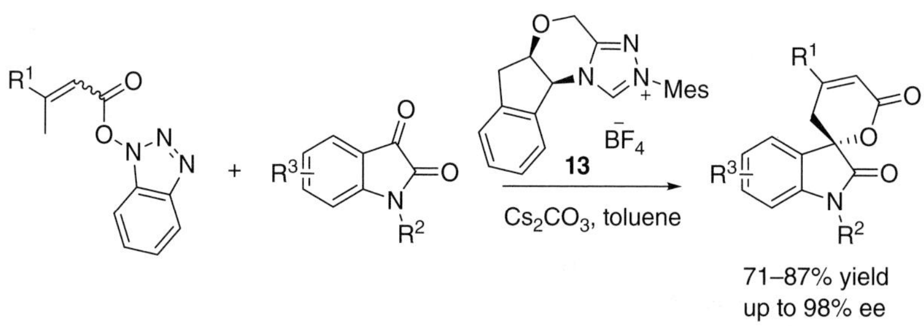

Scheme 8.96

NHC-catalyzed generation of dienoate from α,β-unsaturated carboxylic acid via the *in situ* formed mixed anhydride was developed, and the subsequent [4+2] annulation with hydrazones and isatin-derived imines was realized to allow access to dihydropyridinones and spirocyclic oxindolodihydropyridinones, respectively, in moderate-to-good yields with good-to-excellent enantioselectivities (Scheme 8.97) [89].

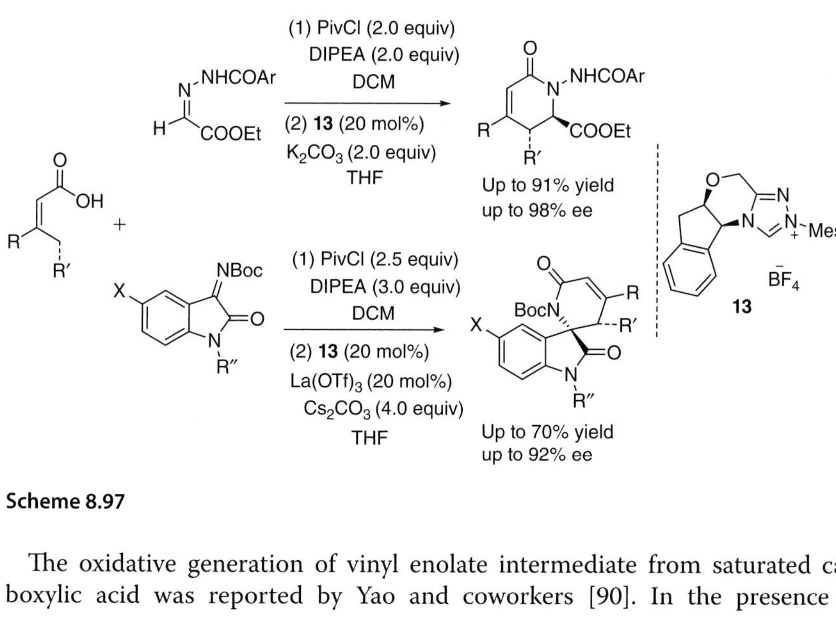

Scheme 8.97

The oxidative generation of vinyl enolate intermediate from saturated carboxylic acid was reported by Yao and coworkers [90]. In the presence of acylhydrazones as the coupling partner, the [2+4] annulation occurs in high yields with excellent enantioselectivity (Scheme 8.98).

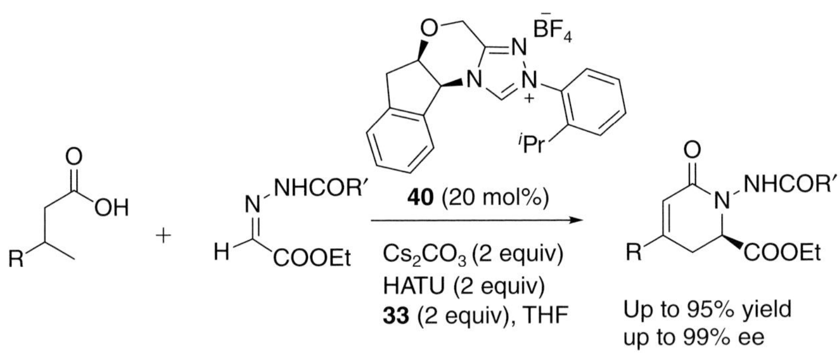

Scheme 8.98

NHCs have been demonstrated as efficient catalysts in the reaction of activation of cyclobutenones [91]. Dienolate was generated through addition of an NHC to cyclobutenones followed by C—C single-bond cleavage (Scheme 8.99). It can readily undergo [4+2] cycloaddition with various sulfonyl imines and isatin imines, providing structurally diverse and multicyclic compounds with high optical purities (Scheme 8.100). This methodology was extended to the [2+4] annulation with azomethine imine using isothiourea as the organocatalyst.

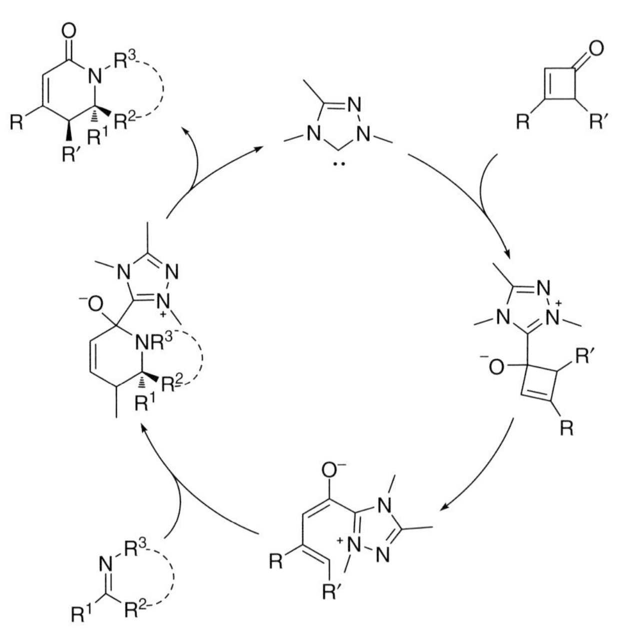

Scheme 8.99

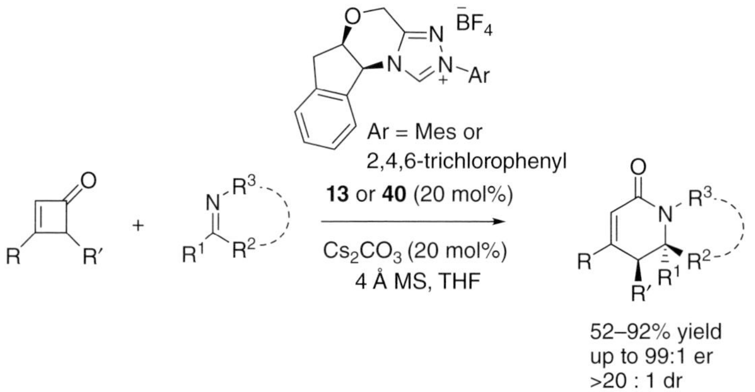

Scheme 8.100

8.9 Conclusions and Outlook

In conclusion, various simple organic molecules can be transformed into a range of meaningful products by NHC catalyst. Various reaction modes have been developed that has led to a large increase in NHC reaction diversity. The NHCs have also been demonstrated as the efficient catalysts to generate azolium enolate and dienolate, which can lead to important transformations. With the rapid development of NHC catalysis, more reaction modes will surely be developed, cooperation with other catalytic systems, and the wide applications in organic synthesis could be expected.

References

1 (a) Sasai, H., Arai, T., and Shibasaki, M. (1994). *J. Am. Chem. Soc.* 116: 1571–1572. (b) Yoshikawa, N., Yamada, Y.M.A., Das, J. et al. (1999). *J. Am. Chem. Soc.* 121: 4168–4178. (c) Trost, B.M. and Ito, H. (2000). *J. Am. Chem. Soc.* 122: 12003–12004. (d) Evans, D.A., Downey, C.W., and Hubbs, J.L. (2003). *J. Am. Chem. Soc.* 125: 8706–8707. (e) Magdziak, D., Lalic, G., Lee, H.M. et al. (2005). *J. Am. Chem. Soc.* 127: 7284–7285. (f) Saito, S. and Kobayashi, S. (2006). *J. Am. Chem. Soc.* 128: 8704–8705.

2 (a) List, B. (2004). *Acc. Chem. Res.* 37: 548–557. (b) Mukherjee, S., Yang, J.W., Hoffmann, S., and List, B. (2007). *Chem. Rev.* 107: 5471–5569.

3 (a) Kolb, H.C., VanNieuwenhze, M.S., and Sharpless, K.B. (1994). *Chem. Rev.* 94: 2483–2547. (b) O'Donnell, M.J. (2004). *Acc. Chem. Res.* 37: 506–517. (c) Marcelli, T., van Maarseveen, J.H., and Hiemstra, H. (2006). *Angew. Chem. Int. Ed.* 45: 7496–7504.

4 (a) Fu, G.C. (2004). *Acc. Chem. Res.* 37: 542–547. (b) Fu, G.C. (2006). *Acc. Chem. Res.* 39: 853–860.

5 Chow, K.Y.-K. and Bode, J.W. (2004). *J. Am. Chem. Soc.* 126: 8126–8127.

6 Reynolds, N.T., Read de Alaniz, J., and Rovis, T. (2004). *J. Am. Chem. Soc.* 126: 9518–9519.

7 Sohn, S.S. and Bode, J.W. (2006). *Angew. Chem. Int. Ed.* 45: 6021–6024.

8 Reynolds, N.T. and Rovis, T. (2005). *J. Am. Chem. Soc.* 127: 16406–16407.
9 (a) Bode, J.W. and Sohn, S.S. (2007). *J. Am. Chem. Soc.* 129: 13798–13799. (b) Vora, H.U. and Rovis, T. (2007). *J. Am. Chem. Soc.* 129: 13796–13797.
10 (a) Vora, H.U. and Rovis, T. (2010). *J. Am. Chem. Soc.* 132: 2860–2861. (b) Wheeler, P., Vora, H.U., and Rovis, T. (2013). *Chem. Sci.* 4: 1674–1679.
11 Kawanaka, Y., Phillips, E.M., and Scheidt, K.A. (2009). *J. Am. Chem. Soc.* 131: 18028–18029.
12 He, M., Uc, G.J., and Bode, J.W. (2006). *J. Am. Chem. Soc.* 128: 15088–15089.
13 He, M., Beahm, B.J., and Bode, J.W. (2008). *Org. Lett.* 10: 3817–3820.
14 Kobayashi, S., Kinoshita, T., Uehara, H. et al. (2009). *Org. Lett.* 11: 3934–3937.
15 Jian, T.-Y., Sun, L.-H., and Ye, S. (2012). *Chem. Commun.* 48: 10907–10909.
16 Zhang, H.-M., Lv, H., and Ye, S. (2013). *Org. Biomol. Chem.* 11: 6255–6257.
17 Chen, K.-Q., Zhang, H.-M., Wang, D.-L. et al. (2015). *Org. Biomol. Chem.* 13: 6694–6697.
18 Wang, D.-L., Liang, Z.-Q., Chen, K.-Q. et al. (2015). *J. Org. Chem.* 80: 5900–5905.
19 Liang, Z.-Q., Wang, D.-L., Zhang, C.-L., and Ye, S. (2016). *Org. Biomol. Chem.* 14: 6422–6425.
20 Lv, H., Mo, J., Fang, X., and Chi, Y.R. (2011). *Org. Lett.* 13: 5366–5369.
21 (a) Davies, A.T., Pickett, P.M., Slawin, A.M.Z., and Smith, A.D. (2014). *ACS Catal.* 4: 2696–2700. (b) Attaba, N., Taylor, J.E., Slawin, A.M.Z., and Smith, A.D. (2015). *J. Org. Chem.* 80: 9728–9739.
22 Taylor, J.E., Daniels, D.S.B., and Smith, A.D. (2013). *Org. Lett.* 15: 6058–6061.
23 Davies, A.T., Slawin, A.M.Z., and Smith, A.D. (2015). *Chem. Eur. J.* 21: 18944–18948.
24 Ni, Q., Zhang, H., Grossmann, A. et al. (2013). *Angew. Chem. Int. Ed.* 52: 13562–13566.
25 Gao, Z.-H., Chen, X.-Y., Cheng, J.-T. et al. (2015). *Chem. Commun.* 51: 9328–9331.
26 Paull, D.H., Weatherwax, A., and Lectka, T. (2009). *Tetrahedron* 65: 6771–6803.
27 Zhang, Y.-R., He, L., Wu, X. et al. (2008). *Org. Lett.* 10: 277–280.
28 Zhang, H.-M., Gao, Z.-H., and Ye, S. (2014). *Org. Lett.* 16: 3079–3081.
29 Douglas, J., Taylor, J.E., Churchill, G. et al. (2013). *J. Org. Chem.* 78: 3925–3938.
30 He, L., Lv, H., Zhang, Y.-R., and Ye, S. (2008). *J. Org. Chem.* 73: 8101–8103.
31 Wang, X.-N., Shao, P.-L., Lv, H., and Ye, S. (2009). *Org. Lett.* 11: 4029–4031.
32 Wang, X.-N., Zhang, Y.-Y., and Ye, S. (2010). *Adv. Synth. Catal.* 352: 1892–1895.
33 Huang, X.-L., Chen, X.-Y., and Ye, S. (2009). *J. Org. Chem.* 74: 7585–7587.
34 Wang, T., Huang, X.-L., and Ye, S. (2010). *Org. Biomol. Chem.* 8: 5007–5011.
35 Lv, H., Zhang, Y.-R., Huang, X.-L., and Ye, S. (2008). *Adv. Synth. Catal.* 350: 2715–2718.
36 Jian, T.-Y., He, L., Tang, C., and Ye, S. (2011). *Angew. Chem. Int. Ed.* 50: 9104–9107.
37 Wang, X.-N., Shen, L.-T., and Ye, S. (2011). *Org. Lett.* 13: 6382–6385.
38 Wang, X.-N., Shen, L.-T., and Ye, S. (2011). *Chem. Commun.* 47: 8388–8390.

39 Namitharan, K., Zhu, T., Cheng, J. et al. (2014). *Nat. Commun.* 5: 3982.
40 Shao, P.-L., Chen, X.-Y., and Ye, S. (2010). *Angew. Chem. Int. Ed.* 49: 8412–8416.
41 Zhang, Y.-R., Lv, H., Zhou, D., and Ye, S. (2008). *Chem. Eur. J.* 14: 8473–8476.
42 Lv, H., Chen, X.-Y., Sun, L.-H., and Ye, S. (2010). *J. Org. Chem.* 75: 6973–6976.
43 Jian, T.-Y., Shao, P.-L., and Ye, S. (2011). *Chem. Commun.* 47: 2381–2383.
44 Lv, H., You, L., and Ye, S. (2009). *Adv. Synth. Catal.* 351: 2822–2826.
45 Jian, T.-Y., Chen, X.-Y., Sun, L.-H., and Ye, S. (2013). *Org. Biomol. Chem.* 11: 158–163.
46 Shao, P.-L., Chen, X.-Y., Sun, L.-H., and Ye, S. (2010). *Tetrahedron Lett.* 51: 2316–2318.
47 Huang, X.-L., He, L., Shao, P.-L., and Ye, S. (2009). *Angew. Chem. Int. Ed.* 48: 192–195.
48 Leckie, S.M., Brown, T.B., Pryde, D. et al. (2013). *Org. Biomol. Chem.* 11: 3230–3246.
49 Wang, X.-N., Lv, H., Huang, X.-L., and Ye, S. (2009). *Org. Biomol. Chem.* 7: 346–350.
50 Concellón, C., Duguet, N., and Smith, A.D. (2009). *Adv. Synth. Catal.* 351: 3001–3009.
51 Douglas, J., Ling, K.B., Concellón, C. et al. (2010). *Eur. J. Org. Chem.* 2010: 5863–5869.
52 He, M., Struble, J.R., and Bode, J.W. (2006). *J. Am. Chem. Soc.* 128: 8418–8420.
53 Kaeobamrung, J., Kozlowski, M.C., and Bode, J.W. (2010). *Proc. Natl. Acad. Sci. U.S.A.* 107: 20661–20665.
54 Fu, Z., Sun, H., Chen, S. et al. (2013). *Chem. Commun.* 49: 261–263.
55 Fang, X., Chen, X., and Chi, Y.R. (2011). *Org. Lett.* 13: 4708–4711.
56 Wadamoto, M., Phillips, E.M., Reynolds, T.E., and Scheidt, K.A. (2007). *J. Am. Chem. Soc.* 129: 10098–10099.
57 Phillips, E.M., Roberts, J.M., and Scheidt, K.A. (2010). *Org. Lett.* 12: 2830–2833.
58 Phillips, E.M., Wadamoto, M., Chan, A., and Scheidt, K.A. (2007). *Angew. Chem. Int. Ed.* 46: 3107–3110.
59 Li, Y., Wang, X.-Q., Zheng, C., and You, S.-L. (2009). *Chem. Commun.* 5823–5825.
60 Wu, Z., Wang, X., Li, F. et al. (2015). *Org. Lett.* 17: 3588–3591.
61 Guo, C., Fleige, M., Janssen-Müller, D. et al. (2015). *Nat. Chem.* 7: 842–847.
62 Zhao, X., Ruhl, K.E., and Rovis, T. (2012). *Angew. Chem. Int. Ed.* 51: 12330–12333.
63 Mo, J., Yang, R., Chen, X. et al. (2013). *Org. Lett.* 15: 50–53.
64 (a) Li, F., Wu, Z., and Wang, J. (2015). *Angew. Chem. Int. Ed.* 54: 656–659. (b) Dong, X., Yang, W., Hu, W., and Sun, J. (2015). *Angew. Chem. Int. Ed.* 54: 660–663.
65 Mo, J., Shen, L., and Chi, Y.R. (2013). *Angew. Chem. Int. Ed.* 52: 8588–8591.
66 Hao, L., Du, Y., Lv, H. et al. (2012). *Org. Lett.* 14: 2154–2157.
67 Chen, S., Hao, L., Zhang, Y. et al. (2013). *Org. Lett.* 15: 5822–5825.

68 Hao, L., Chen, S., Xu, J. et al. (2013). *Org. Lett.* 15: 4956–4959.

69 (a) Fu, Z., Xu, J., Zhu, T. et al. (2013). *Nat. Chem.* 5: 835–839. (b) Jin, Z., Chen, S., Wang, Y. et al. (2014). *Angew. Chem. Int. Ed.* 53: 13506–13509.

70 Lee, A., Younai, A., Price, C.K. et al. (2014). *J. Am. Chem. Soc.* 136: 10589–10592.

71 Lee, A. and Scheidt, K.A. (2015). *Chem. Commun.* 51: 3407–3410.

72 Cheng, J.-T., Chen, X.-Y., and Ye, S. (2015). *Org. Biomol. Chem.* 13: 1313–1316.

73 Xie, Y., Yu, C., Li, T. et al. (2015). *Chem. Eur. J.* 21: 5355–5359.

74 Jin, Z., Jiang, K., Fu, Z. et al. (2015). *Chem. Eur. J.* 21: 9360–9363.

75 Shen, L.-T., Shao, P.-L., and Ye, S. (2011). *Adv. Synth. Catal.* 353: 1943–1948.

76 (a) Shen, L.-T., Sun, L.-H., and Ye, S. (2011). *J. Am. Chem. Soc.* 133: 15894–15897. (b) Shen, L.-T., Jia, W.-Q., and Ye, S. (2013). *Angew. Chem. Int. Ed.* 52: 585–588.

77 Zhao, Y.-M., Cheung, M.S., Lin, Z., and Sun, J. (2012). *Angew. Chem. Int. Ed.* 51: 10359–10363.

78 Chen, X.-Y., Xia, F., Cheng, J.-T., and Ye, S. (2013). *Angew. Chem. Int. Ed.* 52: 10644–10647.

79 Cheng, J.-T., Chen, X.-Y., Gao, Z.-H., and Ye, S. (2015). *Eur. J. Org. Chem.* 2015: 1047–1053.

80 Yang, W., Hu, W., Dong, X. et al. (2016). *Angew. Chem. Int. Ed.* 55: 15783–15786.

81 Mo, J., Chen, X., and Chi, Y.R. (2012). *J. Am. Chem. Soc.* 134: 8810–8813.

82 Chen, X., Yang, S., Song, B.-A., and Chi, Y.R. (2013). *Angew. Chem. Int. Ed.* 52: 11134–11137.

83 Wang, M., Huang, Z., Xu, J., and Chi, Y.R. (2014). *J. Am. Chem. Soc.* 136: 1214–1217.

84 Zhu, T., Zheng, P., Mou, C. et al. (2014). *Nat. Commun.* 5: 5027.

85 Yao, C., Xiao, Z., Liu, R. et al. (2013). *Chem. Eur. J.* 19: 456–459.

86 Xiao, Z., Yu, C., Li, T. et al. (2014). *Org. Lett.* 16: 3632–3635.

87 Xu, J., Jin, Z., and Chi, Y.R. (2013). *Org. Lett.* 15: 5028–5031.

88 Que, Y., Li, T., Yu, C. et al. (2015). *J. Org. Chem.* 80: 3289–3294.

89 Jia, W.-Q., Zhang, H.-M., Zhang, C.-L. et al. (2016). *Org. Chem. Front.* 3: 77–81.

90 Que, Y., Xie, Y., Li, T. et al. (2015). *Org. Lett.* 17: 6234–6237.

91 (a) Li, B.-S., Wang, Y., Jin, Z. et al. (2015). *Nat. Commun.* 6: 6207. (b) Li, B.-S., Wang, Y., Jin, Z., and Chi, Y.R. (2015). *Chem. Sci.* 6: 6008–6012.

9

N-Heterocyclic Carbenes as Brønsted Base Catalysts

Jiean Chen and Yong Huang

Peking University, State Key Laboratory of Chemical Oncogenomics, Key Laboratory of Chemical Genomics, Shenzhen Graduate School, Shenzhen, 518055, China

Following the observation of the thiazolium-mediated benzoin reaction by Ukai et al. [1] and the seminal mechanistic insights proposed by Breslow [2], the chemical reactivity of *N*-heterocyclic carbenes (NHCs) has been investigated intensively [3]. The intrinsic physical characteristics and catalytic properties of such species in organocatalysis failed to receive adequate attention before Arduengo and coworkers isolation of a stable carbene [4]. The partially aromatic structure of NHCs stabilizes and enhances the orbital interactions. The σ-lone-pair in the bent sp^2-carbon and the adjacent π-electrons overlap with the unoccupied p-orbital of this central carbon and are thus orchestrated and form the basis of the carbene's chemical reactivity. NHCs behave as robust σ-donors with medium π-acidity overall. Additionally, the hetero-elements generate a tunable electronic and steric environment ideal for versatile catalysis. In transition metal catalysis, NHCs serve as ligands with properties distinct from those of the more familiar phosphines and nitrogen ligands. They exhibit stronger σ-donating ability and comparable π-electrophilicity (Figure 9.1a) [5]. The C_2-symmetric imidazolylidene structure dominates this series of NHC skeletons and is compatible with mainstream transition metals in numerous catalytic transformations. When a strong σ-donating property encounters an electron-deficient element, a stable complex is formed (Figure 9.1b) [6]. The NHC–borane complex has been shown to have dramatically enhanced stability and orthogonal reactivity. It is thought that a covalent bond exists between the carbene center and the boron atom and that such complexes show a nucleophilic propensity for conjugate addition rather than classical hydroboration [7]. Similar interactions with other semi- and nonmetallic elements have been harvested for new bond-forming reactions. The affinity of NHC for carbon electrophiles is particularly attractive for carbon–carbon and carbon–heteroatom bond formation. Most NHC-catalyzed reactions are initiated by nucleophilic attack of an NHC on an electron-deficient substrate, generating an *umpolung* intermediate (Figure 9.1c) [3]. The aromaticity of NHC is crucial in the modulation of the overall electron flow throughout the entire catalytic circle. The adjacent heteroatom effect enhances the σ-donating ability of NHC in the initial addition reaction

N-Heterocyclic Carbenes in Organocatalysis, First Edition. Edited by Akkattu T. Biju.
© 2019 Wiley-VCH Verlag GmbH & Co. KGaA. Published 2019 by Wiley-VCH Verlag GmbH & Co. KGaA.

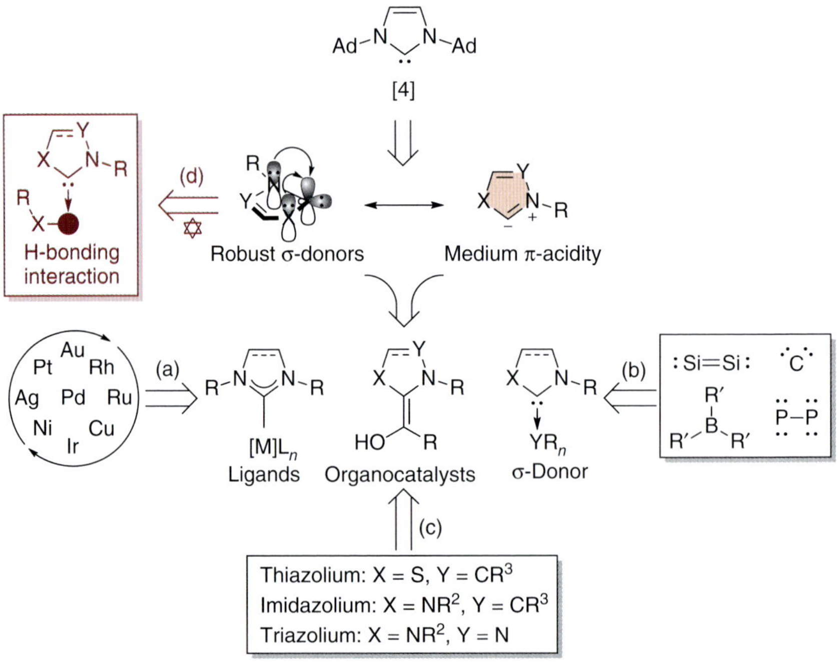

Figure 9.1 NHC catalysis.

and stabilizes the resulting tetrahedral zwitterion. The electron-withdrawing property of the cationic NHC moiety facilitates the hydrogen migration and generation of the Breslow intermediate [8]. The covalent linkage of catalyst and substrate ensures efficient induction and transfer of chirality [9], and the steric nature of the neighboring nitrogen substituent impacts the facial discrimination in the subsequent assembly of chiral centers. Such orchestration of electronic and steric influences expands the versatility of NHC-mediated asymmetric catalysis.

When considering the nucleophilicity of NHCs, a conclusion can be reached that most NHC-catalyzed transformations are a result of synergy between σ-donors and π-acceptors. In organocatalytic systems, the σ-electron lone pair of the NHC takes advantage of either vacant σ-accepting orbitals of the p-block elements or π*-orbitals of carbonyl groups and electron-deficient olefins and forms a covalent bond. The unique electronic structures of heterocycles provide a useful balance between stability and reactivity, and the adjacent nitrogen atoms have a σ-electron-withdrawing and π-donating influence on the carbene center. On the other hand, NHCs may also form hydrogen bonds with acidic substrates (Figure 9.1d) [10]. The aromaticity of NHCs defines them as stronger Brønsted bases than either phosphines or amines. In this chapter, we summarize the recent development of NHC catalysis using such noncovalent interactions, with an emphasis on its stereoselective aspects.

For reactions involving NHCs as Brønsted base catalysts, the predominant factor concerning the catalytic activity is the pK_a value of the corresponding

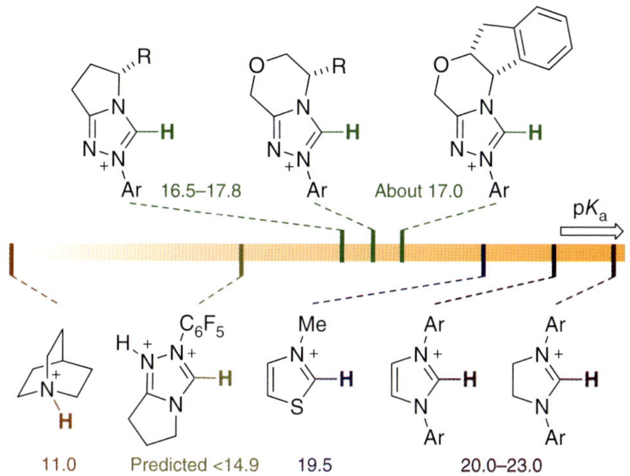

Figure 9.2 Kinetic acidities of common NHC precursors.

precatalysts [11]. The imidazolium and triazolium salts are the most frequently studied NHC precursors. The Brønsted basicity of several such privileged scaffolds has been systematically investigated both experimentally and computationally. The pK_a values of these heteroarenium salts vary from approximately 15.0 to 23.0, and these NHC precursors are therefore orders of magnitude weaker Brønsted acids than typical ammonium salts such as quinuclidinium, whose pK_a is 11.0 in aqueous solution (Figure 9.2). As a result, NHCs derived from heteroarenium precursors are stronger hydrogen-bonding acceptors than amines. The utilization of NHCs as noncovalent templates was initiated and propagated, taking advantage of this chemical feature. Mainstream chiral NHC precursors, most commonly imidazolium and triazolium salts, have pK_a values generally exceeding 17.0, with the exception of those having high electron-withdrawing N-substituents such as $-C_6F_5$. These strong Brønsted base species provide excellent opportunities for asymmetric catalysis using noncovalent, hydrogen-bonding interactions.

The types of reactions catalyzed by NHCs behaving as Brønsted bases fall into two categories: transesterifications and conjugate additions (Figure 9.3). In both cases, strong hydrogen-bonding interactions between NHCs and nucleophiles accelerate the reactions by elevating the highest occupied molecule orbital (HOMO) energy of the nucleophiles, which react with either esters or electron-deficient olefins. Because of the reversibility of these transformations, regeneration of strongly basic NHCs is an important issue. Often proposed is a concerted four-membered-ring transition state (TS) that facilitates proton migration from nucleophile (or NHC) to the preliminary adduct anion with the release of the NHC catalyst.

NHCs are Brønsted bases comparable to alkoxides and enolates, which themselves are versatile reaction intermediates. Strong imidazolylidene NHCs, with a good balance between basicity and stability, are preferred catalysts for reactions involving alcohols and ketones. In 2002, Hedrick and coworkers reported the

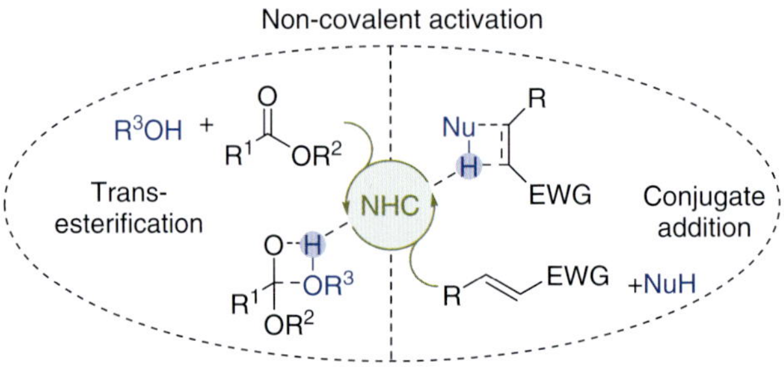

Figure 9.3 Mode of action of NHCs as Brønsted base catalysts.

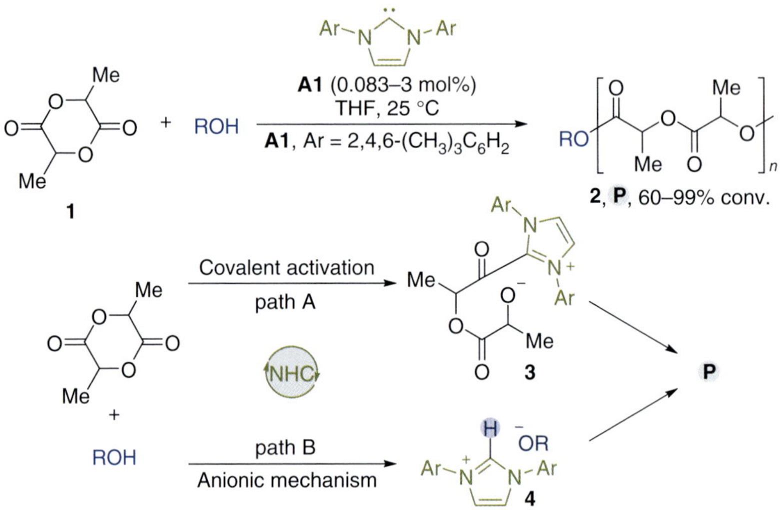

Figure 9.4 First transesterification mediated by NHCs.

first successful example of the NHC-mediated transesterification involving the activation of cyclic esters for ring-opening polymerization (ROP, Figure 9.4) [12]. Similar transesterification-ROP reactions using tertiary amines and phosphines were reported by the same group. In the case of NHC catalysis, a stable NHC (**A1**) was used to activate the lactide monomer (**1**) for alcohol-initiated polymerization affording products with narrow polydispersity. Two possible mechanisms were proposed for this process. The first of these (path A, Figure 9.4) is initiated by nucleophilic attack of **A1** on the carbonyl center of the lactide. This is followed by the generation of a covalent connection between the carbonyl and the NHC catalyst, where the alcohol initiator promotes the catalytic circle by nucleophilic attack on the acyl azolium functionality of the intermediate (**3**). The second mechanism (path B, Figure 9.4) involves deprotonation of the alcohol initiator by **A1** to give the corresponding alkoxide and imidazolium precatalyst. Subsequent nucleophilic addition of the alkoxide to the lactide (**1**) delivers the desired polymeric product. The former pathway is probably favored as the NHC

(**A1**) is not sufficiently basic to fully deprotonate alcohols. The pK_a value of the imidazolium cation is 24 in dimethyl sulfoxide (DMSO), far from that of a typical alcohol whose pK_a in DMSO is 29.

Shortly after this first report, Nolan and coworkers, reported transesterification of vinyl esters with tertiary phosphines as initiators [13]. The authors identified NHCs as effective catalysts for this reaction. Free imidazole derivatives **A1** and **A2** were found to promote facile conversion of various *O*-vinyl esters to alkyl and aryl esters. The reactions are complete within one hour at room temperature. Molecular sieves (MS) were found to absorb the alcohol, accelerating this transesterification. A number of functional groups that are incompatible in traditional Lewis acid systems were found to be tolerated by this mild NHC catalysis. Methyl esters are challenging, and use of the more basic *N*-cyclohexyl substituted catalyst (**A2**) is critical for such substrates (Figure 9.5).

Transesterification between alkyl esters and alcohols is equilibrium driven. Hedrick and Waymouth and coworkers reported the transesterification of methyl benzoate using a large excess of alcohol (20 equiv) [14]. Both the catalytic activity of NHCs and reactivity of alcohols were studied. Imidazolylidenes (**A1** and **A3**) were identified as the most active NHC species in these reactions. The efficiency of transesterification deteriorates rapidly with an increasing size of alcohols. Low conversion was observed for isopropyl alcohol using catalyst **A1**. The smaller, methyl-substituted NHC (**A3**) exhibits higher tolerance for bulky alcohols. Nevertheless, no reaction occurs for *t*-butanol with either NHC catalyst. The primary function of NHC was proposed to be as a nucleophile catalyst. NHC reacts with methyl benzoate to generate the benzoyl azolium intermediate, which subsequently reacts with an alcohol. Concurrent hydrogen-bonding activation of the alcohol by NHC is probably operative. The superior catalytic

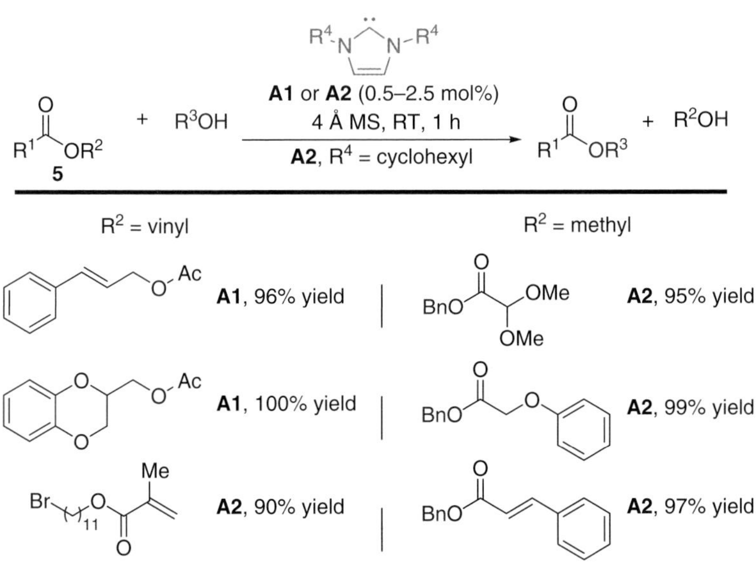

Figure 9.5 Transesterfication of *O*-vinyl and methyl esters.

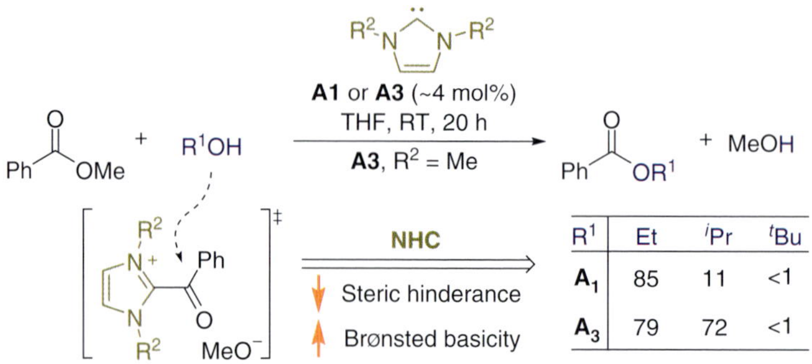

R1	Et	iPr	tBu
A1	85	11	<1
A3	79	72	<1

Figure 9.6 Transesterification of methyl benzoates.

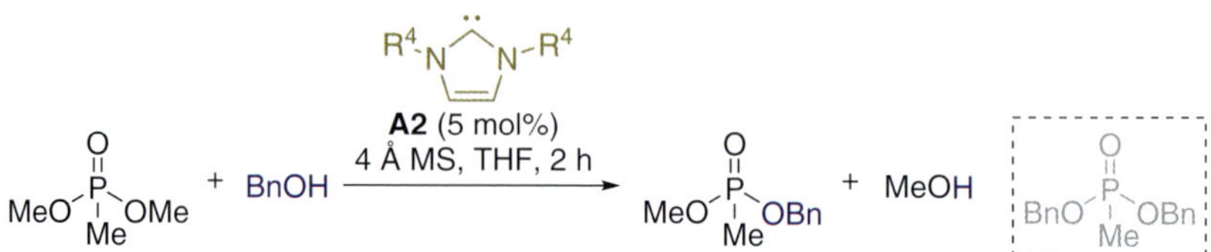

6, 75% yield

7

Figure 9.7 Transesterification of phosphorous esters.

ability of **A3** may be ascribed to the enhanced Brønsted basicity and its small size for a more robust noncovalent interaction with alcohol substrates (Figure 9.6).

A similar catalytic system was utilized by Nolan and coworker for the transesterification of phosphorous esters [15]. By carefully controlling the reaction parameters, such as the stoichiometry of benzyl alcohol and the reaction time, a selective mono-transesterification was accomplished with 75% yield (Figure 9.7). Only one OMe group was replaced by benzyl alcohol. Extending reaction times or increasing the equivalence of the alcohol lead to double transesterification, giving **7**.

The catalytic mechanism for these NHC-catalyzed transesterification reactions was proposed to be initiated by the nucleophilic addition of an NHC to the carbonyl carbon of an ester group. The resulting acyl azolium species were suggested to play a central role in most reported cases. The postulated covalent connection between the NHC catalyst and the carbonyl seems to be a reasonable assumption in view of the high nucleophilicity of NHC toward carbonyls in most reported cases. In 2005, Movassaghi and Schmidt studied NHC-mediated amidation of various ester compounds [16]. Their experimental data brought new insights into the mode of action of NHC in these reactions (Figure 9.8). The interactions of the NHC **A1** with several alcohols were evaluated by NMR analysis, and it was found that the chemical shifts of the proton of the hydroxyl groups increase by 2–5 ppm even when the sterically hindered *t*-BuOH is used. Further evidence of hydrogen-bonding interactions between NHCs and alcohols is supported by X-ray analysis of a stable NHC–alcohol complex (**9**). This noncovalent interaction suggests the likelihood of Brønsted base catalysis by NHC. Tethering a hydroxyl group to the amine was essential for

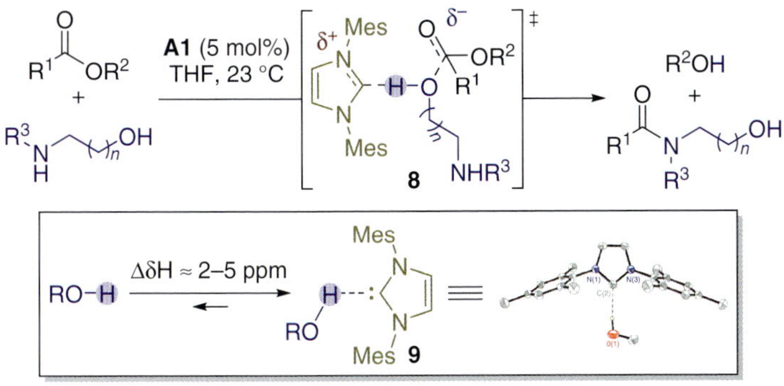

Figure 9.8 NHC-catalyzed amidation and evidence of the involvement of Brønsted base catalysis.

the amidation, and the authors proposed that the reaction proceeds by an initial esterification via noncovalent hydrogen-bonding activation of the alcohol by the NHC catalyst (Figure 9.8, transition state **8**). The ester intermediate undergoes rapid intramolecular O–N transacylation, delivering the amide products.

The role of NHC in transesterification reactions led to further interests in density functional theory (DFT) calculations (Figure 9.9) [17]. Hu and coworkers studied both mechanistic pathways of NHC-catalyzed transesterification: nucleophilic and Brønsted base activation. In the former case, the nucleophilic attack of the NHC on an ester first forms a tetrahedral zwitterion (**10**), which releases an alkoxide anion, generating the key acyl azolium intermediate (**11**). The Brønsted base catalysis pathway involves a concerted four-member-ring transition state

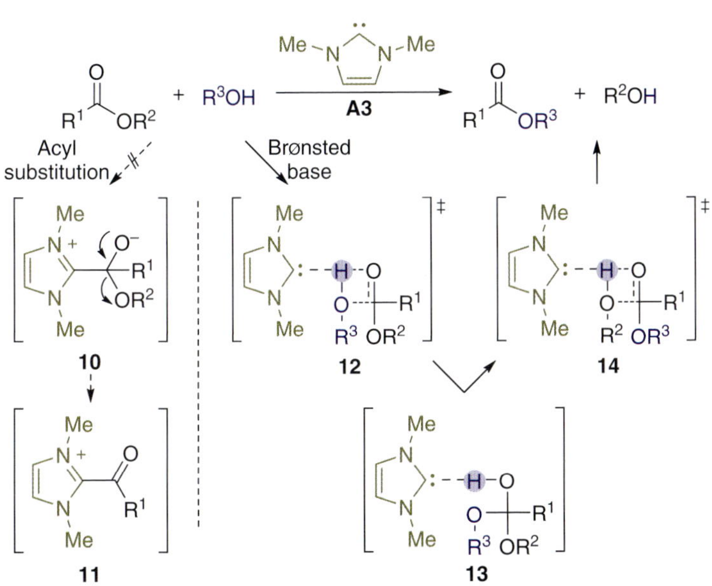

Figure 9.9 DFT calculations for NHC-catalyzed transesterification reactions.

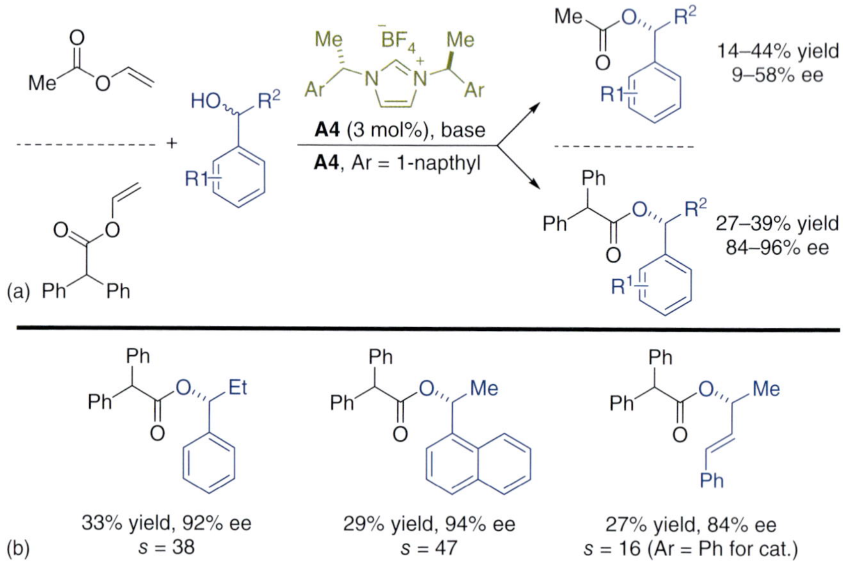

Figure 9.10 Kinetic resolution of secondary alcohols using NHCs.

(**12**) and a neutral tetrahedral intermediate (**13**), which facilitates proton transfer from the incoming alcohol to the dissociating alkoxy group without any ionization. This charge-neutral pathway is noticeably more stable than the former pathway, and calculations also rule out the earlier assumptions that the alcohol is fully deprotonated by NHC because of the unmatching pK_a values. In conclusion, noncovalent, Brønsted base catalysis would appear to be more plausible for the transesterification reactions of alkyl esters.

In addition to the mechanistic investigations, parallel efforts also blossomed on the synthetic applications utilizing NHCs as chiral frameworks. In 2004, Suzuki et al. exploited the kinetic resolution of secondary alcohols with a chiral, C_2 symmetric, imidazolium precatalyst (Figure 9.10a) [18]. Vinyl acetate was speculated to be the acylation reagent and activated by the imidazolylidene catalyst, generated *in situ*. The corresponding acyl transfer products were generated with moderate enantioselectivity. Shortly after this report, Maruoka and coworkers developed a synthesis of chiral esters under kinetic resolution conditions (Figure 9.10b) [19]. In this reaction, a more sterically demanding acyl reagent, vinyl diphenylacetate, was used in combination with a catalyst (**A4**) at a low temperature. The selectivity of various secondary alcohols was evaluated, and the efficiency was dramatically enhanced, compared to reactions using vinyl acetate as the acylating agent, giving up to 96% ee. Despite the nucleophile catalysis mechanism proposed in these reports, these reactions might also operate through the noncovalent Brønsted base catalysis, suggesting a new way of introducing chirality through hydrogen bonding.

Besides activating alcohols and facilitating proton transfer in transesterification processes, the strong Brønsted basicity of NHC was also exploited aimed at activating esters. In 2004, Nolan and coworkers reported the NHC-mediated

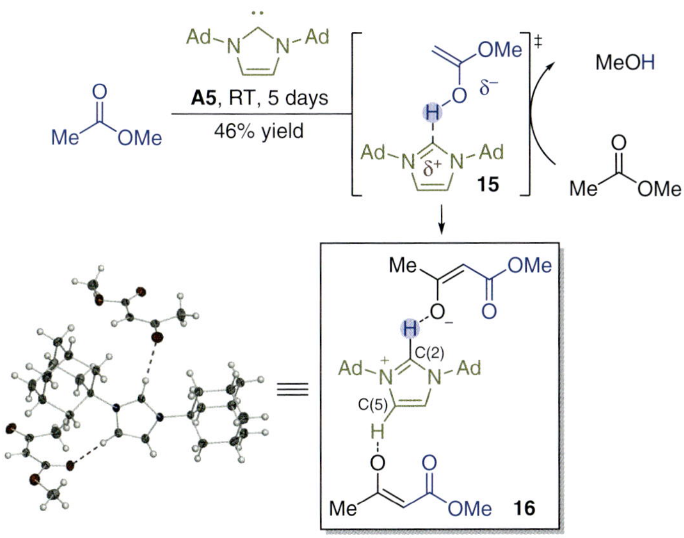

Figure 9.11 Activation of esters in a Claisen condensation.

Claisen condensation of methyl acetate (Figure 9.11) [20]. The author proposed that the NHC (**A5**) activates the enol form of the ester substrate by deprotonation. The role of the NHC catalyst is Brønsted base activation of methyl acetate via complex **15**, whose hydrogen bond stabilizes methyl acetate in its enol form and increases its nucleophilicity. The activated enol is involved in a nucleophilic attack on another methyl ester to give the Claisen condensation product. Interestingly, the product, in its enolate form, forms a stable hydrogen-bonding complex (**16**) with NHC. The structure of **16** shown in Figure 9.11 was established by the analysis of single-crystal X-ray diffraction data. This result indicates that the preliminary adduct anion prefers the enolate form, which is stabilized by a hydrogen-bonding interaction at C2 of the NHC. Another hydrogen bonding at C5 position of the N-heterocycle further stabilizes compound **16** with a discrete charge.

NHC was also used for the activation of ketone in a silyl transfer process (Figure 9.12) [21]. In 2008, Song et al. developed an NHC-mediated synthetic method for silyl enol ethers. The ketone substrate is activated by forming a hydrogen bond with the NHC **A5** (TS **17**) in its enol form. This noncovalent interaction weakens the H—O bond and increases the nucleophilicity of the enol, promoting facile silyl transfer. The silyl transfer product is formed with the release of NHC (**A5**) as the imidazolium precursor. Another side product of the silyl transfer is a highly basic ester enolate, which can deprotonate the imidazolium ion, regenerating the NHC (TS **18**).

The transesterification-type reactions described earlier demonstrate the potential of utilizing NHC's Brønsted basicity as a way of nucleophile activation. Such noncovalent interactions can also mediate another type of Brønsted base-catalyzed reaction, the conjugate addition. In 2009, Coquerel and Rodriguez

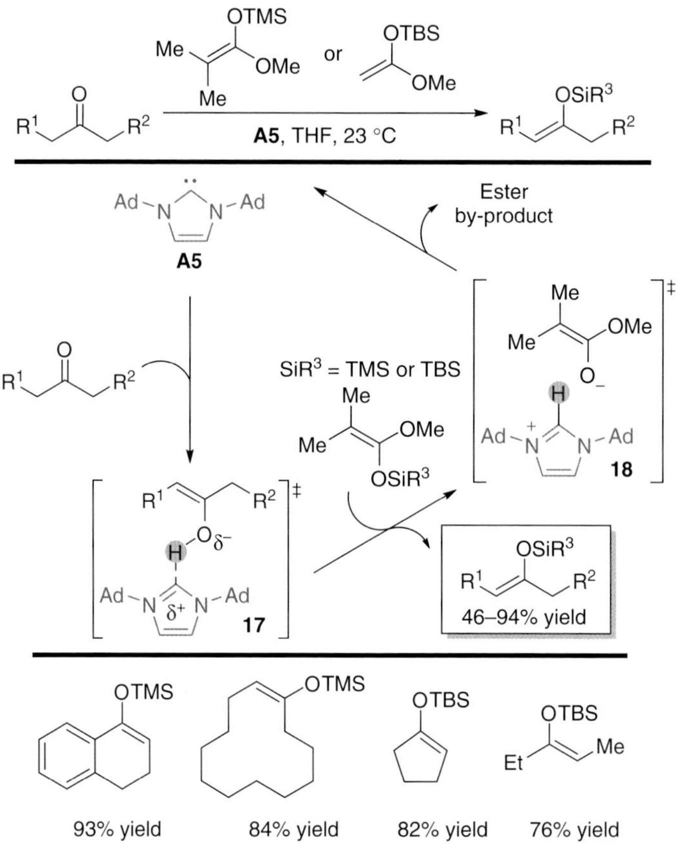

Figure 9.12 NHC-catalyzed silyl transfer reaction.

and coworkers exploited a cascade transformation linking ruthenium-catalyzed cross metathesis and subsequent Michael-type spirocyclization (Figure 9.13) [22]. The dicarbonyl compound (**19**) with an alkene terminus was treated with the second generation of Hoveyda–Grubbs' catalyst (**HGII**) and electron-deficient olefins and the cross metathesis proceeded smoothly. The crude product (**20**) underwent phosphine-mediated intramolecular Michael cyclization to yield an α-spirolactone (**21**) in moderate yield and with excellent diastereoselectivity. Interestingly, no cyclization occurs when purified compound (**20**) and 20 mol% n-Bu_3P were used. The reaction only progressed with crude mixture from the metathesis reaction. This observation led the authors to suspect that the NHC ligand on the **HGII** catalyst might be the actual catalyst, released by ligand exchange with n-Bu_3P. Indeed, when compound (**20**) is treated with NHC (**A6**), a smooth Michael cyclization reaction occurs, confirming that NHC is an efficient catalyst for the Michael spirocyclization.

This transformation represents the first NHC-catalyzed conjugate addition with Brønsted base catalysis. Presumably, the enol form of the 1,3-ketoester (**22**) is activated by the NHC catalyst via hydrogen bonding. Subsequent nucleophilic

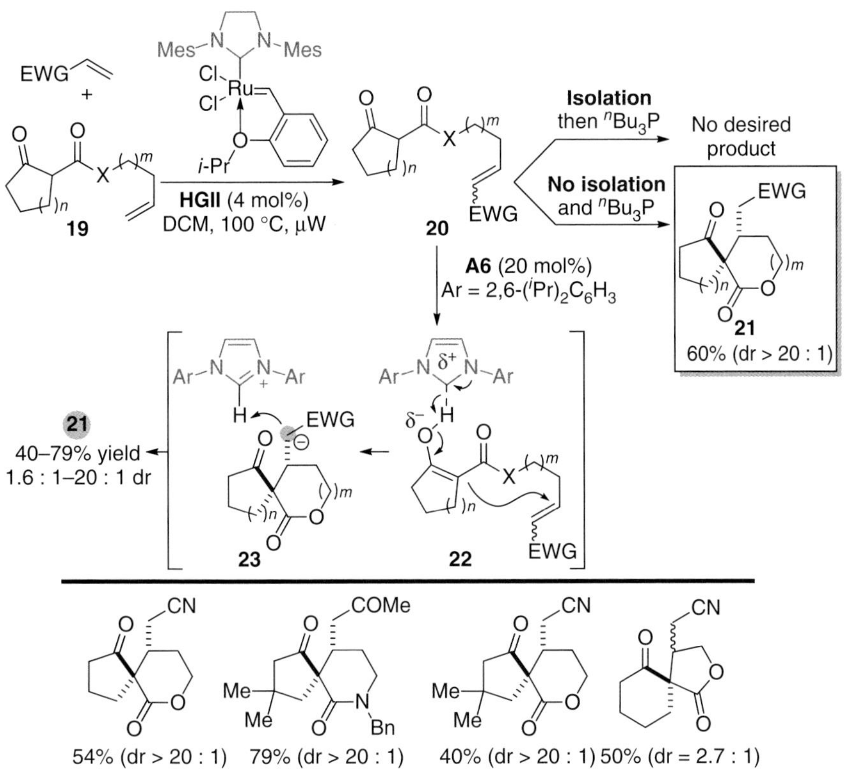

Figure 9.13 NHC-catalyzed intramolecular Michael spirocyclization reaction.

cyclization generates the preliminary adduct anion (**23**), which abstracts the proton from the NHC to complete the reaction loop.

Another report of NHC-mediated intramolecular conjugate addition of enolate was published by Hong and coworkers in 2011 [23]. In this study, the aldehyde (**24**) with a Michael acceptor was chosen as the standard substrate. The function of NHC is highly dependent on the Brønsted basicity of the catalyst. The cyclization of substrate **24** occurs through two distinct reaction pathways (Figure 9.14). When the bulky and strongly basic imidazolylidene precursor (**A3**, $pK_a = 23.0$, aqueous) was used, the NHC catalyst performed nucleophile activation by forming a hydrogen bond with the enol form of the aldehyde moiety. The α-carbon of the aldehyde is activated and functionalized through intramolecular Michael addition with a tethering, electron-deficient olefin to produce a five-membered cyclic product (**27**) in moderate yield. In contrast, when the less basic and hindered triazolium precatalyst (**A7**) was employed, the aldehyde reacted with the NHC catalyst to yield the Breslow intermediate (**28**), which underwent an intramolecular Stetter reaction. The selectivity of Michael vs. Stetter cyclization depends on the NHC catalysts used. The authors concluded that small, nucleophilic NHCs favor the Stetter reaction, whereas bulky, basic NHCs prefer Brønsted base catalysis. For the latter reaction pathway, complete deprotonation of the aldehyde is unlikely because the catalyst

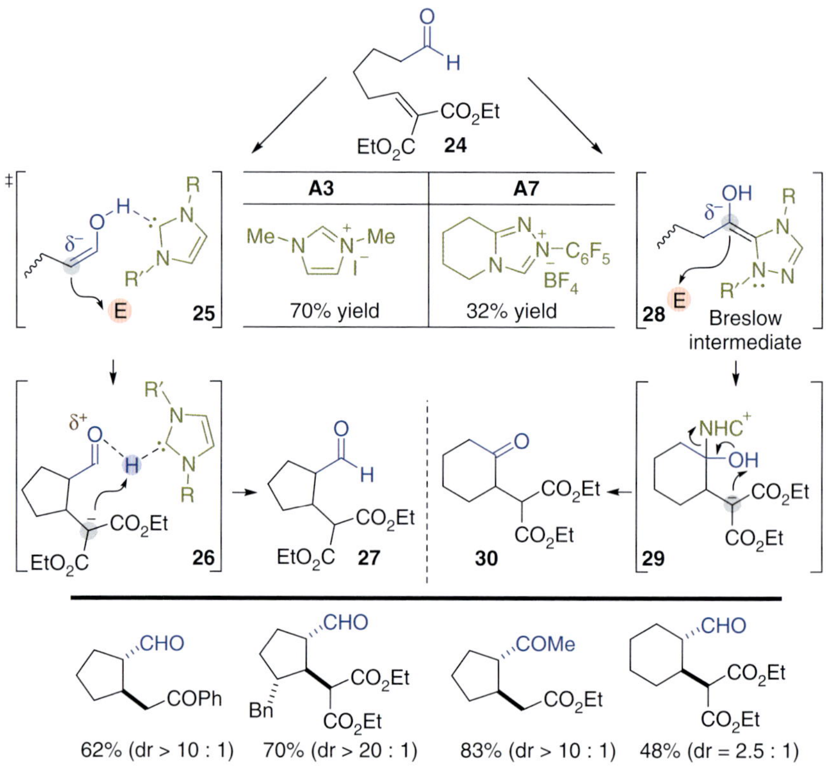

Figure 9.14 Intramolecular cyclization controlled by NHCs.

turnover process would be problematic. The primary Michael adduct anion is not sufficiently basic to re-deprotonate the imidazolium to release the free NHC catalyst. Under this circumstance, NHC appears to function via noncovalent interactions and acts as a proton shuttle, which provides valuable opportunities for asymmetric catalysis using chiral NHCs as a noncovalent template (*vide infra*).

The strategy of enol activation by noncovalent interaction was sequentially extended to the intermolecular Michael addition reactions by Coquerel and Rodriguez and coworkers in 2011 (Figure 9.15) [24]. The catalytic activity of various NHC catalysts was evaluated using conjugate addition reactions of 1,3-dicarbonyl compounds to various Michael acceptors. Highly basic imidazolium-derived NHCs are significantly superior to triazolium salts. The authors proposed an interesting cation activation mechanism. When the NHC catalyst forms a hydrogen bond with the enol form of the 1,3-diketone or 1,3-ketoester (**31**, **32**), a partial positive charge is generated on the N-heterocycle. As a result, this partial cation might act as a Lewis acid to coordinate, by electrostatic interaction, with the Michael acceptor. Such a proposed bifunctional role of the NHC resembles that of the classical tertiary amine-thiourea catalysts. 2-Substituted 1,3-dicarbonyl compounds react smoothly to deliver products containing a quaternary all-carbon center. Asymmetric reactions using chiral

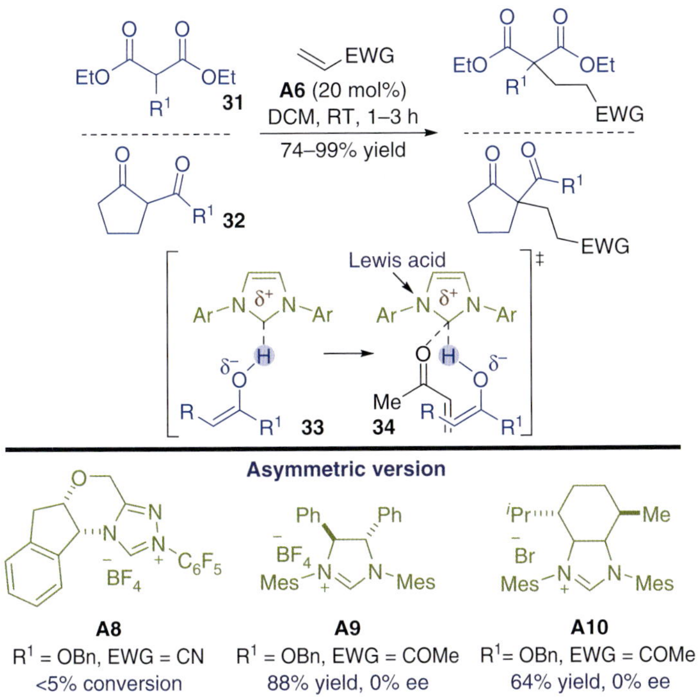

Figure 9.15 NHC-catalyzed intermolecular Michael addition reaction.

NHC precursors were attempted with several chiral NHC scaffolds. The most commonly used aminoindanol-derived triazolium (**A8**), which is a privileged scaffold, failed to promote the reaction. The low Brønsted basicity of this NHC was thought to be the result of poor activity. Both imidazolium-derived NHCs (**A9**, **A10**) promote the conjugate addition with satisfactory efficiency, but no enantioselectivity was observed. The weak noncovalent interaction between the NHC and the nucleophile appears to offer little facial differentiation (*vide infra*).

In 2010, Scheidt and coworkers reported the first oxa-Michael reaction capitalizing on the strong Brønsted basicity of NHC catalysts (Figure 9.16) [25]. By using LiCl as copromoter, a synergistic system for conjugate addition of alcohols to enones was demonstrated. In the control experiments, a prominent lithium effect was observed. When 12-crown-4 was added to scavenge the Li cation, the reaction proceeded with a low yield. In sharp contrast, one equivalent of LiCl was found to effectively promote the reaction with excellent conversion. It was assumed that the reaction is initiated by the noncovalent activation of the alcohol by NHC and Lewis acid activation of the enone by LiCl (intermediates **36**). This concurrent activation of both the nucleophile and the electrophile results in facile oxa-Michael addition at low temperature. In addition, the lithium cation also stabilizes the preliminary adduct anion (**37**). Subsequent protonation leads to the β-alkoxy ketone (**35**) releasing the NHC catalyst. Asymmetric intramolecular oxy-Michael reactions were attempted using various chiral NHC catalysts. Enone **38** was treated with triazolium NHC precursors derived from chiral amino

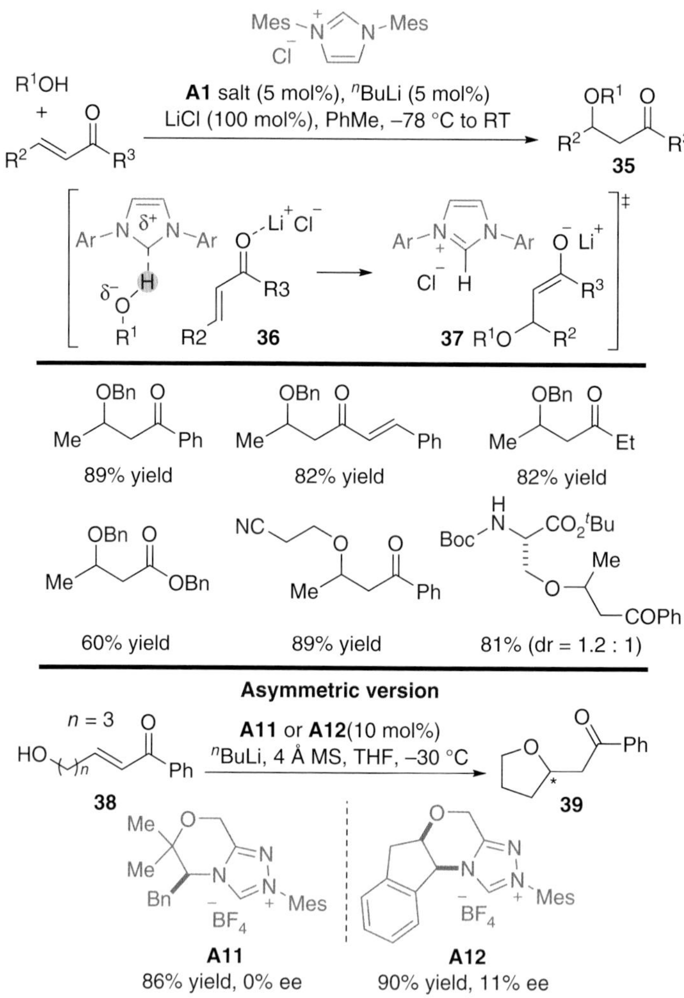

Figure 9.16 NHC-catalyzed oxy-Michael addition reactions.

alcohols (**A11**, **A12**). Despite good conversions, poor ee's were obtained. This report represents the first example of an asymmetric conjugate addition reaction catalyzed by noncovalent activation of NHCs.

Related aza-Michael reactions were developed by Kang and Zhang in 2011 (Figure 9.17) [26]. In this study, both anilines and selected alkyl amines react with enones to give β-aminoketone products under NHC catalysis. In a proposed mechanism similar to that of the oxy-Michael addition, an NHC–amine complex (**40**) is presumably formed, which enhances the nucleophilicity of the amine. The partial cationic charge on the imidazole ring of NHC might act as a Lewis acid activator for the enone. Both primary and secondary alkyl amines were tolerated. Exclusive N-selectivity was observed for the amine with a terminal alcohol (**42**). Addition of aniline to β-disubstituted enones and to α,β-unsaturated esters does not take place under the standard reaction conditions.

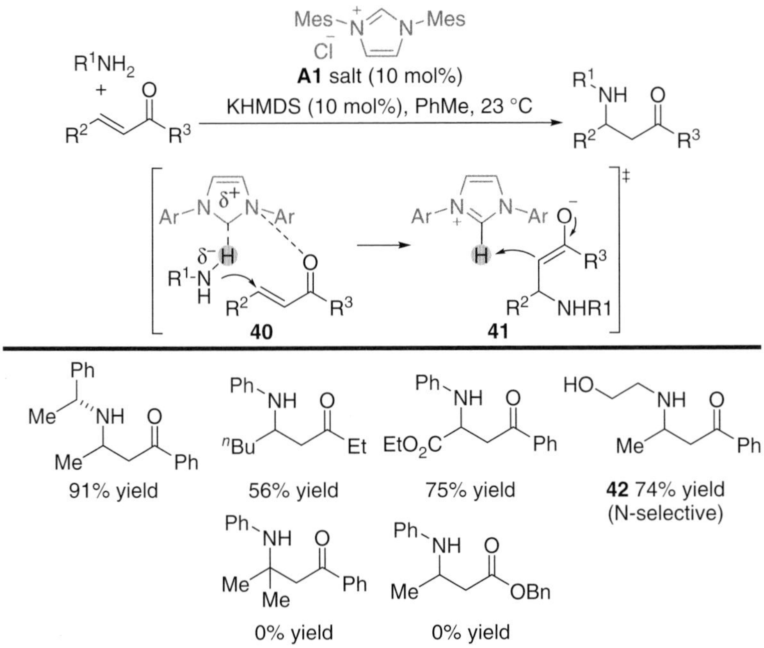

Figure 9.17 NHCs-catalyzed aza-Michael addition reactions.

Close examination of these conjugate addition reaction conditions reveals characteristics unique to noncovalent catalysis by NHCs. The NHC precursors, azolium salts, need to be fully deprotonated by strong bases such as *n*-BuLi or potassium bis(trimethylsilyl)amide (KHMDS). Although classical Breslow type of NHC catalysts often use a combination of NHC precursors and weak bases (e.g. K_2CO_3, DBU, or Et_3N), NMR studies have shown that a large portion of the NHC precursor remains as azolium salts [8b]. This suggests that only a small fraction of the free NHC is formed in these reactions. For noncovalent catalysis, external bases interfere with the hydrogen-bonding activation, and they must be fully consumed before introducing reaction substrates. In 2014, Rodriguez and coworkers used the NHC–CO_2 adduct (**A13**) as a precatalyst, eliminating the need for an external base [27]. Two important features of the NHC–CO_2 adduct were demonstrated. It is more stable than the corresponding free NHC, and it shows similar catalytic efficiency in conjugate addition reactions. Compared to other azolium salt precatalysts, the NHC (**A13**) can be used without any external base. Once in solution, it releases CO_2 and free NHCs.

NHC **A13** was used to catalyze sulfa- and phospha-Michael additions as well. Control experiments were carried out first to confirm the essential role of the carboxylate precatalyst as no desired product was generated in the absence of **A13**. In the presence of 2.5 mol% of **A13**, mercaptans react smoothly with α,β-unsaturated esters, nitriles, sulfones, amides, and ketones to give the desired β-thio products with high yields. 2-Aminobenzenethiol showed complete selectivity for the C—S bond formation (product **43**). Combined with the previous reports, the reactivity order of nucleophiles in NHC-catalyzed hetero-Michael additions is tentatively proposed as thiols > aromatic amines > alcohols. Methyl 2-mercaptoacetate (**44**) reacts with acrolein under standard conditions to give the tetrahydrothiophene (**45**) following a sulfa-Michael–aldol cascade. The dimer of 2-mercaptoacetaldehyde (**46**) reacts with methyl vinyl ketone affording the analogous product (**47**) in excellent yield. In both reactions, competing covalent catalysis by NHC, due to the presence of an aldehyde functionality, did not occur, and no benzoin or Stetter reaction products were isolated. Evidently, the noncovalent interaction between the NHC and the thiols is not affected by aldehydes, and the subsequent cyclization further discourages side reactions from the Breslow intermediates.

The NHC–CO_2 adduct (**A13**) was also tested in phospha-Michael addition reactions. Two phosphorous nucleophiles were investigated with several terminal and internal electron-deficient olefins. Effective C—P bond formation was accomplished using precatalyst (**A13**). In addition to vinyl sulfones, less reactive acrylates were tolerated by this catalyst (Figure 9.18).

These reports of noncovalent catalysis using NHCs involve racemic reactions. As mentioned earlier, several attempts to introduce asymmetric bias were not successful. In the field of Brønsted base catalysis, chiral tertiary amines, often

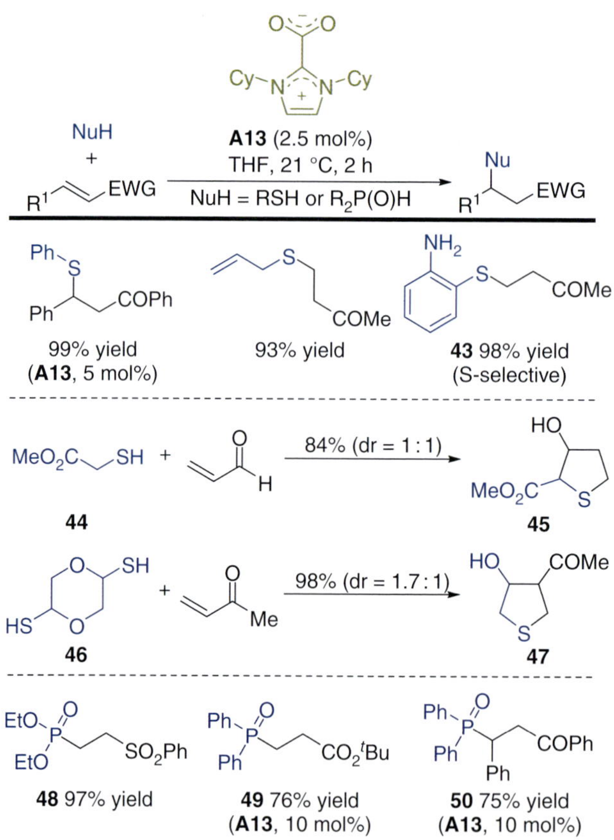

Figure 9.18 NHC–CO_2 adduct-catalyzed sulfa- and phospha-Michael addition.

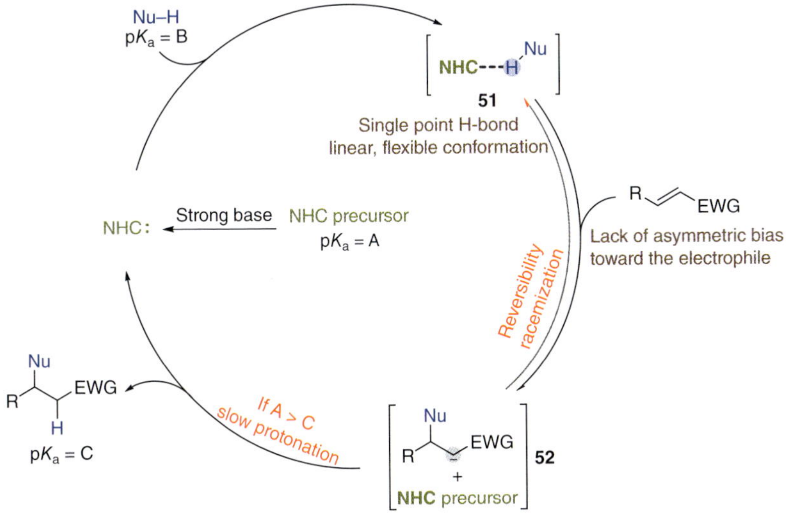

Figure 9.19 Factors contributing to poor asymmetric control in NHC-catalyzed conjugate addition reactions.

containing a hydrogen-bond-donor sidearm, are by far the most prominent catalyst system. Despite the stronger Brønsted basicity of NHCs, enantioselective catalysis via noncovalent catalysis of NHCs has lagged behind. The study of NHC-catalyzed conjugate addition reactions reveals several factors that mitigate asymmetric control (Figure 9.19). First, the key hydrogen bond between the NHC and nucleophile can freely rotate, yielding a flexible complex (**51**). Second, the newly generated chiral center is formed at the electrophile, which calls for intermolecular chirality transfer from the NHC–nucleophile complex. Third, the NHC catalyst needs to return the proton to the conjugate addition product anion (**52**) and, due to the high basicity of the NHC, such a process might be slow. As a result, the retro-Michael reaction would be expected to occur and racemize the β-chiral center.

In 2014, Chen and Huang reported the first enantioselective Michael addition reaction using NHC as a chiral Brønsted base in noncovalent catalysis [28]. In order to limit the conformational freedom of the NHC–nucleophile complex, 1,3-dicarbonyl compounds were chosen as the nucleophiles. These substrates are acidic and primarily exist in their cyclic enol form in solution. The NHC catalyst can form a relatively rigid complex (**53**) via a triply bonded proton. Nitroolefins were selected as the electrophile in view of the fact that the $\mathrm{p}K_a$ value of nitroalkanes is similar to that of 1,3-dicarbonyl compounds. This design was intended to promote facile direct proton transfer from the nucleophile to the adduct anion and the regeneration of the NHC catalyst. Ideally, the C—C bond formation and protonation would occur simultaneously in an ene-type reaction mechanism (Figure 9.20). Despite these efforts, initial results showed very poor enantioselectivity. To further enhance the protonation step, probably the most problematic step, a series of proton shuttles were evaluated (**55**).

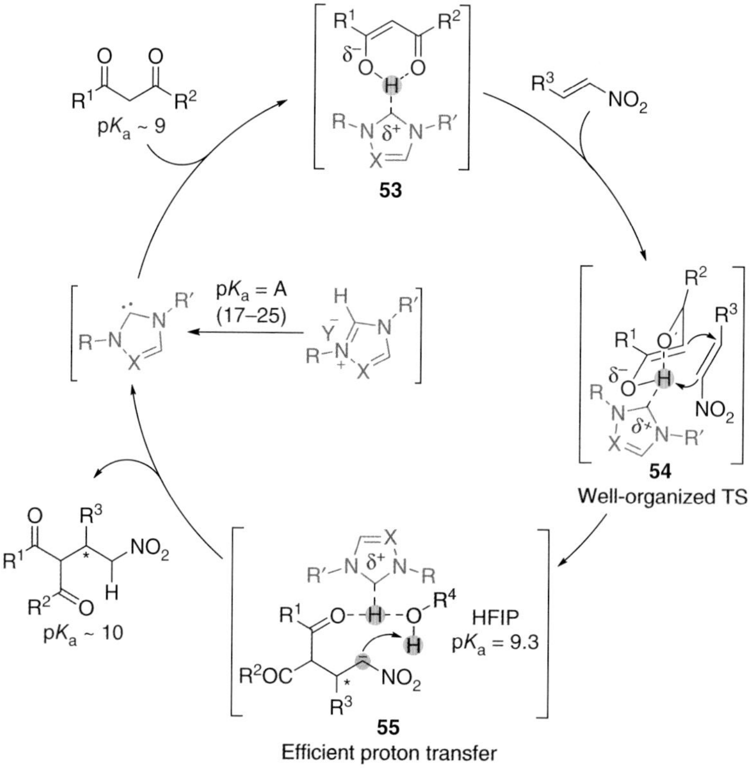

Figure 9.20 Rationale design of substrates for asymmetric Michael reaction using NHC as a chiral Brønsted base catalyst.

Although most alcohols, acids, and thioureas were effective in improving ee, hexafluoroisopropanol (HFIP, pK_a = 9.3, comparable to the pK_a's of both the substrate and the product) was identified as an essential additive for highly enantioselective C—C bond formation. Modest conversions were obtained using the imidazolium precatalyst (**A4**), which is frequently used as a strong Brønsted base catalyst. Because of the large pK_a gap between the NHC precursor and the product, regeneration of the catalyst might prove to be difficult. When a stoichiometric catalyst was used, full conversion was obtained with low ee. The aminoindanol-derived triazolium salt (**A8**) containing a *N*-pentafluorophenyl substituent is a privileged NHC precatalyst in numerous reactions involving covalent catalysis. It fails to promote the desired Michael addition reaction, likely because of its attenuated basicity. On the other hand, the analog with a N-mesityl group (**A14**) delivered excellent yields and ee for this reaction. The base used to generate free NHC also plays an important role. Lithium bis(trimethylsilyl)amide (LiHMDS) was identified as the best cocatalyst, and control experiments using crown ethers suggested that lithium cation might act as a Lewis acid copromoter for this reaction. Various β-aryl nitroolefins with either electron-rich or electron-poor substituents were well tolerated under the optimized reaction conditions. β-Alkyl nitroolefins afforded moderate

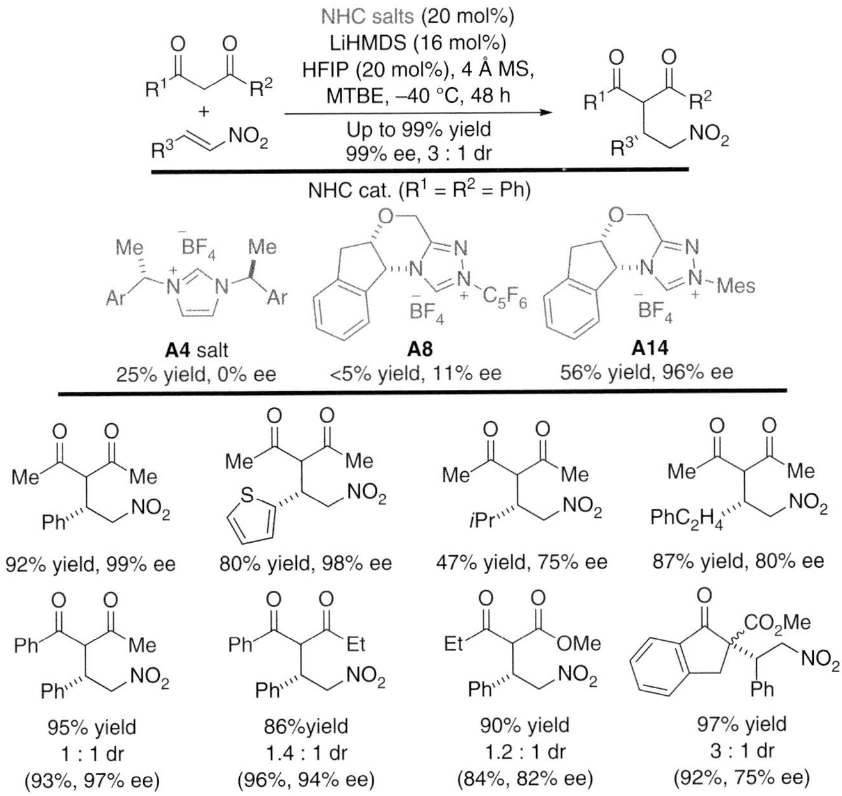

Figure 9.21 The first enantioselective C—C bond formation reaction using NHC as a noncovalent chiral template.

yields and ee. 1,3-Dicarbonyl compounds such as diketones or ketoesters were uniformly compatible with this methodology. β-Ketoesters reacted with various nitroolefins, giving excellent yields and moderate ee and dr (Figure 9.21).

Several control experiments were conducted with the aim of understanding the role of protons and the base in this reaction (Figure 9.22). When the commercially available lithium 2,4-pentanedionate was subjected to the standard reaction conditions, the reaction failed to proceed, suggesting that lack of a proton source led to reaction reversibility (Eq. a1). When HFIP was used in the absence of any NHC catalyst, the reaction proceeded slowly, indicating that a Brønsted base is necessary for the Michael reaction (Eq. a2). As predicted, a combination of catalytic amounts of HFIP and base (lithium 2,4-pentanedionate) greatly accelerated the reaction. In order to confirm the formation of free NHC in the reaction, benzaldehyde was subjected to the standard Michael addition conditions. The benzoin condensation proceeded smoothly and generated the desired product with moderate yield and ee, indicating that the free NHC species was present under the reaction conditions (Eq. b1). The success of this reaction is attributed to a fine balance of pK_a values of the reactant, the catalyst, the proton shuttle, and the product.

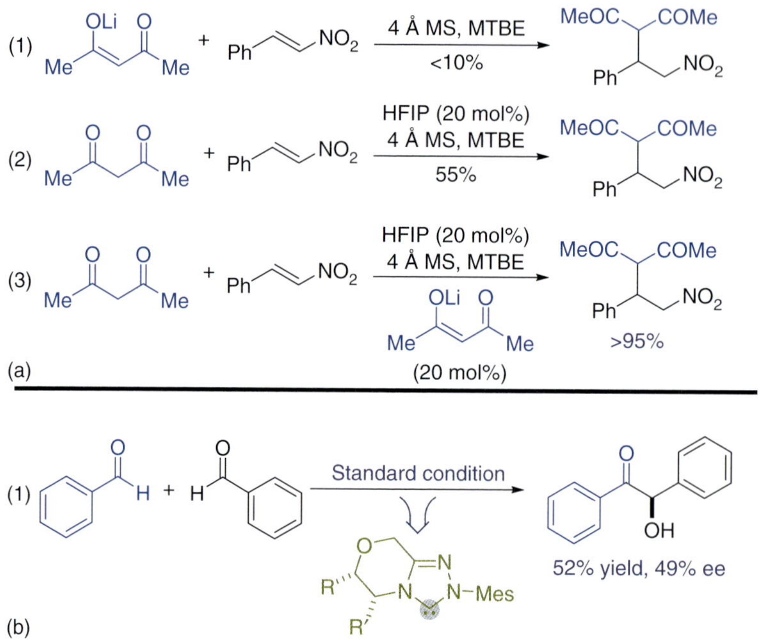

Figure 9.22 Control experiments for the NHC-catalyzed Michael addition reaction. (a) Evidence for HFIP acting as proton shuttle and (b) existence of free carbene species.

In 2015, Huang and coworkers reported the noncovalent activation of NHC catalysts in the asymmetric sulfa-Michael addition (SMA, Figure 9.23) [29]. Chiral centers having both a CF_3 and a sulfur group on the same carbon receive special attentions from medicinal chemists, and asymmetric SMA involving trifluoromethylated substrates are particularly challenging. The authors demonstrated a high degree of catalytic efficiency and enantioselectivity of conjugate addition of aliphatic mercaptan to β-CF_3–β-aryl nitroalkenes, using the NHC precatalyst (**A14**) as a Brønsted base activator. This was the first report of thiol activation by chiral NHCs. Again, HFIP was found as an essential proton shuttle for this reaction, presumably by facilitating the protonation step. The reaction shows a broad substrate scope for alkyl mercaptans including sterically demanding thiols. The pK_a values of thiols are important for enantioselectivity, as the more acidic benzyl mercaptan and thiophenols resulted in reduced ee's. In addition to nitroalkenes, enones are also viable electrophiles for the SMA. Simple acyclic enones are known to be challenging substrates for SMA reactions. Most SMA reactions of enones reported in the literature are limited to cyclic enones. Under the standard Brønsted base catalysis conditions, high ee's were obtained for β-CF_3-β-aryl enones. When the reaction was expanded to a simple alkyl enones, moderate-to-high ee's were also obtained. For alkyl enones, no proton shuttle is required. It was hypothesized that the SMA adduct anion from the simple enones are basic enough (pK_a ~ 20 for ketone vs. pK_a ~ 17 for **A14**) for catalyst turnover. Using the similar strategy, Huang and coworkers subsequently reported the SMA reactions of α,β-unsaturated esters and amides.

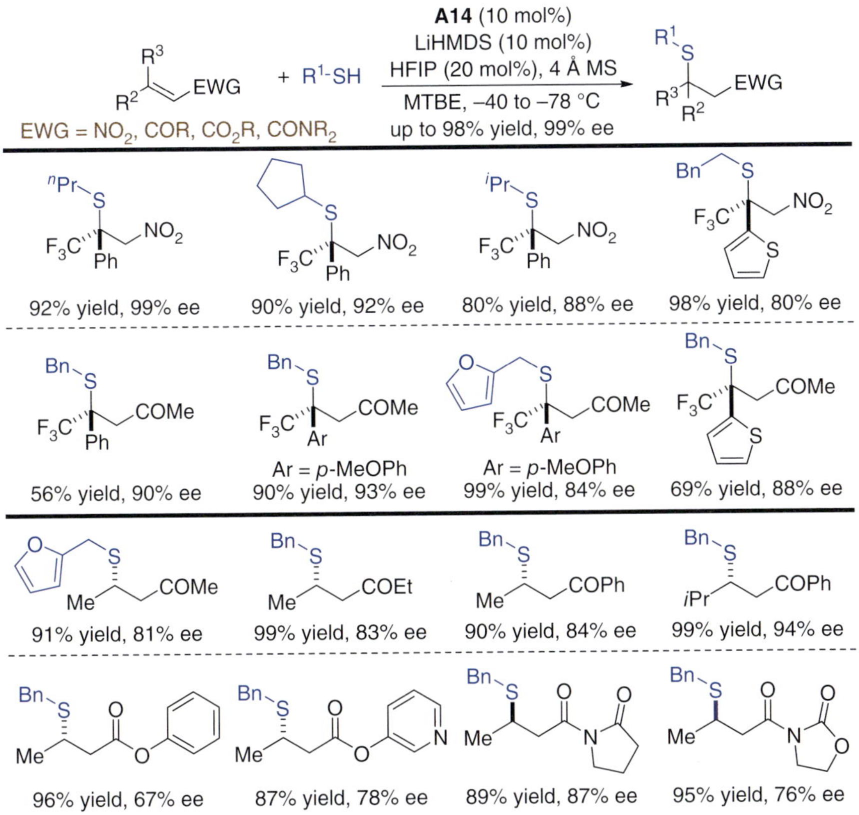

Figure 9.23 Enantioselective SMA reactions using an NHC as a noncovalent chiral template.

Because of the attenuated electrophilicity of α,β-unsaturated esters and amides, asymmetric SMA reactions of these substrates have been problematic. The authors discovered that α,β-unsaturated *O*-phenyl esters, oxazolidinones, and pyrrolidinones were competent Michael acceptors for the NHC-catalyzed SMA reaction. High yields and moderate ee's were obtained (Figure 9.23).

During the examination of the scope of the electrophiles, it was found that the electronic effect of the β-aryl group has an impact on the enantioselectivity. A linear free-energy relationship (LFER) study showed a positive effect of electron-rich substituents on the ee of the products, suggesting a possible π–π interaction between the β-aryl group and the chiral catalyst. Based on these results, a sandwich-like transition state was proposed and is illustrated in Figure 9.24. The free NHC catalyst forms hydrogen bonds with the mercaptan. This noncovalent complex generates a partial positive charge on the NHC and a partial negative charge on the sulfur, enhancing its nucleophilicity. The Michael acceptor approaches the sulfur atom in a way that is controlled by both the chirality of the catalyst and by the π–π stacking. The proposed transition state mimics protein–ligand interactions and both the C—S bond formation and the sequential protonation take place “intramolecularly” within the complex.

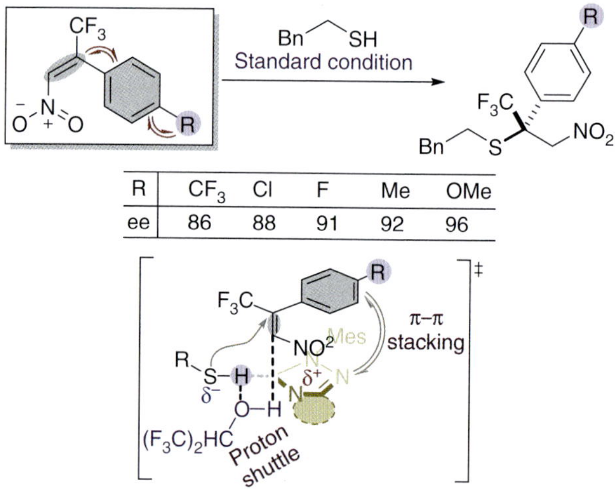

R	CF_3	Cl	F	Me	OMe
ee	86	88	91	92	96

Figure 9.24 Mechanism insight of sulfa-Michael addition. ‡ Means the structures in the brackets represent a transition state.

In 2015, Huang and coworkers demonstrated the activation of aliphatic amines using a chiral NHC as the Brønsted base catalyst [30]. Asymmetric aza-Michael addition reactions are a longstanding challenge for several privileged catalyst systems, and modified nitrogen sources are generally required to temper the basicity of amines. Aliphatic amines interfere with Lewis acid, amine, and Brønsted acid catalysts. Reports on enantioselective aza-Michael addition using aliphatic amines are very rare. The use of NHC as the Brønsted base catalyst opens doors for HOMO-raising amine activation. Researchers found that β-CF_3–β-aryl nitroalkenes were excellent Michael acceptors for alkyl amines under NHC catalysis. A preliminary survey revealed that aminoindanol-derived triazolium NHC precatalysts can catalyze the aza-Michael addition effectively. The *N*-C_6F_5 catalyst (**A8**) afforded the racemic product, probably because of the low basicity of its corresponding NHC that results in weak hydrogen bonding. The mesitylene-derived NHC (**A14**) remained as the most reactive and selective catalyst. Additives play a vital role for both yield and ee. The racemic product was obtained in low conversion in the absence of 4 Å MS. HFIP, as an important proton shuttle, enhances the conversion significantly. Under the optimized conditions, β-CF_3–β-aryl nitroalkenes with different substituents and electronic properties were all tolerated. Both the CF_3 and aryl substituents proved to be indispensable for high stereoselectivity. The primary aliphatic amines, including those with additional basic sites, such as pyridine and tertiary amine, performed well in this system. Secondary amine nucleophiles were not suitable for the reaction because of the instability of the products. The highly functionalized products were converted to 1,2-diamines, which can be cyclized, generating interesting heterocyclic skeletons with a CF_3-substituted chiral center (Figure 9.25).

Based on the preliminary investigations of the mechanism, the transition states analogous to the SMA can be envisaged for the conjugate amine addition.

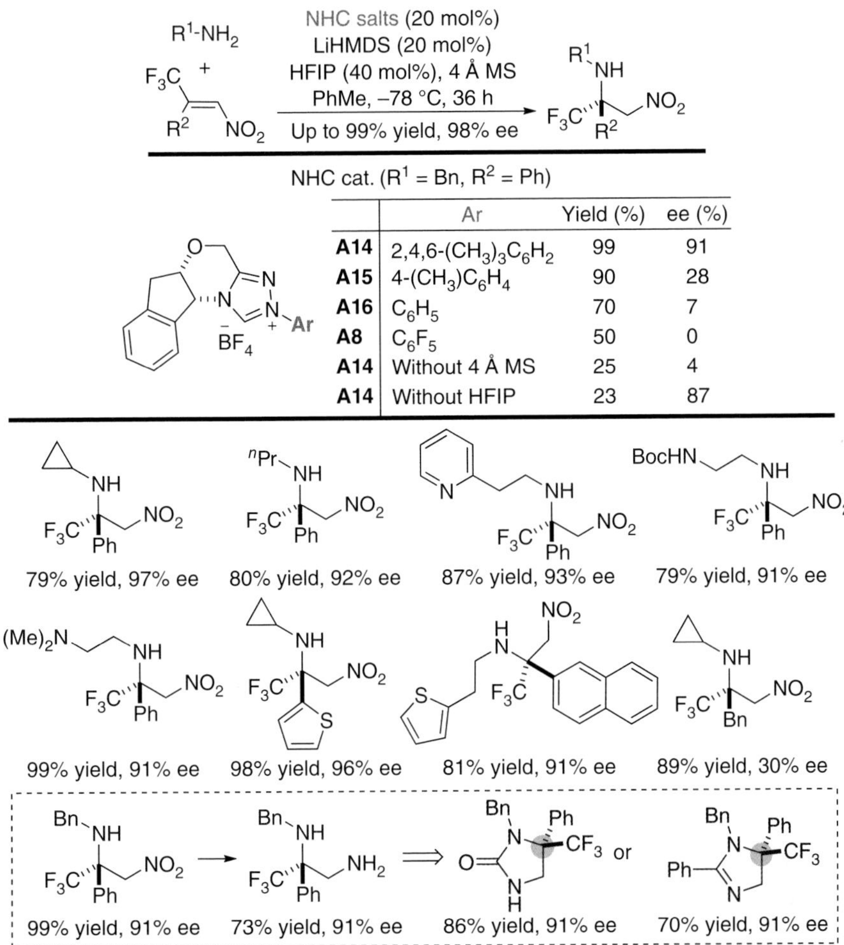

	Ar	Yield (%)	ee (%)
A14	2,4,6-$(CH_3)_3C_6H_2$	99	91
A15	4-$(CH_3)C_6H_4$	90	28
A16	C_6H_5	70	7
A8	C_6F_5	50	0
A14	Without 4 Å MS	25	4
A14	Without HFIP	23	87

Figure 9.25 Asymmetric aza-Michael addition of aliphatic amines using NHC as a HOMO-raising catalyst.

Because of the high pK_a of aliphatic amines, substrate activation by deprotonation is not possible, and instead, the reaction occurs via noncovalent, hydrogen-bonding catalysis. The LFER study continues to confirm the correlation between the electron density of the β-aryl (on the Michael acceptor) and the enantioselectivity. Interactions between the π-electrons of nitroalkenes and the heterocyclic moiety of the NHC catalyst were proposed. The role of 4 Å MS remained unclear, but the function of HFIP was believed to be that of a proton shuttle. Immediately following the C—N bond formation, a zwitterionic intermediate, containing both an ammonium ion and a nitroenolate, is formed. The high Brønsted acidity of the ammonium ion (p$K_a \approx 10$) might deactivate the free NHC catalyst, which is the most basic species in the system. The introduction of HFIP could effectively bridge the acidic proton of the ammonium cation and the adduct anion for a six-membered cyclic proton transfer. As a result,

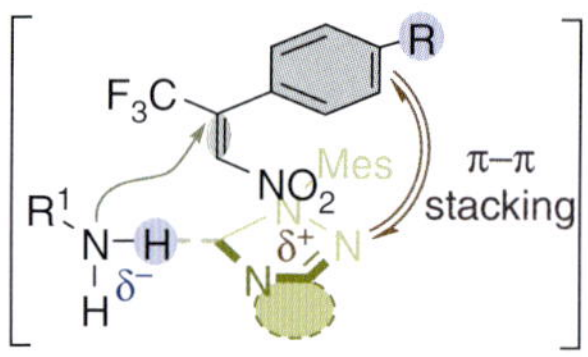

R	ee
CF_3	84
Cl	94
F	90
Me	97
OMe	98

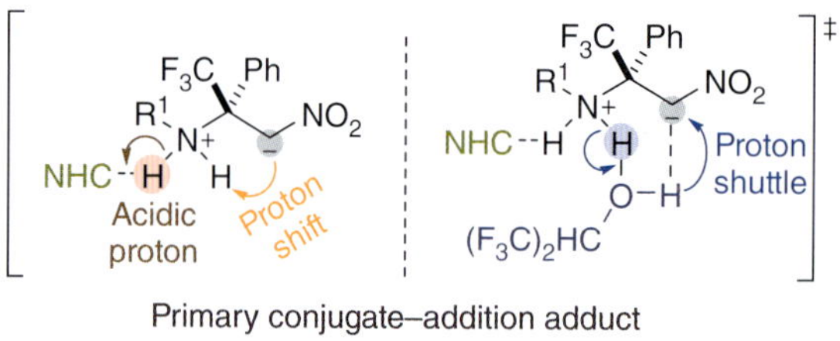

Figure 9.26 Proposed mechanism. ‡ Means the structures in the brackets represent a transition state or a intermediate.

the regeneration of free carbene catalyst is rescued for the next catalytic cycle (Figure 9.26).

In summary, the function of NHCs as organocatalysts occurs naturally and has been exploited extensively for the past two and half decades since Arduengo isolated the first stable NHC. With the reaction scope of a classical covalent activation mechanism involving the Breslow intermediates approaching a bottleneck, chemists turned their attention to the orthogonal noncovalent catalysis. The strong Brønsted basicity of NHCs has been harnessed as a way of nucleophile activation through hydrogen bonding. However, it was not until recently that this new mode of catalysis by NHCs was successfully demonstrated in enantioselective reactions. The limited examples that have been summarized here show how a single, weak hydrogen bond is capable of both enhancing reactivity and relaying chirality in Michael addition reactions. The choices of substrates in combination with proton shuttles suggest that these reactions operate through delicate, enzyme-like transition states. With the design of new catalysts, most likely of multifunctional NHCs, we expect that the concept of Brønsted base catalysis by NHCs will be applied to many other types of reactions and cascades, and we look forward to exciting new developments in this field in the near future.

References

1 Ukai, T., Tanaka, R., and Dokawa, T. (1943). *J. Pharm. Soc. Jpn.* 63: 296–304.

2 Breslow, R. (1958). *J. Am. Chem. Soc.* 80: 3719–3726.

3 (a) Enders, D., Niemeier, O., and Henseler, A. (2007). *Chem. Rev.* 107: 5606–5655. (b) Nair, V., Menon, R.S., Biju, A.T. et al. (2011). *Chem. Soc. Rev.* 40: 5336–5346. (c) Chiang, P.-C. and Bode, J.W. (2011). *TCI MAIL* 149: 2–17. (d) Izquierdo, J., Hutson, G.E., Cohen, D.T., and Scheidt, K.A. (2012). *Angew. Chem. Int. Ed.* 51: 11686–11698. (e) Vora, H.U., Wheeler, P., and Rovis, T. (2012). *Adv. Synth. Catal.* 354: 1617–1639. (f) Bugaut, X. and Glorius, F. (2012). *Chem. Soc. Rev.* 41: 3511–3522. (g) Ryan, S.J., Candish, L., and Lupton,

D.W. (2013). *Chem. Soc. Rev.* 42: 4906–4917. (h) Hopkinson, M.N., Richter, C., Schedler, M., and Glorius, F. (2014). *Nature* 510: 485–496. (i) Flanigan, D.M., Romanov-Michailidis, F., White, N.A., and Rovis, T. (2015). *Chem. Rev.* 115: 9307–9387. (j) Wang, M.H. and Scheidt, K.A. (2016). *Angew. Chem. Int. Ed.* 55: 14912–14922.

4 Arduengo, A.J., Harlow, R.L., and Kline, M. (1991). *J. Am. Chem. Soc.* 113: 361–363.

5 Hahn, F.E. and Jahnke, M.C. (2008). *Angew. Chem. Int. Ed.* 47: 3122–3172.

6 (a) Kuhn, N. and Alsheikh, A. (2005). *Coord. Chem. Rev.* 249: 829–857. (b) Fuchter, M.J. (2010). *Chem. Eur. J.* 16: 12286–12294.

7 Curran, D.P., Solovyev, A., Makhlouf Brahmi, M. et al. (2011). *Angew. Chem. Int. Ed.* 50: 10294–10317.

8 (a) Berkessel, A., Elfert, S., Yatham, V.R. et al. (2012). *Angew. Chem. Int. Ed.* 51: 12370–12374. (b) Mahatthananchai, J. and Bode, J.W. (2012). *Chem. Sci.* 3: 192–197. (c) Collett, C.J., Massey, R.S., Maguire, O.R. et al. (2013). *Chem. Sci.* 4: 1514–1522. (d) Kyan, R., Sato, K., Mase, N. et al. (2017). *Org. Lett.* 19: 2750–2753.

9 DiRocco, D.A., Oberg, K.M., and Rovis, T. (2012). *J. Am. Chem. Soc.* 134: 6143–6145.

10 Fevre, M., Pinaud, J., Gnanou, Y. et al. (2013). *Chem. Soc. Rev.* 42: 2142–2172.

11 (a) Kim, Y.-J. and Streitwieser, A. (2002). *J. Am. Chem. Soc.* 124: 5757–5761. (b) Amyes, T.L., Diver, S.T., Richard, J.P. et al. (2004). *J. Am. Chem. Soc.* 126: 4366–4374; (c) Higgins, E.M., Sherwood, J.A., Lindsay, A.G. et al. (2011). *Chem. Commun.* 47: 1559–1561. (d) Massey, R.S., Collett, C.J., Lindsay, A.G. et al. (2012). *J. Am. Chem. Soc.* 134: 20421–20432.

12 Connor, E.F., Nyce, G.W., Myers, M. et al. (2002). *J. Am. Chem. Soc.* 124: 914–915.

13 Grasa, G.A., Kissling, R.M., and Nolan, S.P. (2002). *Org. Lett.* 4: 3583–3586.

14 Nyce, G.W., Lamboy, J.A., Connor, E.F. et al. (2002). *Org. Lett.* 4: 3587–3590.

15 Singh, R. and Nolan, S.P. (2005). *Chem. Commun.* 5456–5458.

16 Movassaghi, M. and Schmidt, M.A. (2005). *Org. Lett.* 7: 2453–2456.

17 Lai, C.-L., Lee, H.M., and Hu, C.-H. (2005). *Tetrahedron Lett.* 46: 6265–6270.

18 Suzuki, Y., Yamauchi, K., Muramatsu, K., and Sato, M. (2004). *Chem. Commun.* 2770–2771.

19 Kano, T., Sasaki, K., and Maruoka, K. (2005). *Org. Lett.* 7: 1347–1349.

20 Grasa, G.A., Singh, R., Scott, N.M. et al. (2004). *Chem. Commun.* 2890–2891.

21 Song, J.J., Tan, Z., Reeves, J.T. et al. (2008). *Org. Lett.* 10: 877–880.

22 Boddaert, T., Coquerel, Y., and Rodriguez, J. (2009). *Adv. Synth. Catal.* 351: 1744–1748.

23 Kim, H., Byeon, S.R., Leed, M.G.D., and Hong, J. (2011). *Tetrahedron Lett.* 52: 2468–2470.

24 Boddaert, T., Coquerel, Y., and Rodriguez, J. (2011). *Chem. Eur. J.* 17: 2266–2271.

25 Phillips, E.M., Riedrich, M., and Scheidt, K.A. (2010). *J. Am. Chem. Soc.* 132: 13179–13181.

26 Kang, Q. and Zhang, Y. (2011). *Org. Biomol. Chem.* 9: 6715–6720.

27 Hans, M., Delaude, L., Rodriguez, J., and Coquerel, Y. (2014). *J. Org. Chem.* 79: 2758–2764.
28 Chen, J. and Huang, Y. (2014). *Nat. Commun.* 5: 3437–3444.
29 (a) Chen, J., Meng, S., Wang, L. et al. (2015). *Chem. Sci.* 6: 4184–4189. (b) Huang, Y., Chen, J., Yuan, P., and Meng, S. (2016). *Synlett* 27: 1068–1072.
30 Wang, L., Chen, J., and Huang, Y. (2015). *Angew. Chem. Int. Ed.* 54: 15414–15418.

10

NHC-Catalyzed Kinetic Resolution, Desymmetrization, and DKR Strategies

Shenci Lu, Si B. Poh, Jun Y. Ong, and Yu Zhao

National University of Singapore, Department of Chemistry, 3 Science Drive 3, Singapore 117543, Singapore

10.1 Introduction

Scheme 10.1 demonstrates a general schematic summary of methods of kinetic resolution, DKR, and desymmetrization. NHC catalysis has been successfully applied to kinetic resolution of racemic substrates including alcohols, amines, etc. (Scheme 10.1a). If high stereoselectivity is achieved for such processes ($K_R > K_S$), one enantiomer of the racemic substrate will be converted to the *R*-isomer of the product, whereas the other enantiomer of the substrate will be left unreacted. As a resolution process, this approach has the inherent limitation of <50% yield for the recovered starting material ideally in an enantiopure form. Higher conversion of the kinetic resolution can be adopted to enhance the ee of the recovered substrate with a sacrifice in the isolated yield. In order to render a kinetic resolution method practically useful, the choice of readily available racemic starting materials that are difficult to access in an enantioenriched form using alternative methods has served as an important guideline [2].

Compared to simple kinetic resolution, DKR has the power of converting racemic substrates to enantioenriched products in a stereoconvergent manner. Quantitative yield and excellent enantioselectivity can be achieved in these transformations when $K_R > K_S$ is realized. There is one additional key prerequisite that the substrate needs to undergo a facile racemization under the reaction conditions ($K_{rac} \gg K_R$). In fact, many classical DKR methods rely on racemization of a tertiary stereogenic center via a deprotonation–protonation process [3]. This fits NHC catalysis very well, as in many cases, carbonyl-containing compounds bearing an acidic α-proton are used as the substrate, whereas the active NHC catalyst is typically generated *in situ* in the presence of base (Scheme 10.1b).

In contrast to the above two processes, desymmetrization operates on symmetrical, achiral/*meso*-substrates, and delivers products in theoretically high yield and high enantioselectivity (Scheme 10.1c) [4]. In addition to acylation of alcohols/amines, NHC-catalyzed umpolung and annulation processes have also found much application in desymmetrization to deliver a wide array of value-added products.

N-Heterocyclic Carbenes in Organocatalysis, First Edition. Edited by Akkattu T. Biju.
© 2019 Wiley-VCH Verlag GmbH & Co. KGaA. Published 2019 by Wiley-VCH Verlag GmbH & Co. KGaA.

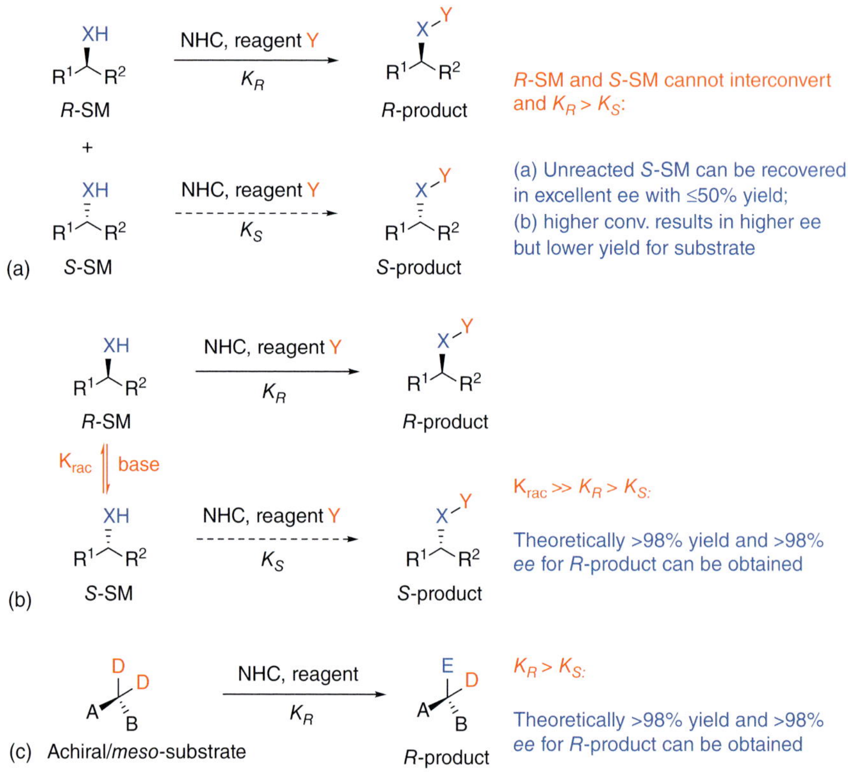

Scheme 10.1 Comparison of (a) kinetic resolution (KR); (b) dynamic kinetic resolution (DKR); and (c) desymmetrization.

10.2 NHC-Catalyzed Acylation

Catalytic acylation of alcohols (and amines) is an extensively explored reaction in organic synthesis [5]. The classical approach utilizes acylating reagents such as acid chloride or anhydride. NHC catalysis has also been applied to such transformations, where an acyl transfer from a vinyl ester **2** to the alcohol substrate **1a** was promoted by imidazolium-based NHC catalyst **A1′** in the pure form or generated *in situ* from azolium **A1** (Scheme 10.2) [6]. The key *acyl azolium* intermediate was generated and attacked by the alcohol in the enantio-determining acylation step. The two systems reported by the Suzuki group and Maruoka group used closely related catalyst. The use of a bulkier acylating reagent and lower temperature led to a much higher level of stereoselectivity (Scheme 10.2b vs. Scheme 10.2a).

In contrast to the classical acylation, NHC catalysis has provided a brand new opportunity for the esterification of alcohols using aldehyde as the acylating reagent. As illustrated in Scheme 10.3, NHC reacts with α,β-unsaturated aldehyde or aldehyde with a α-leaving group to form the functionalized Breslow intermediate, which undergoes elimination of the α-leaving group or isomerization of the allylic anion to generate the key *acyl azolium* intermediate that is attacked by alcohol in the key enantio-determining step. This approach has

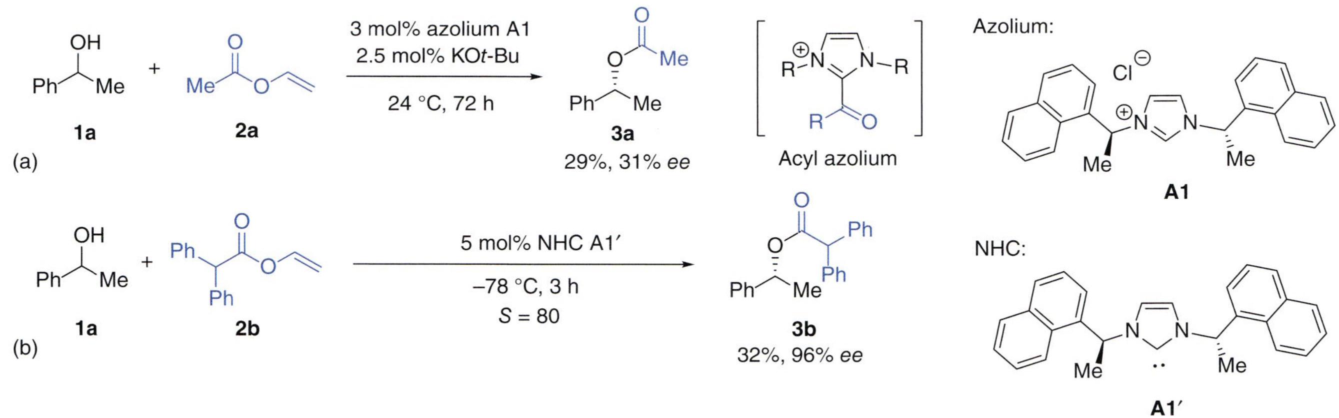

Scheme 10.2 NHC-catalyzed kinetic resolution of alcohols via acyl transfer.

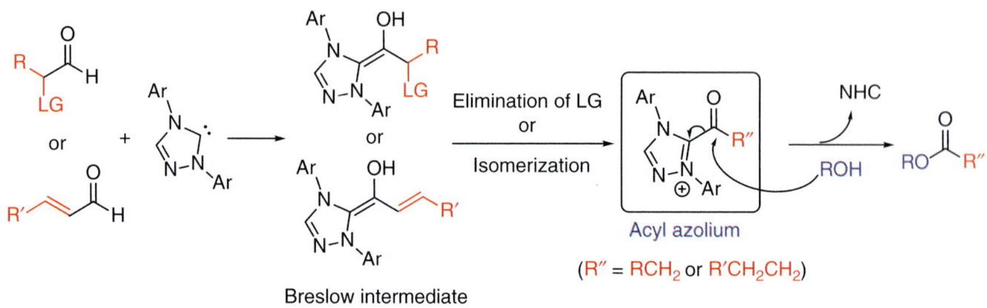

Scheme 10.3 NHC-catalyzed acylation of alcohols using aldehyde as the reagent.

a significant advantage compared to classical acylation: there is no background reaction between the substrate and acylating reagent. The enantioselectivity of the process thus only depends on the effective chirality transfer from the catalyst.

10.2.1 Acylation of Aliphatic Alcohols

10.2.1.1 Acylation of Aliphatic Alcohols

The early work on NHC-catalyzed transformations using functionalized aldehydes was mainly focused on the achievement of novel C—C bond-forming reactions. Some isolated proof-of-principle examples of NHC-catalyzed enantioselective esterification for the resolution of secondary alcohol or desymmetrization of *meso*-diol were also documented (Scheme 10.4) [7]. Although the Rovis group reported the reaction of diol **4a** with α-bromo aldehyde **5a** to yield **6** in 83% ee, the Scheidt group reported resolution of **1a** with **7a** with a moderate selectivity ($S = 4.8$).

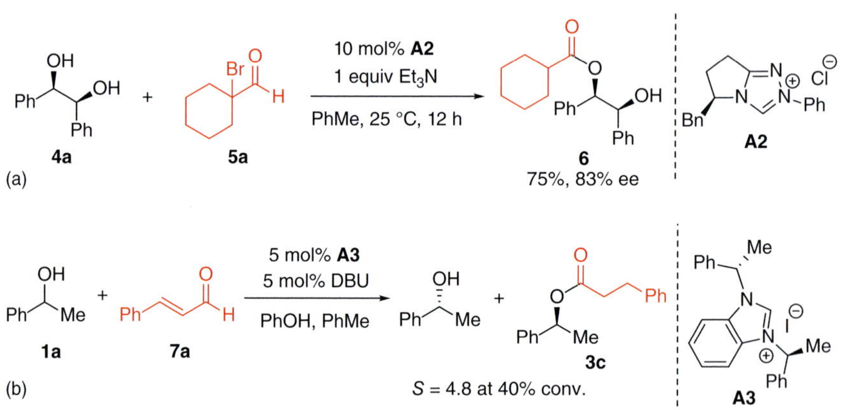

Scheme 10.4 Early examples of NHC-catalyzed acylation of alcohols. (a) Desymmetrization of *meso*-diol using α-bromo aldehyde. (b) Kinetic resolution of sec-alcohol using α,β-unsaturated aldehyde.

With the development of oxidative NHC catalysis [8], more examples for enantioselective acylation of secondary alcohols were reported by the Scheidt group and the Studer group (Scheme 10.5). In these examples, either MnO_2 or

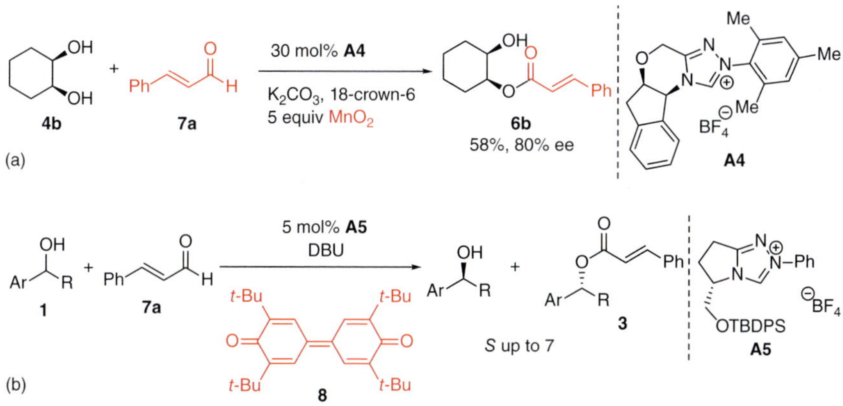

Scheme 10.5 Early examples of acylation of alcohols under oxidative NHC catalysis. (a) Desymmetrization of *meso*-diol. (b) Kinetic resolution of sec-alcohols.

an organic oxidant **8** was used as the additive to deliver the unsaturated ester products [9]. In general, only moderate-to-good level of enantioselectivity was achieved in these early examples (Schemes 10.4–10.5).

In 2013, the Zhao group reported a highly enantioselective kinetic resolution of 3-hydroxy-3-substituted oxindoles, which are readily accessible in the racemic form but difficult to access in excellent enantiopurity. As shown in Scheme 10.6,

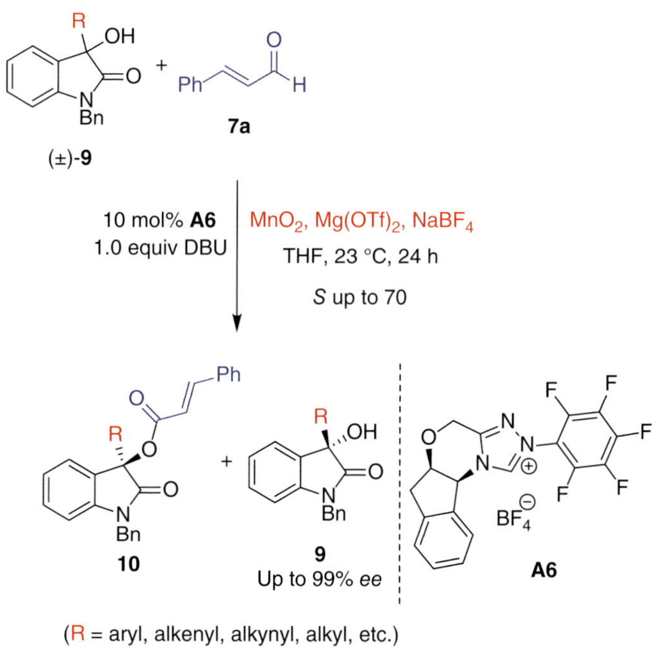

Scheme 10.6 KR of oxindole-derived tertiary alcohols under oxidative NHC catalysis.

high efficiency and enantioselectivity were achieved under oxidative NHC catalysis, and this method can tolerate a wide range of substituent in the 3-position [10]. It is important to note that tertiary alcohols provide much limitation in the reaction efficiency; in this system, only oxidative NHC catalysis using a relatively small aldehyde **7a** was able to convert **9** to the corresponding ester product **10** in good yield. In many other conditions using various types of catalysts, no conversion to the product was observed at all. It is also noteworthy that the addition of Lewis acid cocatalyst proved to be beneficial for the reaction rate and selectivity [11].

In the same year, the Takasu and Yamada group reported that the kinetic resolution and desymmetrization of cyclic diols and amino alcohols **11** were achieved under NHC catalysis with excellent enantioselectivity (up to $S = 218$ for the formation of **12a** and 99% ee for **12b**, respectively) [12]. As shown in Scheme 10.7, it was shown that the rate and enantioselectivity of NHC-catalyzed asymmetric acylation of alcohols with an adjacent H-bond donor functionality (as in diol or amino alcohol) are remarkably enhanced in the presence of a carboxylate cocatalyst. The degree of the enhancement is correlated with the basicity of the carboxylate. Based on the experimental as well as some computed data, a transition state was proposed, in which the carboxylate serves as a key hydrogen bond acceptor to interact with the substrate and accelerate the esterification reaction as a general base.

More recently, the Chi group also reported that oxidative NHC catalysis could be applied to the desymmetrization of chloro-substituted 1,3-propanediols **13** to deliver the mono-ester products **15** in good-to-excellent yield and enantioselectivity (Scheme 10.8a) [13]. These substrates are challenging for desymmetrization as the stereogenic center is distant from the reaction site. The use of a bulky aldehyde **14** together with an organic oxidant **8** was important for achieving good selectivity in this catalytic system.

Very recently, the Wang group reported an elegant intramolecular desymmetrization of 1,3-propanediols **16** bearing a tethered aldehyde moiety. Under oxidative NHC-catalyzed conditions, this intramolecular acylation delivered a range of medium-sized lactone products **17** in excellent yield and enantioselectivity (Scheme 10.8b) [14]. Notably, the addition of phosphoric acid as the additive proved to be beneficial for this NHC-catalyzed reaction to proceed. Although Lewis acid-NHC cooperative catalysis has been reported before, this combination of a Brønsted acid with NHC catalysis represents an interesting example.

10.2.1.2 DKR Involving Acylation of Alcohols

When the alcohol substrates bear an acidic proton, NHC catalysis could be well suited to realize a DKR process. In 2016, the Wang group reported NHC-catalyzed DKR of 6-hydroxypyranone (hemiacetal) **18** with cinnamaldehyde **7a** through an asymmetric redox esterification (Scheme 10.9) [15]. Under basic conditions, the hemiacetal substrates underwent facile racemization. The chiral NHC catalyst promoted an efficient and enantioselective acylation of one isomer preferentially to deliver ester **19** in good-to-high yields and with high levels of enantioselectivity. The derivatization of **19** to functionalized carbohydrate derivatives and natural products was also demonstrated in this report.

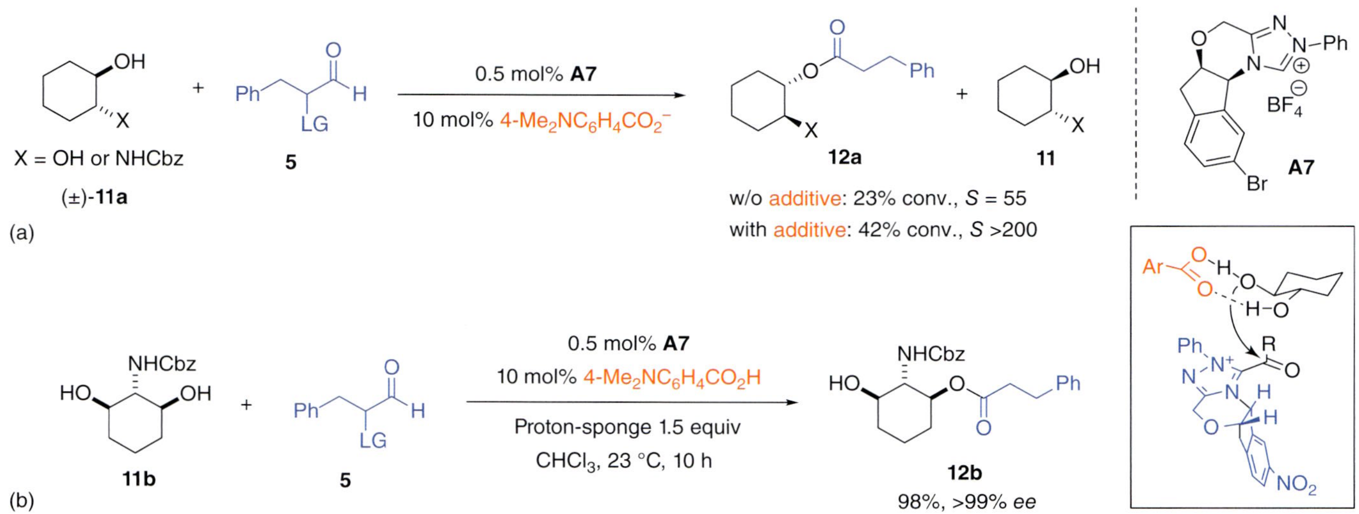

Scheme 10.7 KR of cyclic diols/amino alcohols: carboxylate additive effect.

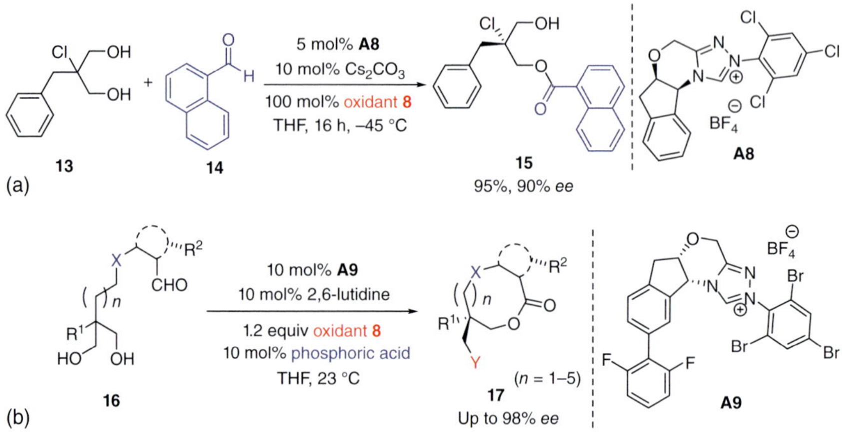

Scheme 10.8 Desymmetrization of prochiral 1,3-propanediols.

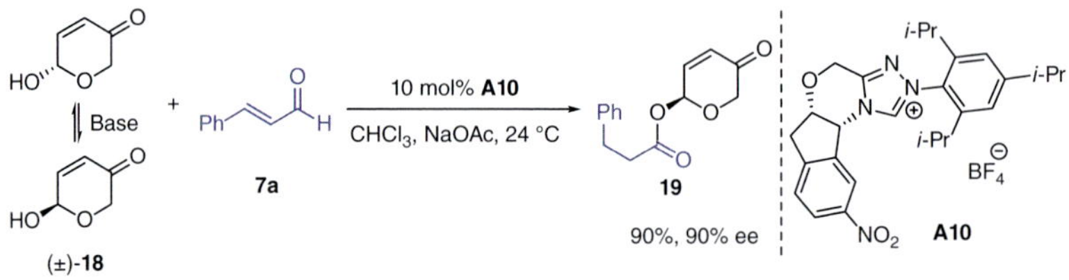

Scheme 10.9 DKR of hemiacetals.

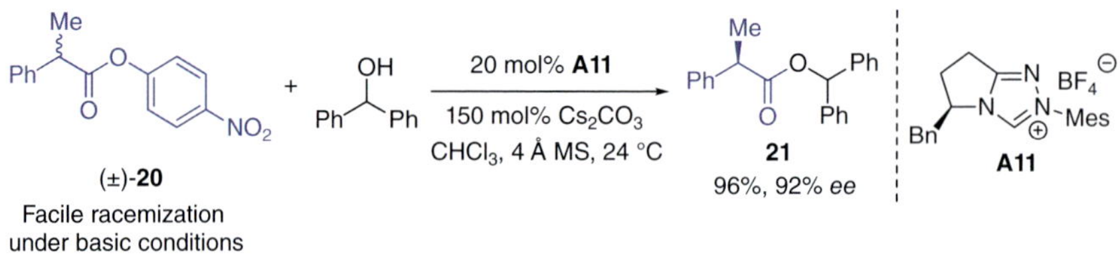

Scheme 10.10 DKR of carboxylic esters.

In 2016, the Chi group reported an elegant application of NHC-catalyzed transesterification of aryl esters with alcohols to a DKR of racemic α,α-disubstituted carboxylic esters **20** (Scheme 10.10) [16]. The readily available racemic aryl esters were converted to highly enantioenriched product **21** in high yield and high ee. This reaction was applied to the preparation of certain important drug molecules such as ibuprofen with high efficiency.

10.2.2 Acylation of Phenols

Compared to the enantioselective acylation of aliphatic alcohols, the analogous transformation of phenols is much less explored. As general base or Lewis base catalysis is typically adopted in catalytic acylation chemistry, the much more acidic phenol substrates may cause much complication in catalytic method development. Aiming toward the development of a general, catalytic access to enantiopure axially chiral diol and amino phenols that are of significant importance in asymmetric catalysis [17], the Zhao group came up with a highly efficient and enantioselective kinetic resolution of BINOL and NOBIN analogs **22** using NHC catalysis in 2014 (Scheme 10.11a) [18]. This system uses a straightforward procedure and is able to deliver a wide range of axially chiral bisphenols and amino phenols in uniformly >99% ee. This reaction represents the first example of highly enantioselective acylation of phenols [19] and allowed the access to a wide range of valuable chiral ligand/catalyst precursors in excellent enantiopurity. The key for the success in this catalytic system includes the use of a bulky aldehyde **5b** bearing a α-leaving group for good reactivity and selectivity. In addition, the use of unprotected diol or amino alcohol was key for the selectivity of the system. It was proposed that the second phenol or carbamate proton might serve as H-bond donor in the enantio-determining step to generate a more organized transition state.

Density functional theory (DFT) calculation of this process as well as other related systems has recently been disclosed by the Wheeler group (Scheme 10.11b) [20]. Intriguingly, their calculation elucidated an important role played by the benzoate leaving group that was correlated with the benzoate additive in the system reported by the Takasu and Yamada group shown in Scheme 10.7. In particular, once NHC attacked aldehyde to form intermediate **I**, the key proton transfer step for the formation of Breslow intermediate is facilitated by a carboxylic acid additive through H-bond interaction with the azolium alkoxide (as in **II**). The acyl azolium intermediate **V** is then formed by the release of benzoate followed by enol-to-ketone tautomerization. In the enantio-determining acylation step, an intramolecular H-bond within the deprotonated substrate was proposed to rigidify the conformation of the phenol in their calculations. These important and informative mechanistic insights may inspire further development in this field of research.

Phosphorus compounds with a P-stereogenic center have found wide applications as ligands and organic catalysts [21]. The development of new and efficient synthesis of these compounds remains an important topic in organic synthesis. As exemplified in Scheme 10.12, the Chi group reported in 2016 that desymmetrization of prochiral bisphenol compounds **24** bearing a remote P-stereogenic center could be achieved under the similar oxidative NHC catalysis conditions as in their previous report shown in Scheme 10.8a [22]. The catalytic reactions can also be performed on gram scale with 1 mol% catalyst, providing efficient access to enantiomerically enriched P-stereogenic phosphinates. The chiral phosphinates **25** were successfully converted to a bidentate P-based Lewis base **26** in a three-step procedure, which showed promising application as a catalyst for reductive aldol reactions.

Also in 2016, the Zhao group reported the application of NHC-catalyzed acylation of phenols to the preparation of enantiopure triarylmethanes and diarylalkanes (Scheme 10.13a) [23]. These products are highly valuable targets in medicinal and material chemistry [24]. The bisphenol substrates **27** are readily available and undergo acylation smoothly to deliver triarylmethanes **28** and diarylalkanes **29**

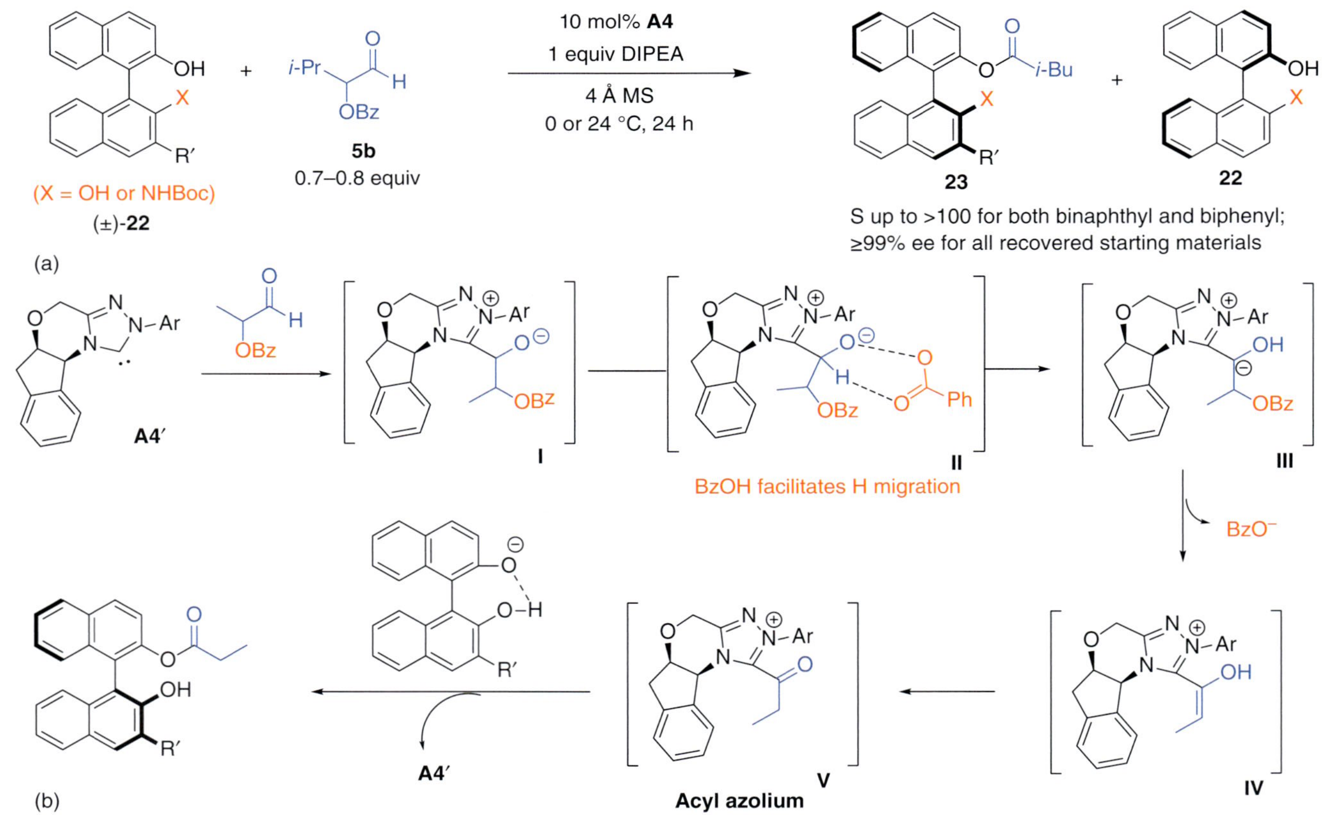

Scheme 10.11 KR of axially chiral bis-phenols and amino phenols.

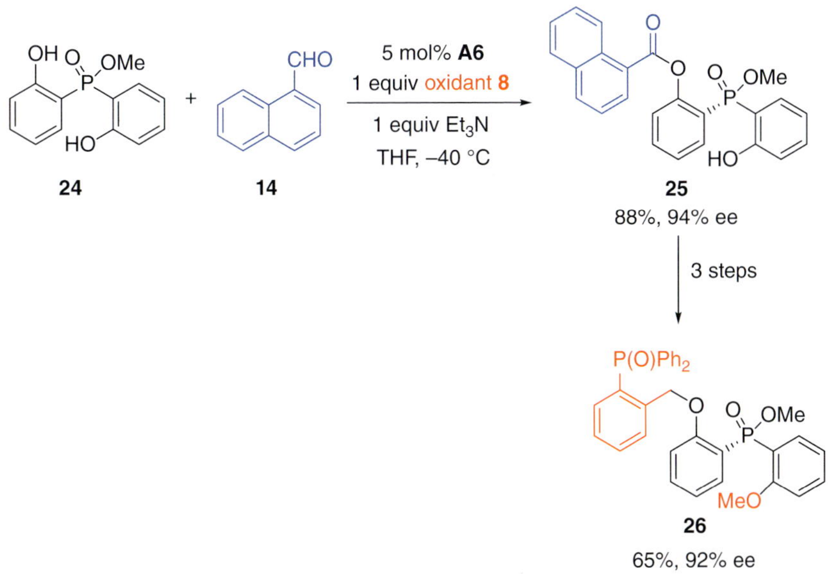

Scheme 10.12 Desymmetrization of P-containing bisphenols.

with uniformly high enantioselectivity. The simple procedure and the wide substrate scope bode well for the application of this catalytic system.

In the same publication, systematic DFT calculations were also included that elucidate the overall catalytic pathway and provide an effective transition state to explain the high enantioselectivity of this process (Scheme 10.13b). The substrate undergoes facile deprotonation in the presence of base to deliver intermediate **II**. Salt metathesis produces the azolium phenoxide ion pair **III**, which reacts simultaneously to deliver the tetrahedral hemiacetal intermediate **IV**. The collapse of **IV** to form ester product **V** then follows that serves as the enantio-determining of this catalytic process. In the key $\mathbf{TS_3}$ that determines the enantioselectivity, there is an important and very strong H-bond interaction between the remaining phenol from the substrate and the tetrahedral intermediate oxygen anion. This interaction, coupled with steric hindrance between the substrate and the chiral backbone of the catalyst, leads to the formation of the major enantiomer of the product with much lower activation barrier. The developed transition state model can readily explain the enantioselectivity of this process; the predicted and experimental selectivity match each other well over a range of substrates.

10.2.3 Acylation of Amines and Sulfoximines

Enantioselective acylation of amines using traditional Lewis base catalysts is much more challenging to achieve than that of alcohols because of the strong background reactivity of the amine substrates. In contrast, the formation of amide by NHC-catalysis is demonstrated to be more difficult than the formation of esters [25]. Nonetheless, important examples of NHC-catalyzed asymmetric acylation of amines/imines have been reported in the literature.

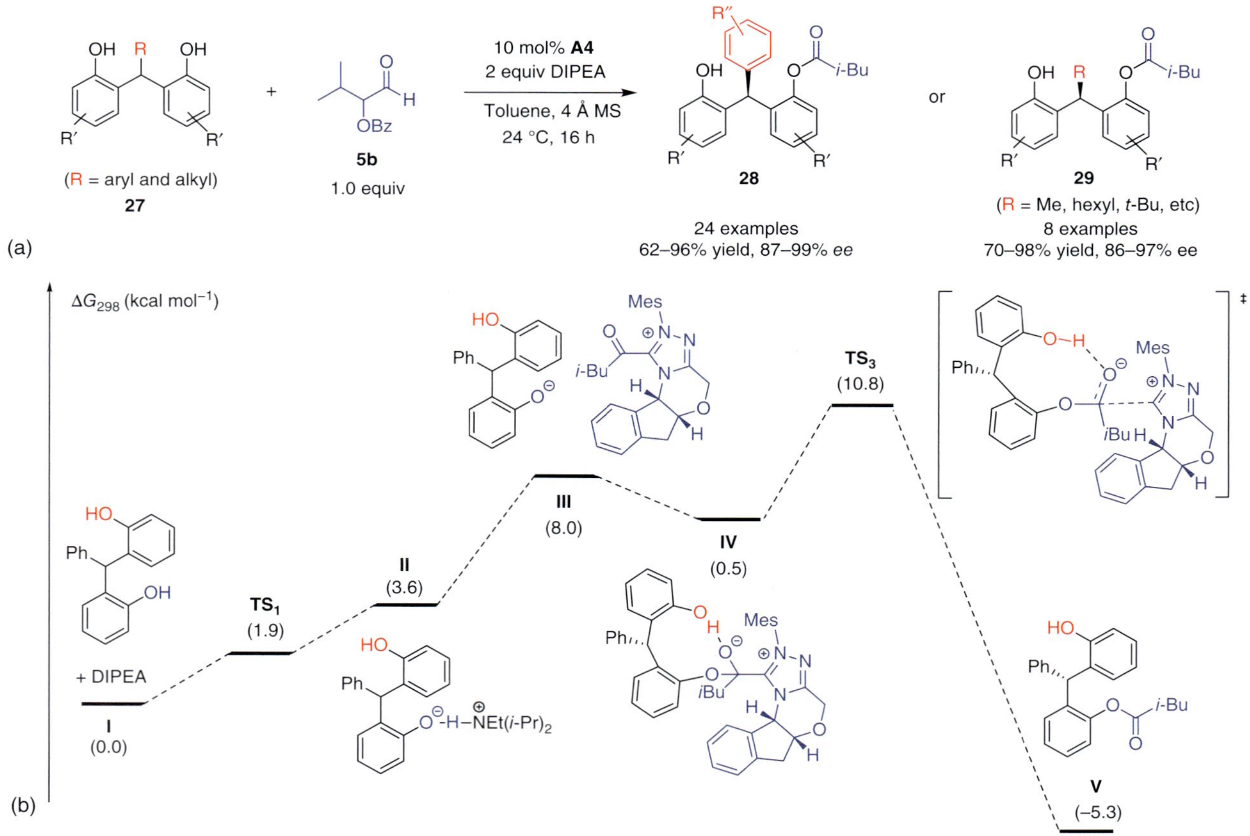

Scheme 10.13 Desymmetrization to access triarylmethanes and diarylalkanes.

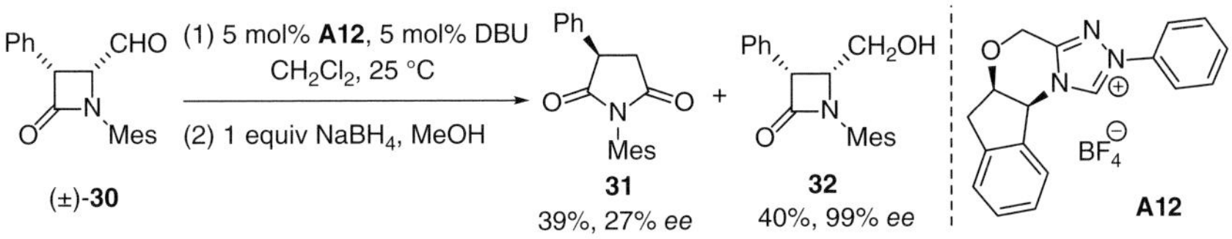

Scheme 10.14 NHC-catalyzed ring expansion of β-lactam.

In 2008, the You group reported an efficient resolution of *cis*-4-formyl-β-lactams **30** by NHC-catalyzed ring expansion (Scheme 10.14) [26]. Although the faster-reacting enantiomer of **30** was converted to succinimide **31**, the other enantiomer of **30** was left unreacted and isolated as the reduced form of **32**. Good yields with moderate-to-good enantioselectivity were obtained for a range of substrates examined. This reaction features the availability of both the substrate and the catalyst as well as mild reaction conditions.

The Bode group developed efficient kinetic resolution of racemic cyclic amines **33** by an enantioselective amidation reaction using the combination of an achiral NHC catalyst **A13** and a new chiral hydroxamic acid cocatalyst **B** (Scheme 10.15a) [27]. This catalytic resolution of amines is unique in that two independent catalytic cycles work concertedly with complementary reactivity and chemoselectivity (Scheme 10.15b). Although NHC reacts with **34** to form the key *acyl azolium* intermediate, the relay of acyl azolium with the hydroxamic acid **B** generates the true acylating reagent to react with the amine substrate. The reaction is operationally simple, proceeds at room temperature, produces no nonvolatile by-products, and affords enantioenriched amines by aqueous extraction. Detailed DFT calculations were also carried out to elucidate the reaction pathway and the origin of enantioselectivity for this catalytic system by the Kozlowski group in collaboration with the Bode group [27c].

In 2016, the Bolm group reported an elegant kinetic resolution of sulfoximines **36** with α,β-unsaturated aldehyde **7b** using oxidative NHC catalysis (Scheme 10.16) [28]. The stereoselective amidation proceeds without an additional acyl transfer agent. A range of new analogs of NHC precursors were prepared and examined in this system, and **A14** catalyzed this reaction with a high level of enantioselectivity and reactivity. The reaction could be performed on a gram scale, and the methodology was also applied to an asymmetric synthesis of FXa inhibitor F.

10.3 Benzoin and Stetter Reactions

Benzoin condensation and Stetter reactions represent important C—C bond formation involving umpolung activity enabled by cyanide or NHC catalyst [29]. As the initial discovery in the nineteenth century, these reactions have been extensively explored because of their intriguing mechanism and the efficient access to 1,2- and 1,4-dicarbonyl compounds. In recent years, a few important examples of Benzoin and Stetter reactions in the form of desymmetrization or DKR have also been reported in the literature.

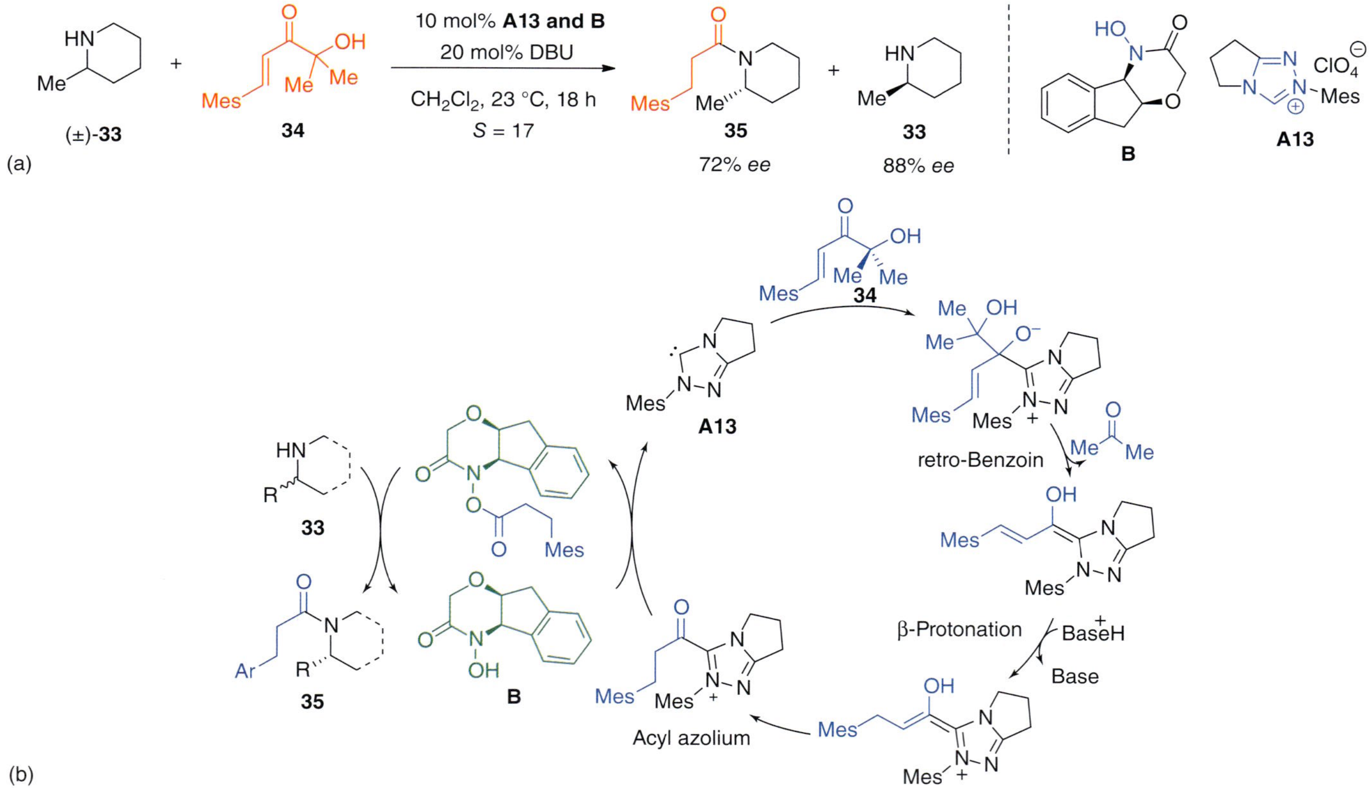

Scheme 10.15 NHC-catalyzed resolution of cyclic amines. (b) Catalytic pathway featuring relay catalysis.

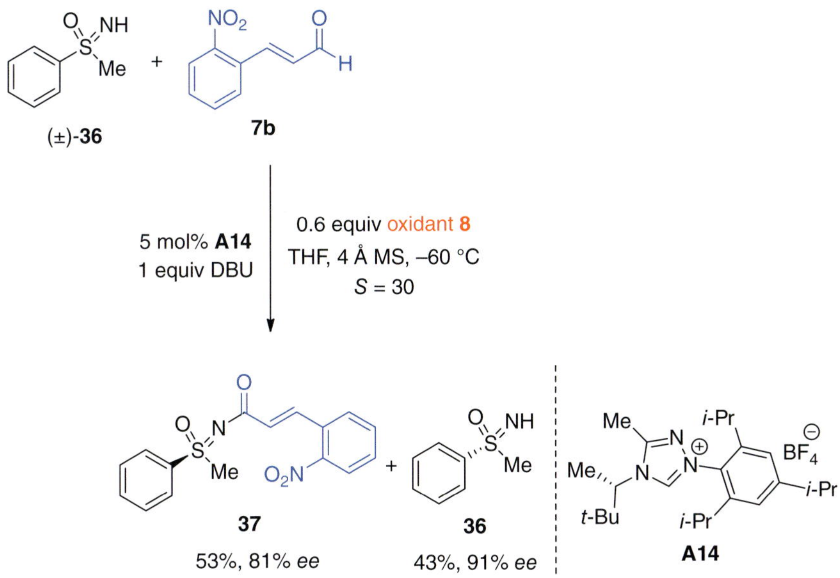

Scheme 10.16 NHC-catalyzed resolution of sulfoximines.

10.3.1 Desymmetrization of Achiral Substrates

In 2009, Sakai and coworkers reported the synthesis of bicyclic tertiary alcohols **39** with two consecutive quaternary centers on the bridgehead carbons through desymmetrization of diketo aldehyde **38** (Scheme 10.17a) [30]. The achiral **38** underwent highly enantioselective intramolecular benzoin reaction to establish the two quaternary stereocenters in **39**. Fused rings of varying sizes were well tolerated to yield the products in moderate yields and high enantioselectivity.

In 2006, the Rovis group reported an elegant desymmetrization of cyclohexadienone **40** to produce hydrobenzofuranone **41** in high yields and excellent diastereo- and enantioselectivity (Scheme 10.17b) [31]. In this case, an intramolecular Stetter reaction took place. By tweaking the electronics on the aryl system of the NHC precursor, the authors found out that aminoindanol-derived triazolium salt **A15** bearing an anisyl substituent produced the optimal results. The reaction could be completed under five minutes. In addition to ether-containing heterocycle, carbocycles were also well tolerated under the same reaction conditions to deliver the analogous products with excellent results, albeit with a relatively longer reaction time.

More recently, the Fan group also developed successful desymmetrization of functionalized 1,4-dienes **42** to deliver bicyclic ketones **43** bearing two continuous stereogenic centers in high yields, excellent enantioselectivity and moderate diastereoselectivity through an intramolecular Stetter reaction (Scheme 10.17c) [32]. This reaction was able to afford a series of tetralone derivatives as well as other classes of cyclic ketone products with two contiguous quaternary stereocenters.

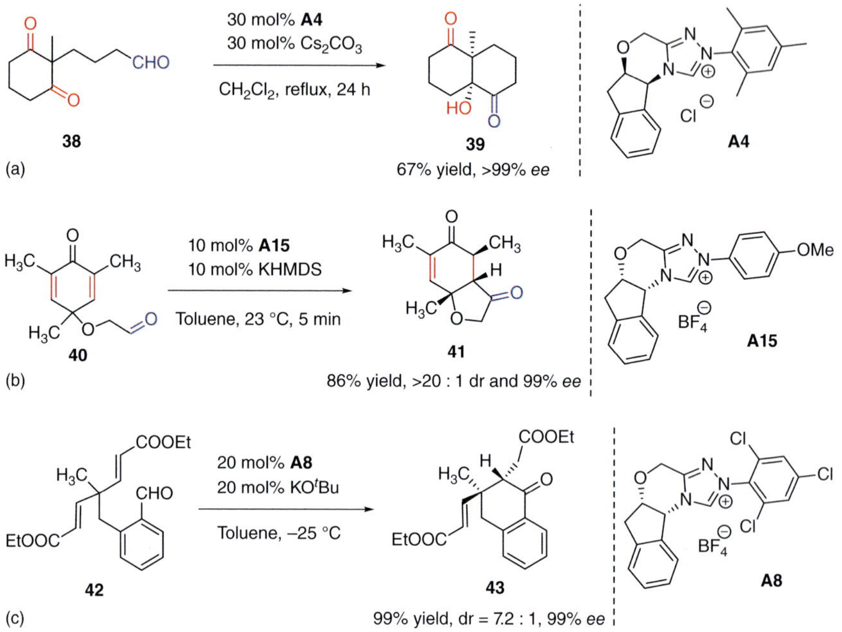

Scheme 10.17 Desymmetrization by (a) benzoin and (b and c) Stetter reactions.

10.3.2 DKR of Racemic Substrates via Benzoin Condensation

As discussed in the introduction part, substrates that can undergo racemization under basic conditions may be well suited for DKR studies under NHC catalysis. As shown in Scheme 10.18, Fan and coworkers reported an impressive example of DKR of β-ketoesters and 1,3-diketones **44** bearing an aldehyde in the same molecule in a highly enantio-, diastereo-, and regioselective manner via intramolecular cross-benzoin reaction (Scheme 10.18a) [33]. The DKR of these substrates was successful because of the relatively acidic α-proton of the 1,3-dicarbonyl compounds. The products **45** bear heavily substituted tetralone backbone. In addition, the authors found out that varying the electronic properties of the substituent on the azolium species helped to speed up the formation of the benzoin product over the corresponding competitive aldol reaction.

In 2014, the Johnson group reported the first DKR of α-ketoesters **46** via an intermolecular cross-benzoin reaction (Scheme 10.18b) [34]. Such a process is highly challenging because of several competing reaction pathways. In addition, the products **47** may also undergo epimerization to destroy the diastereoselectivity of the reaction. Remarkably, the desired cross-benzoin product could be formed in high yields, high enantioselectivity, as well as diastereoselectivity by tweaking the electronics of the catalyst. Because of the faster rate of racemization of the β-stereogenic center of the α-ketoester substrate than that of the cross-benzoin reaction, the desired cross-benzoin product was obtained in high diastereo- and enantioselectivity.

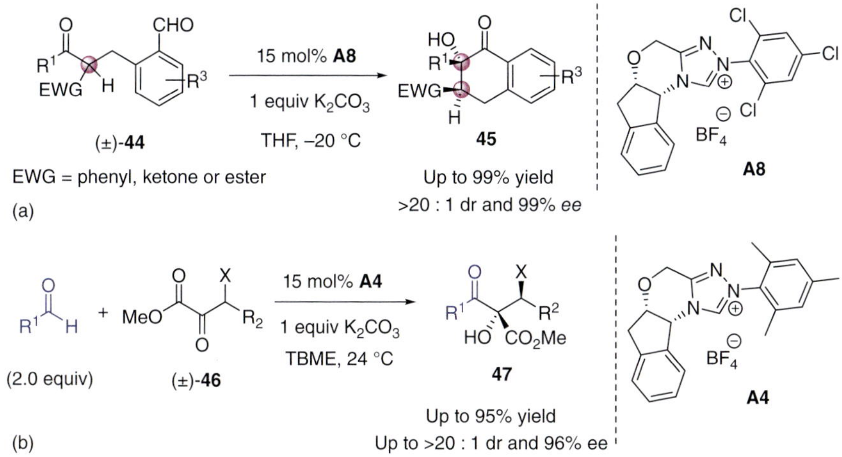

Scheme 10.18 DKR by benzoin condensation.

10.4 Annulation Reactions

As discussed in the previous sections, NHC catalysis provided novel approaches for annulation to deliver various carbo- or heterocyclic compounds in high efficiency and stereoselectivity. Depending on the choice of the system, enolate or homoenolate chemistry as well as γ-addition has been achieved; the corresponding examples for desymmetrization or DKR have also been reported in recent years.

10.4.1 Annulation via Azolium Enolate Addition

In 2007, the Scheidt group reported the desymmetrization of achiral tricarbonyl compounds **48** through an intramolecular aldol/cyclization reaction to produce lactone **49**, which underwent successive decarboxylation to yield enantioenriched α,α-disubstituted cyclopentenes **50** (Scheme 10.19a) [35]. This was made possible through the formation of an azolium enolate intermediate with the reaction between the enal moiety and the chiral NHC catalyst **A16**. Through careful screening of the catalyst, they were able to generate a quaternary carbon stereocenter within the products in high enantioselectivity. In addition, the authors noted the importance of excluding oxygen from the reaction system to avoid the oxidation of the enal into the corresponding carboxylic acid, an unwanted side product.

Subsequently in 2012, the same group developed the first NHC-catalyzed DKR of racemic α-substituted β-keto esters **51** (Scheme 10.19b) [36]. The group envisioned that by making use of the basic environment needed to generate the active carbene species, they could achieve the epimerization of the substrate because of the presence of an acidic proton. By carrying out DFT calculations, the group discovered two possible unconventional mechanisms of either a concerted, asynchronous formal [2+2] aldol-lactonization process, or a stepwise spiro-lactonization mechanism. In addition, it was pointed out that the key

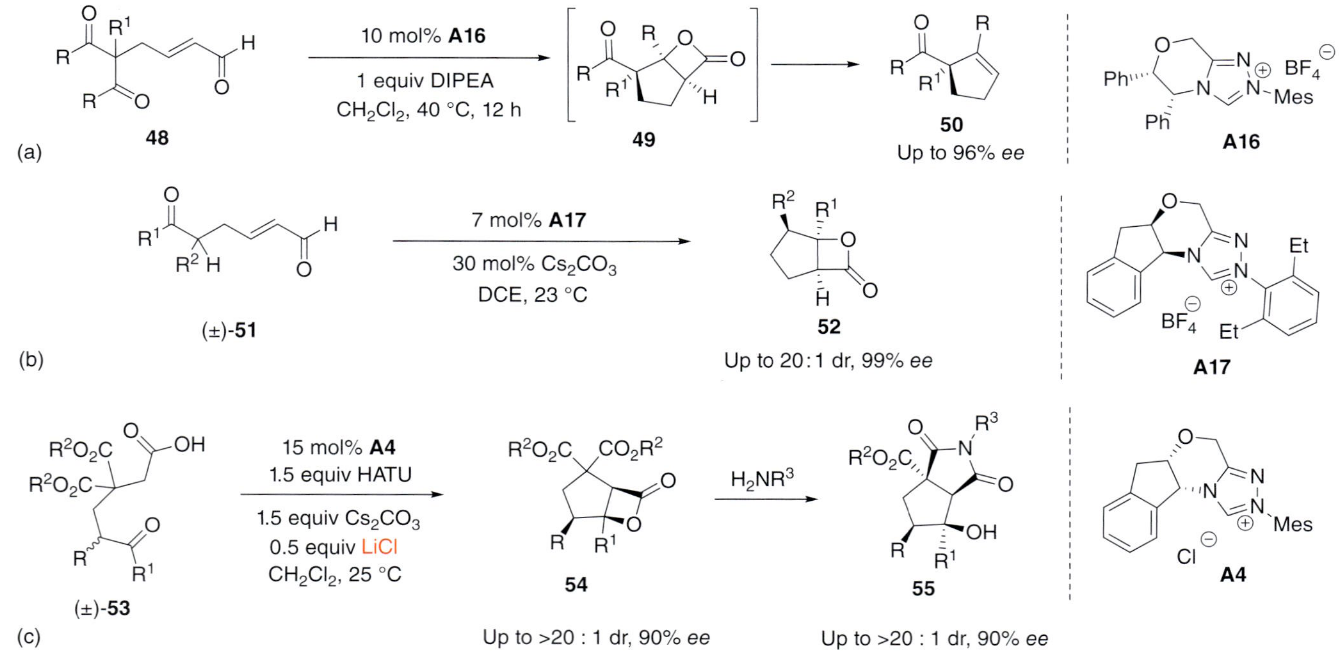

Scheme 10.19 Annulation via enolate addition. (a) Desymmetrization. (b) DKR using ketoenals. (c) DKR using ketoacids.

to achieving high stereoselectivity is both the conjugative stabilization of the electrophile and a nonclassical hydrogen bond between the ketone and the pyranyl C—H on the NHC catalyst **A17**.

In 2017, the Biju group developed another highly enantioselective DKR of ketone acids **53** by the use of HATU and chiral NHC catalyst (Scheme 10.19c) [37]. The β-lactones **54** could be obtained as a single diastereomer and high enantioselectivity. Derivatization of the products to complex succinimides **55** were also demonstrated.

10.4.2 Annulation via Azolium Homoenolate Addition

The concept of NHC-generated azolium homoenolate was also applied for the DKR of β-halo α-keto esters **56** by Johnson and coworkers in 2015 (Scheme 10.20) [38]. Although the two stereocenters were established in the key addition step of homoenolate to the ketone substrate, the chlorine-containing stereocenters in the substrate underwent racemization in the reaction conditions and ended up as a single diastereomer of the product. This example represents the first intermolecular DKR transformation with the construction of three contiguous stereogenic centers in a single step. The resulting γ-butyrolactone products **57** were obtained in high diastereoselectivity and enantioselectivity, as well as good yields amidst the possible competing cross-benzoin and dimerization reactions.

Scheme 10.20 Annulation via homoenolate addition.

10.4.3 Annulation via γ-Addition

In 2014, the Chi group reported an intriguing [3+4] cycloaddition of azomethine imines **58** with a 1,4-dipolarophiles generated by the reaction of **7c** with NHC and an external quinone oxidant (Scheme 10.21a) [39]. The corresponding bicyclic seven-membered heterocycle **59** could be obtained in high yield and ee. In addition, the azomethine imine substrate **58** could also be resolved in high enantioselectivity, with s factors up to 339. It is noteworthy that the resolution was highly effective despite the substrate's reactive site being remote from the NHC catalyst.

By adopting a cooperative catalytic approach with both the NHC catalyst and the Lewis acid $Sc(OTf)_3$, Wang and coworker were able to synthesize enantioenriched δ-lactones **61** in high enantio- and diastereoselectivity through a formal

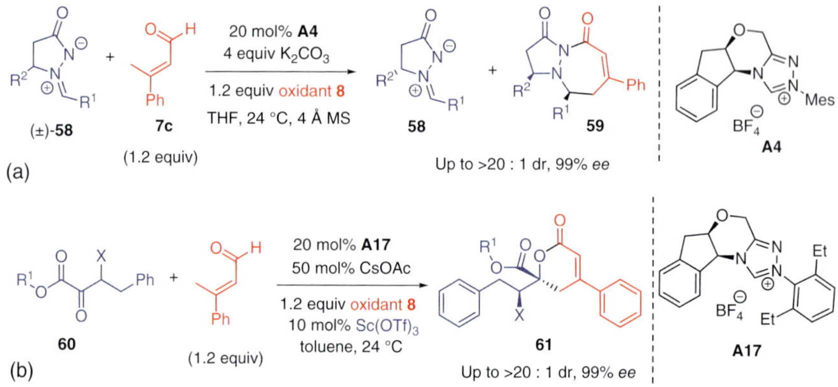

Scheme 10.21 Annulation via γ-addition.

[4+2] cycloaddition (Scheme 10.21b) [40]. The nucleophile generated by the NHC catalyst was the same as that in Scheme 10.21a, and it was hypothesized that the role of $Sc(OTf)_3$ was to bring the ketone electrophile in closer proximity with the nucleophile. This reaction could also be carried out on a gram scale without any erosion in the stereoselectivity.

10.5 Conclusion

In this review, we have presented various approaches and methods enabled by NHC catalysis. These catalytic methods operate on racemic or achiral substrates and proceed in the form of kinetic resolution, DKR, or desymmetrization. With a good understanding of the mechanism and selectivity in NHC-catalyzed processes, we anticipate that further development of NHC-catalyzed kinetic resolution, desymmetrization, and DKR will continue to contribute to synthetic organic chemistry by the introduction of uniquely effective catalytic stereoselective methodologies.

Acknowledgments

The authors are grateful for the generous financial support from the Singapore National Research Foundation (R-143-000-477-281), A*STAR SERC (R-143-000-648-305), and the National University of Singapore.

References

1 (a) Enders, D., Niemeier, D.O., and Henseler, A. (2007). *Chem. Rev.* 107: 5606. (b) Marion, N., Díez-González, S., and Nolan, S.P. (2007). *Angew. Chem. Int. Ed.* 46: 2988. (c) Nair, V., Menon, R.S., Biju, A.T. et al. (2011). *Chem. Soc. Rev.* 40: 5336. (d) Izquierdo, J., Hutson, G.E., Cohen, D.T., and Scheidt,

K.A. (2012). *Angew. Chem. Int. Ed.* 51: 11686. (e) Bugaut, X. and Glorius, F. (2012). *Chem. Soc. Rev.* 41: 3511. (f) Grossmann, A. and Enders, D. (2012). *Angew. Chem. Int. Ed.* 51: 314. (g) Ryan, S.J., Candish, L., and Lupton, D.W. (2013). *Chem. Soc. Rev.* 42: 4906. (h) Bode, J.W. (2013). *Nat. Chem.* (5): 813. (i) Hopkinson, M.N., Richter, C., Schedler, M., and Glorius, F. (2014). *Nature* 510: 485. (j) Flanigan, D.M., Romanov-Michailidis, F., White, N.A., and Rovis, T. (2015). *Chem. Rev.* 115: 9307. (k) Yetra, S.R., Patra, A., and Biju, A.T. (2015). *Synthesis* 47: 1357. (l) Wang, M.H. and Scheidt, K.A. (2016). *Angew. Chem. Int. Ed.* 55: 14912.

2 (a) Keith, J.M., Larrow, J.F., and Jacobsen, E.N. (2001). *Adv. Synth. Catal.* 343: 5–26. (b) Vedejs, E. and Jure, M. (2005). *Angew. Chem. Int. Ed.* 44: 3974.

3 (a) Ebbers, E.J., Ariaans, G.J.A., Houbiers, J.P.M. et al. (1997). *Tetrahedron* 53: 9417. (b) Caddick, S. and Jenkins, K. (1996). *Chem. Soc. Rev.* 25: 447. (c) Huerta, F.F., Minidis, A.B.E., and Bäckvall, J. (2001). *Chem. Soc. Rev.* 30: 321. (d) Pellissier, H. (2003). *Tetrahedron* 59: 8291. (e) Pellissier, H. (2011). *Adv. Synth. Catal.* 353: 659.

4 (a) Willis, M.C. (1999). *J. Chem. Soc., Perkin Trans.* 1: 1765. (b) Zeng, X.-P., Cao, Z.-Y., Wang, Y.-H. et al. (2016). *Chem. Rev.* 116: 7330.

5 (a) France, S., Guerin, D.J., Miller, S.J., and Lectka, T. (2003). *Chem. Rev.* 103: 2985. (b) Fu, G.C. (2004). *Acc. Chem. Res.* 37: 542. (c) Jarvo, E.R. and Miller, S.J. (2004). *Comprehensive Asymmetric Catalysis*, Supplement 1 (ed. E.N. Jacobsen, A. Pfaltz and H. Yamamoto), 189. Berlin: Springer.

6 (a) Suzuki, Y., Yamauchi, K., Muramatsu, K., and Sato, M. (2004). *Chem. Commun.* 2770. (b) Kano, T., Sasaki, K., and Maruoka, K. (2005). *Org. Lett.* (7): 1347.

7 (a) Reynolds, N.T., Read de Alaniz, J., and Rovis, T. (2004). *J. Am. Chem. Soc.* 126: 9518. (b) Chan, A. and Scheidt, K.A. (2005). *Org. Lett.* 7: 905.

8 Knappke, C.E.I., Imami, A., and Wangelin, A.J. (2012). *ChemCatChem* 4: 937.

9 (a) Maki, B.E., Chan, A., Phillips, E.M., and Scheidt, K.A. (2007). *Org. Lett.* 9: 371. (b) De Sarkar, S., biswas, A., Song, C.H., and Studer, A. (2011). *Synthesis* 12: 1974.

10 (a) Lu, S., Poh, S.B., Siau, W.-Y., and Zhao, Y. (2013). *Angew. Chem. Int. Ed.* 52: 1731. (b) Lu, S., Poh, S.B., Siau, W.-Y., and Zhao, Y. (2013). *Synlett* 24: 1165.

11 Raup, D.E.A., Cardinal-David, B., Holte, D., and Scheidt, K.A. (2010). *Nat. Chem.* 2: 766.

12 Kuwano, S., Harada, S., Kang, B. et al. (2013). *J. Am. Chem. Soc.* 135: 11485.

13 Li, B.-S., Wang, Y., Proctor, R.S.J. et al. (2016). *Chem. Commun.* 52: 8313.

14 Wu, Z. and Wang, J. (2017). *ACS Catal.* 7: 7467.

15 Zhao, C., Li, F., and Wang, J. (2016). *Angew. Chem. Int. Ed.* 55: 1820.

16 Chen, X., Fong, J.Z.M., Xu, J. et al. (2016). *J. Am. Chem. Soc.* 138: 7212.

17 (a) Chen, Y., Yekta, S., and Yudin, A.K. (2003). *Chem. Rev.* 103: 3155. (b) Brunel, J.M. (2007). *Chem. Rev.* 107: PR1. (c) Shibasaki, M. and Matsunaga, S. (2011). *Privileged Chiral Ligands and Catalysts* (ed. Q.-L. Zhou), 295. Wiley. (d) Ding, K., Guo, H., Li, X. et al. (2005). *Top. Catal.* 35: 105.

18 Lu, S., Poh, S.B., and Zhao, Y. (2014). *Angew. Chem. Int. Ed.* 53: 11041.

19 Ma, G., Deng, J., and Sibi, M.P. (2014). *Angew. Chem. Int. Ed.* 53: 11818.

20 Maji, R. and Wheeler, S.E. (2017). *J. Am. Chem. Soc.* 139: 12441.
21 (a) Tang, W. and Zhang, X. (2003). *Chem. Rev.* 103: 3029. (b) Xie, J.-H. and Zhou, Q.-L. (2008). *Acc. Chem. Res.* 41: 581.
22 Huang, Z., Huang, X., Li, B. et al. (2016). *J. Am. Chem. Soc.* 138: 7524.
23 Lu, S., Song, X., Poh, S.B. et al. (2017). *Chem. Eur. J.* 23: 2275.
24 (a) Nair, V., Thomas, S., Mathew, S.C., and Abhilash, K.G. (2006). *Tetrahedron* 62: 6731. (b) Mondal, S. and Panda, G. (2014). *RSC Adv.* 4: 28317. (c) Swift, E.C. and Jarvo, E.R. (2013). *Tetrahedron* 69: 5799. (d) Nambo, M. and Crudden, C.M. (2015). *ACS Catal.* 5: 4734.
25 De Sarkar, S., Grimme, S., and Studer, A. (2010). *J. Am. Chem. Soc.* 132: 1190.
26 Li, G.-Q., Li, Y., Dai, L.-X., and You, S.-L. (2008). *Adv. Synth. Catal.* 350: 1258.
27 (a) Binanzer, M., Hsieh, S.-Y., and Bode, J.W. (2011). *J. Am. Chem. Soc.* 133: 19698. (b) Hiseh, S.-Y., Binanzer, M., Kreutuss, I., and Bode, J.W. (2012). *Chem. Commun.* 48: 8892. (c) Allen, S.E., Hsieh, S.-Y., Gutierrez, O. et al. (2014). *J. Am. Chem. Soc.* 136: 11783.
28 Dong, S., Frings, M., Cheng, H. et al. (2016). *J. Am. Chem. Soc.* 138: 2166.
29 Menon, R.S., Biju, A.T., and Nair, V. (2016). *Beilstein J. Org. Chem.* 12: 444.
30 (a) Ema, T., Oue, Y., Akihara, K. et al. (2009). *Org. Lett.* 11: 4866. (b) Ema, T., Akihara, K., Obayashi, R., and Sakai, T. (2012). *Adv. Synth. Catal.* 354: 3283.
31 Liu, Q. and Rovis, T. (2006). *J. Am. Chem. Soc.* 128: 2552.
32 Lin, Q., Li, Y., Das, D.K. et al. (2016). *Chem. Commun.* 52: 6459.
33 Zhang, G., Yang, S., Zhang, X. et al. (2016). *J. Am. Chem. Soc.* 138: 7932.
34 Goodman, C.G. and Johnson, J.S. (2014). *J. Am. Chem. Soc.* 136: 14698.
35 Wadamoto, M., Phillips, E.M., Reynolds, T.E., and Scheidt, K.A. (2007). *J. Am. Chem. Soc.* 129: 10098.
36 Cohen, D.T., Eichman, C.C., Phillips, E.M. et al. (2012). *Angew. Chem. Int. Ed.* 51: 7309.
37 Mondal, S., Mukherjee, S., Das, T.K. et al. (2017). *ACS Catal.* 7: 3995–3999.
38 Goodman, C.G., Walker, M.M., and Johnson, J.S. (2015). *J. Am. Chem. Soc.* 137: 122.
39 Wang, M., Huang, Z., Xu, J., and Chi, Y.R. (2014). *J. Am. Chem. Soc.* 136: 1214.
40 Wu, Z., Li, F., and Wang, J. (2015). *Angew. Chem. Int. Ed.* 54: 1629–1633.

11

N-Heterocyclic Carbenes for Organopolymerization: Metal-Free Polymer Synthesis

Romain Lambert, Joan Vignolle, and Daniel Taton

CNRS-Université de Bordeaux-Bordeaux INP, Laboratoire de Chimie des Polymères Organiques (LCPO UMR 5629), 16 Avenue Pey-Berland, 33607 Pessac Cedex, France

11.1 Introduction

In the past decade, organocatalysts have been introduced in the macromolecular synthesis for producing polymer materials free of any metallic residues and a potential use in high-value applications, such as biomedical and personal beauty care applications, microelectronic devices, or food packaging [1–4]. Several classes of organic activators, behaving as true catalysts or as direct initiators (see the distinction further), including Brønsted/Lewis acids or bases, and mono- or bicomponent bifunctional catalytic systems have been utilized, both for step-growth and chain-growth polymerizations, and for depolymerization reactions as well. Several general reviews or highlights on organocatalyzed polymerizations are already available [1–4], while other reviews have dealt with more specific aspects of this topic, such as, for instance, the ring-opening polymerization (ROP) [5, 6], the group-transfer polymerization (GTP) of acrylics and methacrylics [7], the zwitterionic polymerization [8], or the organocatalyzed polymerizations induced by H-bond components [9].

In this context, *N*-heterocyclic carbenes (NHCs) have been among the first organocatalysts employed for polymer synthesis, providing a straightforward and metal-free synthetic strategy to a wide range of polymers [10–12]. This is obviously related to the near-unlimited structural diversity of NHCs, which allows finely tuning their steric and electronic properties and using them either as Brønsted and/or Lewis bases [13–16]. Cyclic esters (e.g. D,L-lactide [LA] and lactones) have been by far the most investigated monomer substrates for NHC-organocatalyzed ring-opening polymerization (OROP). However, the range of monomers amenable to polymerization by NHC catalysis has been largely expanded to the ROP of other heterocycles, including *O*-carboxyanhydrides, N-substituted carboxyanhydrides, phosphoesters, carbonates, siloxanes and carbosiloxanes, lactams, epoxides, and N-activated aziridines, to the polymerization of alkyl (meth)acrylates, and finally to some polymerizations proceeding by a step-growth process, such as polyurethane (PU), polyester, or polyepoxide synthesis [10–12].

N-Heterocyclic Carbenes in Organocatalysis, First Edition. Edited by Akkattu T. Biju.
© 2019 Wiley-VCH Verlag GmbH & Co. KGaA. Published 2019 by Wiley-VCH Verlag GmbH & Co. KGaA.

Besides their role as true catalysts, i.e. in truly organocatalyzed polymerizations, some NHCs can operate as direct nucleophilic initiators in the absence of an initiator, i.e. in organoinitiated polymerization, for instance, through ring-opening of heterocyclic monomers or through 1,4-conjugate addition of some (meth)acrylics. In many cases of chain-growth processes, NHCs provide high polymerization rates, excellent chain-end fidelity, control over molar masses of resulting polymers, and also allow achieving cyclic polymer structures [10–12]. The present chapter aims at giving a general view of NHC-mediated organopolymerization, including both organocatalyzed and organoinitiated polymerization reactions. Main monomer substrates and general polymerization mechanisms are first briefly discussed.

11.2 Main NHCs and Fundamental Mechanisms of NHC-Induced Polymerization

Owing to their σ-donor properties, NHCs can behave either as Brønsted bases ($pK_a \sim 15–30$ in H_2O) [17, 18] or as nucleophilic type catalysts, which depend on the monomer and/or the initiator – if present – to activate. The different mechanisms found in NHC-mediated polymerization as a function of substrate(s) are depicted in Schemes 11.1–11.3. It is worth mentioning that there is still a lack of rationalization as for which mechanism truly operates in some NHC-mediated polymerizations, which arguably depends on various parameters, including the nature of both the NHC and the initiator, and the overall experimental conditions as well (solvent polarity and temperature). In this regard, calculations by density functional theory (DFT) prove often useful to explain some effects observed experimentally [19, 20].

NHCs eventually behave as nucleophiles in direct reaction with the neutral monomer. This mode of action has been proposed, not only in NHC-mediated of heterocycles (e.g. lactide [8, 21–23], epoxides [24–26], and phosphoesters [27]) but also for methacrylate monomers involving selective 1,4-conjugate addition reactions (Scheme 11.1) [28–31]. In these cases, the nucleophilic attack of the monomer by the NHC yields a zwitterionic intermediate that is sometimes regarded as an "activated monomer"; hence, the related mechanisms also refer to as the activated monomer mechanism. Subsequent protonation of this zwitterionic species, for instance, by an excipient alcohol initiator (ROH), and concomitant release of the NHC catalyst, gives rise to a monoadduct (extended alcohol). A similar catalytic cycle utilizing the latter monadduct will form the bis-adduct, and so on so forth, so that polymerization can proceed in a controlled manner.

In the absence of any (ROH) initiator, the NHC nucleophile is no more a catalyst and instead plays the role of a true initiator. Thus, it will be incorporated in α-position of the polymer, the chain growth occurring by a zwitterionic mechanism. In this way, macrocyclic polymers can be achieved, providing a new synthetic route to these architectures under relatively mild conditions (see Scheme 11.1; see also Section 2.1.2).

NHCs would also behave as nucleophilic catalysts in the specific context of the GTP of (meth)acrylic monomers, in the presence of a silyl ketene acetal (SKA)

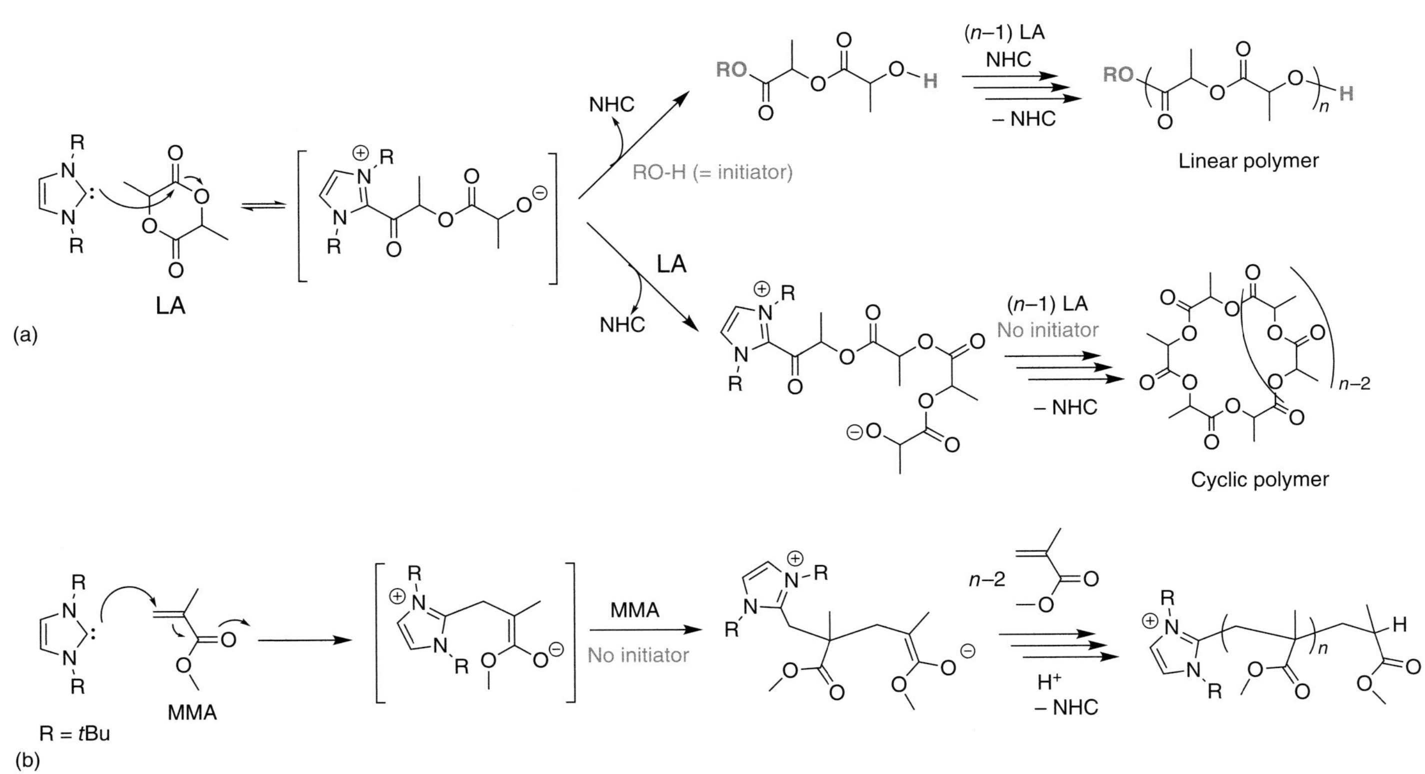

Scheme 11.1 Nucleophilic mechanism for the ROP of lactide (LA) (a) and for the polymerization of MMA (b).

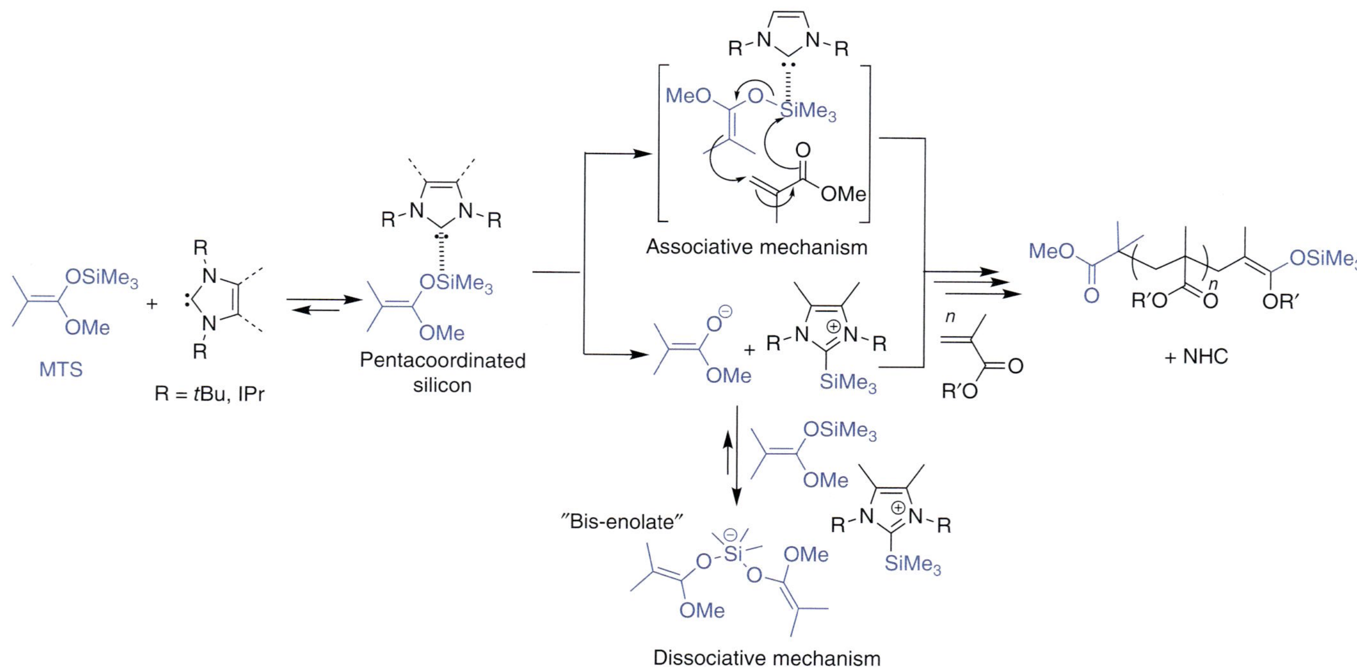

Scheme 11.2 Associate and dissociative NHC-catalyzed group transfer polymerization of methyl methacrylate.

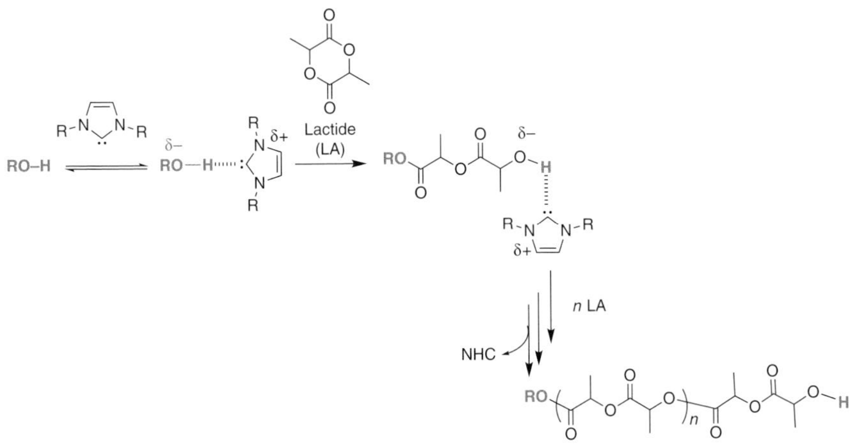

Scheme 11.3 Basic mechanism for the ROP of lactide.

employed as an initiator (Scheme 11.2) [32, 33]. GTP is eventually an extension of the Mukaiyama–Michael reaction, which has been developed in the 1980s at Dupont as a convenient method to control the polymerization of methacrylic monomers at room temperature (RT) [7, 34, 35]. However, the absence of a single catalytic system that could be generalized for both acrylic and methacrylic monomers has limited the development of this polymerization method. NHCs have been shown to catalyze the GTP of both families of monomers and have also enabled the synthesis of structurally well-defined block copolymers by NHC-catalyzed sequential GTP [32, 33, 36–38]. In this case, activation of the SKA initiator – and not the monomer – would take place by nucleophilic interaction of the NHC onto the silicon atom of the SKA (see Scheme 11.2). The ensuing hypervalent SKA-based intermediate would then add to the monomer following a 1,4-addition pathway.

Polymerization reactions involving selective activation of the initiator by the NHC – and later on activation of the polymer chain end – refer to as the basic mechanism, also sometimes called the activated chain-end mechanism. In such a case, the NHC plays the role of a Brønsted base interacting with – or even deprotonating – the initiator, and then with the polymer chain ends, favoring the addition of the latter to the neutral monomer substrate (Scheme 11.3). The ensuing monoadduct can then propagate an anionic-type polymerization by the same catalytic cycle, with the protonated NHC, i.e. a soft and generally bulky noninnocent azolium-type counter ion. Such a mechanism has been postulated in many situations, where alcoholic initiators are employed, not only for OROP [10–12, 39] but also for the NHC-catalyzed polymerization of methyl methacrylate (MMA) [19]. The interactions developing between the selected NHC and the protic component are obviously correlated with the pK_a values of both reaction partners.

Another situation where NHCs would operate as Brønsted bases is encountered when the monomer itself contains an acidic proton that can be readily abstracted. This has been proposed in step-growth polymerizations involving OH-containing monomers, for instance, to achieve polyurethanes [40–42] and polysiloxanes [43], from a diol and a diisocyanate and from a disilanol, respectively (see Section 11.4). In these cases, indeed, the NHC catalyst is thought to selectively interact with the hydroxyl function, before the attack of the as-activated alcohol.

Finally, a basic mechanism has also been reported for the ROP of ε-caprolactam and laurolactam (LL), forming polyamide-6 and polyamide-12, respectively, when directly initiated by peculiar NHC precursors [44, 45]. Deprotonation of the lactam monomer by the NHC would generate an activated monomer in the form of an amide anion that could initiate the ROP (Scheme 11.4).

As already mentioned, the precise mechanism by which NHC-mediated polymerizations operate has not been systematically established. Some investigations have eventually revealed possible coexistence of the two competitive nucleophilic and Brønsted basic mechanisms, for instance, for the OROP of cyclic esters [9–11, 46], epoxides [24–26, 47], or siloxanes [39, 48] as well as for the polymerization of alkyl (meth)acrylates [19, 29].

The rest of this chapter discusses representative cases of the use of NHCs as organic nucleophiles and/or Brønsted bases in chain-growth or step-growth polymerization reactions. The discussion is organized according to the polymerization method considered and the type of polymer synthesized (e.g. linear vs. cyclic). For a more complete description of these works, the reader can refer to review articles and highlights already mentioned [10–12].

11.3 NHC-Mediated Chain-growth Polymerization

11.3.1 Ring-opening Polymerization

Main stable carbenes used in the context of the ROP are shown in Figure 11.1, whereas monomers and related polymers obtained from a NHC activation are overviewed in Figure 11.2. Cyclic esters (lactides, ε-caprolactone [ε-CL], and β-propiolactone, Figure 11.2) have been by far the most studied, which is in part due to the importance of corresponding biodegradable and *in vivo* bioresorbable aliphatic polyesters [10–12]. Cyclic carbonates have also received a special attention, for which related polycarbonates exhibit a lower degradation rate in water and allow introducing reactive side-chain functionalities [5].

11.3.2 NHC-OROP (in the Presence of an Initiator)

The first report by the Hedrick group at IBM Almaden on NHC-OROP of cyclic esters (e.g. D,L-lactide and lactones) dates back to 2002. Exploiting the ability of NHCs to catalyze transesterification reactions in molecular chemistry, metal-free linear aliphatic polyesters have thus been achieved [21].

Under certain conditions, NHC-OROP reactions provide high polymerization rates, high chain-end fidelity, and controlled molar masses, hence showing the features of "controlled/living" polymerizations (Scheme 11.1, path a) [10–12]. In a typical experiment, an α-benzyloxy poly(L-lactide) with narrowly distributed molar masses (dispersity, $Đ$ = 1.15) can be synthesized in 92% yield after

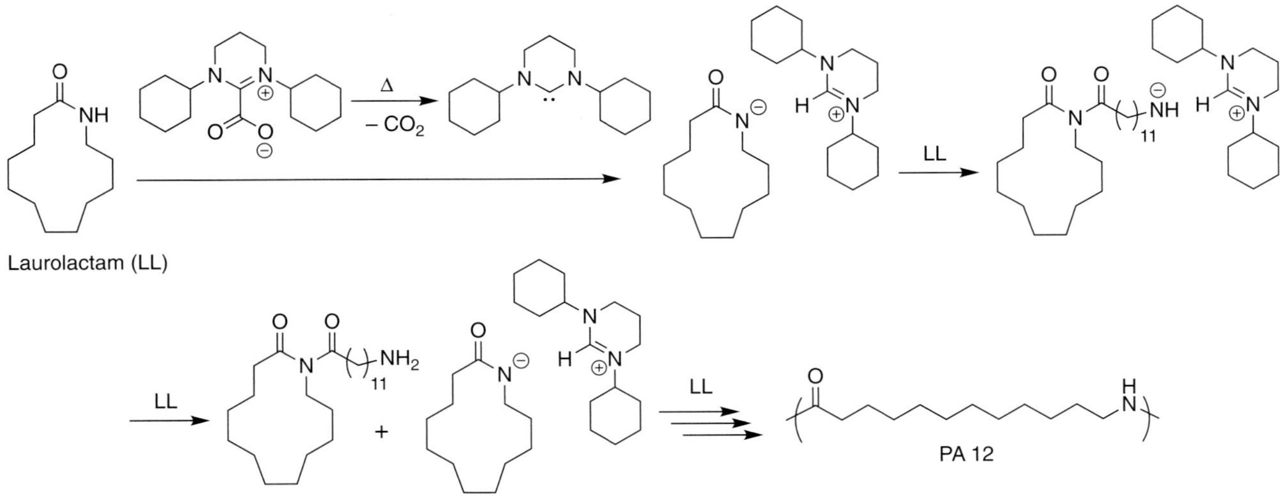

Scheme 11.4 Synthesis of polyamide-12 (PA 12) by NHC-mediated anionic-like polymerization of laurolactam (LL); the NHC here is thermally generated from a NHC-CO_2 precursor.

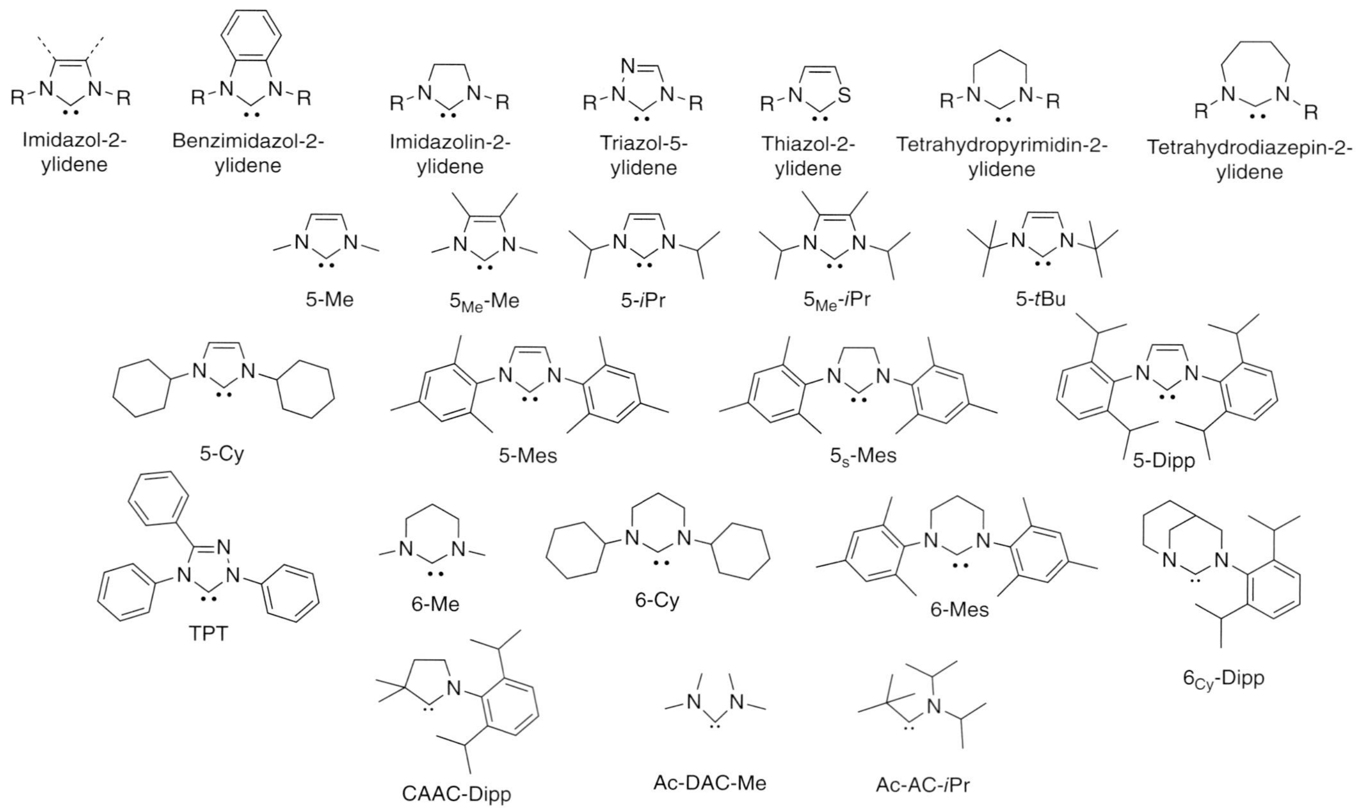

Figure 11.1 Main singlet carbenes investigated for the activation/polymerization of monomer substrates.

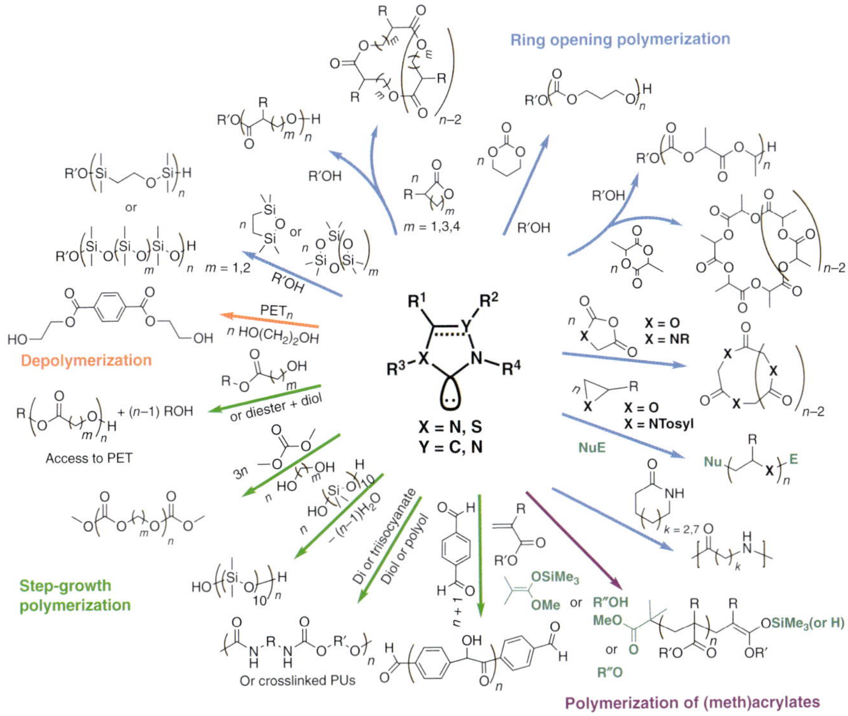

Figure 11.2 Scope of NHC-mediated polymerization reactions.

two hours of reaction, using initial molar ratios of LA/BnOH/NHC = 100/1.5/1 [21]. Under similar conditions, the NHC-OROP of ε-CL requires 24 hours to reach completion because of its lesser reactivity compared to LA. As expected, the catalytic activity of the NHC closely depends on its substituent pattern. Generally, small-size *N*-substituents on the NHC favors faster polymerization. Besides, less electron-rich thiazol-2-ylidenes are less efficient ROP organocatalysts [30].

Although initial experimental observations have led to the proposition of the aforementioned nucleophilic mechanism, i.e. involving the formation of a zwitterionic intermediate species, it has soon become apparent that the basic mechanism can also operate; this is summarized in Scheme 11.5.

Another interest in NHC-OROP of lactide relates to the possibility to induce a stereoselective process from chiral NHC precursors. Consequently, the polymer tacticity might be controlled, which has a crucial influence in the final thermal and mechanical properties of polylactide (PLA). Yet, only a few reports have addressed that aspect, and there is probably room for improvement. First attempt has been reported in 2004 by Hillmyer and coworkers who have used **5-Mes**, **5-Dipp**, and **5-*t*Bu** (Figure 11.1), in the presence of benzyl alcohol as an initiator [49]. Only moderately isotactic enriched PLAs have been obtained, however, from the racemic LA. Further studies by Dove et al. have established that the PLA tacticity is governed by a chain-end control mechanism [50].

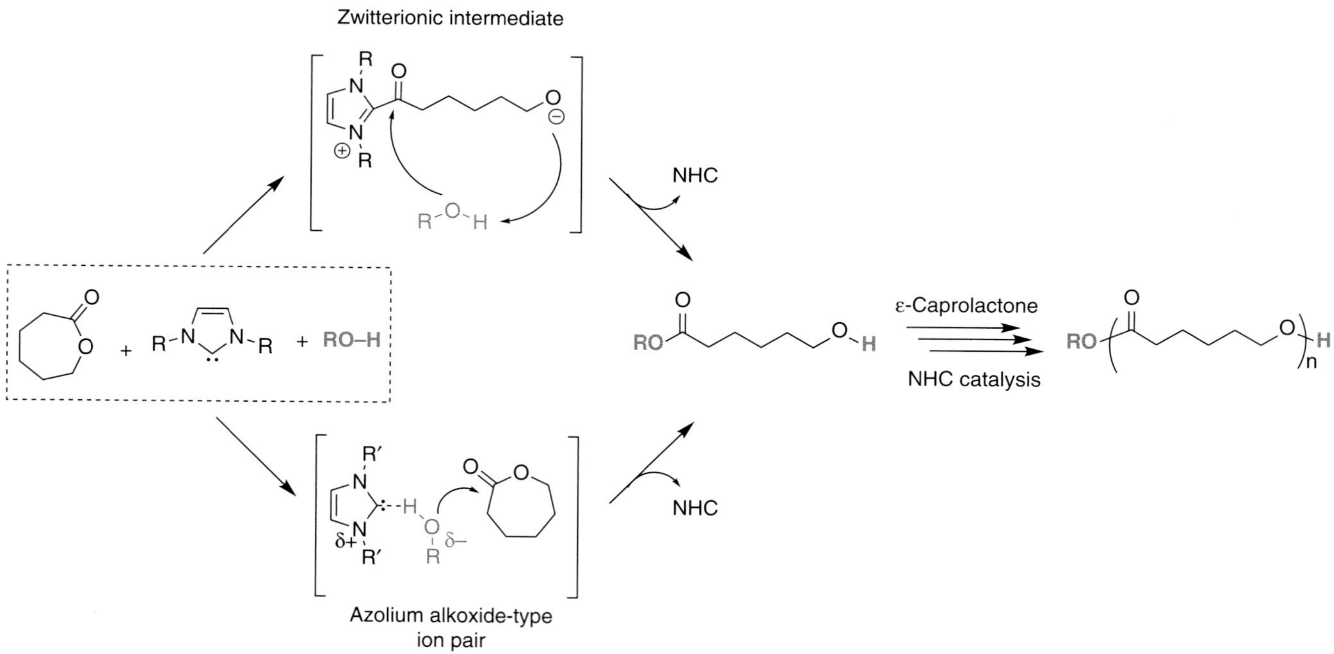

Scheme 11.5 Nucleophilic *versus* basic mechanism in competition, here for the NHC-OROP of ε-caprolactone.

As mentioned, the range of monomers amenable to ROP by a NHC catalysis has been expanded to many other heterocycles than cyclic esters, as highlighted in Figure 11.2. For instance, three-membered-ring heterocycles, such as oxiranes and aziridines, have been successfully polymerized by an organocatalytic pathway utilizing NHCs (Scheme 11.6). The NHC-OROP of both ethylene oxide (EO) and propylene oxide (PO) provides metal-free and well-defined poly(ethylene oxide)s (PEOs) and poly(propylene oxide)s (PPOs) in the presence of different types of initiators, including functionalized and nonfunctionalized alcohols, trimethylsilyl azide, and trimethyl silyl ethers [25, 26]. At 50 °C, no polymerization has been observed in THF in contrast to DMSO, although requiring three days for the reaction to reach completion with EO, whereas PO leads to 30–40% conversion only, evidencing a much slower process in comparison to cyclic esters. Both the nucleophilic and the basic mechanisms have been postulated, similarly to the NHC-mediated polymerization of cyclic esters.

Three-membered-ring monomers such as *N*-*p*-toluenesulfonyl (*N*-tosyl = *N*-Ts) aziridines, namely, 2-alkyl-*N*-*p*-toluenesulfonyl aziridines, can also be subjected to a controlled NHC-OROP [51]. Typically, catalytic amounts of 1,3-bis(isopropyl)-4,5(dimethyl)imidazol-2-ylidene (**5_{Me}-*i*Pr**) as organocatalyst and THF as a solvent, at 50 °C, have been successfully employed. In this case, a designed *N*-tosyl secondary amine serves as an initiator mimicking the growing chain. Metal-free polyaziridines (PAz) have thus been achieved, with an excellent control over molar masses (up to 20 000 g mol^{-1}) and low dispersities ($Đ < 1.15$), as depicted in Scheme 11.6. In addition, all PAz-based block copolymers have been achieved by sequential **5_{Me}-*i*Pr**-mediated OROP of two different *N*-sulfonylaziridines (2-methyl- and 2-benzyl-*N*-*p*-toluenesulfonyl aziridines).

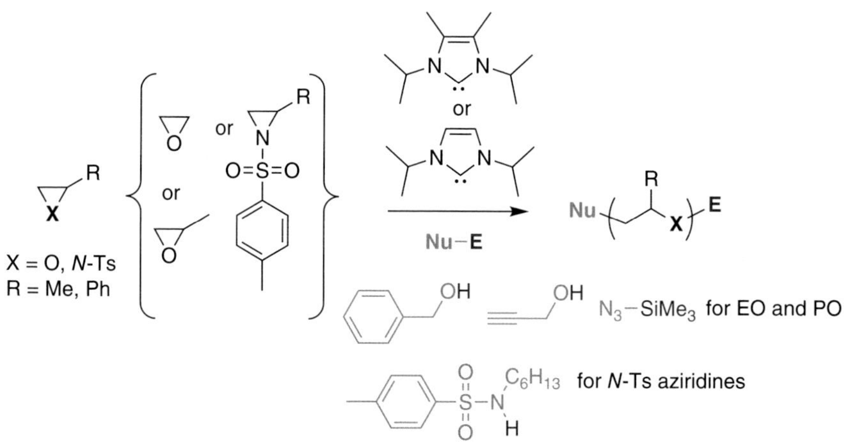

Scheme 11.6 NHC-OROP of three-membered ring hetrocycles such as ethylene and propylene oxides and 2-alkyl-*N*-*p*-toluenesulfonyl aziridines.

Many reports have focused on free NHCs (or bare NHCs) derived from imidazol(in)ium salts, in most cases, in the presence of OH-containing initiators. Miscellaneous examples of block copolymer synthesis via NHC catalysis by sequential polymerization have also been reported [10–12]. However, despite their efficiency as catalysts, the air and moisture sensitivity of free NHCs, their limited storability, and, in some cases, synthetic requirements have limited their widespread adoption in (macro)molecular chemistry. Thus, some studies have focused on the development of protected carbenes, as more air-tolerant precursors [10, 52]. These NHC progenitors are indeed generally easier to handle than bare NHCs, avoiding the use of controlled inert atmospheres. This *modus operandi* happens to be an important step toward large-scale application.

Typical protected or masked NHCs include adducts, such as imidazol(in)ium carboxylates (NHC–CO_2), NHC–alcohol adducts, or NHC–metal complexes, or some azolium hydrogen carbonates ([NHC(H)·HCO_3]) (see Scheme 11.7). Masked NHC thus operates as latent catalysts, the release of the free carbene being triggered, for instance, upon heating or by a simple solvent effect. An NHC showing a photoswitchable catalytic for the OROP of lactones has also been reported, the photocyclized NHC–alcohol adduct being catalytically inactive [53]. It must be pointed out, however, that the protected NHCs generally exhibit lower catalytic performances in comparison to free NHCs, likely because of a slow or incomplete generation of active species under polymerization conditions [10, 52, 54].

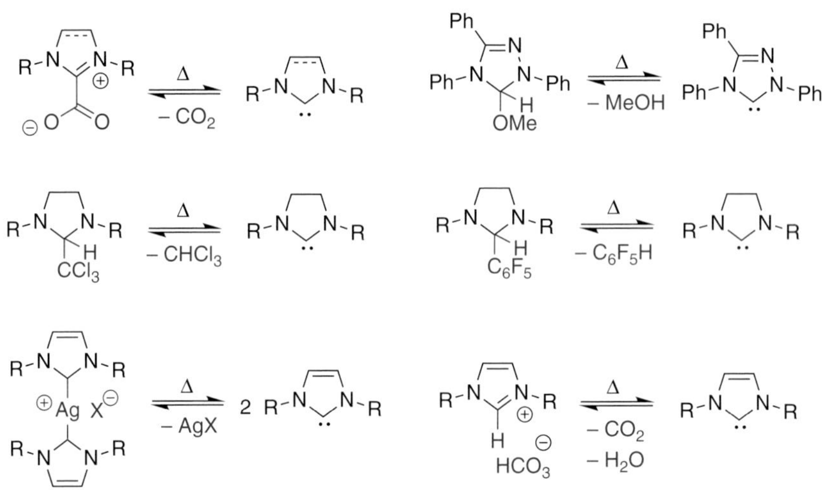

Scheme 11.7 Examples of masked NHCs used for polymer synthesis.

Different groups have thus resorted to NHC–CO_2 adducts to trigger the NHC-OROP of ε-CL, β-butyrolactone, or trimethylene carbonate [46, 54–57] via a thermal generation of the free NHCs *in situ*. A study by the BASF researchers has focused on the bulk NHC-OROP of PO at 120 °C, from imidazolium (e.g. **5-Me-CO$_2$**, **5-*i*Pr-CO$_2$**, **5-*t*Bu-CO$_2$**, and **5-Mes-CO$_2$**), imidazolinium (**5$_s$-Mes-CO$_2$**), or benzimidazolium carboxylates, using diethylene glycol (DEG) as an initiator [47]. Diol-functionalized PPO oligomers (M_n = 300–1200 g mol^{-1} and $Đ$ = 1.08–1.23) have thus been obtained after four hours, with 40–74% monomer conversion. Consistently with previous findings by Taton and coworkers [26], imidazolium precatalysts prove the most active, with the least bulky substituents affording the higher catalytic activity. DFT calculations and *in situ* IR spectroscopy have suggested that decarboxylation is slow, enabling a continuous delivery of the free NHC and a better control over the NHC-OROP process. In other words, NHC–CO_2 adducts would serve as a reservoir for free NHCs. Interestingly, this protocol has also allowed synthesizing PPO-*b*-polyester copolymers (M_n = 750–3700 g mol^{-1} and $Đ$ = 1.38–1.77), by sequential ROP of PO and ε-CL (or L-LA). The authors have then discussed the nucleophilic mechanism initially reported by Taton et al. by proposing a free anionic-type mechanism, as highlighted in Scheme 11.8, based on their DFT calculations. In the latter anionic mechanism, the zwitterionic imidazolium alkoxide would evolve into an imidazolium alkoxide ion pair (route A_1) [47].

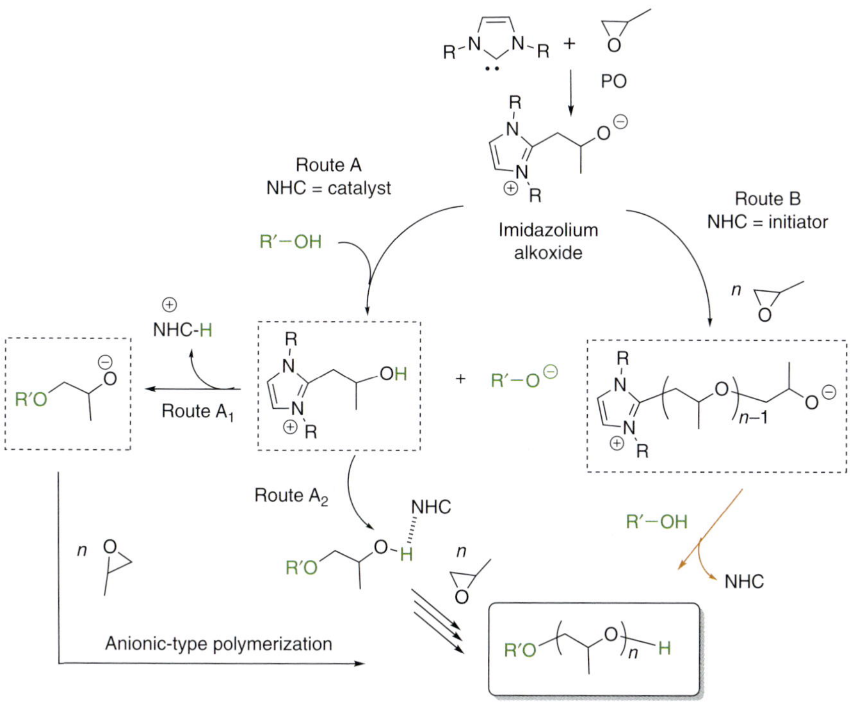

Scheme 11.8 Proposed mechanisms for the NHC-OROP of propylene oxide in the presence of an alcohol as the initiator.

Finally, it is worth mentioning that NHC-OROP can be applied to derive not only linear homopolymers and block copolymers but also branched polymer architectures, including star polymers [21], dendrimer-like structures [58], and "H-shaped" polymers [59].

11.3.3 Directly NHC-Mediated ROP (in the Absence of an Initiator): Synthesis of Cyclic vs. Linear Polymers

As emphasized above, NHCs can serve as direct nucleophiles, and no more as organocatalysts, i.e. in the absence of any external initiator, to initiate chain-growth polymerization reactions, either through ring-opening or through 1,4-conjugate addition, for cyclic and (meth)acrylic monomers, respectively (see Scheme 11.1 and Figure 11.2). Most studies have employed "bare" NHCs (made before synthesis or commercially available), but NHC precursors in their protected form have also been evaluated for that purpose [10–12].

For instance, the ROP of D,L-LA can be directly initiated by 1,3-bis(2,4,6-trimethylphenyl)imidazol-2-ylidene (**5-Mes**, Figure 11.1) via the so-called zwitterionic ring-opening polymerization (ZROP) mechanism [60]. The reaction reaches completion within minutes at RT, yielding cyclic PLAs (see Scheme 11.1 and Figure 11.2) of high molar masses, and a dispersity <1.4, provided monomer conversion does not exceed 90%, otherwise the dispersity can increase to about 1.6. Kinetic modelings have shown that this ZROP process features slow initiation, fast propagation, and side reactions such as a depropagation step and slow termination [61]. In other words, NHC-induced ZROP of LA is clearly not a "controlled/living" process.

Landis and coworker have recently reinvestigated the kinetics of **5-Mes**-mediated ZROP of LA in toluene at RT, using a stopped flow NMR probe, which allows analyzing reactions on very fast time scales [62]. The latter conditions thus slightly differ from those previously reported by Waymouth, Hedrick and coworkers [60, 61] because THF has been used in the latter case vs. toluene in Landis's works. In particular, the authors have evidenced a rapid rac-meso **5-Mes**-induced interconversion of the lactide monomer, i.e. lactide epimerization, that is kinetically in competition with the polymerization process. The tacticity and the molar mass distribution of the PLAs have also been assessed by ^{13}C NMR and size exclusion chromatography (SEC) analysis, respectively. Analysis by ^{13}C NMR spectroscopy of the microstructure of the PLAs synthesized from L-lactide, rac-lactide, and meso-lactide has established that the ZROP occurs by a chain-end control of polymer tacticity. These data support the earlier findings by Dove et al. [50] Kinetic models have also allowed the authors to propose that the mechanism of **5-Mes**-mediated ZROP eventually includes reversible and unimolecular termination reactions and a first-order catalyst death term. These two events had not been considered previously by Hedrick, Waymouth and coworkers [60, 61] The catalyst deactivation process, although not clearly established, may be the result of a protonation of **5-Mes** by the PLA forming an enolate-type anion.

The ZROP of ε-CL and δ-VL in the presence of 1-isopropyl-3-benzylimidazol-2-ylidene or by methyl-3-menthoxymethyl imidazol-2-ylidene allows achieving cyclic polylactones of $M_n = 31\,900\,g\,mol^{-1}$ in THF at RT, but again with broad dispersities ($Đ = 2.27$) [63, 64]. Cyclic PCLs with M_n values up to $114\,000\,g\,mol^{-1}$ have also been synthesized in toluene at RT, from 1,3,4,5-tetramethylimidazol-2-ylidene (**5_{Me}-Me**, Figure 11.1) and 1,3-bis(isopropyl-4,5-dimethyl)imidazol-2-ylidene (**5_{Me}-*i*Pr**) [65]. Interestingly, cyclization can be totally suppressed by adding LiCl during the 5Me-*i*Pr-ZROP of δ-VL in THF, thus yielding linear polyester instead [66].

Modeling experiments have suggested that initiation of lactones is also a slow and reversible process, whereas propagation is rapid, but complicated both by intra- and intermolecular backbiting reactions, and by the formation of cyclized chains (Scheme 11.9) [65, 67]. The latter event would be the key step to the high molar mass cyclic PCLs.

Waymouth and coworkers have described the NHC-initiated ZROP of 2,2,5,5-tetramethyl-2,5-disila-1-oxacyclopentane (TMOSC) in THF at RT using different NHCs [68]. **5_{Me}-Me** and **5_{Me}-*i*Pr** are the most efficient, with >95% of monomer conversion reached only after five seconds (Scheme 11.10a). Cyclic poly(carbosiloxane)s of molar masses as high as M_n = 587 000 g mol^{-1} ($Đ$ = 1.74) and 942 000 g mol^{-1} ($Đ$ = 3.18), respectively, have been obtained in this way.

Recently, the Waymouth group has extended the ZROP method to the synthesis of *N*-functionalized cyclic polycarbonates derived of eight-membered *N*-functionalized cyclic carbonates (8CCs), i.e. azacarbonates [69]. Such heterocycles are obtained from diethanolamine derivatives. The polymerization of 8CCs has been initiated at RT by **5_{Me}-*i*Pr** in THF (Scheme 11.10b), yielding *N*-substituted cyclic polycarbonates with molar masses in the range 30–100 000 g mol^{-1} and a dispersity in the range 1.5–1.9. The azacarbonate polymerizability proves to strongly depend on the substitution on the N-atom of the 8CC. Thus, the *N*-benzyl-substituted 8CC polymerizes readily, whereas the *N*-phenyl-substituted 8CC undergoes a dimerization reaction catalyzed by the same NHC, affording a crystalline cyclic dimer.

As mentioned, the NHC-induced ZROP of monomers such as LA, ε-CL, 8CCs, and TMOSC can lead to cyclic polymers, but in a noncontrolled manner owing to the occurrence of side reactions. In contrast, peculiar cyclic monomers can yield very well-defined cyclic polymers under rather mild conditions by "controlled/living" ZROP. For instance, β-BL can be converted into cyclic polyester by the **5s-Mes** NHC, with the experimental degree of polymerization matching the initial NHC-to-monomer ratio [70]. The control over molar masses and dispersity is attributed, in this case, to the formation of a dormant spirocyclic structure, which enables to minimize side reactions and the reshuffling of molar masses (Scheme 11.11).

The NHC-mediated ZROP of *N*-substituted *N*-carboxyanhydrides (^{N}R-NCAs) has been intensively investigated by Zhang and coworkers (see Scheme 11.12) [71–74]. Among the various imidazol-2-ylidenes (**5-*i*Pr**, **5-*t*Bu**, **5-Mes**, and **5-Dipp**) evaluated, small-size *N*-substituents on the NHC have been found to favor faster polymerization. Different types of cyclic polypeptoids, including cyclic brushes [72] and amphiphilic diblock copolypeptoid cycles [73] (M_n = 3000–30 000 g mol^{-1} and $Đ$ = 1.2), have been synthesized in this way. In contrast to all examples discussed previously, these polymerizations proceed with the release of CO_2 at each propagation step. As emphasized, well-defined cyclic polypeptoids can be achieved, i.e. with low dispersities ($Đ$ = 1.04–1.12) and molar masses ranging from 3000 and 40 000 g mol^{-1}, using THF as the solvent, at monomer concentrations between 0.4 and 0.8 mol l^{-1}. This high level of control is due to the fact that the main chain of polypeptoids – in contrast to aliphatic polyesters – is not subjected to a nucleophilic attack by the NHC and/or by the growing chains, which avoids side reactions and molar mass reshuffling. Moreover, formation of dormant spirocycles has also been postulated in this case, contributing to the controlled chain growth [74].

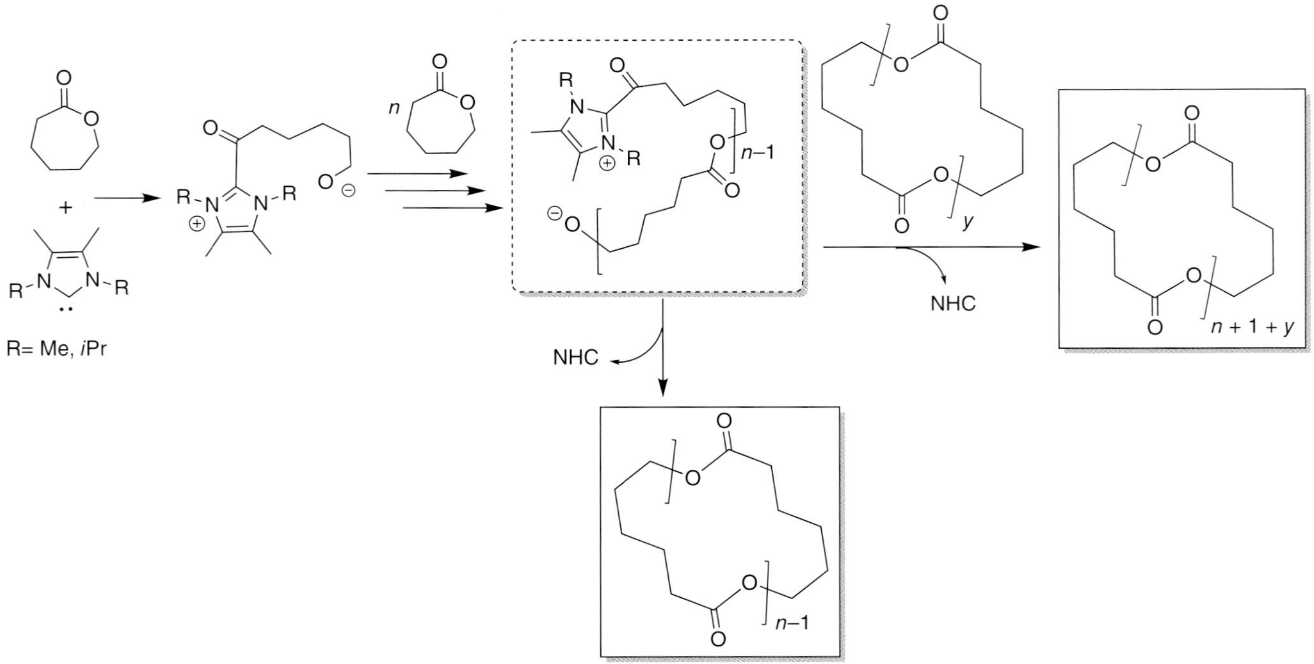

Scheme 11.9 Simplified mechanism forming cyclic polylactone by NHC-mediated ZROP (here of ε-caprolactone).

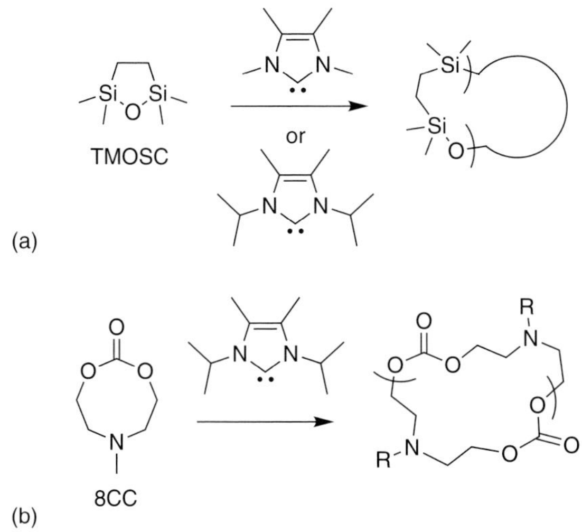

Scheme 11.10 Synthesis of (a) cyclic poly(carbosiloxane) and (b) poly(*N*-functionalized carbonate).

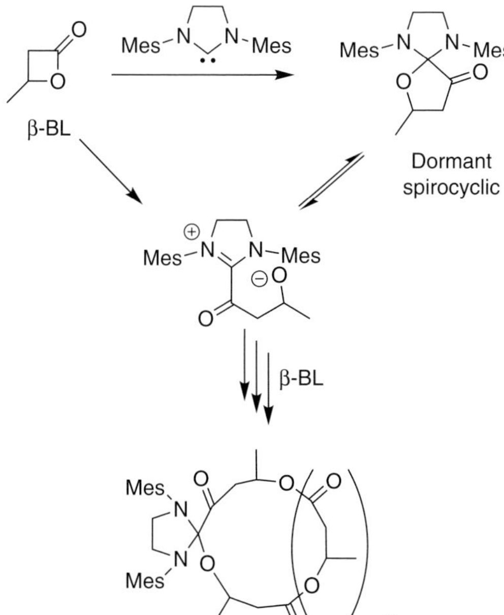

Scheme 11.11 Synthesis of cyclic poly(β-butyrolactone) by NHC-initiated ZROP via the formation of a dormant spirocyclic structure.

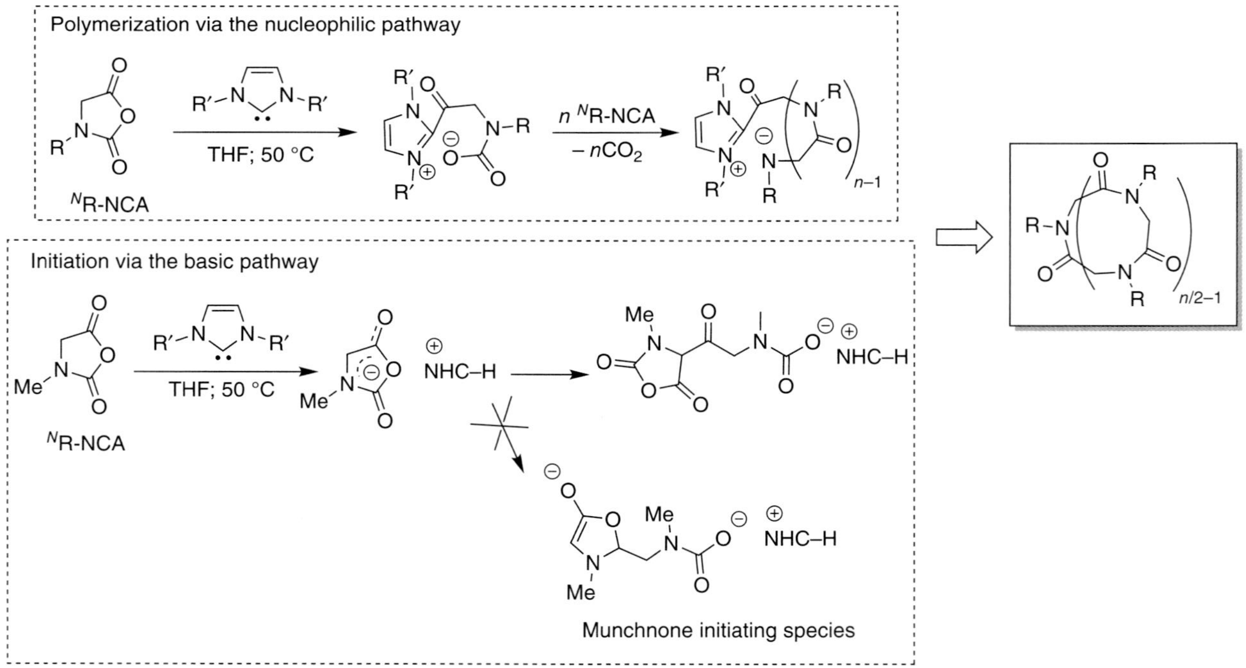

Scheme 11.12 Nucleophilic or basic mechanism for the NHC-mediated ZROP of ^{N}R-NCA.

Recent computational investigations by Cavallo and coworkers into the ZROP process of ^{N}Me-NCA have supported the existence of the two competing basic and nucleophilic path*w*ays for the initiation step (Scheme 11.12) [75]. Although the nucleophilic mechanism occurs similarly as for cyclic esters, i.e. by generating a zwitterionic azolium carbamate adduct, the basic mechanism here involves the formation of an azolium carbamate ion-pair-type adduct, by the abstraction of a proton from the methylene group of the monomer by the NHC. The authors have also proposed that the mechanism involving the “Munchnone initiating species,” formed by addition of the monomer to the N-atom, is not possible as it is too high in energy (Scheme 11.12). Both the zwitterionic and ion-pair-type adducts can eventually propagate this NHC-organopolymerization by repeated nucleophilic additions of the carbamate chain end to the monomer and elimination of CO_2. The two ROP processes show similar energetic barriers, consistently with experimental observations, including a narrow molar mass distribution. However, the zwitterionic mechanism has been found kinetically more favorable and thermodynamically reversible, whereas the basic mechanism is preferred thermodynamically. Finally, calculations have suggested that the termination pathway forming the cyclic polypeptoids occurs by a nucleophilic attack of the N-atom of the carbamate chain end to the carbonyl group bound to the azolium moiety derived from the NHC.

In a more systematic study, Falivene and Cavallo have reported on DFT calculations of the elementary direct reaction between various monomers (1 equiv), including MMA, ethylene oxide, glycolide, a ^{N}R-NCA and valerolactone, and 1 equiv of a given NHC [20]. A total of nine different NHCs have been evaluated, so that 45 NHC/monomer combinations have been examined. With the aforementioned monomers studied, the possibility for the zwitterionic intermediate species to form a spirocyclic (neutral) compound has been taken into account. Results have indicated that the relative stability of both the spirocyclic and the zwitterionic adducts between the monomer and the NHC is governed by the singlet–triplet energy gap of the free NHC. Low-energy gaps have been found to favor the spirocycle adduct, whereas preference for the zwitterionic adduct is observed for high-energy gaps. Overall, these calculations prove very useful to rationalize some effects observed experimentally and could serve as guidelines for properly selecting the NHC for a given monomer.

Polymerization of ε-caprolactam and laurolactam directly mediated by a NHC does not follow a ZROP nucleophilic mechanism. Instead, a basic mechanism operates, first by deprotonation of the lactam monomer by the NHC, resulting in the formation of polyamide-6 and polyamide-12 (see Scheme 11.4). This has been reported by Buchmeiser and coworkers who have employed various NHC–CO_2 adducts for this purpose [44, 45], reactions being conducted in bulk at 180 °C, i.e. by delivering the free NHC *in situ*. Tetrahydropyrimidinium-based (**6-Cy-CO$_2$**) carbenes prove more active than imidazol(in)ium compounds, giving after 45 minutes PA-6 of molar masses $M_n = 140\,000–420\,000\ g\ mol^{-1}$ (as determined by viscometric measurements). The broadened dispersity observed and the high molar masses obtained have suggested incomplete initiation or other side reactions, such as chain transfer, as expected for the anionic-like ROP of ε-caprolactame (ε-CLA) [44]. Nevertheless, the inhomogeneity of the system caused by the precipitation of PA-6 from the monomeric melt may also explain these results.

The activity of tetrahydropyrimidinium carboxylates is related to the basicity of the corresponding free carbenes, suggesting a basic mechanism where the monomer is anionically polymerized. Thus, polymerization starts by deprotonation of the lactam by the NHC, forming an activated monomer amide anion that can initiate the ROP (Scheme 11.4).

NHC-CO_2 adducts are also effective precatalysts for the bulk homopolymerization of LL, leading to PA-12 of molar masses M_n of 14 000 g mol^{-1} ($Đ$ = 2.2–3.1) [45]. The copolymerization of LL and ε-CLA gives a gradient copolymer, ε-CLA being incorporated preferentially at the onset of the polymerization.

11.4 Reaction with Alkyl (meth)acrylates

As mentioned, the reactivity of NHCs is directly correlated with their electronic structure. Although the nucleophilic basic reactivity of NHCs toward cyclic monomers solely relies on the presence of a sp^2-hybridized lone pair of electrons at carbon, direct reaction of NHCs with α,β-unsaturated monomers involves their ambiphilic character, which results from both the presence of a filled sp^2 and a formally p empty orbitals at carbon.

Scheme 11.13 highlights the various reaction outcomes of NHCs toward α,β-unsaturated substrates. Although the nucleophilicity of NHCs favor a selective 1,4-addition onto alkyl (meth)acrylate (M(M)A) derivatives, their basic properties allow alcohols, as well as crotonates, to be activated by hydrogen bonding and deprotonation. SKAs can also be activated, thanks to the silicophilicity of NHCs; depending on the carbene, the pentavalent silicate resulting from the addition of the carbene onto the silicon center may further evolve toward an imidazolium enolate upon rupture of the Si—O bond.

The ambiphilic reactivity of NHCs has early been recognized with the "Umpolung reaction" of carbonyl moieties, which leads their polarity to be inverted [13–16]. Thus, the 1,2-addition of an NHC to an (hetero)aromatic aldehyde generates a 1,1-imidazolium alcoolate, which evolves via proton transfer, to a Breslow intermediate. This nucleophilic species may then undergo several transformations, such as the benzoin condensation or the Stetter reaction, by reacting with aldehydes or α,β-unsaturated compounds, respectively. These two elementary reactions have been adapted for polymer synthesis by step-growth polymerization (see Section 11.4) [76–78].

The ambiphilic reactivity of NHCs is also operative in the presence of α,β-unsaturated compounds. In this case, the imidazolium enolate resulting from the 1,4-addition of the NHC onto the alkyl (meth)acrylate (e.g. MMA; R = R_1 = Me in Scheme 11.13 in red) undergoes a proton transfer to generate a nucleophilic deoxy-Breslow intermediate. Because proton transfer reactions compete with the 1,4-addition of the enolate moiety, direct polymerization of MMA triggered by NHCs has only rarely been observed [31, 79].

11.4.1 Basic Nucleophilic Reactivity of Stable Carbenes in the Absence of Initiator

Studies by Chen and coworkers have revealed that the strongly nucleophilic **5-*t*Bu** can directly initiate the zwitterionic polymerization of MMA in DMF

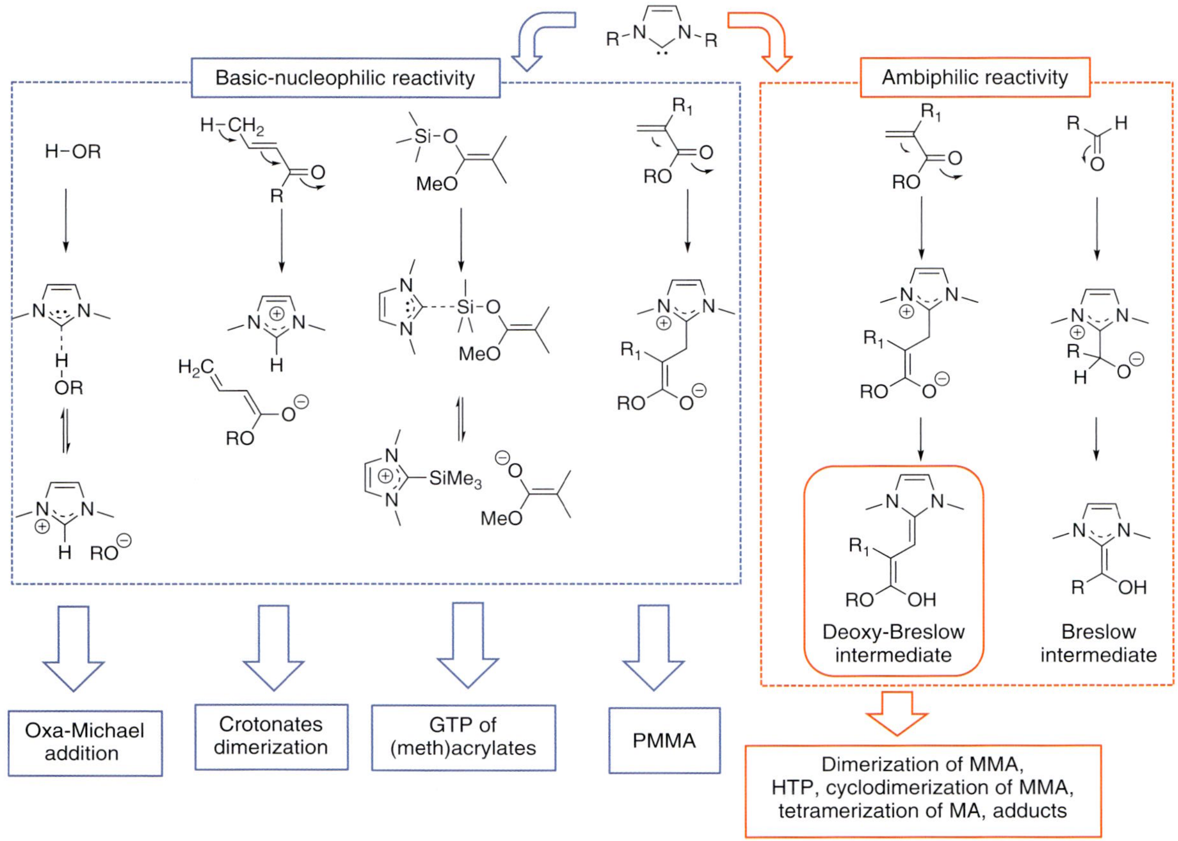

Scheme 11.13 Basic nucleophilic and ambiphilic reactivities of NHCs.

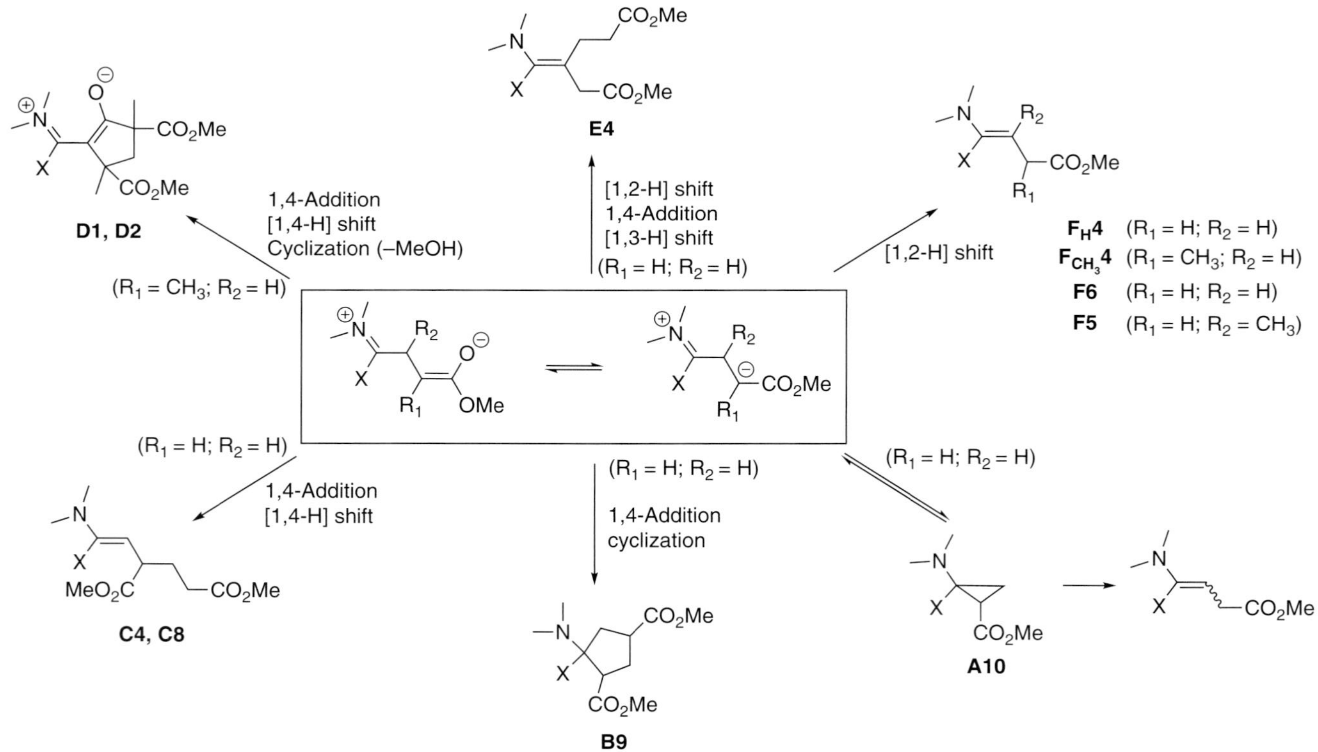

Scheme 11.14 Noncatalytic reactivity of stable carbenes toward M(M)A.

at RT in the absence of any other initiators [79, 80]. In this way, PMMA of 10.6–84.7 × 10^3 g mol^{-1} ($Đ$ = 1.4–2.1) can be obtained. The corresponding masked **5-*t*Bu-CO_2** adduct showing a thermally latent behavior can also successfully initiate the MMA polymerization at 85 °C, upon the release of CO_2 [29]. Remarkably, this polymerization reaction appears to be selectively triggered by **5-*t*Bu**, different reaction outcomes being experimentally observed with other NHCs (*vide infra*). Although the formation of macrocycles from interaction between the enolate and imidazolium moieties is conceivable, the high ΔS–T of **5-*t*Bu** precludes such a behavior, as described by Cavallo and coworker [20].

In the presence of DMSO, a different initiation step occurs, involving the deprotonation of DMSO by the NHC, which affords α-functional DMSO-PMMA [29]. In this case, the basicity of the carbene is the key to generate the $CH_3SOCH_2^-$ (dimsyl)anion. Tetrahydropyrimidinium-2-carboxylates (e.g. the **6-Cy-CO_2** or **6-Cy-CO_2** adduct; Figure 11.1) proves well suited for the preparation of PMMA at 60 or 85 °C (M_n = 11 000–80 000 g mol^{-1}; $Đ$ = 1.7–4.4). In contrast, the more reactive cyclic renewable α-methylene-γ-butyrolactone (MBL) can be polymerized within minutes in DMF at RT by different NHCs, including **TPT** and **5-Mes**, to afford poly(MBL) of M_n = 10 600–84 700 g mol^{-1} ($Đ$ = 1.36–2.11). Vinyl-substituted MBL and MMBL can be chemoselectively polymerized at the α,β-unsaturation by **5-*t*Bu** and **5-*i*Pr**, leading to unsaturated polyesters with M_n up to 73 800 g mol^{-1} ($Đ$ = 3), with 50 ppm of **5-*t*Bu**, a turn over frequency (TOF) as high as 80 000 per hour can be reached [80].

A completely different reaction pathway has been evidenced with methyl crotonates, which only differs from MMA by the position of the methyl group. Instead of polymerizing, methyl crotonates indeed undergo a quantitative dimerization process, in the presence of a catalytic amount of **5_{Me}-*i*Pr**, yielding the corresponding 2-ethylidene-3-methylpentanedioates [81]. Although both nucleophilic and basic mechanisms can take place in this dimerization reaction, a basic mechanism is favored to account for the reaction catalyzed by *t*BuOK. In this case, the first step would consist in the deprotonation of the methyl group of the crotonate to generate the corresponding conjugated enolate (Scheme 11.13). The final product would then form by sequential 1,4-addition of this enolate onto an incoming crotonate, followed by proton transfer and isomerization. Overall, only a few transformations involving the nucleophilic basic properties of NHCs toward α,β-unsaturated monomers have been described.

11.4.1.1 Ambiphilic Reactivity of Stable Carbenes

In contrast, the ambiphilic reactivity of stable carbenes, including NHCs, cyclic aminoalkyl carbenes (CAACs), and acyclic carbenes, toward α,β-unsaturated monomers (e.g. MMA) has also led to a variety of reaction pathways, which mainly depends on the structure of the carbene, the monomer, and the number and sequence of 1,4-addition proton transfer. In those transformations, which all involve the formation of deoxy-Breslow-type intermediates, carbenes can behave either as (sub)-stoichiometric reagents or as catalysts.

11.4.1.2 Noncatalytic Reactivity

The noncatalytic reactivity of carbenes has been observed with both MMA and methyl acrylate (MA), and a variety of multisubstituted aminoalkenes have been isolated depending on the nature of the carbene. This is summarized in Scheme 11.14. All these transformations involve a common zwitterionic enolate intermediate, which results from the initial 1,4-addition of the carbene onto M(M)A. Depending on the number and sequence of 1,4-additions and the type of proton transfer reactions ([1,2-H], [1,3-H], or [1,4-H]) undergone by the common zwitterionic enolate, various adducts have indeed been isolated.

In the case of acyclic (di)aminocarbenes **Ac-DAC-Me** and **Ac-AC-*i*Pr** (Figure 11.1), no proton transfer has been observed with MA as the substrate; instead, C-nucleophilic intramolecular addition of the enolate onto the iminium moiety occurred, yielding to the cyclopropyl and cyclopentyl derivatives **A10** and **B9**, in the case of the aminoalkylcarbene **Ac-AC-*i*Pr** and diamino-carbene **Ac-DAC-Me**, respectively [82]. This unusual behavior has to be paralleled with the enhanced electrophilicity of the iminium moiety in comparison with cyclic unsaturated imidazolium. The amino-substituted cyclopropane **A10** is not stable and has been found to isomerize over a week at RT to the corresponding aminoalkene. With NHCs, it has been shown that the deoxy-Breslow intermediate can be isolated with both MMA and MA and **5-Mes** (**F_{CH_3}4** and **F_H4**) but also with MA and the anti-Bredt bicyclic carbene **6_{Cy}-Dipp** (**F_H6**) [28, 82]. In this reaction, the initial imidazolium enolate, resulting from the conjugate addition of the carbene onto the monomer, undergoes a [1,2-H] shift to generate the corresponding single-addition product. A similar single-addition product **F_H5** has also been observed with crotonates as substrates and **TPT** as the catalyst [81]. In the presence of 2 equiv of MA, a 1 : 2 mixture of two isomeric double-addition products **C4** and **E4** has been observed with **5-Mes** [82]. Although the first compound results from the conjugate addition of the aminoalkene **C4** onto another equivalent of MA followed by proton transfer, the second compound **E4** is accessed from a different pathway, involving two consecutive 1,4-addition of the carbene onto MA, followed by proton transfer. Despite the structural and electronic differences between the **5-Mes** and the **CAAC-Dipp** (Figure 11.1), a similar aminoalkene **C8** has been isolated upon reaction with 2 equiv of MA. The reactivity between **5-*i*Pr** and 2 equiv of MMA, as well as **5_{Me}-*i*Pr** and 2 equiv of MMA, likely proceeds through a similar diaminoalkene. However, in contrast to **C4**, this deoxy-Breslow compound is not stable and undergoes a cyclization onto one ester group to generate a diastereoisomeric mixture of imidazolium enolate cyclodimers **D1** and **D2** upon the release of MeOH [31, 81]. The inability of **5-*t*Bu** to cyclodimerize MMA seems related to the higher energetic barriers (relative to those calculated for **5-*i*Pr**) computed for the ring closure and the loss of MeOH according to the DFT calculation [31].

11.4.1.3 Catalytic Reactivity

Besides stoichiometric reactions, NHCs have also been demonstrated to catalyze several transformations, including polymerizations, with α,β-unsaturated monomers, such as MMA, *n*-BuMA, and MAN [83–86]. All those transformations involve the formation of a deoxy-Breslow intermediate at some point.

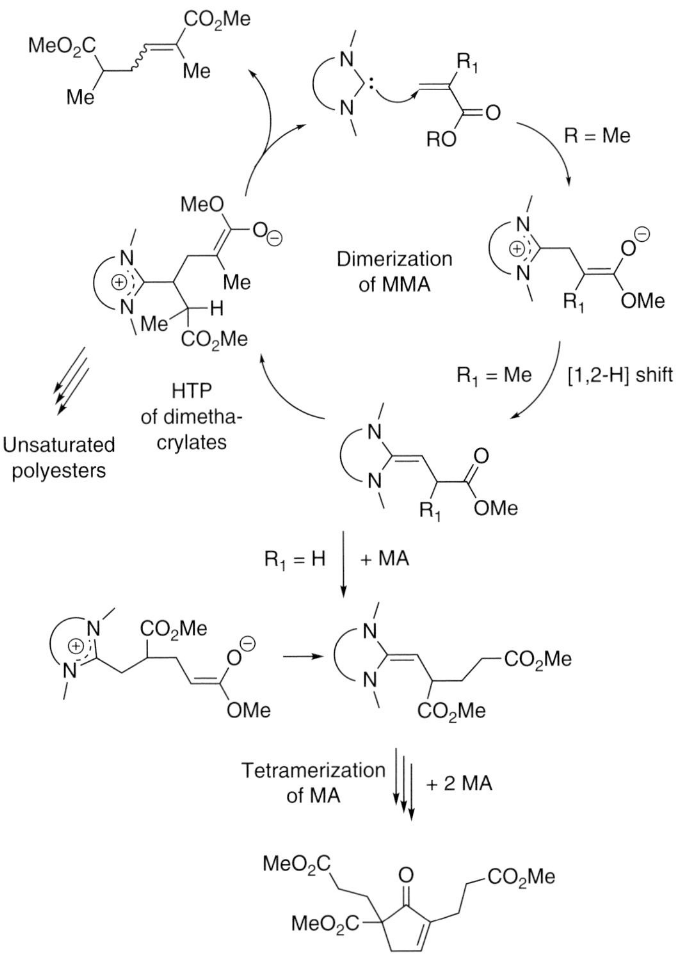

Scheme 11.15 Dimerization of MMA and proton transfer polymerization (HTP) of dimethacrylates catalyzed by TPT; tetramerization of MA catalyzed by 5-Mes.

TPT, which is one of the weakest basic nucleophilic NHC, is particularly well suited for catalytic purposes because it behaves as a good leaving group. With MMA as the substrate, no polymerization occurs; instead, a head-to-tail dimerization has been observed (Scheme 11.15), leading to the corresponding dimethyl 2,5-dimethyl-2-hexenedioate (86%, E/Z: 95/5). The scope of this transformation has also been extended to various functional acrylate derivatives (R = $CH_2CH_2NMe_2$, allyl,…; R_1 = alkyl, aryl,…), using 10 mol% of triazolium perchlorate and 1 equiv of DBU at 80 °C for 24 hours in 1,4-dioxane [86].

From a mechanism point of view, it has been proposed that the deoxy-Breslow intermediate, resulting from the first 1,4-addition/proton transfer sequence, reacts with a second equivalent of MMA to generate the corresponding imidazolium enolate. After [1,4-H] shift, the carbene and the final product are concomitantly released. Similarly, dimerization of MAN can be selectively catalyzed by **TPT** or the corresponding 2-OMe adduct, in the presence of *iso*-propanol [87]. Conversions up to 82% have been achieved using a catalytic quantity of the phenol derivative. Although the reaction is sluggish in the absence of alcohols, it has been reported that those protic species can promote the intermolecular proton transfer to generate the key deoxy-Breslow intermediate. Dimerization of *n*-BuMA and codimerization of *n*-BuMA and MAN has also been demonstrated using **TPT**, in the presence of 5 equiv of *n*-BuOH.

This NHC-catalyzed dimerization of a methacrylate substrate has been further applied for the preparation of unsaturated polyesters using, in this case, bis-methacrylates (Scheme 11.15) [88]. Thus, unsaturated polyesters of M_w = 10 900–27 300 g mol^{-1} ($Đ$ = 1.7–1.9) can be prepared using **TPT** or the corresponding methanol adduct, in toluene at 80–120 °C. In these reactions, a small amount of methoxyphenol is beneficial, not only to suppress the radically induced chain-growth polymerization of the vinyl moieties but also to enhance monomer conversion. Post-functionalization of the alkene moieties can also be performed using thiol-ene-type reaction [88].

With MA, a process involving 4 equiv of MA has been observed (Scheme 11.15) [84]. Although a variety of carbenes, including **TPT**, have appeared inactive, only **5-Mes** has been found to catalyze the tetramerization of MA, yielding substituted cyclopentenone derivatives. This transformation would involve a deoxy-Breslow intermediate resulting from a sequential addition of 4 equiv of MA, followed by [1,4-H] shift. The final product would then be formally obtained by the addition of 2 equiv of MA, followed by ring closure and loss of MeOH.

11.4.2 Reactivity of NHCs Toward α,β-Unsaturated Esters in the Presence of Initiators

Catalytic polymerization of alkyl (meth)acrylate monomers has been described in the presence of different types of initiators. With SKAs, the NHC-organocatalyzed GTP has been independently reported by Taton, Hedrick, Waymouth and coworkers (see Scheme 11.2) [32, 33, 36–38]. As already mentioned, GTP is based on the Mukaiyama–Michael reaction, which is the conjugate addition of a SKA onto an alkyl (meth)acrylate. In this process, the NHC activates the Si—O bond of the SKA by interacting with the silicon center, thanks to its silicophilicity. Remarkably, unlike other catalysts, NHCs not only enables the homopolymerization of MMA and MA but also the block copolymerization of these monomers to generate valuable all acrylate-based block copolymers. From a mechanism point of view, an associative and a dissociative mechanism have been both proposed to account for experimental observations; although the GTP mechanism is still under debate, it seems that both mechanisms may operate depending on the nature of the carbene catalyst.

Very recently, controlled polymerization of MMA and MA has been achieved with simple alcohols as initiating species, in the absence of any other additives (Scheme 11.4) [19]. In this way, α-alkoxy-PMMA with M_n up to 8000 g mol^{-1} ($Đ$ = 1.1–1.9) has been prepared in DMF at 25 °C, in the presence of a catalytic amount of **5-*t*Bu**. It is noteworthy that, in contrast to conventional methods, this methodology allows the direct access to valuable amphiphilic PEO-*b*-PMMA

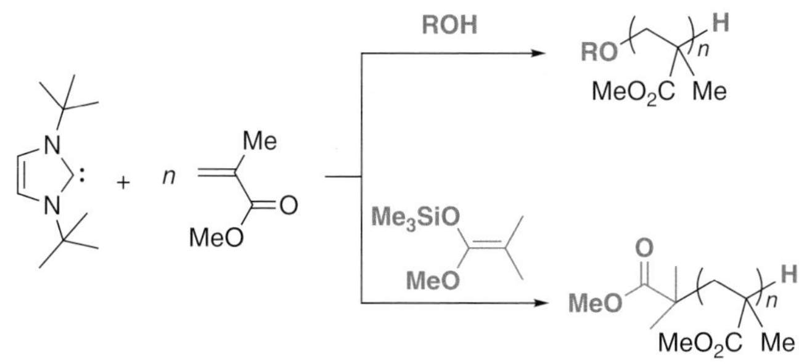

Scheme 11.16 Catalytic polymerization of MMA by oxa-Michael addition polymerization (top) and group transfer polymerization (GTP; bottom).

block copolymers, using a hydroxyl-functional MeO–PEO$_{33}$–OH macroinitiator. According to the DFT calculation, the nucleophilic mechanism can compete with the activated alcohol (basic) mechanism, both initiation and propagation steps proceeding through a concerted low-energy pathway (Scheme 11.16).

Hydroxyl-functional acrylate monomers have also been shown to polymerize with **TPT** or IDipp catalysts (Scheme 11.17); however, in contrast to the polymerization of MMA initiated by alcohols, saturated polyesters were obtained instead of polyacrylates [89]. Although a similar "formal" conjugate addition of the alcohol moiety can account for this oxa-Michael addition polymerization, the authors involve the initial 1,4-addition of **TPT** followed by multiple proton transfer between the growing chain and the monomers. Although molecular weights are rather modest, monomers incorporating an alkene moiety can be derived into poly(ol)-functional poly(ester) from acrylate monomers by thio-ene click reaction, using mercaptohaxanol.

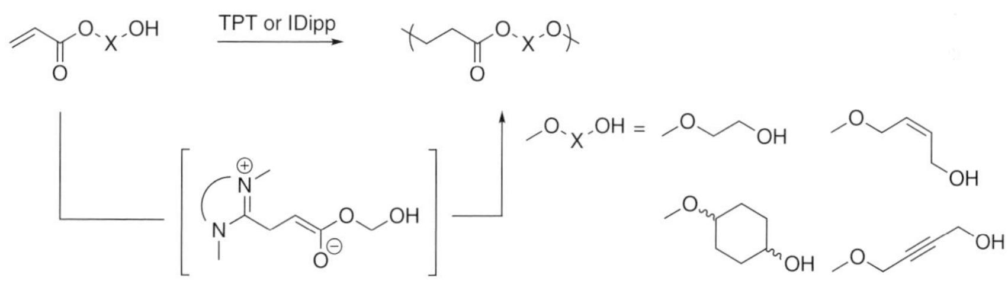

Scheme 11.17 Polymerization of hydroxyl-functionalized acrylates with NHCs catalysts.

11.4.3 Reactivity of NHCs in Conjunction with a Lewis Acid: Frustrated Lewis Pair-Type Reactivity

The association between a bulky Lewis base and a bulky Lewis acid may result in the formation of a frustrated Lewis pair, where the interaction between the basic and acidic centers is precluded for steric reasons [90, 91]. Such a frustrated system displays enhanced reactivity toward a variety of small molecules, including H_2. Accordingly, the combination between a bulky nucleophilic phosphine, such as $P(tBu)_3$ and a bulky electrophilic borane, such as $B(C_6F_5)_3$, leads to the activation of H_2 by heterolytic cleavage of the enthapically strong H—H bond and the formation of the corresponding phosphonium borate. In the context of polymer chemistry, this FLP-like reactivity has successfully been applied to some α,β-unsaturated monomers by Chen and coworkers, using NHCs as Lewis bases [92–95]. In particular, the association between **5-*t*Bu**, **5-Mes**, or **TPT** and $Al(C_6F_5)_3$ affords a highly active system (TOF up to 48 000 per hour) for the polymerization of MMA in toluene or dichloromethane at 25 °C. Although the polymerization is not well controlled, molar masses M_n = 26 600–600 000 g mol^{-1} have been obtained within 1–15 minutes under optimized conditions, depending on the nature of the carbene [94, 95].

The activity of such systems is eventually not directly correlated with their frustration as both **5-*t*Bu** and **5-Mes** form classical Lewis pair with $Al(C_6F_5)_3$ at RT. In the presence of MMA, the carbene is displaced to yield the corresponding MMA:$Al(C_6F_5)_3$ adduct, which is more electrophilic than the free monomer because of the coordination of the oxygen atom to the Al center. Such an activated monomer undergoes a 1,4-conjugate addition of the carbene to yield the corresponding imidazolium enolaluminate (Scheme 11.18). It is noteworthy that a second equivalent of $Al(C_6F_5)_3$ is required to enable the propagation of the polymerization by 1,4-addition of the zwitterionic species onto an activated monomer. This Lewis pair polymerization (LPP) has also been extended to other monomers, such as the renewable methacrylic derivatives MBL and MMBL, but also dimethylacrylamide and diethylvinylphosphonate. The bifunctional allyl methacrylate can also be selectively polymerized at the α,β-unsaturation, affording allyl-functional polymethacrylate of M_n = 140 000 g mol^{-1} ($Đ$ = 1.37) in three minutes at 25 °C in toluene [96].

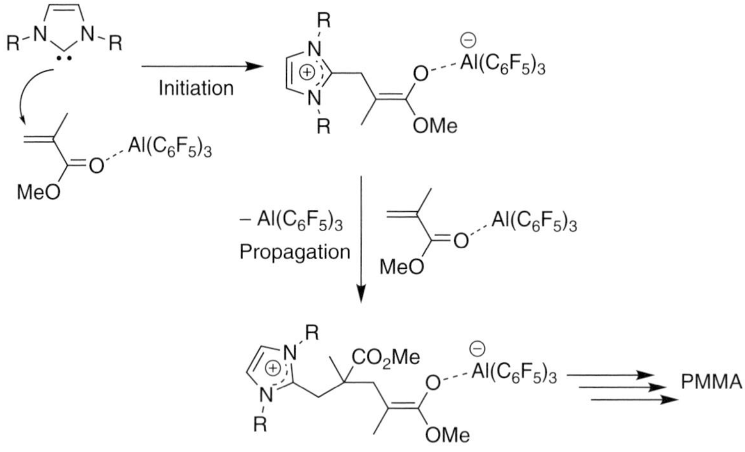

Scheme 11.18 Polymerization of methacrylates by NHC/$Al(C_6F_5)_3$ Lewis pair.

11.5 NHC-Mediated Step-growth Polymerization

For a large majority, examples of polymer synthesis by NHCs have concerned chain-growth polymerization – and most often ROP-type reactions. As summarized in Scheme 11.19, NHCs have also proven particularly efficient in step-growth polymerization, for instance, for the synthesis of polyurethanes, aliphatic and aromatic polyesters, polysiloxanes, or polyepoxides [1–4, 10–12].

Some other polymers have also been synthesized by taking inspiration of NHC-catalyzed elementary reactions in molecular chemistry, e.g. "Umpolung" reactions such as the Stetter reaction and the benzoin condensation [14, 15] applied to difunctional monomers. The first example of polymer synthesis by NHC catalysis has been actually reported in the 1990s [76, 77], that is, long before the seminal paper by Herick and coworkers on NHC-mediated ROP [21]. Bis-formylarenes (=bis-aldehydes, Scheme 11.19) and bis-α,β-unsaturated ketones have thus been reacted to form polymeric1,4-diketones at 85 °C in DMF, via the Stetter route, in the presence of a thiazolylidene catalyst (route A). Further ring closure of the polyketones via the Paal–Knorr reaction has then led to conjugated alternating copolymers constituted of pyrroles, furyls, or thienyl moieties.

In a similar manner, terephthaldehyde has been self-polymerized following the NHC-catalyzed benzoin condensation mechanism (route B, Scheme 11.19) [78]. Among NHCs tested, the typical "Umpolung" triazolydene **TPT** catalyst has been found the most efficient, leading to oligobenzoin-type polymers.

Synthesis of cross-linked polyurethanes by NHC catalysis, from polyisocyanates and polyols, has been reported first by Buchmeiser and coworkers NHC–CO_2 adducts have been used as precatalysts for that purpose, reactions being carried out at 60–70 °C in CH_2Cl_2, allowing thermally activating precatalysts (route C, Scheme 11.19) [40, 41]. A basic mechanism involving the activation of the alcohol by the NHC catalyst has been postulated, on the basis of NMR investigations. Later on, Destarac and coworkers have suggested the same mechanism in their study utilizing the free **5-*t*Bu** NHC for the organocatalyzed step-growth polymerization of aliphatic diisocyanates and diols, at 30–50 °C, using a catalyst loading <1 mol%) [42]. Linear PUs with $M_n = 3000–5000\ g\ mol^{-1}$ have thus been achieved by adding the NHC organocatalyst to a solution of both monomers.

The ability of NHCs to catalyze transesterification reactions has also been used for step-growth polymerization, through a few examples. For instance, polytransesterification of 6-hydroxyhexanoate and of ethyl glycolate in the presence of **5-Me** at RT and under reduced pressure gives poly(6-hydroxyhexanoate) and poly(glycolate) with M_n ranging from 8000 to 20 000 g mol^{-1} (route D, Scheme 11.19) [97].

The synthesis of poly(ethyleneterephthalate) (PET) by NHC organocatalysis has also been reported by Hedrick and coworkers [97]. For that purpose, a two-step synthetic procedure involving the preparation of bis(2-hydroxyethyl) terephthalate (BHET), followed by self-condensation of this precursor within one hour at 250 °C, has been implemented (route E, Scheme 11.19).

Similarly, aliphatic polycarbonates have been synthesized by NHC-catalyzed polytransesterification between dimethylcarbonate (DMC) and various diols (route F, Scheme 11.19) [98]. To this end, 1 mol% NHC–CO_2 adducts serving as precatalysts have been employed in bulk at 150 °C under reduced pressure for one hour. Homopolymers and random copolymers with average molar

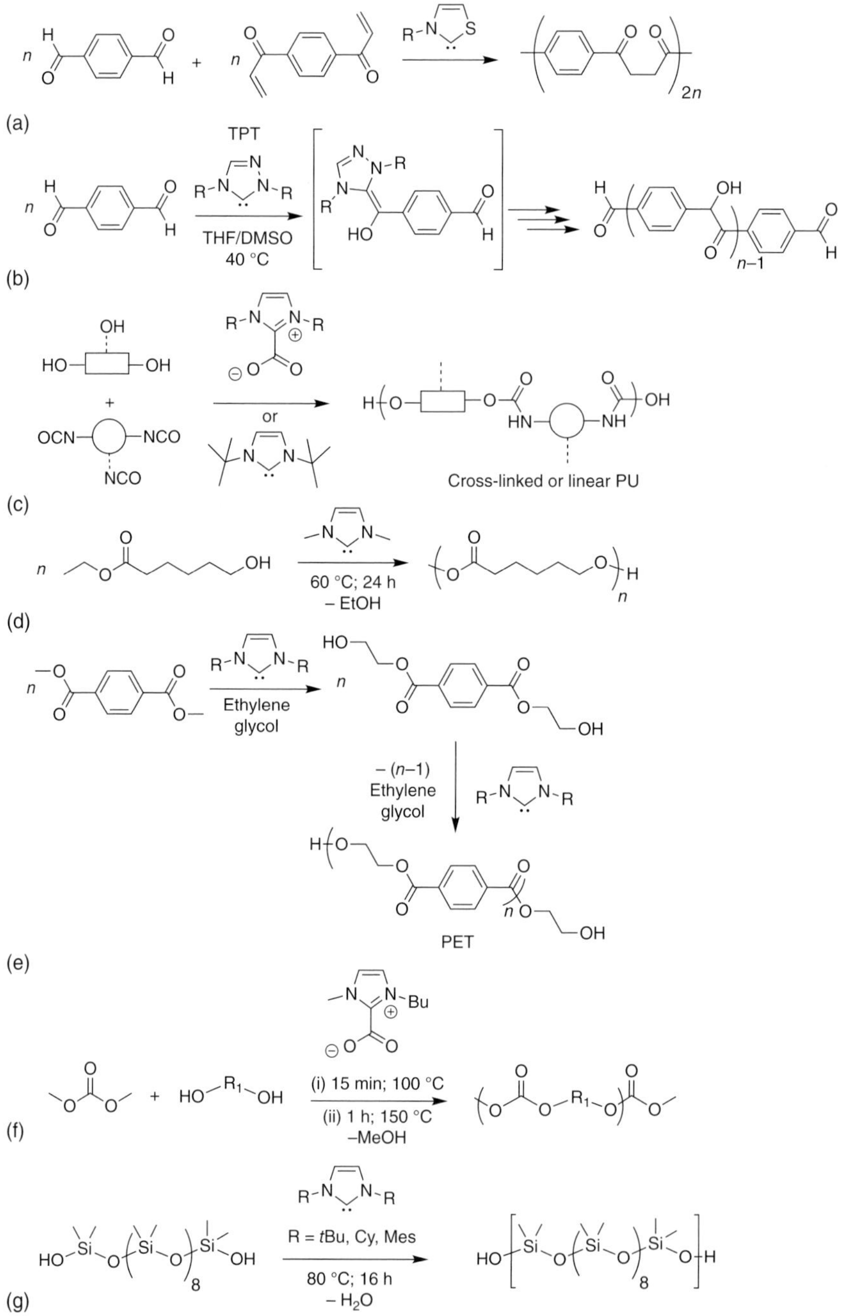

Scheme 11.19 Representative examples of NHC-catalyzed step-growth polymerization.

masses (M_n) in the range 1500–20 000 g mol^{-1} and a dispersity between 1.4 and 2.8 have thus been obtained.

Synthesis of polysiloxanes has been described following a NHC-catalyzed polycondensation of a α,ω-dihydroxy disilanol oligomer (route F, Scheme 11.19) [43]. Although water is released as the by-product, no loss in the catalytic activity of NHC has been noted, reactions being carried out 80 °C for 16 hours. This has enabled the formation of polysiloxanes with molar masses ranging from 50 000 to 300 000 g mol^{-1}. The authors have proposed an activation of silanols by the basic NHC to catalyze the dehydration reaction.

Buchmeiser and coworkers have described the NHC-mediated thermal curing of anhydride hardened epoxy resins (Scheme 11.20), e.g. based on bisphenol A diglycidyl ether (BADGE) and on phthalic acid anhydride (PhA) or hexahydrophthalic anhydride (HHPA) [99, 100]. The authors have resorted to thermally latent NHC–CO_2 adducts as well as to metal-protected NHCs in a process that has been shown compatible with vacuum-assisted resin infusion (VARI). Mixtures of the epoxy compound, the anhydride, and protected NHC form homogeneous and stable formulations. Hence, such monocomponent mixtures can be handled and stored under air without any particular precautions. Upon heating to 140–160 °C, free NHCs (0.1–1 mol%) are released and prove more efficient in comparison with commonly used accelerators (e.g. tertiary amines or *N*-alkylimidazoles) for warm-curing epoxy resins. Therefore, use of NHCs allows reducing catalyst loadings, while affording fast polymerization kinetics and milder curing temperatures. Interestingly, the polymerization rate can be modulated by the chemical structure of the NHC. A NHC-triggered ring-opening of cyclic anhydrides, rather

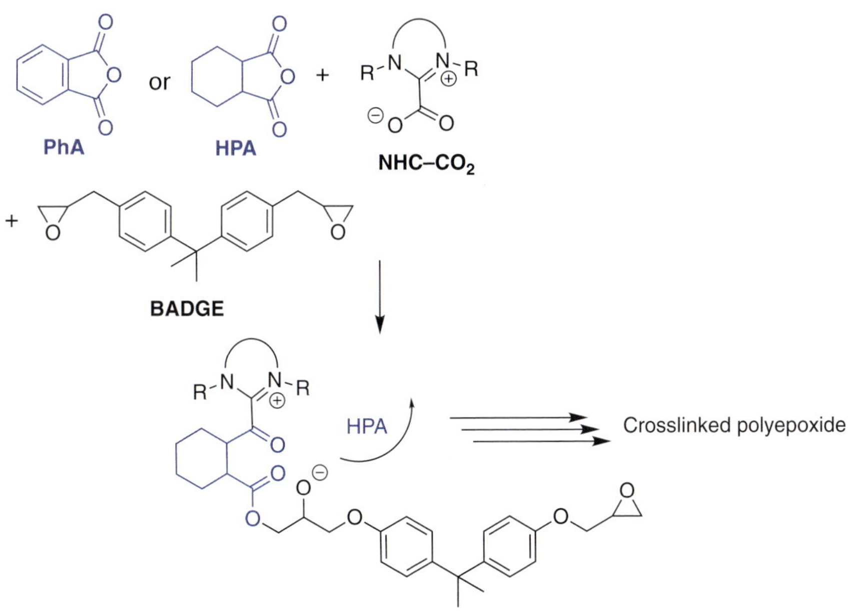

Scheme 11.20 Curing of epoxide/anhydride resins using NHCs.

than the generally admitted ring-opening of oxirane moieties by conventional imidazole-based accelerators, has been postulated (Scheme 11.20) [99, 100].

11.6 Conclusion

The diverse reactivity of stable carbenes found in organocatalyzed reactions of molecular chemistry has been successfully applied to polymer synthesis and macromolecular engineering. NHCs have grown significantly in organopolymerization in the past 15 years, through the activation of a wide range of monomer substrates and involving various reaction mechanisms, including anionic-like and zwitterionic chain-growth processes and step-growth reactions as well. The very high structural modularity of NHCs and their tunable steric and electronic properties have allowed using them either as Brønsted and/or Lewis bases. Subtle changes in their chemical structure sometimes cause dramatic differences in reaction with specific substrates. In this regard, calculations can prove very worthwhile to rationalize experimental results and could even serve as guidelines for properly selecting the NHC for a particular monomer to be polymerized.

Their potential has been originally exploited for ROP of cyclic esters, mainly lactones, and lactide. The scope of NHCs has been, however, significantly extended to other chain-growth polymerization reactions, including the ROP of *N*-substituted *N*-carboxyanhydrides, phosphoesters, carbonates, (carbo)siloxanes and carbosiloxanes, lactams, epoxides and *N*-activated aziridines, the polymerization of (meth)acrylic monomers, and finally to some step-growth polymerization reactions such as the synthesis of industrially important polymers (e.g. polyuretanes, polysiloxanes, PET, or epoxy resins). NHCs exhibit an adapted selectivity for polymer synthesis, and they sometimes exhibit performances that compare with some common organometallic catalysts, making NHC valuable alternatives for the synthesis of polymers free of any metallic residues under relatively mild conditions. They can now be readily generated from readily accessible precursors serving, in some cases, as air-stable and easy-handling masked NHCs showing a latent catalytic behavior. This provides more convenient means of NHC delivery and allows applying NHCs in more robust formulations for on-demand reactivity.

There are still several challenges to tackle in the utilization of NHCs in organopolymerization. For instance, enantioasymmetric polymerization using chiral NHCs has probably been overlooked. Another challenge concerns the activation of nonpolar monomers (e.g. styrenics, dienes, and also olefins). This will probably require the design of specific carbenes and/or the assistance of a coactivator. By associating the nucleophilic properties of NHCs with the activation of some (organic or metal-based) Lewis acids, each partner can display a specific activity toward the targeted substrate and the desired transformation, via a dual/cooperative pathway. This concept of LPP utilizing NHCs has already been translated in polymer chemistry and allows dramatically increasing both the polymerization activity and selectivity.

Overall, one can thus reasonably expect new exciting developments in the use of carbene chemistry for macromolecular synthesis.

References

1 Kamber, N.E., Jeong, W., Waymouth, R.M. et al. (2007). *Chem. Rev.* 107: 5813–5840.
2 Kiesewetter, M.K., Shin, E.J., Hedrick, J.L., and Waymouth, R.M. (2010). *Macromolecules* 43: 2093–2107.
3 Fèvre, M., Vignolle, J., Gnanou, Y., and Taton, D. (2012). *Polymer Science: A Comprehensive Reference*, vol. 4, 67–115. Elsevier.
4 Ottou, W.N., Sardon, H., Mecerreyes, D. et al. (2016). *Prog. Polym. Sci.* 56: 64–115.
5 Dove, A.P. (2012). *ACS Macro Lett.* 1: 1409–1412.
6 Mespouille, L., Coulembier, O., Kawalec, M. et al. (2014). *Prog. Polym. Sci.* 39: 1144–1164.
7 Fuchise, K., Chen, Y.G., Satoh, T., and Kakuchi, T. (2013). *Polym. Chem.* 4: 4278–4291.
8 Brown, H.A. and Waymouth, R.M. (2013). *Acc. Chem. Res.* 46: 2585–2596.
9 Thomas, C. and Bibal, B. (2014). *Green Chem.* 16: 1687–1699.
10 Fèvre, M., Pinaud, J., Gnanou, Y. et al. (2013). *Chem. Soc. Rev.* 42: 2142–2172.
11 Naumann, S. and Dove, A.P. (2015). *Polym. Chem.* 6: 3185–3200.
12 Naumann, S. and Dove, A.P. (2016). *Polym. Int.* 65: 16–27.
13 Bourissou, D., Guerret, O., Gabbai, F.P., and Bertrand, G. (2000). *Chem. Rev.* 100: 39–92.
14 Enders, D., Niemeier, O., and Henseler, A. (2007). *Chem. Rev.* 107: 5606–5655.
15 Hopkinson, M.N., Richter, C., Schedler, M., and Glorius, F. (2014). *Nature* 510: 485–496.
16 Wang, M.H. and Scheidt, K.A. (2016). *Angew. Chem. Int. Ed.* 55: 14912–14922.
17 Magill, A.M., Cavell, K.J., and Yates, B.F. (2004). *J. Am. Chem. Soc.* 126: 8717–8724.
18 Higgins, E.M., Sherwood, J.A., Lindsay, A.G. et al. (2011). *Chem. Commun.* 47: 1559–1561.
19 Nzahou Ottou, W., Bourichon, D., Vignolle, J. et al. (2015). *Chem. Eur. J.* 21: 9447–9453.
20 Falivene, L. and Cavallo, L. (2017). *Macromolecules* 50: 1394–1401.
21 Connor, E.F., Nyce, G.W., Myers, M. et al. (2002). *J. Am. Chem. Soc.* 124: 914–915.
22 Kamber, N.E., Jeong, W., Gonzalez, S. et al. (2009). *Macromolecules* 42: 1634–1639.
23 Shin, E.J., Jeong, W., Brown, H.A. et al. (2011). *Macromolecules* 44: 2773–2779.
24 Raynaud, J., Absalon, C., Gnanou, Y., and Taton, D. (2009). *J. Am. Chem. Soc.* 131: 3201–3209.
25 Raynaud, J., Absalon, C., Gnanou, Y., and Taton, D. (2010). *Macromolecules* 43: 2814–2823.

26 Raynaud, J., Nzahou Ottou, W., Gnanou, Y., and Taton, D. (2010). *Chem. Commun.* 46: 3203–3205.
27 Stukenbroeker, T.S., Solis-Ibarra, D., and Waymouth, R.M. (2014). *Macromolecules* 47: 8224–8230.
28 Zhang, Y. and Chen, E.Y. (2012). *Angew. Chem. Int. Ed.* 51: 2465–2469.
29 Naumann, S., Schmidt, F.G., Schowner, R. et al. (2013). *Polym. Chem.* 4: 2731–2740.
30 Nyce, G.W., Glauser, T., Connor, E.F. et al. (2003). *J. Am. Chem. Soc.* 125: 3046–3056.
31 Nzahou Ottou, W., Bourichon, D., Vignolle, J. et al. (2014). *Chem. Eur. J.* 21: 3989–3997.
32 Scholten, M.D., Hedrick, J.L., and Waymouth, R.M. (2008). *Macromolecules* 41: 7399–7404.
33 Raynaud, J., Ciolino, A., Baceiredo, A. et al. (2008). *Angew. Chem. Int. Ed.* 47: 5390–5393.
34 Webster, O.W., Hertler, W.R., Sogah, D.Y. et al. (1983). *J. Am. Chem. Soc.* 105: 5706–5708.
35 Webster, O.W. (2000). *J. Polym. Sci., Part A: Polym. Chem.* 38: 2855–2860.
36 Raynaud, J., Gnanou, Y., and Taton, D. (2009). *Macromolecules* 42: 5996–6005.
37 Raynaud, J., Liu, N., Gnanou, Y., and Taton, D. (2010). *Macromolecules* 43: 8853–8861.
38 Raynaud, J., Liu, N., Fèvre, M. et al. (2011). *Polym. Chem.* 2: 1706–1712.
39 Lohmeijer, B.G.G., Dubois, G., Leibfarth, F. et al. (2006). *Org. Lett.* 8: 4683–4686.
40 Bantu, B., Pawar, G.M., Decker, U. et al. (2009). *Chem. Eur. J.* 15: 3103–3109.
41 Bantu, B., Pawar, G.M., Wurst, K. et al. (2009). *Eur. J. Inorg. Chem.* 2009: 1970–1976.
42 Coutelier, O., El Ezzi, M., Destarac, M. et al. (2012). *Polym. Chem.* 3: 605–608.
43 Marrot, S., Bonnette, F., Kato, T. et al. (2008). *J. Organomet. Chem.* 693: 1729–1732.
44 Naumann, S., Epple, S., Bonten, C., and Buchmeiser, M.R. (2013). *ACS Macro Lett.* 2: 609–612.
45 Naumann, S., Schmidt, F.G., Speiser, M. et al. (2013). *Macromolecules* 46: 8426–8433.
46 Naumann, S., Schmidt, F.G., Frey, W., and Buchmeiser, M.R. (2013). *Polym. Chem.* 4: 4172–4181.
47 Lindner, R., Lejkowski, M.L., Lavy, S. et al. (2014). *Chem. Cat. Chem.* 6: 618–625.
48 Rodriguez, M., Marrot, S., Kato, T. et al. (2007). *J. Organomet. Chem.* 692: 705–708.
49 Jensen, T.R., Breyfogle, L.E., Hillmyer, M.A., and Tolman, W.B. (2004). *Chem. Commun.* 40: 2504–2505.
50 Dove, A.P., Li, H., Pratt, C. et al. (2006). *Chem. Commun.* 42: 2881–2883.
51 Bakkali-Hassani, C., Rieger, E., Vignolle, J. et al. (2016). *Chem. Commun.* 52: 9719–9722.

52 Naumann, S. and Buchmeiser, M.R. (2014). *Cat. Sci. Technol.* 4: 2466–2479.
53 Neilson, B.M. and Bielawski, C.W. (2013). *Chem. Commun.* 49: 5453–5455.
54 Fèvre, M., Vignolle, J., and Taton, D. (2013). *Polym. Chem.* (4): 1995–2003.
55 Hoppe, A., Sadaka, F., Brachais, C. et al. (2013). *Beilstein J. Org. Chem.* 9: 647–654.
56 Brulé, E., Guérinea, V., Vermaut, P. et al. (2013). *Polym. Chem.* 4: 2414–2423.
57 Reitz, A.K., Sun, Q., Wilhelm, R., and Kuckling, D. (2017). *J. Polym. Sci., Part A: Polym. Chem.* 55: 820–829.
58 Coulembier, O., Lohmeijer, B.G.G., Dove, A.P. et al. (2006). *Macromolecules* 39: 5617–5628.
59 Coulembier, O., Mespouille, L., Hedrick, J.L. et al. (2006). *Macromolecules* 39: 4001–4008.
60 Culkin, D.A., Jeong, W., Csihony, S. et al. (2007). *Angew. Chem. Int. Ed.* 46: 2627–2630.
61 Jeong, W., Shin, E.J., Culkin, D.A. et al. (2009). *J. Am. Chem. Soc.* 131: 4884–4891.
62 Dunn, A.L. and Landis, C.R. (2017). *Macromolecules* 50: 2267–2275.
63 Vishwa Prasad, A. and Yinghuai, Z. (2013). *J. Appl. Polym. Sci.* 128: 3411–3416.
64 Zhang, L., Wang, Z., Wang, J. et al. (2012). *J. Macromol. Sci., Pure Appl. Chem.* 49: 528–532.
65 Brown, H.A., Xiong, S., Medvedev, G.A. et al. (2014). *Macromolecules* 47: 2955–2963.
66 Chang, Y.A. and Waymouth, R.M. (2015). *Polym. Chem.* 6: 5212–5218.
67 Acharya, A.K., Chang, Y.A., Jones, G.O. et al. (2014). *J. Phys. Chem. B* 118: 6553–6560.
68 Brown, H.A., Chang, Y.A., and Waymouth, R.M. (2013). *J. Am. Chem. Soc.* 135: 18738–18741.
69 Chang, Y.A., Rudenko, A.E., and Waymouth, R.M. (2016). *ACS Macro Lett.* 5: 1162–1166.
70 Jeong, W., Hedrick, J.L., and Waymouth, R.M. (2007). *J. Am. Chem. Soc.* 129: 8414–8415.
71 Guo, L. and Zhang, D. (2009). *J. Am. Chem. Soc.* 131: 18072–18074.
72 Lahasky, S.H., Serem, W.K., Guo, L. et al. (2011). *Macromolecules* 44: 9063–9074.
73 Lee, C., Smart, T.P., Guo, L. et al. (2011). *Macromolecules* 44: 9574–9585.
74 Guo, L., Lahasky, S.H., Ghale, K., and Zhang, D. (2012). *J. Am. Chem. Soc.* 134: 9163–9171.
75 Falivene, L., Al Ghamdi, M., and Cavallo, L. (2016). *Macromolecules* 49: 7777–7784.
76 Jones, R.A., Karatza, M., Voro, T.N. et al. (1996). *Tetrahedron* 52: 8707–8724.
77 Jones, R.A. and Civcir, P.U. (1997). *Tetrahedron* 53: 11529–11540.
78 Pinaud, J., Vijayakrishna, K., Taton, D., and Gnanou, Y. (2009). *Macromolecules* 42: 4932–4936.
79 Zhang, Y., Schmitt, M., Falivene, L. et al. (2013). *J. Am. Chem. Soc.* 135: 17925–17942.
80 Gowda, R.R. and Chen, E.Y.-X. (2016). *ACS Macro Lett.* 5: 772–776.

81 Flanagan, J.C.A., Kang, E.J., Strong, N.I., and Waymouth, R.M. (2015). *ACS Catal.* 5: 5328–5332.
82 Martin, D., Canac, Y., Lavallo, V., and Bertrand, G. (2014). *J. Am. Chem. Soc.* 136: 5023–5030.
83 Kato, T., Ota, Y., Matsuoka, S.-I. et al. (2013). *J. Org. Chem.* 78: 8739–8747.
84 Matsuoka, S.-I., Ota, Y., Washio, A. et al. (2011). *Org. Lett.* 13: 3722–3725.
85 Hans, M., Delaude, L., Rodriguez, J., and Coquerel, Y. (2014). *J. Org. Chem.* 79: 2758–2764.
86 Biju, A.T., Padmanaban, M., Wurz, N.E., and Glorius, F. (2011). *Angew. Chem. Int. Ed.* 50: 8412–8415.
87 Kato, T., Matsuoka, S.-I., and Suzuki, M. (2014). *J. Org. Chem.* 79: 4484–4491.
88 Hong, M. and Chen, E.Y.-X. (2014). *Angew. Chem. Int. Ed.* 53: 11900–11906.
89 Matsuoka, S.-I., Namera, S., and Suzuki, M. (2015). *Polym. Chem.* 6: 294–301.
90 Welch, G.C., San Juan, R.R., Masuda, J.D., and Stephan, D.W. (2006). *Science* 314: 1124–1126.
91 Stephan, D.W. and Erker, G. (2015). *Angew. Chem. Int. Ed.* 54: 6400–6441.
92 Zhang, Y., Miyake, G.M., John, M.G. et al. (2012). *Dalton Trans.* 41: 9119–9134.
93 Chen, E.Y.X. (2013). *Frustrated Lewis Pairs II*, vol. 334 (ed. G. Erker and D.W. Stephan), 239–260. Berlin and Heidelberg, Germany: Springer.
94 Zhang, Y., Miyake, G.M., and Chen, E.Y.X. (2010). *Angew. Chem. Int. Ed.* 49: 10158–10162.
95 He, J., Zhang, Y., Falivene, L. et al. (2014). *Macromolecules* 47: 7765–7774.
96 Jia, Y.-B., Ren, W.-M., Liu, S.-J. et al. (2014). *ACS Macro Lett.* 3: 896–899.
97 Nyce, G.W., Lamboy, J.A., Connor, E.F. et al. (2002). *Org. Lett.* 4: 3587–3590.
98 Naik, P.U., Refes, K., Sadaka, F. et al. (2012). *Polym. Chem.* 3: 1475–1480.
99 Naumann, S., Speiser, M., Schowner, R. et al. (2014). *Macromolecules* 47: 4548–4556.
100 Buchmeiser, M.R., Kammerer, J.A., Naumann, S. et al. (2015). *Macromol. Mater. Eng.* 300: 937–943.

12

N-Heterocyclic Carbene Catalysis in Natural Product and Complex Target Synthesis

M. Todd Hovey, Ashley A. Jaworski, and Karl A. Scheidt

Northwestern University, Department of Chemistry, 2145 Sheridan Road, Evanston, IL 60208, USA

12.1 Introduction

Natural product chemistry encompasses a vast chemical space with almost a limitless variety in the structure and biological activity of newly discovered molecules. In response, chemical research is inspired and driven to develop a new chemical methodology to access compounds of continuously increasing complexity. One of the challenges currently being surmounted in this field is the implementation of new tactics and synthetic strategies by harnessing catalytic bond-forming reactions. These new reactions have the propensity to enhance synthetic efficiency and quickly provide access to compounds of greater complexity and functionality. One facet of this work centers on *N*-heterocyclic carbene (NHC) catalysis, a subset of organocatalysis, which has emerged as a powerful tactic to construct new bonds with a high degree of stereoselectivity through both Umpolung and normal polarity-based reactions. Historically, few total syntheses implementing carbene strategies have been realized, presumably because the substrate scope of known benzoin and Stetter NHC technology was not fully developed. In the past decade, NHC catalysis has moved beyond the acyl anion equivalent operator to access enolate, homoenolate, and redox relay type transformations, greatly expanding the breadth of different reaction classes that are now rapidly emerging. Furthermore, with the discovery of NHC catalysts that promote highly efficient and enantioselective transformations, the implementation of strategies mediated by NHCs in total synthesis has rapidly increased. This chapter aims to summarize NHC catalysis in the construction of natural products.

12.2 NHC-Catalyzed Benzoin Condensations

The coupling between two aldehydes or an aldehyde and ketone through the formation of acyloins, also known as the benzoin condensation, is the most venerable of Umpolung reactions [1–8]. Investigation into this reaction began in 1832

N-Heterocyclic Carbenes in Organocatalysis, First Edition. Edited by Akkattu T. Biju.
© 2019 Wiley-VCH Verlag GmbH & Co. KGaA. Published 2019 by Wiley-VCH Verlag GmbH & Co. KGaA.

by von Liebig and Woehler with the discovery that benzoin was produced upon exposure of aldehydes to cyanide [9]. Later in 1943, it was demonstrated by Ugai that upon exposure to base, thiazoliums could serve as effective catalysts for this unusual reaction, which marks one of the earliest examples of NHC catalysis [10]. In the years following, research into the reactivity of NHCs has produced numerous benzoin condensation reactions promoted by these small molecules, including examples of enolizable aldehydes and ketones, as well as mixed aldehyde systems being productive in this reaction regime [11–13]. Furthermore, continued research has yielded asymmetric variants of the benzoin reaction with reports by Sheehan [14, 15], Enders [16–18], and others [19–22], disclosing the synthesis and application of chiral NHCs as new powerful catalysts.

12.2.1 Synthesis of *trans*-Resorcylide

Isolated from *Penicillium* sp in 1978, the 12-membered macrocyclic lactone *trans*-resorcylide [7] belongs to a family of naturally occurring benzoic-acid-derived macrolide plant growth inhibitors [23]. In 2007, Miller reported a formal synthesis of *trans*-resorcylide through a benzoin-mediated NHC macrocyclization (Scheme 12.1) [24]. Activation of aliphatic aldehyde of **3** by NHC **4** was preferred over the *ortho*-substituted benzaldehyde, giving rise to macrolactones of type **5** as the only product, albeit in a low yield. This strategy was chosen over more traditional macrocyclization methods because Miller was investigating the potential of thiamine pyrophosphate to promote similar macrocyclizations in biosynthetic pathways. To begin the synthesis, thioether bis-aldehyde **3** was prepared through a five-step sequence from alcohol **2** and benzoic acid **1**. From bis-aldehyde **3**, an intramolecular benzoin condensation afforded macrolactone **5** as a diastereomeric mixture in 21% yield. The benzylic alcohol was subsequently removed through an acylation/deoxygenation sequence, and oxidation of phenyl sulfide to the sulfone followed by elimination led to the bis-benzylated *trans*-resorcylide precursor **6**, which could be converted by following the procedure of Couladouros to *trans*-resorcylide (**7**), thus completing the formal synthesis [25].

12.2.2 Synthesis of (+)-Sappanone B

(+)-Sappanone B (**11**) is a homoisoflavanoid isolated by Saitoh in 1986 from the heartwood of *Caesalpinia sappan* Leguminosae [26, 27]. In 2005, it was discovered that this 3-hydroxy chromanone possesses an inhibitory activity toward xanthine oxidase [28]. Subsequently, an elegant total synthesis of (+)-sappanone B was reported in 2007 by Suzuki (Scheme 12.2) [29]. They envisioned that an intramolecular benzoin condensation could yield the necessary α-hydroxy-keto moiety present in this family of natural products. Ketoaldehyde **8** was obtained in five synthetic transformations starting from 3-methoxysalicylic acid. The exposure of aldehyde **8** to a mixture of triazolium salt **9** and triethylamine promoted the intramolecular benzoin reaction in 12 hours, affording methylated (+)-sappanone B (**10**) in 92% yield and 95% ee. The incorporation of trifluoromethyl substituents on the triazolium *N*-aryl

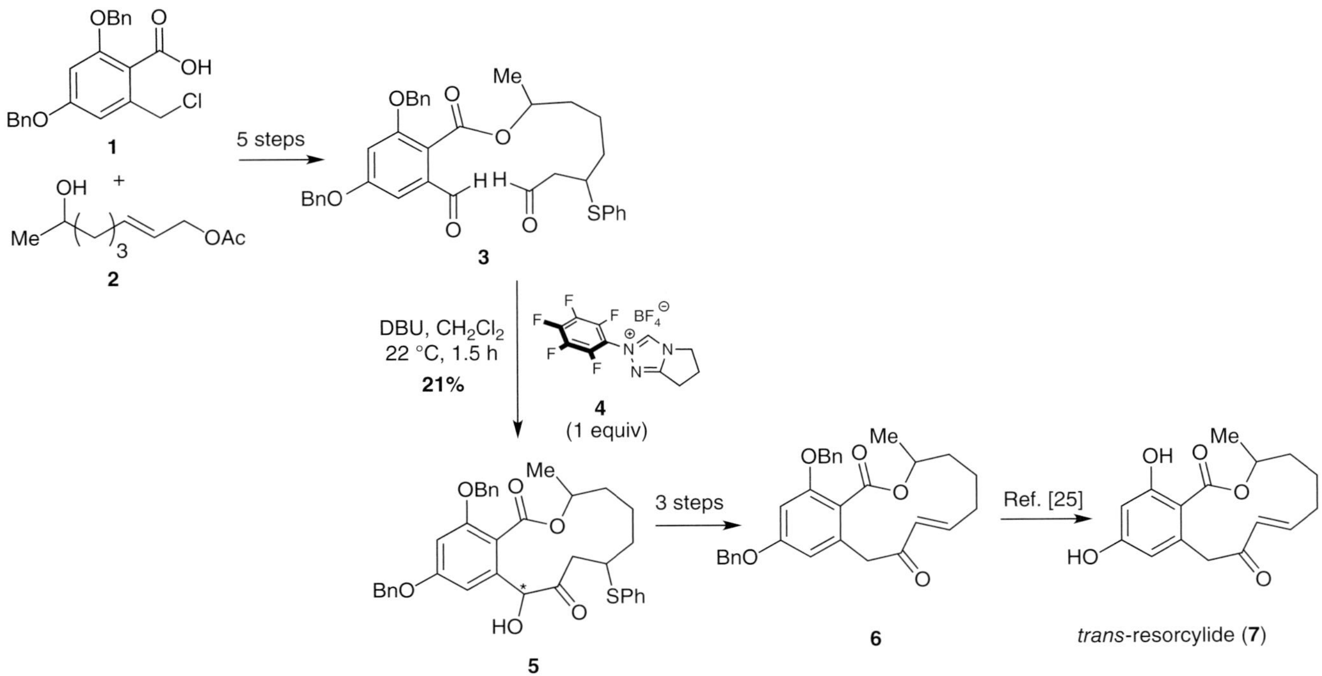

Scheme 12.1 Formal synthesis of *trans*-resorcylide.

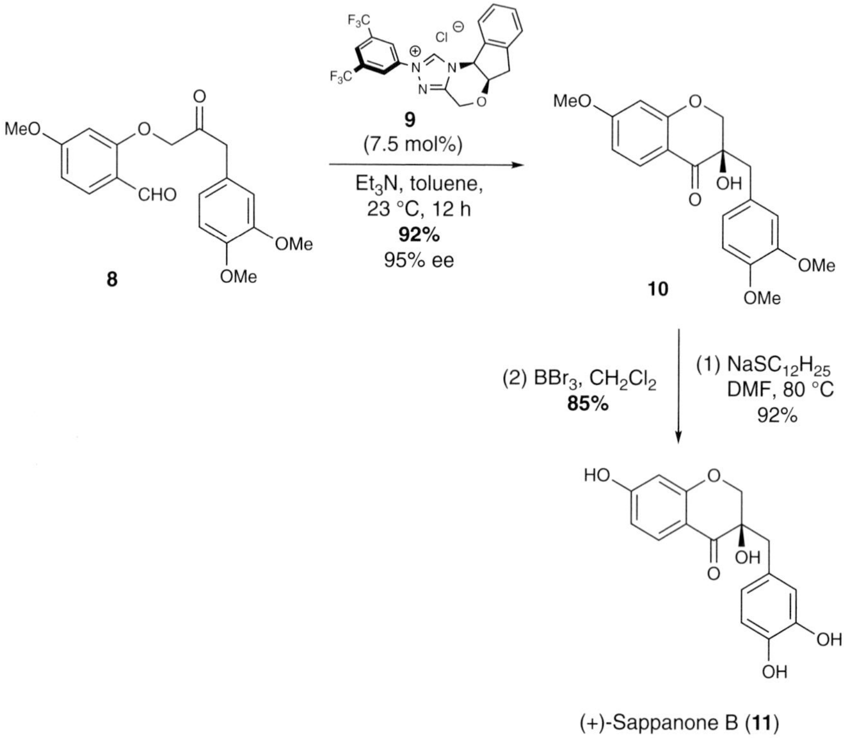

Scheme 12.2 Synthesis of (+)-sappanone B.

component is notable because the NHC is tuned for an optimal asymmetric induction. In comparison, the pentafluoro *N*-aryl triazolium variant of catalyst **9** gave diminished enantioselectivity, providing precedent to employ **9** as a viable alternative when more common NHC catalysts fail. To finish their synthesis, global demethylation of **10** gave (+)-sappanone B (**11**) in an excellent overall yield of 59% for the eight-step route.

12.2.3 Synthesis of Cassialoin

Cassialoin (**17**) is an anthrone C-glycoside found in the plant extracts of *Cassia garrettiana* [30, 31], the roots of *Rheum emodi* [32], and the Chinese herb *Rumex dentatus* [33]. Cassialoin (**17**) is characterized by a phenolic anthraquinone skeleton attached to a stereogenically defined β-glycoside. *In vivo* biological studies have shown that cassialoin (**17**) possesses tumor growth inhibition as well as inhibition of metastasis in mice [34]. In 2007, Suzuki reported the first asymmetric total synthesis of cassialoin beginning from nonracemic, chiral starting materials [35]. The diastereoselective assembly of the anthroquinone skeleton was accomplished by an intramolecular NHC-catalyzed benzoin condensation. The exposure of ketoaldehyde **12** to thiazolium salt **13** and 1,8-diazabicyclo[5.4.0]undec-7-ene (DBU) at 40 °C for two hours afforded ketol **14** in 99% yield with an excellent diastereoselectivity (Scheme 12.3). Ketone

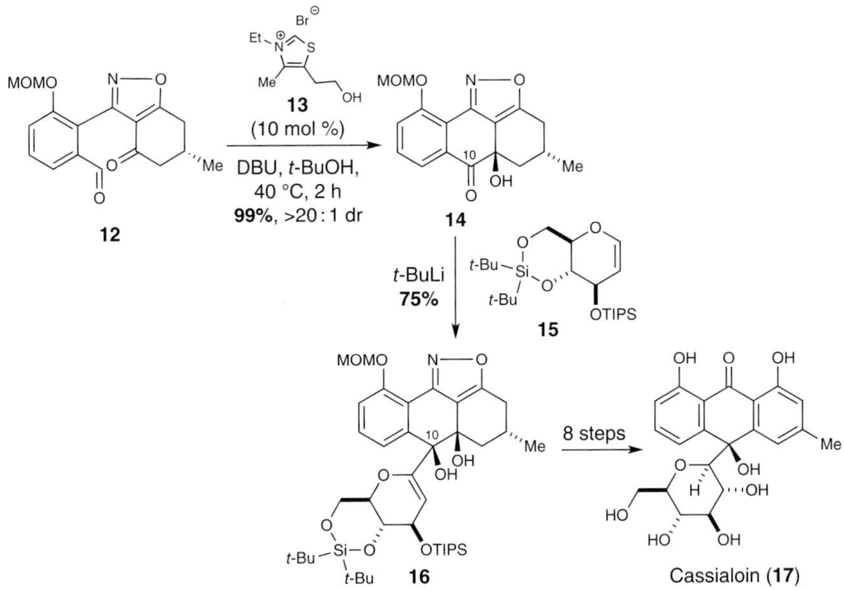

Scheme 12.3 Synthesis of anthrone C-glycoside cassialoin.

14 was further elaborated to cassialoin (**17**) by the addition of the lithium anion of glycal **15** to yield *cis*-diol **16** as a single diastereomer. Subsequently, eight synthetic steps produced the natural product cassialoin (**17**). Upon the completion of this synthetic effort, they confirmed the originally proposed structure of cassialoin [30, 31].

12.2.4 Synthesis of the Kinamycins and the Monomeric Unit of Lomaiviticin Aglycon

The kinamycins [36–39] and lomaiviticins A and B [40] are structurally related natural products that exhibit antimicrobial and cytotoxic activities. Both families of natural products possess structurally novel diazofluorene groups as well as densely oxygenated cyclohexane rings and a poly-aromatic core. The lomaiviticins are interesting dimeric structures composed of two similar diazofluorene-containing polycyclic components related to the kanamycin backbone. In 1970, Omura isolated kinamycins A–D from the fermentation broth of *Streptomyces murayamaensis* [36–39]. In 2007, Nicolaou reported a synthesis of kinamycins C, F, and J (Scheme 12.4) [41]. Central to their synthetic route is an Ullmann coupling to bring together bromide **18** and iodide **19** to yield aldehyde **20** on a gram scale. This coupling reaction was followed by a benzoin condensation to fuse the fourth ring by allowing aldehyde **21** to react in the presence of azolium **4** to yield cyclopentanone **22** in 78% yield and 3 : 1 dr (Scheme 12.4). This strategy installed the necessary functionality to promote the synthesis of diazofluorene **22** in 10 additional synthetic steps. The selective hydrolysis or acetylation of **22** using complimentary conditions gave each of the three kinamycins (**23–25**) with overall yields between 9 and 11% from the key Ullmann coupling product **20**.

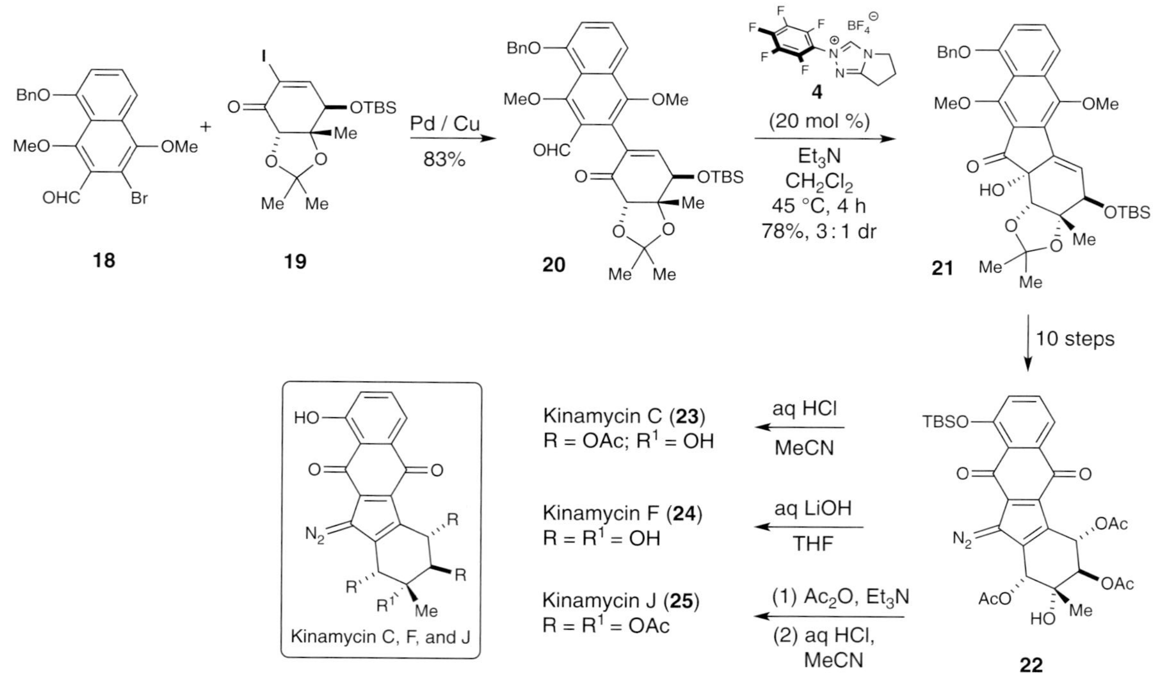

Scheme 12.4 Synthesis of kinamycins C, F, and J.

Similar to kinamycins, lomaiviticins A (**31**) and B (**32**) are dimeric natural products containing the structurally unique diazofluorene moiety. These highly cytotoxic and antimicrobial compounds were originally isolated from *Micromonospora lomaivitiensis* in 2001 by Carter [40]. Of particular note, lomaiviticin A has IC_{50} values ranging from 0.01 to 98 μM against several cancer cell lines [40]. In 2009, Nicolaou reported an asymmetric synthesis of the lomaiviticin aglycon (**30**) and a monomeric subunit of the lomaiviticins [42]. Because of the structural homology between lomaiviticin aglycon and the kinamycins, they employed an analogous synthetic strategy toward the aglycon. When the NHC-catalyzed benzoin condensation was attempted on an aldehyde derived from the Ullmann coupling of a benzyl-protected version of **27** (not depicted) and iodide **26**, the major observed product was the undesired Stetter annulation product. To disfavor the undesired Stetter reaction pathway, the aldehyde component was redesigned to arise from bromide **27**, which contains 2-(trimethylsilyl)ethoxymethyl ether (SEM) protecting groups at C-7 and C-10 instead of benzyl ethers. The SEM groups direct the methoxy group at C-11 into the plane of the C-4 carbonyl groups, thus affecting a conformational change between both cyclic systems and thereby allowing the acyl anion equivalent to favor the benzoin condensation pathway. The reaction of ketoaldehyde **28** with NHC precatalyst **4** in the presence of triethylamine yielded benzoin condensation product **29** with excellent chemo- and regioselectivity (20 : 1) (Scheme 12.5). Notably, in this particular case, sterically encumbered protecting groups lowered chemoselectivity in the benzoin condensation or were incompatible with the previous Ullmann coupling conditions. An additional five synthetic transformations yielded lomaiviticin aglycon (**30**) in 16.5% yield over nine steps.

12.2.5 Synthesis of (−)-Seragakinone A

In 1999, seragakinone A (**39**) was characterized after isolation from a marine fungal symbiote rhodophyta *Ceratodictyon spongiosum* [43, 44]. It has gained attention for its unique poly-aromatic structure and for possessing antifungal and antibacterial activities [43, 44]. The synthesis of this natural product class presents several challenges. Stereoselective construction of a densely oxygenated pentacyclic core and installation of a sterically hindered angular prenyl group are nontrivial synthetic tasks. In 2011, Suzuki reported an enantioselective synthesis of (−)-seragakinone A (**39**), harnessing the power of the benzoin condensation to fashion its tetracyclic core [45]. Their synthesis began with a six-step sequence to prepare aldehyde **34** from aryl bromide **33**. The treatment of aldehyde **34** with *N*-(3,5-trifluoromethylphenyl) triazolium salt **9** [29] and triethylamine yielded tetracycle **35** in high yield (86%) with an excellent enantioselectivity (99% ee, Scheme 12.6). An additional 13 synthetic steps provided ketoaldehyde **36**, which underwent a second benzoin condensation using triazolium salt **33** [46] to furnish ketol **38** in 90% yield with an excellent diastereoselectivity. X-ray crystallographic analysis verified the stereochemistry of ketol **38**. Four additional steps furnished seragakinone A (**39**) in 26 steps with a 2% overall yield.

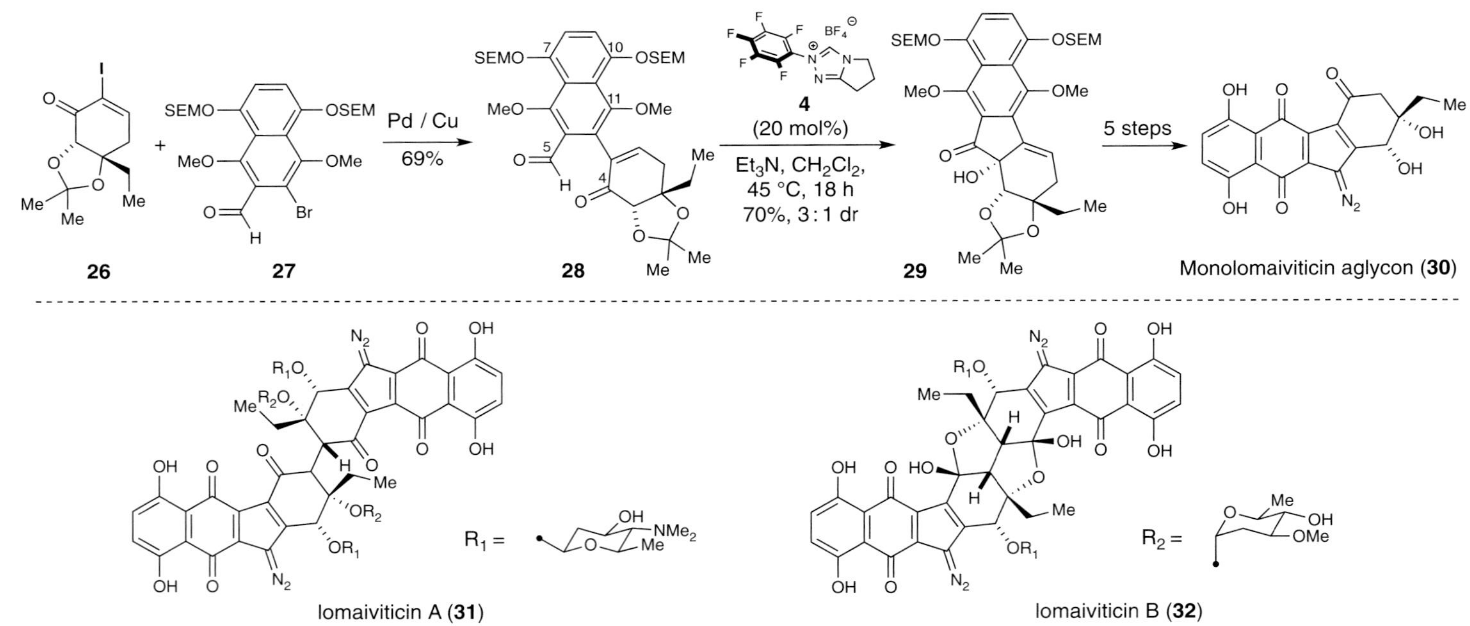

Scheme 12.5 Synthesis of monolomaiviticin aglycon.

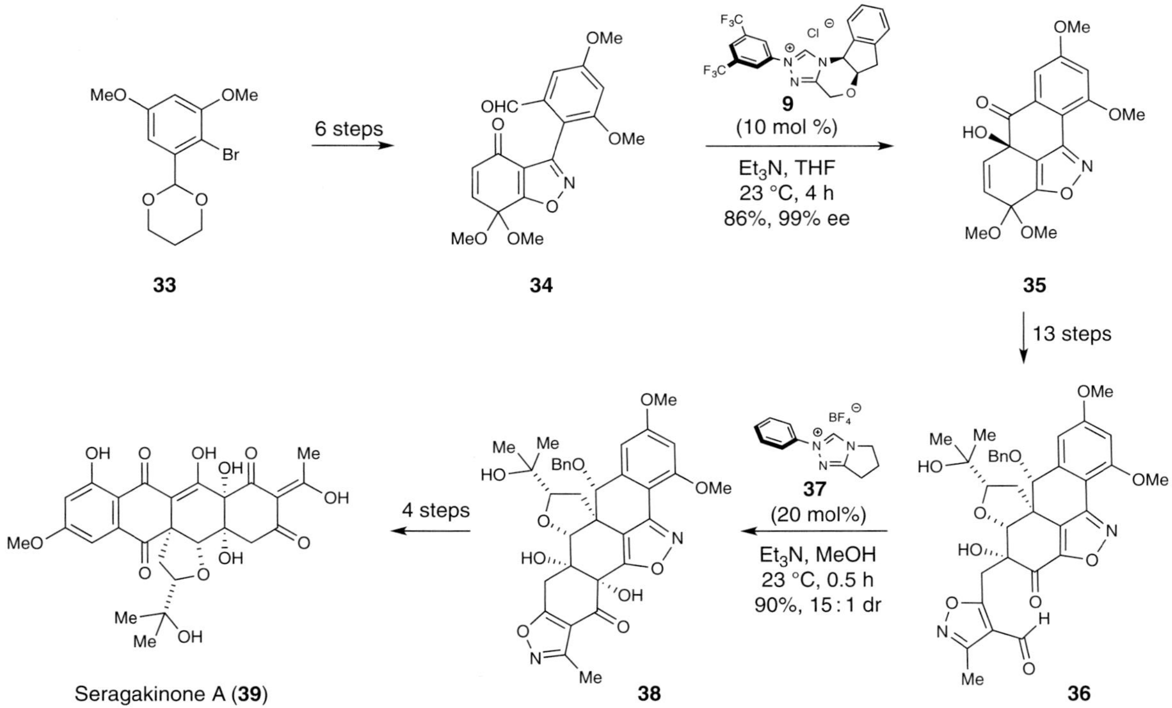

Scheme 12.6 Synthesis of (−)-seragakinone A.

12.2.6 Synthesis of Originally Assigned Structure of Pleospdione

The polyketide synthase found in the type-II polyketide biosynthesis assembles densely functionalized polyclic aromatic polyketides encompassing an entire family of biologically active natural products [47]. The repetitive reactions of the involved enzymes prepare linear polyketide precursors to be assembled into linearly and angularly fused polycyclic frameworks [47]. Suzuki set out to devise a general synthetic strategy toward these natural products through sequential isoxazole formation and benzoin condensation to yield the densely functionalized core skeletons [48]. Originally isolated in 2005 from the *Pleospora* sp. IFB-E006, tricyclic octaketide-derived Pleospdione (**44**) was chosen as a proving ground for this synthetic strategy [49]. Starting from previously reported nitrile oxide **40** [45] and racemic cyclohexenone **41** [50], oxo-ketone **42** was synthesized in three transformations through isoxazole formation and functional group manipulations (Scheme 12.7). When exposed to pentafluoro-aryl trizazolium **4** in the presence of triethylamine, a highly efficient and diastereoselective benzoin condensation occurred between the aryl aldehyde moiety and ketone moiety of **42** to yield tetracyclic intermediate **43**. Suzuki was able to elaborate this ketone to the reported structure of pleospdione (**44**) in eight synthetic transformations. The researchers determined the structure of an advanced intermediate by X-ray crystallography to confirm the relative stereochemistry, which matched the

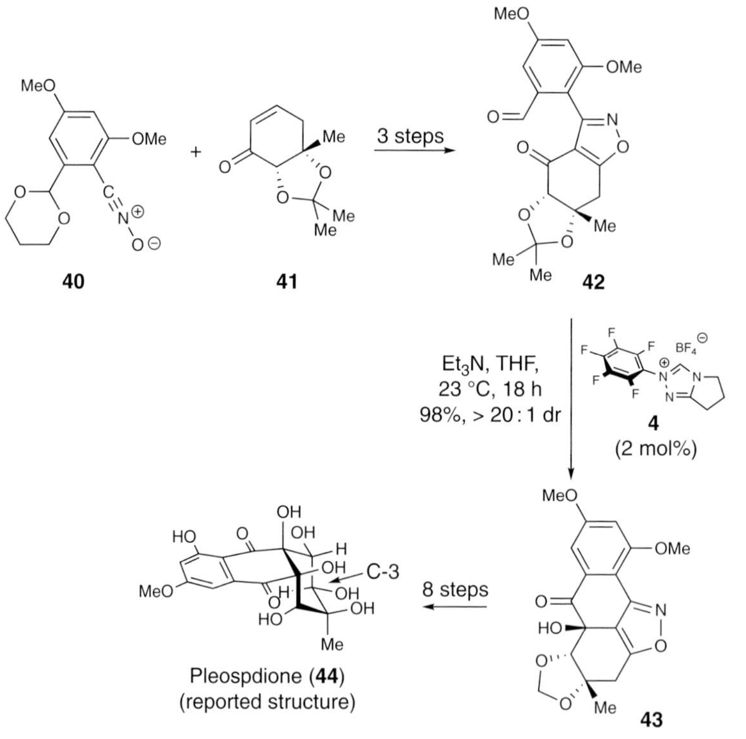

Scheme 12.7 Synthesis of pleospdione.

reported structure, but upon further elaboration of the intermediate to the natural product, they obtained spectral data that did not match the isolation data. Additionally, the C-3 epimer did not match the reported spectral data, thus posing a new challenge for characterization as well as highlighting the role that natural product synthesis still plays in structure elucidation.

12.2.7 Formal Synthesis of Natural Inositols

The cyclitols are a class of poly-hydroxylated cycloalkanes that have gained much attention for both their biological activities and their potential as synthetic building blocks in natural product synthesis. The inositol subclass of the cyclitols is characterized by a hexa-hydroxy cyclohexane structure, mimicking the shape and chemistry of isoelectronically similar hexoses [51]. Of the inositols, several naturally occurring diastereomers are involved in metabolic processes, whereas other stereoisomers are not found in nature [51–53]. In particular, Yamada sought to develop a general strategy for the asymmetric synthesis of the inositol family of cyclitols from nonracemic starting material [54]. Specifically, their executed strategy utilized a benzoin condensation between C_2-symmetrical dialdoses to prepare the cyclohexane core, as well as a ketone moiety, which could be selectively differentiated into hydroxyl epimers. Starting from tetra-benzyl-protected C_2-symmetrical dialdose **45**, exposure to amino-indanol-derived triazolium **46** with triethylamine yielded the desired keto-hexose **47** in excellent yield and diastereoselectivity (Scheme 12.8). Subsequent protection of the free hydroxyl group was necessary to overcome the preference for equatorial hydride delivery during reduction to hexa-hydroxyl cycloalkane **48**. A palladium on carbon hydrogenolysis yielded the naturally occurring D-*chiro*-inositol (**49**) [52]. In order to reverse the selectivity of the benzoin condensation for the epimeric product, acetonide formation and silyl protection of the α-hydroxy substituents were necessary on the dialdose **50**. Fortunately, they found that achiral triazolium **4** promoted the desired benzoin condensation of dialdehyde substrate **50** to yield the epimeric product **51**. Attempts to use *ent*-**46** with substrate **45** generated no product, and with acetonide **50**, *ent*-**46** yielded a 1 : 1 mixture of epimers, suggesting a match/mismatch scenario when employing the chiral catalysts for this transformation. Again, a triethylsilane (TES) protection, reduction, and deprotection sequence of **51** yielded desired hexose **52**, which constitutes the formal synthesis of *neo*-inositol through the procedure of Fernández-Mayoralas [55].

12.2.8 Synthesis of (+)-7,20-Diisocyanoadociane

The isocyanoterpene natural products are generally found in soft-bodied marine species and have potent and highly selective antimalarial activity against *Plasmodium falciparum* at low-nanomolar concentrations [56–58]. These activities have prompted a broad range of synthetic efforts toward the construction of these unique marine terpenoids [59]. Shenvi has reported a succinct strategy toward these molecules using a dendralene cycloaddition approach, which quickly produces the complex framework of the isocyanoterpenes [60]. In 2016,

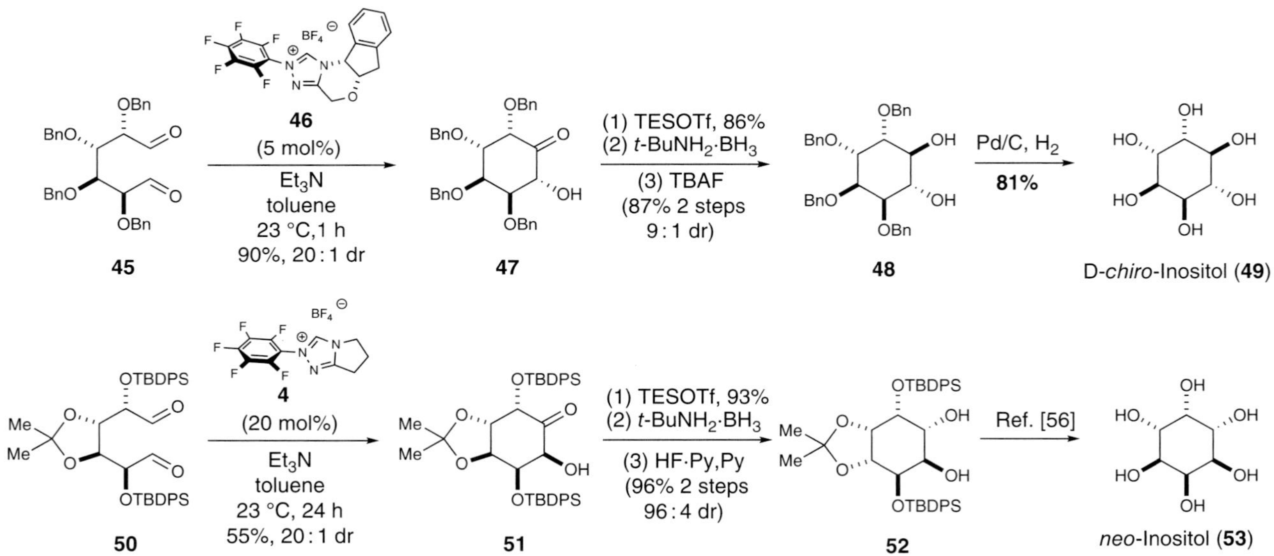

Scheme 12.8 Synthesis of naturally occurring inositols.

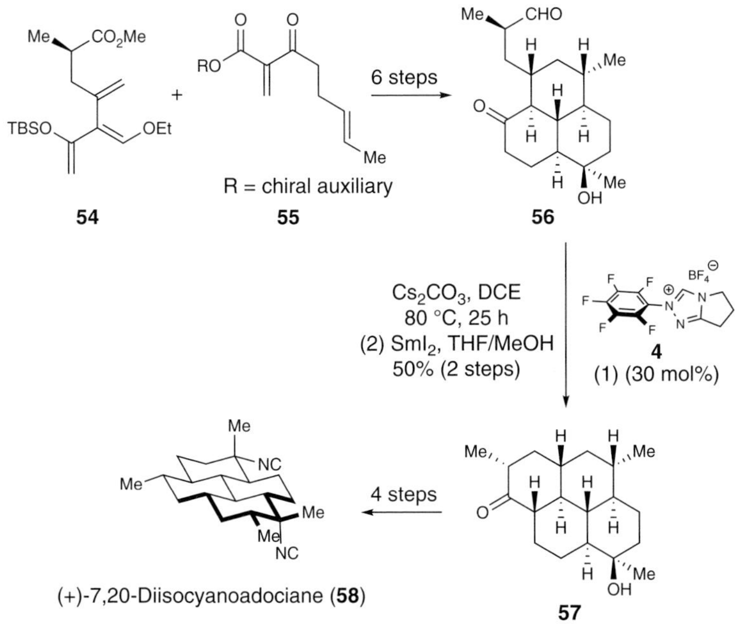

Scheme 12.9 Synthesis of (+)-7,20-diisocyanoadociane.

Shenvi's dendralene cycloaddition strategy was successfully implemented in a total synthesis of (+)-7,20-diisocyanoadociane (**58**) (Scheme 12.9) [61]. A benzoin condensation reaction and α-hydroxy reduction sequence were central to their strategy, appending the final carbocycle of the tetracyclic framework of **57**. Starting from simple dendralene precursor **54** and dienophile **55**, six synthetic transformations yield advanced intermediate aldehyde **56**, which is primed for the benzoin condensation. Aldehyde **56** reacts with precatalyst **4** in the presence of inorganic base (to prevent undesired epimerization of the aldehyde) in 1,2-dichloroethane to yield intermediate α-hydroxy ketone, which was reduced using a samarium diiodide protocol, to provide ketone **57**. This two-step protocol is a unique method to afford functionalized polycyclic ketones like **57**, and it has potential for application to other synthetic annulation strategies. They were able to convert ketone **57** to (+)-7,20-diisocyanoadociane (**58**) in four additional steps, totaling 11 steps from dendralene **54** with 2% overall yield.

12.3 The Stetter Reaction

The Setter reaction is the conjugate addition of acyl anion equivalents to α,β-unsaturated carbonyls and related conjugated acceptors [62–65]. This reaction produces 1,4-dicarbonyl compounds, which are useful synthetic intermediates for a variety of further transformations [66–70]. Significant advances of the asymmetric intramolecular Stetter reaction have been made by Enders [71–73], Rovis [74], and Hamada [75]. In contrast, the intermolecular Stetter reaction variant poises a greater fundamental challenge, thus prompting further exploration in this area. Typical α,β-unsaturated carbonyls are still not viable in many selective intermolecular Stetter reactions; however, there has been success in using more activated substrates, such as alkylidene malonates [76–78], nitroalkenes [79, 80], or β,γ-ketoesters [81]. The most common issues in these intermolecular variants are self-condensation or benzoin by-products complicating the reaction mixture. Several methods exist to suppress these undesired reactions, which include using thiazolium-derived NHC catalysis with α-keto carboxylates or acylsilanes [82] as precursors to the acyl anion equivalents [83]. This section aims to highlight some total and formal syntheses that use an NHC-catalyzed Stetter reaction as a central tactic to execute the synthetic strategy [84].

12.3.1 Annulation Reactions

The construction of cyclic frameworks through annulation reactions is a synthetic strategy ubiquitous across the field of total synthesis. New tactics to implement annulations continue to attract the attention of the synthetic community due to a variety of ways that (poly)-cyclic targets can be strategically disconnected using a wide diversity of chemical transformations. Within the field of NHC catalysis, utilization of intramolecular Stetter reactions to yield compounds of greater complexity with dissonant charge affinity patterns is a powerful method for total synthesis. This subsection describes the known examples of the Stetter reaction applied in this manner.

12.3.1.1 Synthesis of Hirsutic Acid C

Derived from the humulene biosynthetic pathway, hirsutic acid C (**65**) is a tricyclic sesquiterpenoid characterized by stereogenic complexity and possessing seven chiral centers, six of which are contiguous. The hirsutic acids have been a venerable proving ground for new synthetic methods and strategies because of their complexity as well as their observed antibiotic and antitumor activities [85–87]. In 1979, Trost reported the first stereocontrolled synthesis of (±)-hirsutic acid C (**65**) [88]. This disclosure served as the seminal demonstration of the Stetter reaction as a powerful annulation strategy to form poly-carbocyclic frameworks. The route began by the conversion of cyanocyclohexane **59** to polycyclic aldehyde **60** in 10 synthetic transformations (Scheme 12.10). Intermediate aldehyde **62** was treated with 2.3 equiv of thiazolium salt **61** and a large excess of triethylamine to yield the densely functionalized ketone **62** in 67% yield. A four-step sequence encompassing a reduction of the ketone in tricycle **62** and γ-lactonization followed by ozonolysis of the olefin moiety and reduction of the resultant carbonyls to the pendant-methyls afforded lactone **63**, thereby delivering the core of hirsutic acid. Further elaboration through a methylation and aldol condensation yielded enone **64**. A subsequent four-step reaction sequence, following the procedure of Matsumoto, yielded hirsutic acid C (**65**) with 3% overall yield [89].

12.3.1.2 Formal Synthesis of Platensimycin

The polyphenol platensimycin (**70**) possesses *in vitro* activity against *Staphylococcus aureus* and *Enterococcus faecium* [90, 91]. The formal synthesis of

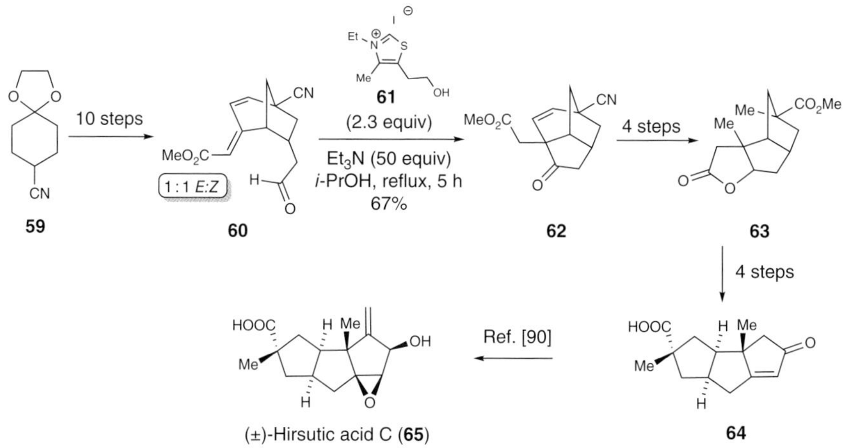

Scheme 12.10 Synthesis of (±)-hirsutic acid C.

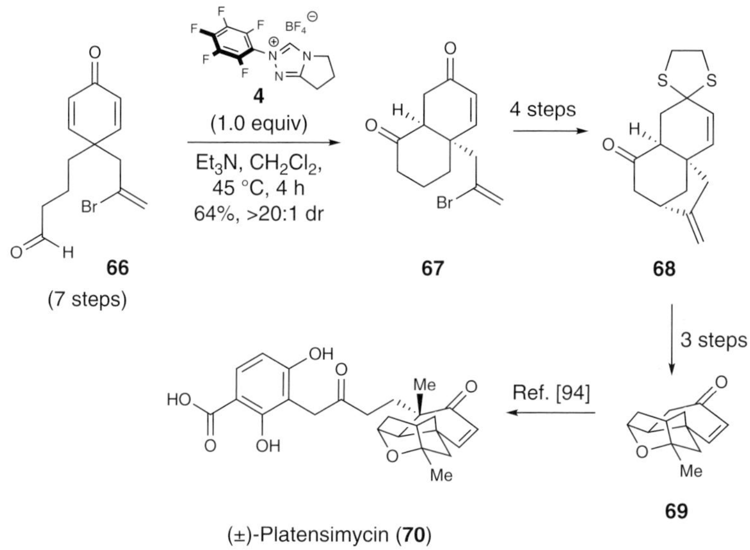

Scheme 12.11 Formal synthesis of (±)-platensimycin.

(±)-platensimycin [92] was reported in 2007 by Nicolaou. Central to their synthetic strategy was an intramolecular Stetter reaction, a tin-mediated radical annulation, and a trifluoroacetic acid (TFA)-promoted cyclization (Scheme 12.11). Ketoaldehyde **66** was reacted with 1.0 equiv of azolium **4** and triethylamine to yield the corresponding Stetter adduct **67** in 64% yield and >20 : 1 diastereoselectivity. Cyclohexenone **67** was selectively transformed into tetracycle **68** as described: a chemoselective dithiane formation protected the ketone of the enone moiety, which allowed for selective oxidation of the cyclohexanone to the enone (not depicted). This intermediate was then converted into **68** through functional group manipulations and a tin-promoted radical annulation. Three subsequent steps furnished the core structure **69** of

(±)-platensimycin (**70**), which is an intermediate in the previously published synthesis of **70** by Nicolaou [93].

12.3.2 Fragment Coupling

In contrast to the syntheses that utilize the Stetter reaction as a tactic to implement annulation strategies, it is far more common to see the Stetter reaction employed in an intermolecular manner to couple two molecular fragments. The advantages of this approach are twofold. First, coupling two fragments increases the convergency of the synthetic route. Second, the intermolecular Stetter reaction affords access to 1,4-dicarbonyl compounds and their functional equivalents. The employment of the intermolecular Stetter reaction in the context of total synthesis (see Section 12.3.1 for the intramolecular variant) is exemplified in the following section.

12.3.2.1 Synthesis of *cis*-Jasmon and Dihydrojasmon

The naturally occurring compounds, *cis*-jasmon (**73**) and dihydrojasmon (**74**) from the jasmine plant, are used by the fragrance and cosmetic industry as scent additives [94, 95]. The total syntheses of these two fragrance compounds were first reported in 1975 by Stetter and Kuhlmann utilizing a Stetter reaction to quickly construct the cyclopentenone core (Scheme 12.12) [96]. The key reaction utilizes thiazolium **13** to catalyze an intermolecular addition of saturated aldehyde **70** or **71** to methyl vinyl ketone **72**, followed by an intramolecular aldol condensation to prepare *cis*-jasmon (**73**) and dihydrojasmon (**74**). This disclosure by Stetter and Kuhlmann appear to be the first reported total synthesis that employs the Stetter reaction or similar transformations.

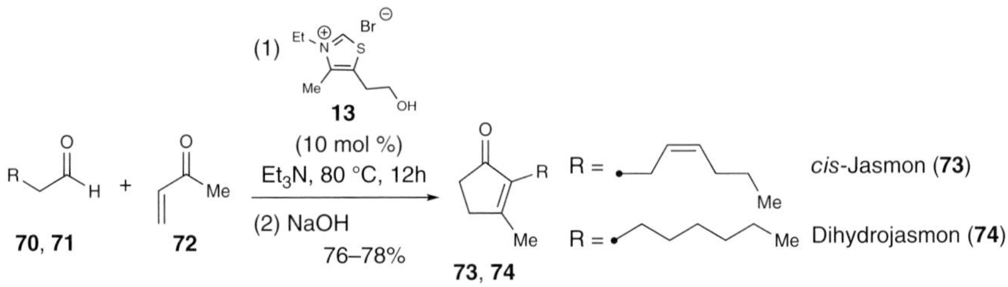

Scheme 12.12 Synthesis of *cis*-jasmon and dihydrojasmon.

12.3.2.2 Synthesis of the Core of Atorvastatin

Atorvastatin calcium (**79**), known by the trade name Lipitor, is a drug developed and marketed by Pfizer since 1997. Lipitor is used to moderate blood cholesterol levels through the inhibition of the HMG-CoA enzyme. The efficacy of Lipitor has made it one of the most successful drugs in history, in addition to becoming a flagship compound of the statin drug class that now contains many complimentary drugs. Lipitor is characterized by a tetrasubstituted pyrrole possessing a *N*-alkyl chain bearing a chiral 1,3-diol as well as a terminal carboxylic acid (Scheme 12.13). Quantitative structure activity relationship (QSAR) studies identified that phenyl and phenylamide moieties at the C-3 and C-4 positions improved the drug efficacy but also complicated the synthesis

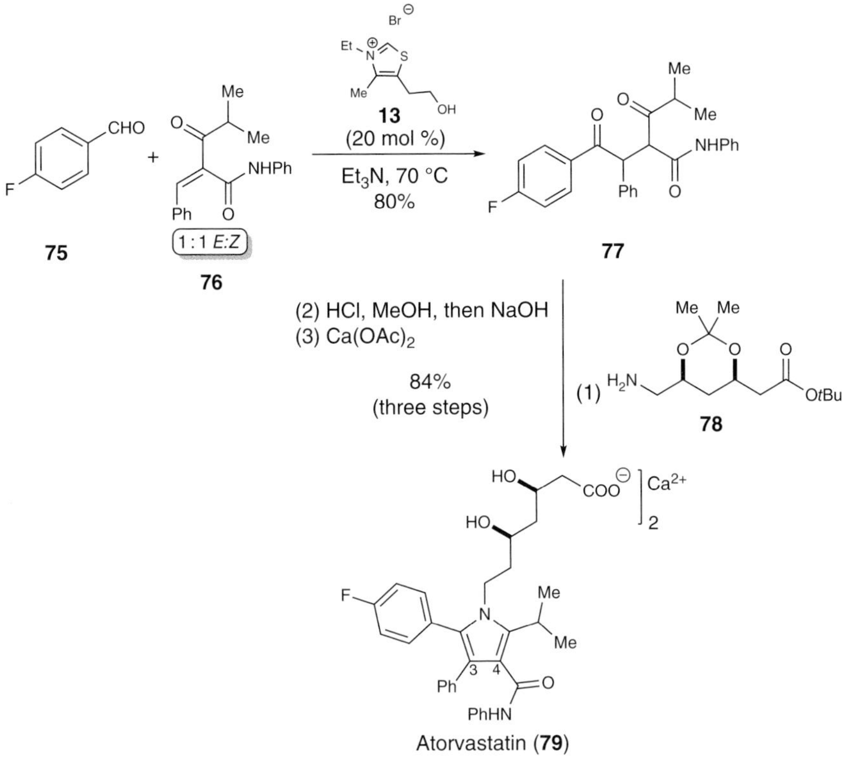

Scheme 12.13 Synthesis of atorvastatin (Lipitor).

[97]. The synthetic route reported by Roth employs a Stetter reaction catalyzed by *N*-alkyl thiazolium **13**, between *p*-fluorobenzaldehyde **75** and conjugate acceptor carboxamide **76** to furnish the required 1,4-diketone **77** in 80% yield. Roth noted that *N*-alkyl thiazolium catalysts are preferred to the *N*-benzyl thizaoliums (such as **82**) because only the Stetter product is observed as opposed to the competing benzoin condensation (which was not reversible under these reaction conditions, although earlier reports suggest the reversibility from benzoin, see previous section). The route is completed through a subsequent Paal–Knorr reaction between heptanoate amine **78** and 1,4-diketone **77** to produce atorvastatin **79** [98]. The overall synthetic route is notable for two reasons: a) the significant scale of the Umpolung Stetter reaction and b) the choice of thiazolium precatalyst to prevent the formation of the undesired benzoin condensation product from aldehyde **75**. Overall, this synthetic approach exemplifies the complexity inducing power of the Stetter reaction in synthetic planning.

12.3.2.3 Synthesis of Roseophilin

In 1992, Seto isolated roseophilin (**85**) from the fermentation broth of *Streptomyces griseoviridis*, which was measured to have cytotoxic activity against K562 human erythroid leukemia cells [99]. Roseophilin is characterized by an unusual azafulvene conjugated to a pyrrolylfuran as well as an ansa macrocycle. Because of the report of this structurally unique natural product, several considerable efforts have been made toward its synthesis [100]. In 1999, Tius reported a 12-step formal synthesis of roseophilin by intercepting the known racemic core **84** found in the previously reported route by Fürstner (Scheme 12.14) [100, 101]. Two years later, Tius reported an updated enantioselective route where the key synthetic step was the formation of cyclopentenone **81** through a Nazarov cyclization. Central Tius' strategy is a Stetter reaction in which unsaturated ketone **81** and 2 equiv of 6-heptenal (**80**) were allowed to react with thiazolium **82** and triethylamine to yield the Stetter product **83** in 77% yield. The resulting *trans*-isomeric product was favored (Scheme 12.14). Ring-closing metathesis, a selective olefin hydrogenation, and subsequent Knorr pyrrole synthesis yielded the enantiopure core of roseophilin **84**. An alkylation of the necessary pyrrolylfuran component with the ansa-core followed by functional group interconversions completed the synthesis of roseophilin (**85**) in 15 steps and 7% overall yield [102].

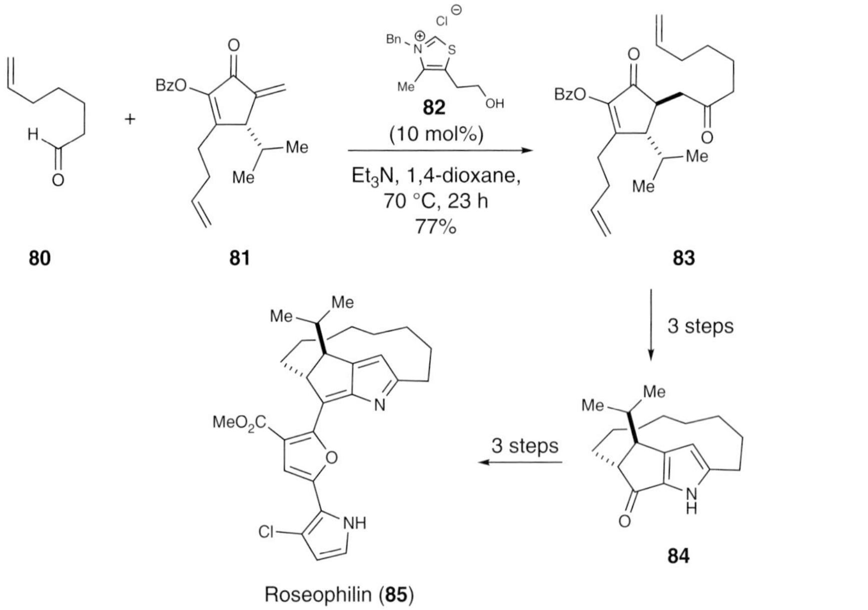

Scheme 12.14 Synthesis of roseophilin.

12.3.2.4 Synthesis of *trans*-Sabinene Hydrate

The natural product *trans*-sabinene hydrate (**90**) is a flavor found in a variety of essential oils and is generally isolated from mint and other herbs [103–105]. In an effort to facilitate the use of *trans*-sabinene hydrate as a food additive and flavorant, Galopin developed and executed a short and efficient synthesis of this natural product [106]. A Stetter reaction is used as a fragment coupling in the key synthetic step toward the carbocyclic framework. Isovaleraldehyde **86** was allowed to react with methyl vinyl ketone **87** in the presence of thiazolium catalyst **82** and triethylamine to yield the dicarbonyl compound **88** in 82% yield (Scheme 12.15). The sequence is completed through a base-promoted

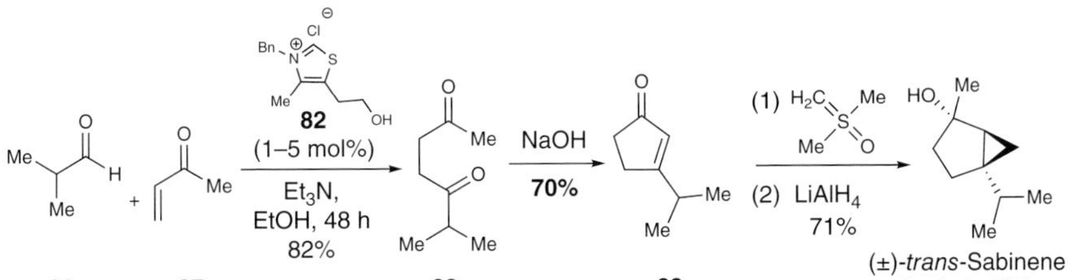

Scheme 12.15 Synthesis of (±)-*trans*-sabinene hydrate.

intramolecular aldol cyclization to afford cyclopentenone **89** followed by a subsequent Corey–Chaykovsky reaction and a lithium aluminum hydride reduction to install both the cyclopropyl moiety and the methyl group with the desired relative stereochemistry, yielding (±)-*trans*-sabinene hydrate (**90**) in 28% yield over four steps. Galopin notes that the starting materials are readily available in bulk, and the catalytic nature and high efficiency of the Stetter reaction employed affords a scalable and economically viable method for the synthesis of **90**.

12.3.2.5 Synthesis of (+)-Monomorine I and Related Natural Products

In 2003, Blechert reported a general strategy toward the synthesis of indolizidine natural products in which a cross-metathesis and reductive cyclization cascade were facilitated by substrates prepared through Stetter reaction fragment couplings. To exemplify the power of this strategy, the indolizidine alkaloid (+)-monomorine I (**99**) isolated from the pharaoh ant *Monomorium pharaonis L.* [107], and (3R, 5S, 9S)-3-ethyl-5-methylindolizidine **100** isolated from the venom of the ant *Solenopsis (Diplorhoptrum) conurata* were chosen as synthetic targets [108]. Blechert reported the successful syntheses of both (+)-monomorine I (**99**) and (3R, 5S, 9S)-3-ethyl-5-methylindolizidine (**100**). This approach features a concise route consisting of a five-step linear sequence utilizing a sequential cross-metathesis and double-reductive cyclization transformations [109]. The synthetic routes commence with intermolecular Stetter reactions catalyzed by *N*-alkyl thiazolium (**94**) between acrolein, protected as 5-norbornen-2-carbalde hyde **93** (through Diels–Alder cyclization with cyclopentadiene), and vinyl ketone **91** or **92** to afford 1,4-diketones **95** and **96** (Scheme 12.16). To prepare the reductive cyclization substrate, a retro-Diels–Alder cycloaddition removes the norbornene moiety to reveal an enone that undergoes an intermolecular cross-metathesis with Cbz-protected (S)-4-penten-2-amine, affording carbamates **97** and **98**. The reductive amination/cyclization of carbamates **97** and **98** yielded (+)-monomorine I (**99**) and (3R, 5S, 9S)-3-ethyl-5-methylindolizidine (**100**) in 35% and 29% overall yield, respectively.

12.3.2.6 Synthesis of Haloperidol

Haloperidol (**103**), also known as Haldol, was developed in 1958 by Paul Janssen and is now a widely used antipsychotic with therapeutic properties believed to arise from its activity as a dopamine receptor D2 antagonist [110–116]. In 2004,

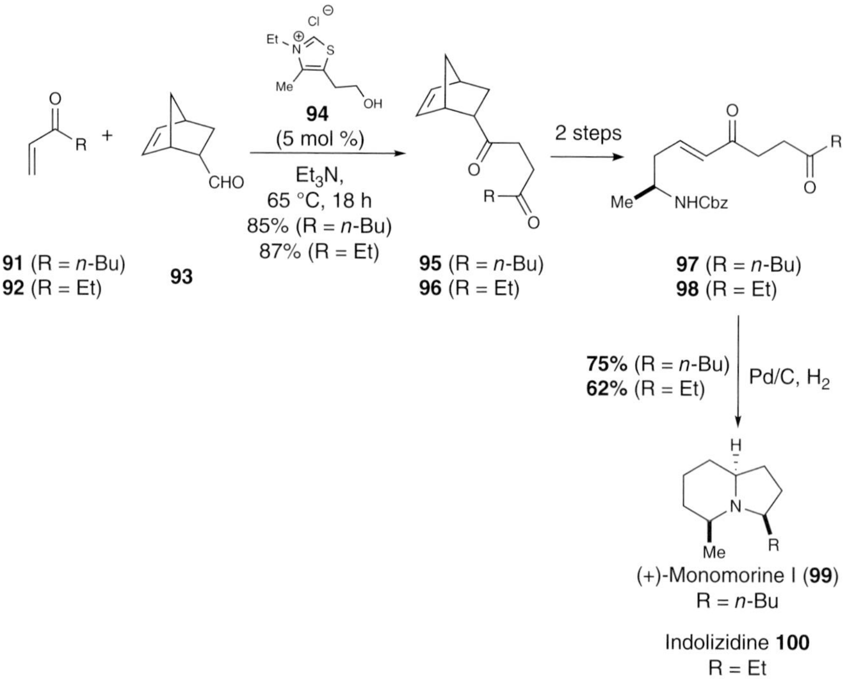

Scheme 12.16 Synthesis of (+)-monomorine I and indolizidine.

during the investigation of ionic liquids as solvents for Stetter reactions, Grée reported a concise synthetic route to haloperidol, which utilizes a Stetter reaction in an Umpolung approach. The 1,4 dicarbonyl compound **102** was prepared through a Stetter reaction between *p*-fluorobenzaldheyde **75** and methyl acrylate (**101**) promoted by thiazolium salt **82** in butylmethylimidazolium tetraflouroborate ionic liquid (Scheme 12.17). In a four-step sequence where ketone **102** was protected as a ketal, methyl ester was selectively reduced to an aldehyde, the piperidine fragment was appended via a reductive amination, and then deprotection of the ketal yielded haloperidol (**103**) in 30% overall yield [117].

12.3.2.7 Synthesis of (−)-Englerin A

In an effort to find new small-molecule treatments for renal cell carcinoma, englerin A (**108**) gained considerable interest because it has shown highly potent and selective cytotoxicities against various renal cancer cell lines at low nanomolar levels [118]. Englerin A (**108**) is characterized as a guaiane-type sesquiterpene natural product isolated from the Tanzanian plant *Phyllanthus engleri* [118]. In 2010, Theodorakis disclosed a succinct asymmetric formal synthesis of englerin A (**108**). Central to the synthetic strategy is a rhodium (II)-catalyzed [4+3] cycloaddition reaction, which couples chiral diazo silyl enol ether **104** and 2-isopropyl-5-methylfuran to establish the oxo-bridging bicyclic framework of the target compound (Scheme 12.18). Further elaboration over three steps yielded intermediate α-hydroxyl enone **105** [119]. A straightforward silyl protection of the free alcohol primed the substrate for a Stetter reaction. The

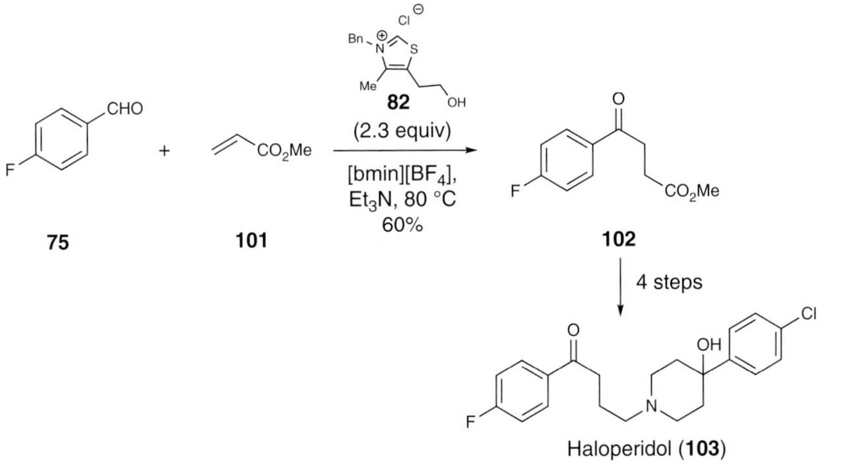

Scheme 12.17 Synthesis of haloperidol (Haldol).

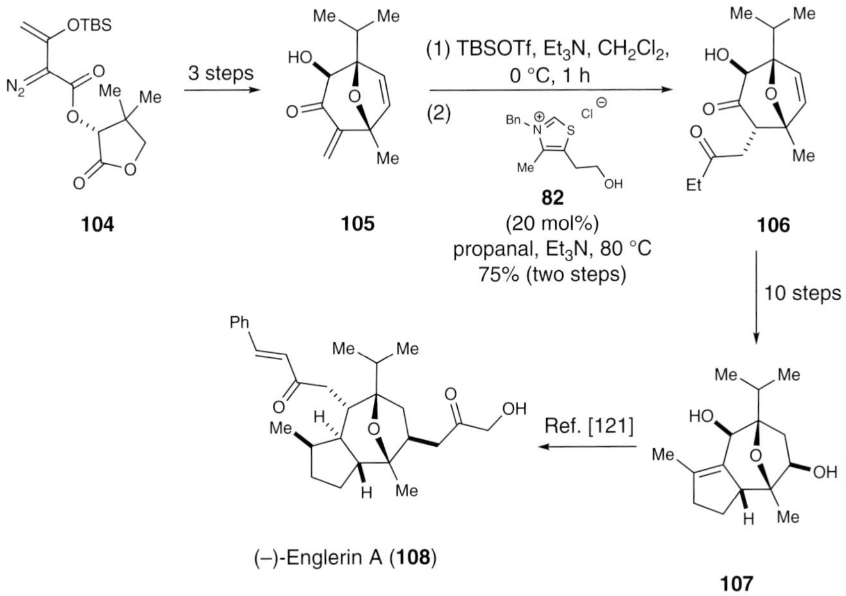

Scheme 12.18 Formal synthesis of (–)-englerin A.

intermediate was allowed to react with propanal in the presence of thiazolium precatalyst **82** and triethylamine, yielding diketone **106** in 75% yield over two steps. This transformation only produces a single diastereomer, suggesting a thermodynamic preference for the observed product **106**. A subsequent 10-step synthetic sequence, which includes substituted cyclopentene formation, yielded the tricyclic framework **107**. Compound **107** is an intermediate in Ma's synthesis of englerin A, completing Theodorakis' formal synthesis of the compound [120].

Theodorakis' route to englerin A was achieved in 15 synthetic steps with 5% overall yield.

12.3.2.8 Synthesis of Piperodione

The pepper family of pantropical plants contains thousands of species each exhibiting a wide range of metabolites with varying biological activities [121]. Plants of the *Piper* genus have gained attention because many of the secondary metabolites isolated from these plants have physiological activities ranging from antineurodegenerative to selective cytotoxicity *in vitro* and *in vivo* [122, 123]. In 2013, piperodione (**111**) was isolated from *Piper retrofractum* in extremely low yield by Fukuyama [124]. This compound is of interest because it enhances the outgrowth of neurites of nerve growth factor (NGF)-mediated PC12 cells at concentrations less than 1 μM [124]. To facilitate the continued investigation of the peculiar biological activity of this compound, Csuk reported an expedient and modular total synthesis of piperodione (**111**) in 2014. Using a Stetter reaction to couple the key fragments was central to their synthetic strategy (Scheme 12.19) [125]. Enone **109** was prepared through aldol condensation from the precursor aryl methyl ketone. Aldehyde **110** was prepared through amidation of diethyl-tartrate followed by oxidative cleavage using sodium periodate. This sequence was used to prepare other aldehydes, enabling analog synthesis in addition to the synthesis of the natural product. The critical reaction of the synthetic sequence is between enone **109** and aldehyde **110**, which is facilitated by the thiazolium precatalyst **82** in the presence of triethylamine to form natural piperodione **111** in 92% yield.

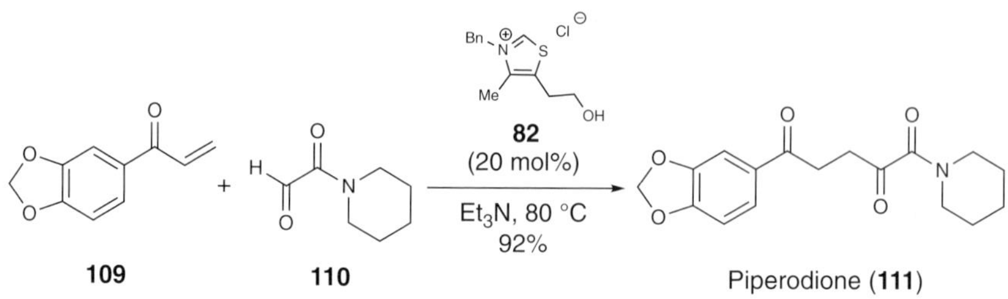

Scheme 12.19 Synthesis of piperodione.

12.4 NHC-homoenolate Equivalents

Homoenolates are broadly defined as reactive species in which the nucleophilic carbon is beta- to a carbonyl or carbonyl equivalent. Classically, these species have been used to produce both linear and cyclic frameworks through anionic metal-mediated reactions. In contrast to the enolate (which is characterized by a nucleophilic carbon being alpha- to a carbonyl or carbonyl equivalent and can be accessed through a large variety of catalytic and stoichiometric reaction conditions), few general methods exist to access homoenolates, especially under mild or catalytic conditions. In general, classic protocols to prepare homoenolate equivalents utilize harsh anionic reaction conditions, thus limiting the use of this potentially powerful operator in synthetic sequences [126–130].

By extending the acyl anion equivalent generated under NHC catalysis conditions through the incorporation of a vinyl group, a homoenolate equivalent can be accessed. Glorius and Bode realized this potential independently in 2004 when it was disclosed that NHC catalysts could prepare γ-butyrolactones through annulation reactions between enals and aldehydes or ketones [131, 132]. An additional report by Scheidt in 2005 also revealed the ability to access NHC-homoenolate equivalents, which were protonated to give saturated esters [133]. These methods to catalytically generate the homoenolate equivalent rely on the use of an NHC precatalyst in the presence of a strong amine or alkoxide base in order to suppress unwanted side reactions and acyl anion pathways. After these preliminary reports on catalytically generated NHC-homoenolate equivalents, the homoenolate synthon has since been harnessed to assemble a myriad of structural motifs as well as important heterocycle classes, solidifying this method as a general strategy to enantioselectively construct the frameworks in Umpolung manner [134–147].

12.4.1 Synthesis of Salinosporamide A

Isolated in 2003 from the marine actinomycete bacteria Salinospora strain CNB-392, Salinosporamide A (**115**) has attracted much attention because of its *in vitro* cytotoxic activity against several tumor cell lines, including HCT-116 human colon carcinoma (IC_{50} = 1 μM). Additional exciting biological activities include 20S proteasome inhibition and proteasomal chymotrypsin-like proteolytic activity (IC_{50} = 1.3 nM) [148]. In 2009, Bode disclosed the formal synthesis of salinosporamide A through an intramolecular homoenolate annulation to yield a γ-butyrolactone with the required functionality toward the target compound [149]. Optimization of the annulation reaction was complicated by many side reactions, such as NHC-enolate pathways and competing intramolecular aldol annulations. Following extensive experimentation, the undesired reaction pathways were minimized and the yield of the desired product was increased. Unfortunately, the diastereoselectivity of the annulation could not be improved even with the use of chiral NHC catalysts. Annulation precursor **112** was prepared in five steps from a previously reported chiral amine (Scheme 12.20). Under the optimized reaction conditions, the desired bicycle **114** was afforded in high yield, albeit with no diastereoselectivity. Bode found that the NHC catalyst prepared from azolium precursor **113** provided the desired diastereomer in slight excess. All attempts to improve this product distribution through further catalyst optimization and modification of reaction parameters failed to improve the reaction outcome (**116–118**). The product of the annulation, bicyclic lactone **114**, intercepts the synthetic sequence toward salinosporamide A reported by Lam [150]. Preparation of bicyclic lactone **114** constitutes a formal synthesis of salinosporamide A **115** with eight additional transformations.

12.4.2 Synthesis of Bakkenolides I, J, and S

Kitahara first reported the initial isolation of the bakkanes in 1968 [151, 152]. Since this initial disclosure, more than 50 related compounds have been isolated

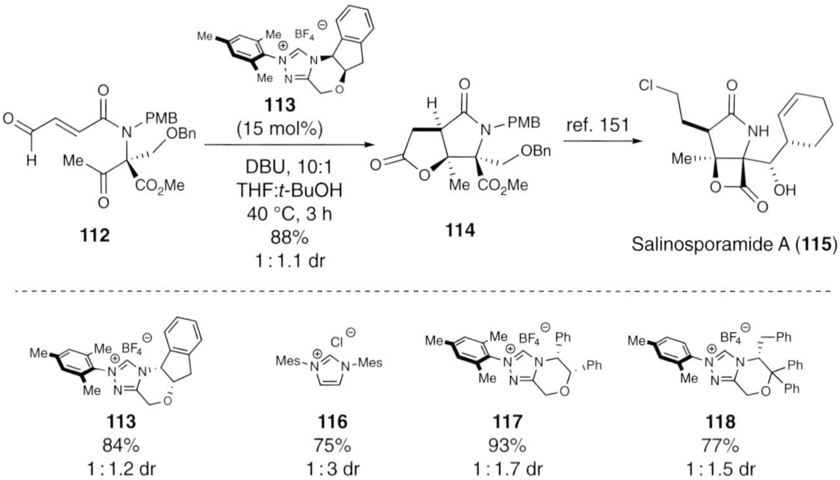

Scheme 12.20 Formal synthesis of salinosporamide A.

[151, 152]. Bakkanes are sesquiterpenoid natural products characterized by a *cis*-fused 5,6-bicyclic core displaying five contiguous stereocenters centered on the hydrindane core [153, 154]. Furthermore, the backbone possesses two quaternary carbons and a spiro γ-butyrolactone [153, 154]. The bakkenolides have gained considerable interest because they have shown biological properties ranging from antifeedant activity to inhibition of platelet aggregation as well as *in vitro* cytotoxicity against several tumor cell lines [155, 156].

In 2010, Scheidt disclosed a highly diastereoselective and enantioselective NHC-catalyzed desymmetrization of 1,3-diketones through NHC-homoenolate chemistry, wherein β-protonation of a homoenolate yields an enolate equivalent [157]. It was identified that this methodology could construct the indane core of bakkenolides. Subsequently, bakkenolides I, J, and S were prepared employing this desymmetrization in the key step of the synthetic strategy (Scheme 12.21). Enal **119** was prepared in two synthetic transformations from the requisite diketone by a palladium-catalyzed Tsuji–Trost allylation and allylic alcohol oxidation. Aldehyde **119** was allowed to react in the presence of chiral triazolium precatalyst **120** and 1 equiv of Hunig's base to furnish lactone **121** in good yield and excellent diastereoselectivity and enantioselectivity. Propargyl ester **122** was prepared through a 13-step sequence from tricyclic lactone **121**. Spiro lactone **123** was prepared through a manganese(III)-acetate-promoted spiro γ-butyrolactonization, followed by a samarium-diiodide-directed ketone reduction. Unfortunately, this reaction sequence furnished the undesired epimer **123** of the bakkenolides. This conundrum was rectified using a flouride-initiated retro-aldol/aldol sequence first described by Deprés [158, 159], allowing for the construction of bakkenolide S (**124**) directly. Bakkenolides I and J could be accessed by acylation of **124** with their corresponding acid chlorides. In summary, bakkenolides S, I, and J were prepared in 20 steps in 2% overall yield. Furthermore, this strategy represented the first complete total synthesis utilizing an NHC-catalyzed homoenolate reaction.

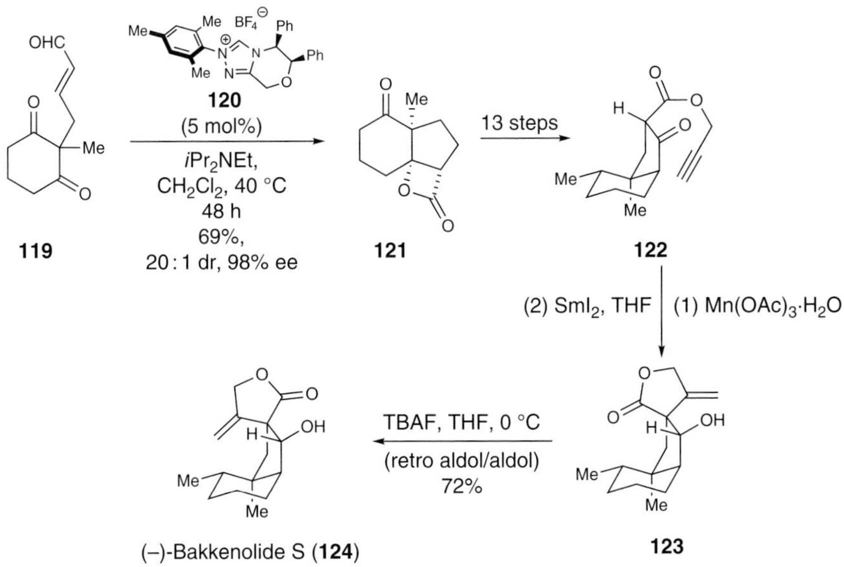

Scheme 12.21 Synthesis of (−)-bakkenolide S.

12.4.3 Synthesis of Maremycin B

The diketopiperazine alkaloid maremycin B (**128**) was isolated by Laatsch in 1995 from the marine *Streptomyces* species B 9173 [160, 161]. The maremycin family of natural products has been of synthetic interest because of their unique structures and anticancer activities [162–164]. In 2012, Scheidt reported a short enantioselective synthesis of maremycin B employing an NHC-homoenolate equivalent addition of a β-alkyl substituted α,β-unsaturated aldehyde (Scheme 12.22) [165]. The key step in the synthetic route was the homoenolate annulation between crotonaldehyde (**126**) and *N*-methyl-isatin (**125**) to form the desired spiro-oxindole **127** in the presence of precatalyst **117** and DBU. The desired lactone **127** was prepared in high yield and diastereoselectivity. The enantioselectivity of 78% excess could be improved to 99% through recrystallization of the product from a mixture of ethyl acetate and hexanes. It is noteworthy that lithium chloride was not utilized in the annulation with crotonaldehyde. This is in contrast to the optimized system with cinnamaldehyde, which required lithium chloride as a Lewis acid for the best selectivity. From the spiro-oxindole **127**, a diastereoselective bromination, amide formation, and then diketopiperazine formation yielded maremycin B (**128**) in five synthetic steps from **125**. This six-step sequence gives the target compound in 17% overall yield, with no utilization of protecting group chemistry.

12.4.4 Synthesis of Clausenamide

In 1984, the γ-lactam natural product (±)-clausenamide (**133**) was isolated by the extraction of the dry-leaf material from *Clausena lansium* [166, 167]. Interest in the extracts of this plant arises from the observation that its teas are used in traditional Chinese folk medicine to treat acute and chronic hepatitis [166, 167].

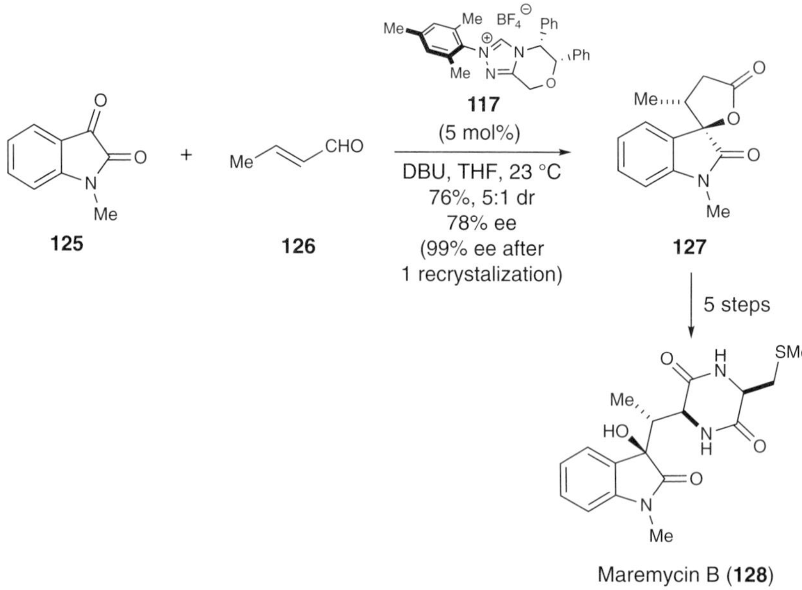

Scheme 12.22 Synthesis of maremycin B.

Biological testing of (±)-clausenamide (**133**) showed hepatoprotective activity against toxins [166, 167]. Further biological investigation has revealed neuroprotective and nootropic effects, making clausenamide an attractive synthetic target [168]. Furthermore, (±)-clausenamide is isolated as a racemic mixture, which is unusual for chiral, naturally occurring compounds, making its enantioselective synthesis a valuable endeavor to organic chemists. In 2012, Bode disclosed an Umpolung formal synthesis of **133**, which utilizes an NHC-homoenolate equivalent in a lactam forming annulation reaction (Scheme 12.23) [169]. The unsaturated protected imine **129** was allowed to react with cinnamaldehyde **130** in the presence of imidazolium precatalyst **116** and DBU to afford phenyl γ-lactam **131**. A deprotection of the sulfonamide, amide methylation, ozonolysis, and addition of phenyl magnesium bromide yielded methyl-lactam **132** in four steps. From intermediate **132**, following the procedure of Hartwig, three further transformations gave clausenamide **133**, which completed the formal synthesis of the natural product [170]. Bode also attempted to optimize an enantioselective variant of the homoenolate annulation; however, it was observed that enantiopurities no greater than 53% ee or yields above 16% could be achieved.

12.4.5 Synthesis of (−)-Paroxetine and (−)-Femoxetine

The selective serotonin reuptake inhibitor (SSRI) (−)-paroxetine (**138**), also known under the trade name Paxil, was discovered in 1970 and released into the clinic in 1992 as a therapeutic to treat psychological disorders ranging from depression to panic disorders [171]. Concomitant to this development, (−)-femoxetine (**142**) was discovered but was not pursued as a clinical agent [172]. Much research has been done to develop efficient and catalytic approaches

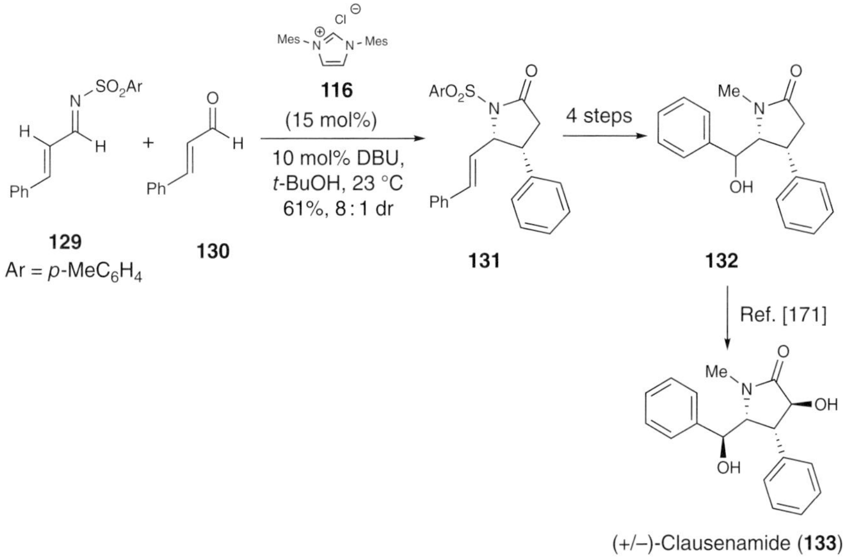

Scheme 12.23 Formal synthesis of (±)-clausenamide.

to these compounds. In 2014, Rovis reported a concise synthetic strategy to these molecules that installed all of the necessary carbons of the molecular framework in a single enantioselective transformation [173]. To accomplish this, Rovis employed an NHC-catalyzed homoenolate addition to a nitro-alkene compound that possessed the necessary functionality of the target molecule. In the synthetic sequence outlined in Scheme 12.24, enal **135** was allowed to react with nitro-alkene **134** in the presence of modified triazolium catalyst **136** to yield piperidone **137** after reduction of the nitro moiety with zinc dust in a mixture of ethanol and acetic acid (Scheme 12.24). A reduction of lactam **137** with lithium aluminum hydride gave (−)-paroxetine (**138**) in good yield. Synthesis of (−)-femoxetine (**142**) commenced with a similar transformation in which enal **140** and nitro-alkene **139** reacted under the same reaction conditions. A methylation and reduction of lactam **141** yielded (−)-femoxetine (**142**). Rovis' synthetic strategy offers a complimentary approach to normal polarity synthetic strategies to these compounds, providing target compounds (−)-paroxetine **138** and (−)-femoxetine **142** in just three and four synthetic steps, respectively.

12.4.6 Synthesis of (*S*)-Baclofen and (*S*)-Rolipram

In 2013, Chi disclosed a new strategy by which NHC catalysts could access homoenolate synthetic operators (Scheme 12.25) [174]. Activated phenolic esters can generate homoenolate equivalents when exposed to chiral NHC precatalysts of type **145** and a large excess of base in a reverse β-protonation pathway. This method greatly simplifies NHC-homoenolate chemistry by eliminating the requirement that enals be used as precursors to access the extended Breslow intermediate. In Chi's study, several reaction systems were developed. Specifically, they applied the addition of homoenolate equivalents to

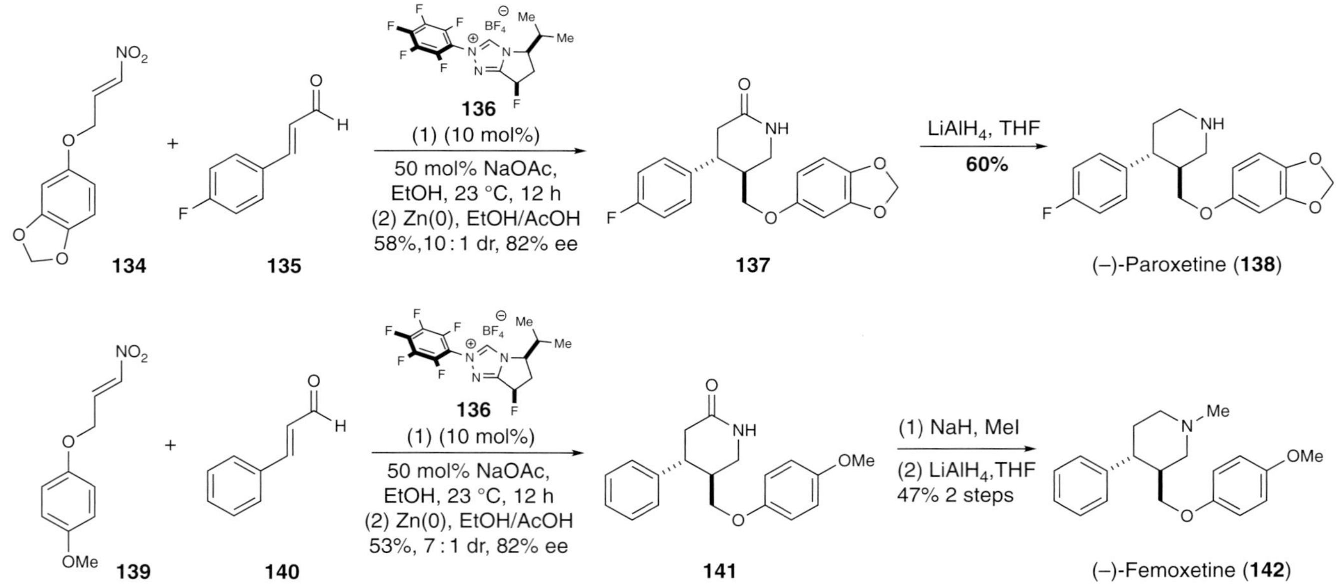

Scheme 12.24 Synthesis of (–)-paroxetine and (–)-femoxetine.

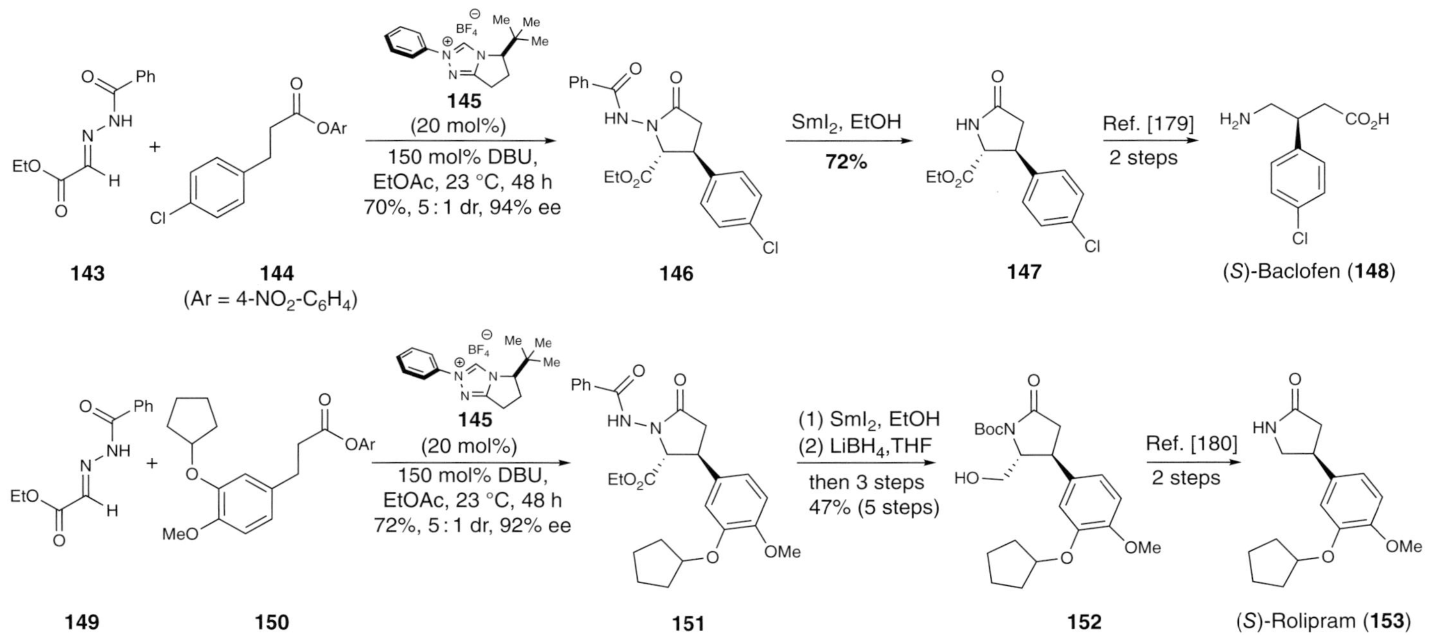

Scheme 12.25 Synthesis of *(S)*-baclofen and *(S)*-rolipram.

acyl hydrazones in the syntheses of two pharmaceutical compounds. The first, *(S)*-baclofen (**148**), is used to treat spasticity and the second, *(S)*-rolipram (**153**), was originally developed as an atypical antidepressant, although it failed in clinical trials because of its narrow therapeutic window [175–177]. Chi's synthesis of *(S)*-baclofen commenced with the addition of the homoenolate equivalent derived from ester **144** and acyl hydrazone **143** to yield lactam **146** in a very good 70% yield and 94% enantiomeric excess. Cleavage of the N–N bond using samarium diiodide yielded aryl lactam **147** in 72% yield. Following the procedure of Oba, *(S)*-baclofen (**148**) could be prepared in two steps, constituting a formal synthesis of the target compound [178]. In a similar manner, Chi allowed ester **150** and acyl hydrazone **149** to react in the presence of precatalyst **145** and DBU to yield the functionalized lactam **151**. Cleavage of the N–N bond followed by selective reduction of the ester and protecting group manipulations yielded the carbamate-protected lactam **152**. Following the procedure of Alvarez-Builla, lactam **152** could be converted into *(S)*-rolipram (**153**) in two steps, completing the formal synthesis of the desired target, exemplifying the power of this new method to access homoenolate equivalents with NHC catalysis [179].

12.4.7 Synthesis of 3-Dehydroxy Secu'amine A

The securinega alkaloids (**154**–**156**) are a collection of secondary metabolites found in the *Euphorbiaceae* plant family [180]. These compounds are structurally unique because they typically contain a bridged, tetracyclic framework possessing an α,β,γ,δ-unsaturated lactone moiety. Several members of the family exhibit noteworthy biological activities, with securinine (**156**) being a potent γ-aminobutyric acid receptor (GABA) antagonist [181]. Beginning with the total synthesis of securinine (**156**) in 1966 by Horii, a significant amount of synthetic research has been conducted toward this family of compounds [182]. Several strategies have been explored to construct the α,β,γ,δ-unsaturated lactone moiety. In 2013, Snyder disclosed a succinct strategy toward this family of molecules, which harnesses an unprecedented homoenolate equivalent to directly access the deceptively difficult unsaturated lactone framework [183]. Starting from Boc-piperidine **157**, a series of alkylations and homologations yielded ynal **158** (Scheme 12.26). Substrate **158** was allowed to react with precatalyst **159** in the presence of DBU and titanium(IV) tetra-isopropoxide to yield tricyclic lactone **160** in one synthetic transformation. The Lewis acid additive as well as slow addition of the substrate was essential for the success of this reaction. The use of the ynal as a allenolate precursor has since been utilized in a related annulation reported by Scheidt [184]. Further elaboration of lactone **160** in a two-step hydroboration/oxidation and alcohol activation sequence proceeded with concomitant carbamate deprotection under the acidic mesylation conditions followed by intramolecular alkylation of the resultant secondary amine, yielding 3-deshydroxy secu'amine A (**161**) in a short order.

12.5 NHC-Catalyzed Aroylation Reactions

In 1990, Miyashita disclosed the NHC-catalyzed nucleophilic aroylation of 4-chloro-1H-pyrazolo[3,4d]pyrimidines [185]. This reaction employs an

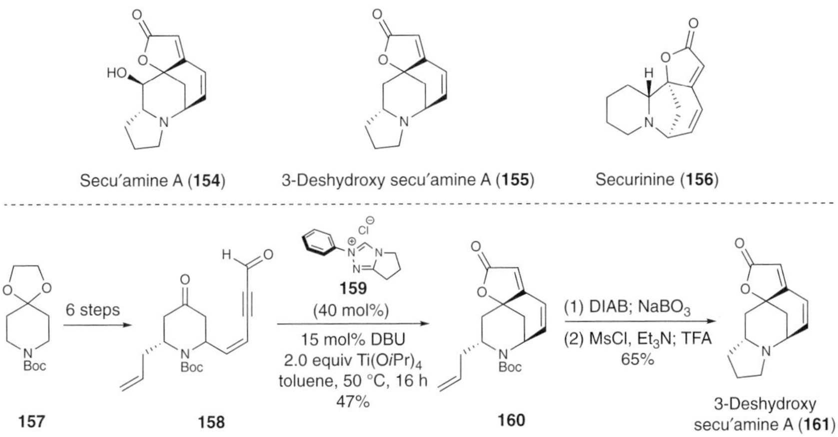

Scheme 12.26 Synthesis of 3-deshydroxy secu'amine A.

NHC-generated acyl anion as the nucleophile in an aromatic substitution reaction. In the previous two decades, Miyashita and Suzuki have expanded the scope of this method into a general strategy for the synthesis of arenyl compounds [186–190]. In 2008, Miyashita and Suzuki disclosed the aroylation of fluorobenzenes using NHC precatalyst concentrations as low as 1 mol%, which eventually instructed a pathway to prepare xanthone natural products as discussed in the following section [191].

12.5.1 Synthesis of Atroviridin

The oxygenated xanthone atroviridin (**166**) was originally isolated from *Garcinia atroviridis* in 1998 [192]. In 2011, Suzuki disclosed the total synthesis of atroviridin through an NHC-catalyzed aroylation strategy to construct its tetracyclic framework (Scheme 12.27) [190]. The benzaldehyde derivative **162**

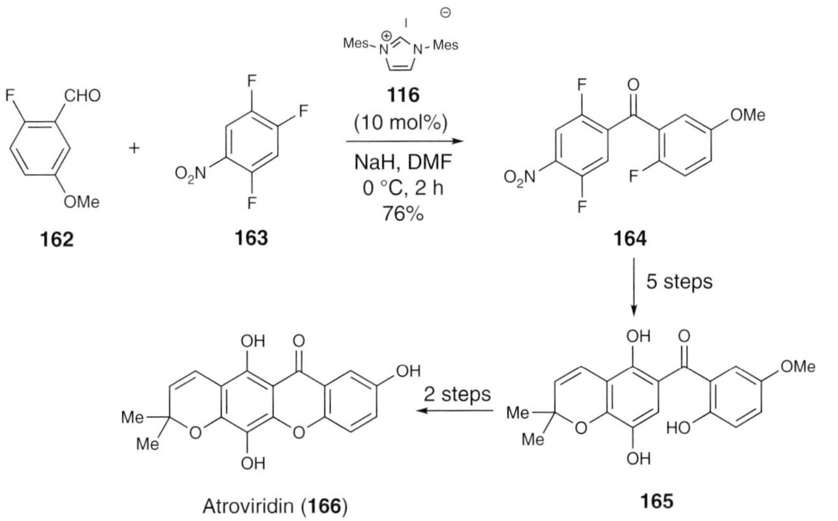

Scheme 12.27 Synthesis of atroviridin.

and poly-functionalized arene **163** were treated with precatalyst **116** and sodium hydride to yield benzophenone **164** in 76% yield. The product was converted into triphenol **165** over five steps by the reduction of the nitro substituent, chromene synthesis, conversion of the aryl fluorides to allyl ethers, and global allyl ether deprotection using catalytic palladium. Triphenol **165** was subjected to a manganese dioxide oxidation to generate an intermediate quinone, which spontaneously underwent an intramolecular oxo-conjugate addition. A final demethylation with boron tribromide yielded atroviridin (**166**) in nine overall steps with a 14% combined yield.

12.6 NHC-Catalyzed Redox and Oxidative Processes

In addition to the previously outlined Umpolung reactions catalyzed by NHCs such as benzoin, Stetter, and homoenolate addition reactions, NHCs can also promote unique redox (internal or external) reactions that terminate with the addition of heteronucleophiles in a formal acylation step [3, 4, 193–198]. Both Scheidt [199] and Studer [200] have reported that activated alcohols and aldehydes can be directly converted to the ester oxidation state via the coupling of metal or organic oxidants with NHC catalysts. Aldehydes activated with α-substitution (e.g. chloride) have also been employed in a redox conversion to the ester oxidation state [201–203]. Furthermore, additional functional groups can be accessed, such as amides [204–207] and acyl azides [208, 209], using redox NHC catalysis. In 2009, a report by Lupton pioneered the use of α,β-unsaturated acyl azoliums as conjugate acceptors, adding to the growing repertoire of NHC-catalyzed redox reactions [210]. Further studies have demonstrated the generality of α,β-unsaturated acyl azoliums to be excellent conjugate acceptors for a variety of transformations [211–215]. These normal polarity reaction manifolds have extended the scope of NHC-catalyzed reactions and continue to inspire new applications. This section includes reports of total and formal syntheses completed using NHC redox catalysis [216, 217].

12.6.1 Redox Esterifications

The internal redox reaction offers a clever means by which to convert aldehydes that contain unsaturation or additional sites of oxidation (e.g. α-substitution or an α,β-epoxide) to be converted into saturated esters with additional functionality. Although the number of reactions that employ this redox esterification strategy is numerous and beyond the scope of this chapter, only a few total syntheses have utilized an NHC-catalyzed internal redox esterification in their synthetic strategy. The following section outlines the implementation of redox esterification reactions.

12.6.1.1 Synthesis of (+)-Davanone

The sesquiterpene natural product davanone (**170**) is the major component of davana oil. It was originally isolated in 1968 from *Artemisia pallens* and showed promise as an antifungal and antispasmodic agent [218, 219]. In 2009, Vosburg disclosed the concise total synthesis of (+)-davanone employing a redox NHC-catalyzed reaction in which an epoxy aldehyde is converted to a β-hydroxy ester in a stereoselective manner (Scheme 12.28) [220]. When treated with thiazolium salt **168** and Hünig's base, the α,β-epoxyaldehyde **167** underwent a redox esterification in which the epoxide opened and the aldehyde was converted to the ester oxidation state to afford the *anti*diastereomeric product **169** in good yield. This intermediate was then converted to the target compound through a palladium-catalyzed allylic *O*-alkylation, conversion of the ester to a Weinreb amide, and a Grignard alkylation to furnish (+)-davanone (**170**) in seven synthetic transformations from geranyl acetate.

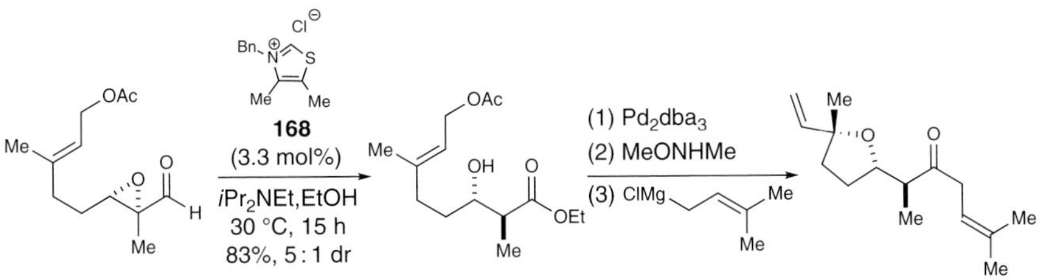

167 **169**

Scheme 12.28 Synthesis of (+)-davanone.

12.6.1.2 Synthesis of Gelsemoxonine

In 2016, Wang reported a dynamic kinetic resolution (DKR) of hemiacetals using an NHC-catalyzed diastereoselective redox esterification reaction employing enals as a precursor to the ester oxidation state [221]. This work was preceded in 2012 by the intramolecular NHC-catalyzed DKR of β-ketoesters reported by Scheidt, in which enals were reacted with epimerizable β-ketoesters through an NHC-enolate pathway [222]. Johnson disclosed in 2014 a DKR reaction catalyzed by NHCs between β-halo-α-ketoesters and aldehydes through a selective benzoin condensation [223]. Furthermore Wang, along with Johnson, have reported DKR reactions involving [3+2] and [4+2] formal cycloadditions [224, 225]. Additional studies have applied NHC catalysis to the DKR of activated esters, as reported by Chi in 2016 [226]. In the context of total synthesis, Wang's 2016 report on the DKR of hemiacetals sought to harness 6-hydroxy-3-pyranones specifically for applications in glyochemical synthesis. This work culminated not only in the formal synthesis of several common sugar frameworks but also the formal synthesis of gelsemoxonine (**176**) (Scheme 12.29) [221]. Gelsemoxonine (**176**) was isolated in 1991 from *Gelsemium elegans*, and it has shown potency as a cytotoxic agent. Consequently, this target has garnered much attention from the synthetic chemistry community [227, 228]. Starting from 6-hydroxypyranone **171** and enal **172**, the DKR reaction promoted by chiral NHC precatalyst **173** in the presence of sodium acetate afforded the acylated derivative **174** in good yield and enantioselectivity. A cyclopropanation followed by an ionic reduction yielded cyclopropyl-ketone **175** in two steps. From this intermediate, the group

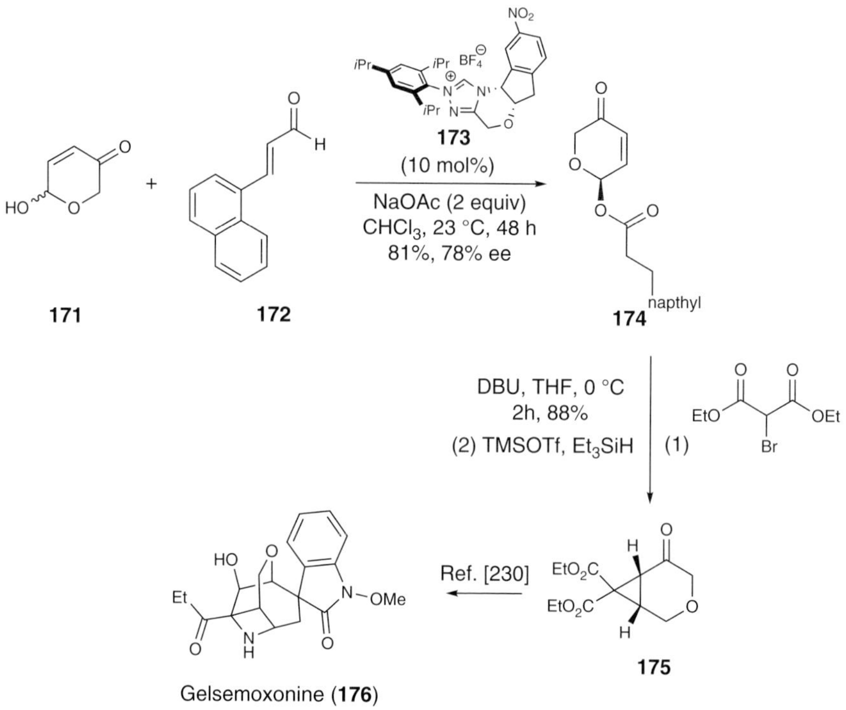

Scheme 12.29 Formal synthesis of gelsemoxonine.

of Fukuyama has prepared gelsemoxonine (**176**) in several synthetic transformations [229]. Access to chiral ketone **175** constitutes the formal synthesis of gelsemoxonine (**176**).

12.6.1.3 Synthesis of (+)-Tanikolide

(+)-Tanikolide (**182**) was isolated and characterized by Gerwick in 1999 from *Lyngbya majuscula*, collected in Madagascar on Tanikeli Island [230]. The lactone natural product gained attention from the synthetic community because it showed activity against *Candida albicans*, the common source of thrush in humans [230]. In 2016, Yakura reported the total synthesis of (+)-tanikolide (**182**), harnessing the power of a rhodium (II)-catalyzed oxonium ylide formation-[2,3]-sigmatropic rearrangement sequence. This transformation was coupled with an NHC-catalyzed stereoselective ring-expanding lactonization to prepare the critical intermediate **181** (Scheme 12.30) [231]. This NHC-catalyzed method to convert aldehydes to lactones was first reported by Gravel in 2009 and serves as an unconventional strategy to generate chiral lactone targets [232]. Yakura's synthesis commences by converting chiral pool starting material malic acid (**177**) to protected alcohol **178** following the procedure of Carreira [233]. Thirteen synthetic transformations, including the rhodium (II)-catalyzed rearrangement, yielded the α-hydroxy-aldehyde substrate **179**, which is primed for the NHC-catalyzed ring-expansion lactonization reaction. Aldehyde **179** was allowed to react with NHC precatalyst **180**, affording lactone **181** in good

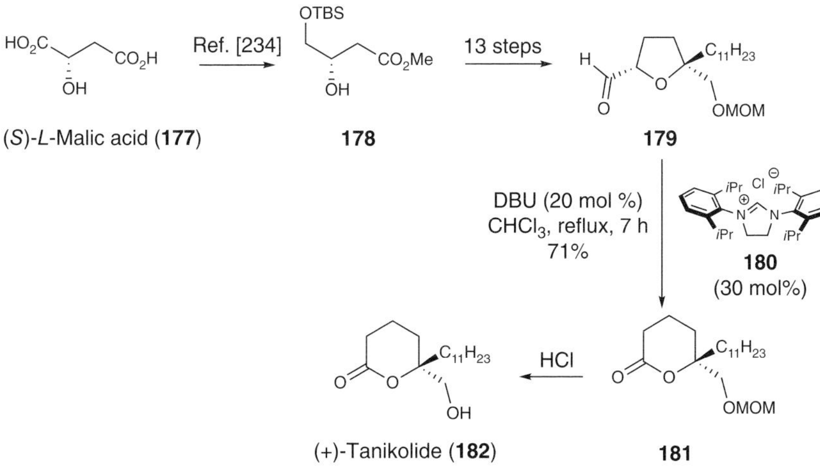

Scheme 12.30 Synthesis of (+)-tanikolide.

yield. Acidic deprotection of the primary hydroxyl group completed the total synthesis of (+)-tanikolide (**182**).

12.6.2 Oxidative Esterification

Classically, the oxidation of aldehydes and alcohols directly to the carboxylic acid oxidation state (in either the free acid or ester form) requires either multistep procedures or the use of harsh reaction conditions [234, 235]. One solution for the direct conversion of aldehydes to esters is the Corey–Gilman oxidation, which utilizes stoichiometric cyanide to mediate this conversion [236]. Using this transformation as an inspiration, Scheidt reported in 2007 the direct conversion of activated alcohols to esters mediated by NHC catalysis via the oxidation of the intermediate acyl anion equivalent with the mild oxidant, manganese (IV) dioxide [198]. Scheidt later extended this method to unactivated alcohols [199]. Many other oxidative esterification methods have been disclosed [237], employing organic oxidants [200], anodic electrochemical oxidations [238], and organohalogen oxidants [239]. This section highlights the application of this facile functional group manipulation in the area of total synthesis.

12.6.2.1 Synthesis of (+)-Dactylolide

Isolated in 2001 by Riccio from the Vanuatu sponge *Dactylospongia* sp, (+)-dactylolide (**188**) shows an activity against several cancer cell lines including leukemia and ovarian cancers. This cytotoxic compound possesses a 20-membered macrolide framework, making it an attractive target for synthesis [240]. In 2012, Kim and Hong disclosed an elegant total synthesis of (+)-dactylolide (**188**), in which an NHC-catalyzed oxidative macrolactonization is used as the key transformation to form the 20-membered macrocyclic core (Scheme 12.31) [241]. To prepare the substrate for macrolactonization, an 11-step sequence was employed starting from 1,3-dithiane-2-ethanol **183** to

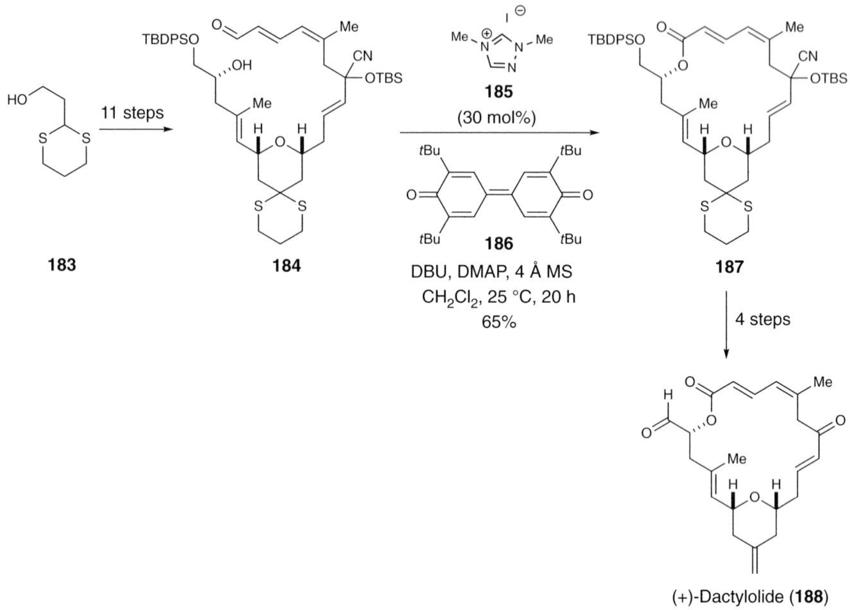

Scheme 12.31 Synthesis of (+)-dactylolide.

yield the substrate ω-hydroxy aldehyde **184**. Dithiane **184** was allowed to react in the presence of triazolium precatalyst **185**, 4-(dimethylamino)pyridine (DMAP), and 3,3′,5,5′-tetra-*tert*-butyldiphenoquinone **186** and promoted the impressive oxidative macrolactonization to yield the desired 20-membered lactone **187** in 65% yield [242]. This protocol combines the conditions previously reported by Scheidt [237] and Studer [200] for oxidative NHC-catalyzed esterification. Furthermore, this macrolactonization is the first example of its kind to use an NHC-catalyzed oxidation to append ω-hydroxy aldehydes for the formation of macrocycles. From macrocycle **187**, an additional four-synthetic transformations are required to yield (+)-dactylolide (**188**) in 16 total steps and a 2% yield overall.

12.6.2.2 Synthesis of Cyanolide A and Clavosolide A

Cyanolide A (**191**) was isolated and characterized in 2010 by Gerwick using a molluscicidal bioassay to screen marine cyanobacterial extracts to identify potential agents to control the animal vector of the *Schistosoma* flatworm parasite [243]. Schistosomiasis is a chronic and debilitating disease caused by this parasitic flatworm, and its rate of infection is second only to malaria, which necessitates the development of new methods to curb rates of infection [244]. Cyanolide A (**191)** is defined by a 16-membered macrocycle as well as two appended modified xyloses [243]. Hong reported a succinct total synthesis of Cyanolide A (**191**) in 2010 in which the dimeric nature of the target was exploited in designing a convergent synthetic strategy (Scheme 12.32) [245]. Central to their successful route is an oxidative oxo-conjugate addition/esterification sequence mediated by NHC catalysis. Allylic alcohol **189** (prepared via a dithiane addition to the required complementary epoxide) was oxidized with

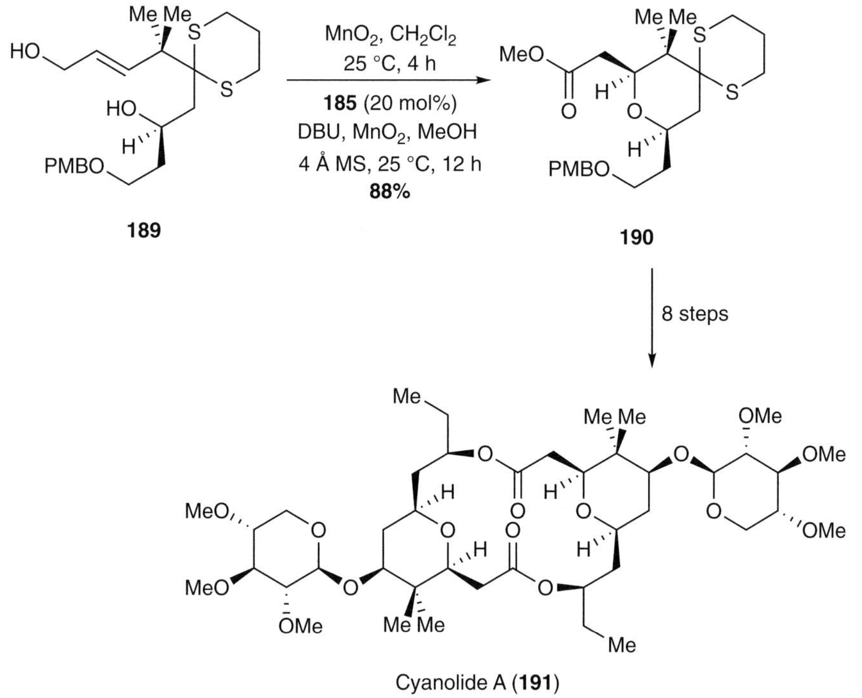

Scheme 12.32 Synthesis of cyanolide A.

manganese (IV) dioxide, followed by the addition of triazolium **185** in the presence of additional oxidant. These reaction conditions promote tetrahydropyran formation as well as conversion of the allylic alcohol to the saturated methyl ester **190** in a one-pot protocol in an excellent 88% yield. Eight additional synthetic transformations, including dimerization and glycosylation, afforded the target compound, cyanolide A (**191**).

Using a similar strategy to the synthesis of cyanolide A (**191**), Hong reported the total synthesis of the structurally related clavosolide A (**194**) in 2015 (Scheme 12.33) [246]. The 16-membered macrolide clavosolide A (**194**) was isolated from *Myriastra clavosa* in the Philippines by Faulkner in 2002 [247]. Allylic alcohol **192** was prepared via a dithiane coupling with the requisite partner epoxide to yield the secondary alcohol substrate **192** [248, 249]. Alcohol **192** was exposed to the oxo-conjugate addition/oxidative esterification protocol, analogous to the procedure used in the synthesis of cyanolide A (**191**), to yield saturated ester **193** in an excellent 84% yield in a single direct sequence. This oxo-conjugate addition is of note because the tetrahydropyran heterocycle was formed as a single diastereomer. Eight additional synthetic steps were necessary to arrive at the target compound, clavosolide A (**194**).

12.6.2.3 Synthesis of Bryostatin 7

Bryostatin 7 (**201**) was isolated from *Bugula neritina* among 20 related polyacetate macrolides characterized from bryozoan [250, 251]. The complex

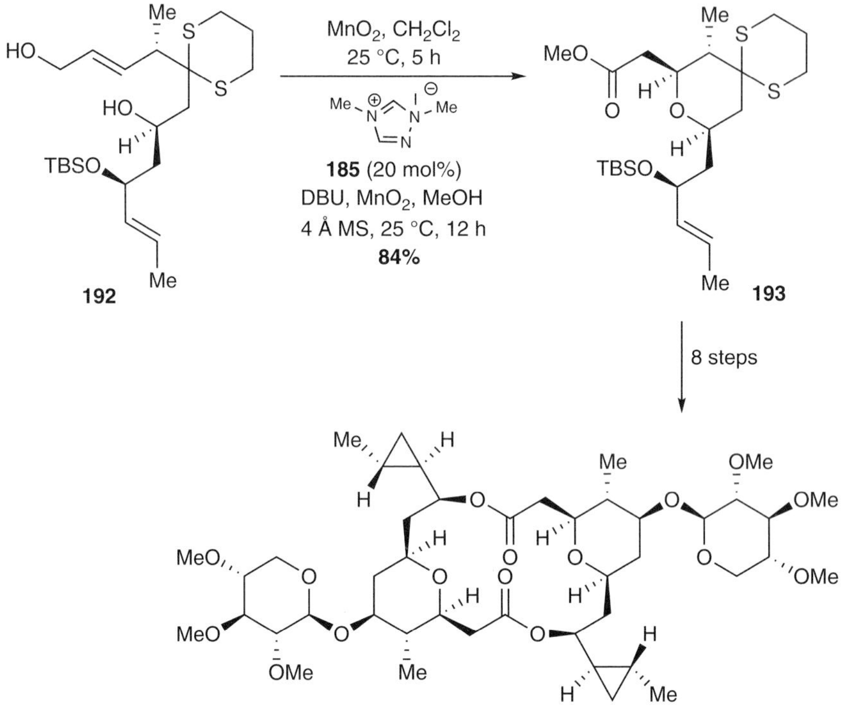

Clavosolide A (**194**)

Scheme 12.33 Synthesis of clavosolide A.

polyketide macrocyclic structure of these natural products has been the focus of intense synthetic research. Additionally, biological activities ranging from synergistic behavior with chemotherapeutics for cancer treatment [252] to the reversal of stroke damage [253], as well as the activity against Alzheimer's disease [254], make the bryostatins an attractive family of synthetic targets. In 2011, Krische reported the brief and elegant total synthesis of bryostatin 7 (**201**) in 20 steps (longest linear sequence) and 36 steps overall (Scheme 12.34) [255]. Central to their successful synthetic strategy are sequential iridium-catalyzed allylations of aldehydes and alcohols using transfer-hydrogenative coupling reactions developed in their laboratory [256–259]. Krische's strategy avoids unnecessary functional group manipulations and multiple protection/deprotection sequences, which can greatly reduce the efficiency of more traditional synthetic approaches toward polyketide natural products. Starting from 1,3-propanediol **195**, the polyfunctionalized tetrahydropyran fragment **196** was prepared in 10 steps utilizing three iridium-catalyzed transfer-hydrogenative C–C bond forming reactions. Starting from isovaleraldehyde and allylic cyanide **197**, polyfunctionalized linear fragment **198** was prepared in 10 steps in the longest linear sequence. Fragment **196** and fragment **198** were coupled through a Keck–Yu annulation to form the required A-ring tetrahydropyran ring [260, 261]. Six additional transformations yielded the macrocyclic enal **199** (full compound not

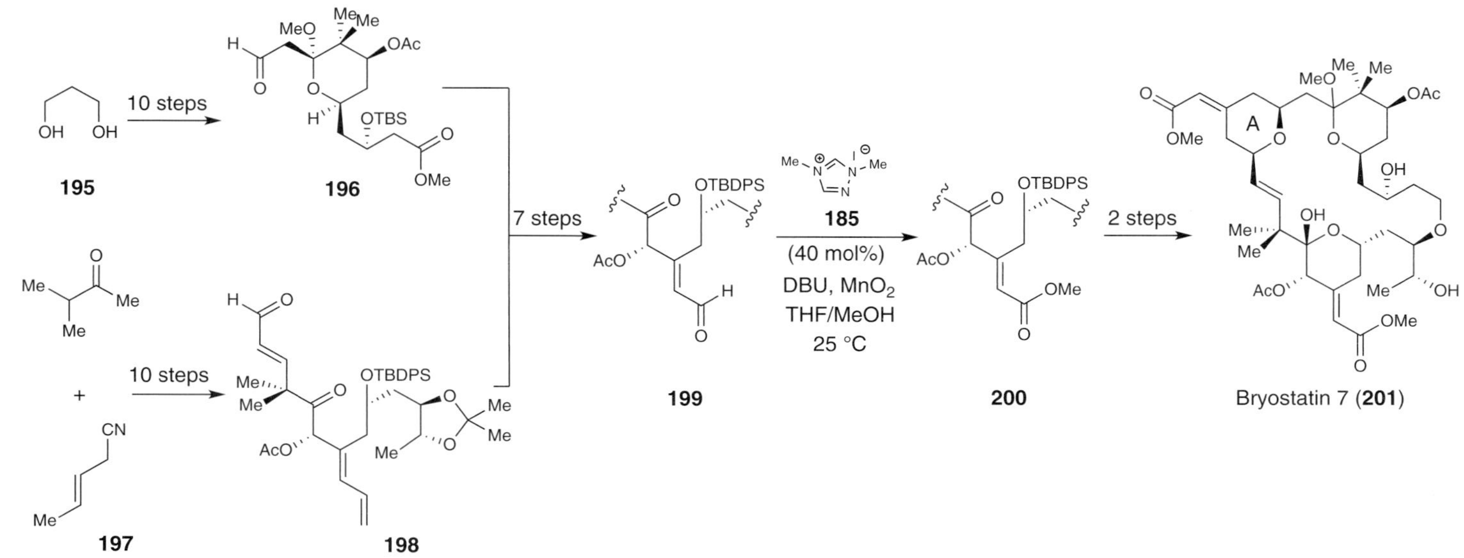

Scheme 12.34 Synthesis of bryostatin 7.

shown), which was oxidized to the enoate using an NHC-catalyzed oxidative esterification with triazolium precatalyst **185** to give ester **200** [199]. Two additional steps completed their synthesis of bryostatin 7 (**201**).

12.6.3 Carbon–Carbon Bond Formation

In addition to accessing acyl anion equivalents, enolates, and homoenolate equivalents, NHCs are able to activate α,β-unsaturated ester equivalents to undergo conjugate addition by nucleophiles. This reactivity has been exploited in several methods for C–C bond formation by Lupton, Studer, and others but has scarcely been applied to the synthesis of natural products. One example of this mode of activation is known and is outlined below.

12.6.3.1 Synthesis of (−)-7-Deoxyloganin

(−)-7-Deoxyloganin (**206**) is a member of the monoterpenoid iridoid family of natural products, and it is considered one of the only non-nitrogen containing alkaloid compounds because of the biosynthetic relationship between (−)-7-deoxyloganin (**206**) and the majority of known plant secondary metabolite alkaloids [262–264]. The majority of iridoid natural products possess a *cis*-fused [6.5.0] bicyclic framework in which the *cis*-relationship is maintained at C-5 and C-9 of the cyclopenta-pyran core (Scheme 12.35) [262–264]. This family of natural products is characterized by broad biological activities, and many studies have been undertaken to show their role as key intermediates in indole and alkaloid natural product biosynthesis [265–267]. In 2010 and 2011, Lupton disclosed the total synthesis of (−)-7-deoxyloganin (**206**) by harnessing a unique NHC-catalyzed annulation to establish the *cis*-fused cyclopenta-pyranone core. In Lupton's original synthesis published in 2010, α,β-unsaturated enol ester substrate **203** was prepared from 2,5-dimethoxytetrahydrofuran in 13 synthetic transformations. Substrate **203** was treated with the preformed stable carbene **204** to directly facilitate a conjugate addition to the α,β-unsaturated enol of **203** to give dihydropyranone **205** [268]. Although NHCs are known to activate methyl esters for transesterifications, this specific transformation yielded the desired product **205** without significant decomposition of its methyl ester [269]. The temperature profile of the NHC-catalyzed annulation (–78 to 23 °C) was essential for success, as it was proposed that the rearrangement may proceed through a tetrahedral intermediate at the lactone carbonyl carbon, resulting in the observed diastereoselectivity. A set of clever crossover experiments revealed that the rearrangement probably proceeds through an intramolecular pathway [270]. From bicycle **205**, four synthetic transformations yielded (−)-7-deoxyloganin (**206**) in 18 steps with a 0.7% overall yield. In 2011, Lupton reported an improved synthesis of substrate **203** from chiral starting material (*S*)-citronellal (**202**) in only five steps, which increased the efficiency of their route to 4% overall yield in just 10 steps (Scheme 12.35) [270].

12.6.4 Brønsted Base Catalysis

In addition to acting as Lewis base catalysts, free NHCs have been shown to possess a modest range of Brønsted basic reactivities. Beginning with the

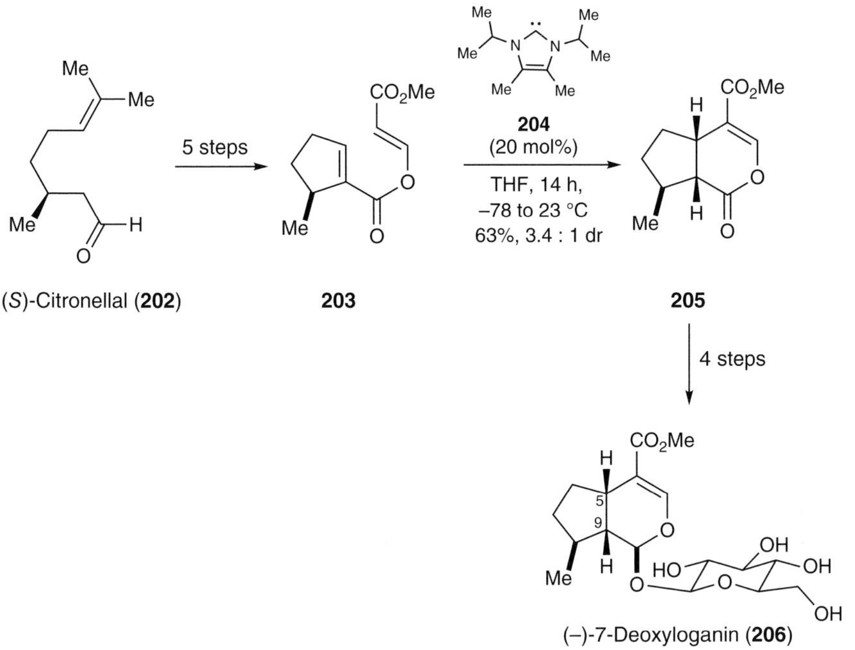

Scheme 12.35 Synthesis of (–)-7-deoxyloganin.

observation that free NHCs promote transesterification reactions in a catalytic manner, several additional Brønsted basic NHC-promoted reactions have been identified [271–273]. Free NHCs have been used to initiate polymerizations [274], acylate symmetrical diols in an asymmetric manner [275], promote the oxo-conjugate addition reaction of alcohols to enones [276], and promote the asymmetric addition of malonate nucleophiles to nitroalkenes [277]. Despite these advances, only one example of Brønsted basic NHC catalysis has been reported in total synthesis thus far, and it is summarized below.

12.6.4.1 Synthesis of (1R)-Suberosanone

Suberosanone (**213**) was isolated in 2000 by Sheu et al. from the gorgonian *Isis hippuris* [278]. It garnered the attention for its unique polycyclic architecture as well as its cytotoxic activity against P-388, A549, and HT-29 cancer cell lines [278]. In 2016, Rodriguez and Coquerel reported the organocatalytic enantioselective total synthesis of (1R)-suberosanone (**213**) using a tandem secondary amine-catalyzed Michael addition/Brønsted base-catalyzed aldol annulation to form the carbocyclic core of the target compound (Scheme 12.36) [279]. β-ketoester **207** and crotonaldehyde (**208**) were treated first with secondary amine catalyst **209**, followed by the addition of free carbene **211** buffered with acetic acid, to promote the desired aldol annulation to produce compound **212**. Coquerel found that using free carbene alone without acetic acid favored a retro-Michael addition, which leads to the racemization of intermediate **210**. From aldol product **212**, 14 synthetic steps were required to finish the synthesis of (1*R*)-suberosanone (**213**) in 9% overall yield.

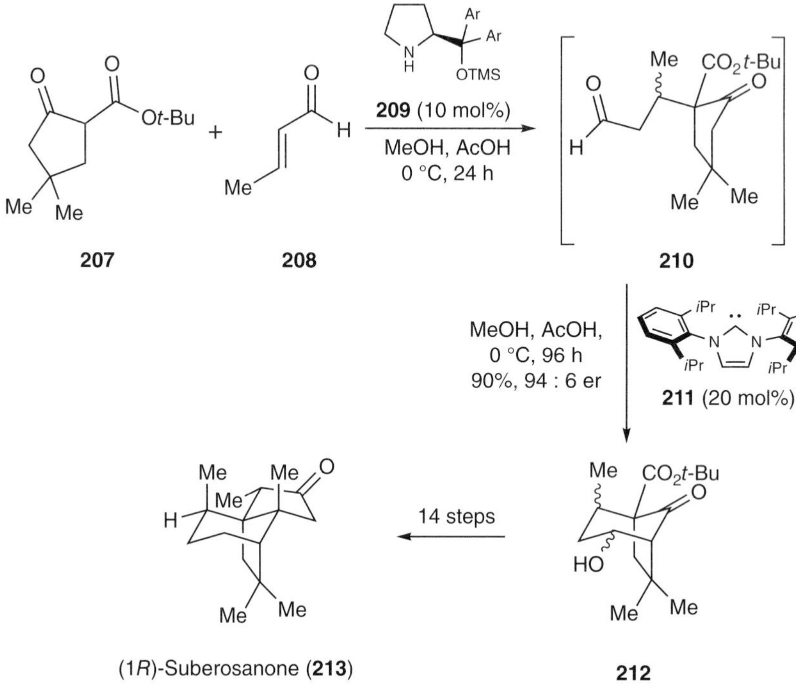

Scheme 12.36 Synthesis of (1*R*)-suberosanone.

12.7 Summary

NHC catalysis has emerged as a powerful and versatile tool to construct new bonds with a high degree of stereoselectivity through both Umpolung and normal polarity-based reactions. With a continuously growing repertoire of powerful NHC-catalyzed chemical transformations becoming available to synthetic chemists, the potential of NHC catalysis to enable the synthesis of complex targets of biological interest has only just started to be realized. As new innovations to the scope and variety of new reactive intermediates that can be accessed using NHC catalysis continue to emerge, so will new strategies that leverage these advances for complex chemical synthesis.

References

1 Enders, D. and Balensiefer, T. (2004). Nucleophilic carbenes in asymmetric organocatalysis. *Acc. Chem. Res.* 37 (8): 534–541.
2 Johnson, J.S. (2004). Catalyzed reactions of acyl anion equivalents. *Angew. Chem. Int. Ed.* 43 (11): 1326–1328.
3 Enders, D., Niemeier, O., and Henseler, A. (2007). Organocatalysis by *N*-heterocyclic, carbenes. *Chem. Rev.* 107 (12): 5606–5655.

4 Marion, N., Diez-Gonzalez, S., and Nolan, I.P. (2007). N-heterocyclic carbenes as organocatalysts. *Angew. Chem. Int. Ed.* 46 (17): 2988–3000.
5 Enders, D. and Narine, A.A. (2008). Lessons from nature: biomimetic organocatalytic carbon–carbon bond formations. *J. Org. Chem.* 73 (20): 7857–7870.
6 de Maria, P.D. and Shanmuganathan, S. (2011). Umpolung catalysis in Benzoin-type and Stetter-type reactions: from enzymatic performances to bio-mimetic organocatalytic concepts. *Curr. Org. Chem.* 15 (13): 2083–2097.
7 Grossmann, A. and Enders, D. (2012). N-heterocyclic carbene catalyzed domino reactions. *Angew. Chem. Int. Ed.* 51 (2): 314–325.
8 Bugaut, X. and Glorius, F. (2012). Organocatalytic umpolung: N-heterocyclic carbenes and beyond. *Chem. Soc. Rev.* 41 (9): 3511–3522.
9 Wöhler, F. and Liebig, J. (1832). Untersuchungen über das Radikal der Benzoesäure. *Ann. Pharmacother.* 3: 249–282.
10 Ugai, T., Tanaka, S., and Dokawa, S. (1943). *J. Pharm.* 63: 269–300. https://www.jstage.jst.go.jp/article/yakushi1881/63/6/63_6_296/_article/-char/ja/.
11 Dunkelmann, P., Kolter-Jung, D., Nitsche, A. et al. (2002). Development of a donor-acceptor concept for enzymatic cross-coupling reactions of aldehydes: the first asymmetric cross-benzoin condensation. *J. Am. Chem. Soc.* 124 (41): 12084–12085.
12 Enders, D., Grossmann, A., Huang, H., and Raabe, G. (2011). Dual secondary amine/N-heterocyclic carbene catalysis in the asymmetric Michael/cross-benzoin cascade reaction of β-oxo sulfones with enals. *Eur. J. Org. Chem.* 4 (23): 4298–4301.
13 O'Toole, S.E., Rose, C.A., Gundala, S. et al. (2011). Highly chemoselective direct crossed aliphatic-aromatic acyloin condensations with triazolium-derived carbene catalysts. *J. Org. Chem.* 76 (2): 347–357.
14 Sheehan, J. and Hunnemann, D.H. (1966). Homogeneous asymmetric catalysis. *J. Am. Chem. Soc.* 88: 3666–3667.
15 Sheehan, J. and Hara, T. (1974). Asymmetric thiazolium salt catalysis of the benzoin condensation. *J. Org. Chem.* 39: 1196–1199.
16 Enders, D., Breuer, K., and Teles, J.H. (1996). A novel asymmetric benzoin reaction catalyzed by a chiral triazolium salt – preliminary communication. *Helv. Chim. Acta* 79 (4): 1217–1221.
17 Enders, D. and Kallfass, U. (2002). An efficient nucleophilic carbene catalyst for the asymmetric benzoin condensation. *Angew. Chem. Int. Ed.* 41 (10): 1743–1745.
18 Enders, D., Niemeier, O., and Balensiefer, T. (2006). Asymmetric intramolecular crossed-benzoin reactions by N-heterocyclic carbene catalysis. *Angew. Chem. Int. Ed.* 45 (9): 1463–1467.
19 Tagaki, W., Tamura, Y., and Yano, Y. (1980). Asymmetric benzoin condensation catalyzed by optically-active thiazolium salts in micellar 2-phase media. *Bull. Chem. Soc. Jpn.* 53 (2): 478–480.
20 Takikawa, H., Hachisu, Y., Bode, J.W., and Suzuki, K. (2006). Catalytic enantioselective crossed aldehyde-ketone benzoin cyclization. *Angew. Chem. Int. Ed.* 45 (21): 3492–3494.

21 Chiang, P.C., Kaeobamrung, J., and Bode, J.W. (2007). Enantioselective, cyclopentene-forming annulations via NHC-catalyzed benzoin-oxy-cope reactions. *J. Am. Chem. Soc.* 129 (12): 3520–3521.

22 Baragwanath, L., Rose, C.A., Zeitler, K., and Connon, S.J. (2009). Highly enantioselective benzoin condensation reactions involving a bifunctional protic pentafluorophenyl-substituted triazolium precatalyst. *J. Org. Chem.* 74 (23): 9214–9217.

23 Oyama, H., Sassa, T., and Ikeda, M. (1978). Structures of new plant-growth inhibitors, *trans*-resorcylide and *cis*-resorcylide. *Agr Biol Chem Tokyo.* 42 (12): 2407–2409.

24 Mennen, S.M. and Miller, S.J. (2007). Development of a bio-inspired acyl-anion equivalent macrocyclization and synthesis of a trans-resorcylide precursor. *J. Org. Chem.* 72 (14): 5260–5269.

25 Couladouros, E.A., Mihou, A.P., and Bouzas, E.A. (2004). First total synthesis of trans- and *cis*-resorcylide: remarkable hydrogen-bond-controlled, stereospecific ring-closing metathesis. *Org. Lett.* 6 (6): 977–980.

26 Siddalah, V., Rao, C.V., Venkateswarlu, S. et al. (2006). Synthesis, stereochemical assignments, and biological activities of homoisoflavonoids. *Bioorg. Med. Chem.* 14 (8): 2545–2551.

27 Saitoh, T., Sakashita, S., Nakata, H. et al. (1986). 3-Benzylchroman derivatives related to Brazilian from Sappan Lignum. *Chem. Pharm. Bull.* 34 (6): 2506–2511.

28 Nguyen, M.T.T., Awale, S., Tezuka, Y. et al. (2005). Xanthine oxidase inhibitors from the heartwood of Vietnamese Caesalpinia sappan. *Chem. Pharm. Bull.* 53 (8): 984–988.

29 Takikawa, H. and Suzuki, K. (2007). Modified chiral triazolium salts for enantioselective benzoin cyclization of enolizable keto-aldehydes: synthesis of (+)-sappanone B. *Org. Lett.* 9 (14): 2713–2716.

30 Hata, K., Kozawa, M., and Baba, K. (1976). Structure of cassialoin, a new anthrone C-glycoside from heartwood of Cassia-garrettiana Craib. *Chem. Pharm. Bull.* 24 (7): 1688–1689.

31 Hata, K., Baba, K., and Kozawa, M. (1978). Chemical studies on the heartwood of Cassia-garrettiana Craib. 1. Anthraquinones including cassialoin, a new anthrone C-glycoside. *Chem. Pharm. Bull.* 26 (12): 3792–3797.

32 Krenn, L., Pradhan, R., Presser, A. et al. (2004). Anthrone C-glucosides from Rheum emodi. *Chem. Pharm. Bull.* 52 (4): 391–393.

33 Zhu, J.J., Zhang, C.F., Zhang, M.A. et al. (2010). Separation and identification of three epimeric pairs of new C-glucosyl anthrones from *Rumex dentatus* by on-line high performance liquid chromatography-circular dichroism analysis. *J. Chromatogr. A* 1217 (33): 5384–5388.

34 Kimura, Y., Sumiyoshi, M., Taniguchi, M., and Baba, K. (2008). Antitumor and antimetastatic actions of anthrone-C-glucoside, cassialoin isolated from Cassia garrettiana heartwood in colon 26-bearing mice. *Cancer Sci.* 99 (11): 2336–2348.

35 Koyama, Y., Yamaguchi, R., and Suzuki, K. (2008). Total synthesis and structure assignment of the anthrone C-glycoside cassialoin. *Angew. Chem. Int. Ed.* 47 (6): 1084–1087.

36 Ito, S., Matsuya, T., Omura, S. et al. (1970). A new antibiotic, kinamycin. *J. Antibiot.* 23 (6): 315–317.

37 Hata, T., Omura, S., Iwai, Y. et al. (1971). New antibiotic, kinamycin – fermentation, isolation, purification and properties. *J. Antibiot.* 24 (6): 353–359.

38 Furusaki, A., Watanabe, T., Hata, T. et al. (1972). Crystal and molecular structure of kinamycin-C *p*-bromobenzoate. *Israel J. Chem.* 10 (2): 173–183.

39 Omura, S., Nakagawa, A., Yamada, H. et al. (1973). Structures and biological properties of kanamycin a, B, C, and D. *Chem. Pharm. Bull.* 21 (5): 931–940.

40 He, H.Y., Ding, W.D., Bernan, V.S. et al. (2001). Lomaiviticins a and B, potent antitumor antibiotics from Micromonospora lomaivitiensis. *J. Am. Chem. Soc.* 123 (22): 5362–5363.

41 Nicolaou, K.C., Li, H.M., Nold, A.L. et al. (2007). Total synthesis of kinamycins C, F, and J. *J. Am. Chem. Soc.* 129 (34): 10356–10357.

42 Nicolaou, K.C., Nold, A.L., and Li, H.M. (2009). Synthesis of the monomeric unit of the lomaiviticin aglycon. *Angew. Chem. Int. Ed.* 48 (32): 5860–5863.

43 Shigemori, H., Komatsu, K., Mikami, Y., and Kobayashi, J. (1999). Seragakinone a, a new pentacyclic metabolite from a marine-derived fungus. *Tetrahedron* 55 (52): 14925–14930.

44 Komatsu, K., Shigemori, H., Shiro, M., and Kobayashi, J. (2000). Revised stereochemistry and biosynthesis of seragakinone A. *Tetrahedron* 56 (45): 8841–8844.

45 Takada, A., Hashimoto, Y., Takikawa, H. et al. (2011). Total synthesis and absolute stereochemistry of seragakinone A. *Angew. Chem. Int. Ed.* 50 (10): 2297–2301.

46 Thomson, J.E., Rix, K., and Smith, A.D. (2006). Efficient *N*-heterocyclic carbene-catalyzed O- to C-acyl transfer. *Org. Lett.* 8 (17): 3785–3788.

47 Zhou, H., Li, Y., and Tang, Y. (2010). Cyclization of aromatic polyketides from bacteria and fungi. *Nat. Prod. Rep.* 27 (6): 839–868.

48 Takikawa, H., Ishikawa, Y., Yoshinaga, Y. et al. (2016). Pleospdione, a tricyclic natural product with dense oxygenation at the A-ring: total synthesis and incongruity of the originally assigned structure and its C3-epimer. *Bull. Chem. Soc. Jpn.* 89 (8): 941–954.

49 Ge, H.M., Song, Y.C., Shan, C.Y. et al. (2005). New and cytotoxic anthraquinones from Pleospora sp. IFB-E006, an endophytic fungus in *Imperata cylindrica. Planta Med.* 71 (11): 1063–1065.

50 Sha, C.-K., Huang, S.-J., and Zhan, Z.-P. (2002). Anionic cyclization approach toward perhydrobenzofuranone: stereocontrolled synthesis of the hexahydrobenzofuran subunit of avermectin. *J. Org. Chem.* 67 (3): 831–836.

51 Brown, N.W., Marmelstein, A.M., and Fiedler, D. (2016). Chemical tools for interrogating inositol pyrophosphate structure and function. *Chem. Soc. Rev.* 45 (22): 6311–6326.

52 Larner, J. (2002). D-chiro-inositol – its functional role in insulin action and its deficit in insulin resistance. *Int. J. Exp. Diabetes Res.* 3 (1): 47–60.

53 Michell, R.H. (2008). Inositol derivatives: evolution and functions. *Nat. Rev. Mol. Cell Biol.* 9 (2): 151–161.

54 Kang, B., Sutou, T., Wang, Y. et al. (2015). N-heterocyclic carbene-catalyzed benzoin strategy for divergent synthesis of cyclitol derivatives from alditols. *Adv. Synth. Catal.* 357 (1): 131–147.

55 Carpintero, M., Fernández-Mayoralas, A., and Jaramillo, C. (1997). Protecting group-directed, diastereoselective samarium diiodide-promoted carbocyclization: application to the synthesis of cyclitols. *J. Org. Chem.* 62 (7): 1916–1917.

56 Schnermann, M.J. and Shenvi, R.A. (2015). Syntheses and biological studies of marine terpenoids derived from inorganic cyanide. *Nat. Prod. Rep.* 32 (4): 543–577.

57 Emsermann, J., Kauhl, U., and Opatz, T. (2016). Marine isonitriles and their related compounds. *Mar. Drugs* 14 (1): 16–91.

58 Angerhofer, C.K., Pezzuto, J.M., König, G.M. et al. (1992). Antimalarial activity of sesquiterpenes from the marine sponge Acanthella klethra. *J. Nat. Prod.* 55 (12): 1787–1789.

59 Newton, C.G., Drew, S.L., Lawrence, A.L. et al. (2015). Pseudopterosin synthesis from a chiral cross-conjugated hydrocarbon through a series of cycloadditions. *Nat. Chem.* 7 (1): 82–86.

60 Pronin, S.V. and Shenvi, R.A. (2012). Synthesis of a potent antimalarial amphilectene. *J. Am. Chem. Soc.* 134 (48): 19604–19606.

61 Lu, H.-H., Pronin, S.V., Antonova-Koch, Y. et al. (2016). Synthesis of (+)-7,20-diisocyanoadociane and liver-stage antiplasmodial activity of the isocyanoterpene class. *J. Am. Chem. Soc.* 138 (23): 7268–7271.

62 Stetter, H. and Schrecke, M. (1973). New method for addition of aldehydes to activated double-bonds. *Angew. Chem. Int. Ed. Engl.* 12 (1): 81.

63 Stetter, H., Schreckenberg, M., and Wiemann, K. (1976). Addition of aldehydes to activated double-bonds. 9. Addition of aromatic and heterocyclic aldehydes to α,β-unsaturated esters. *Chem. Ber. Recl.* 109 (2): 541–545.

64 Stetter, H., Ramsch, R.Y., and Kuhlmann, H. (1976). Synthesis of thiazolium salt catalyzed acyloin and benzoin. 1. Synthesis of simple acyloins and benzoins. *Synthesis-Stuttgart* (11): 733–735.

65 Stetter, H. and Kuhlmann, H. (1976). Addition of aldehydes to activated double-bonds. 11. Addition of aliphatic, heterocyclic and aromatic-aldehydes to butenone. *Chem Ber-Recl.* 109 (10): 3426–3431.

66 Jones, T.H., Blum, M.S., and Fales, H.M. (1980). Chrysomelidial by chromyl chloride oxidation – a revised structure for gastrolactone. *Tetrahedron Lett.* 21 (18): 1701–1704.

67 Elhajj, T., Martin, J.C., and Descotes, G. (1983). Derivatives of 5-hydroxymethyl-furfural. 1. Synthesis of difuran and terfuran derivatives. *J. Heterocyclic Chem.* 20 (1): 233–235.

68 Wynberg, H. and Metselaar, J. (1984). A convenient route to polythiophenes. *Synth. Commun.* 14 (1): 1–9.

69 Perrine, D.M., Kagan, J., Huang, D.B. et al. (1987). 4 Novel phenyldithienoindole isomers from the oxidative photocyclization of dithienylpyrroles. *J. Org. Chem.* 52 (11): 2213–2216.

70 Raghavan, S. and Anuradha, K. (2003). Solid-phase synthesis of heterocycles from 1,4-diketone synthons. *Synlett* (5): 711–713.

71 Enders, D. (1993). *Stereoselective Synthesis*, 63. Heidelberg: Springer-Verlag.

72 Enders, D., Bockstiegel, B., Dyker, H. et al. (1993). *Dechema-Monographies*, vol. 129, 209. Weinheim: Wiley-VCH.

73 Enders, D. and Breuer, K. (1999). *Comprehensive Asymmetric Catalysis*, 1093–1104. Berlin: Springer.

74 Moore, J.L., Kerr, M.S., and Rovis, T. (2006). Enantioselective formation of quaternary stereocenters using the catalytic intramolecular Stetter reaction. *Tetrahedron* 62 (49): 11477–11482.

75 Nakamura, T., Hara, O., Tamura, T. et al. (2005). A facile synthesis of chroman-4-ones and 2,3-dihydroquinolin-4-ones with quaternary carbon using intramolecular Stetter reaction catalyzed by thiazolium salt. *Synlett* (1): 155–157.

76 Liu, Q., Perreault, S., and Rovis, T. (2008). Catalytic asymmetric intermolecular Stetter reaction of glyoxamides with alkylidenemalonates. *J. Am. Chem. Soc.* 130 (43): 14066–14067.

77 Liu, Q. and Rovis, T. (2009). Enantio- and diastereoselective intermolecular Stetter reaction of glyoxamide and alkylidene ketoamides. *Org. Lett.* 11 (13): 2856–2859.

78 Jousseaume, T., Wurz, N.E., and Glorius, F. (2011). Highly enantioselective synthesis of α-amino acid derivatives by an NHC-catalyzed intermolecular Stetter reaction. *Angew. Chem. Int. Ed.* 50 (6): 1410–1414.

79 DiRocco, D.A., Oberg, K.M., Dalton, D.M., and Rovis, T. (2009). Catalytic asymmetric intermolecular Stetter reaction of heterocyclic aldehydes with nitroalkenes: backbone fluorination improves selectivity. *J. Am. Chem. Soc.* 131 (31): 10872–10874.

80 DA, D.R. and Rovis, T. (2011). Catalytic asymmetric intermolecular Stetter reaction of enals with nitroalkenes: enhancement of catalytic efficiency through bifunctional additives. *J. Am. Chem. Soc.* 133 (27): 10402–10405.

81 Sanchez-Larios, E., Thai, K., Bilodeau, F., and Gravel, M. (2011). Highly enantioselective intermolecular Stetter reactions of β-aryl acceptors: α-ketoester moiety as handle for activation and synthetic manipulations. *Org. Lett.* 13 (18): 4942–4945.

82 Mattson, A.E., Bharadwaj, A.R., and Scheidt, K.A. (2004). The thiazolium-catalyzed Sila-Stetter reaction: conjugate addition of acylsilanes to unsaturated esters and ketones. *J. Am. Chem. Soc.* 126 (8): 2314–2315.

83 Myers, M.C., Bharadwaj, A.R., Milgram, B.C., and Scheidt, K.A. (2005). Catalytic conjugate additions of carbonyl anions under neutral aqueous conditions. *J. Am. Chem. Soc.* 127 (42): 14675–14680.

84 Orellana, A. and Rovis, T. (2008). Towards the total synthesis of FD-838: modular enantioselective assembly of the core. *Chem. Commun.* (6): 730–732.

85 Comer, F.W. and Trotter, J. (1966). Structure of hirsutic acid. X-ray analysis of P-bromophenacyl ester. *J. Chem. Soc. B* (1): 11–18.

86 Comer, F.W., Mccapra, F., Qureshi, I.H., and Scott, A.I. (1967). Structure and chemistry of hirsutic acid. *Tetrahedron* 23 (12): 4761–4768.

87 Feline, T.C., Mellows, G., Jones, R.B., and Phillips, L. (1974). Biosynthesis of hirsutic acid C using C-13 nuclear magnetic-resonance spectroscopy. *J. Chem. Soc. Chem. Commun.* (2): 63–64.

88 Trost, B.M., Shuey, C.D., Dininno, F., and Mcelvain, S.S. (1979). Stereocontrolled total synthesis of (+/−)-hirsutic acid-C. *J. Am. Chem. Soc.* 101 (5): 1284–1285.

89 Hashimoto, H., Tsuzuki, K., Sakan, F. et al. (1974). Total synthesis of d1-hirsutic acid. *Tetrahedron Lett.* 15 (43): 3745–3748.

90 Singh, S.B., Jayasuriya, H., Ondeyka, J.G. et al. (2006). Isolation, structure, and absolute stereochemistry of platensimycin, a broad spectrum antibiotic discovered using an antisense differential sensitivity strategy. *J. Am. Chem. Soc.* 128 (36): 11916–11920.

91 Wang, J., Soisson, S.M., Young, K. et al. (2006). Platensimycin is a selective FabF inhibitor with potent antibiotic properties. *Nature* 441 (7091): 358–361.

92 Nicolaou, K.C., Tang, Y.F., and Wang, J.H. (2007). Formal synthesis of (+/−)-platensimycin. *Chem. Commun.* (19): 1922–1923.

93 Nicolaou, K.C., Li, A., and Edmonds, D.J. (2006). Total synthesis of platensimycin. *Angew. Chem. Int. Ed.* 45 (42): 7086–7090.

94 Treff, W. and Werner, H. (1933). The constitution of jasmon. *Ber. Dtsch. Chem. Ges.* 66: 1521–1527.

95 Ruzicka, L., Lardon, F., and Treadwell, P. (1943). Jasmine smelling substances. (2. Announcement). Synthesis of lactones with similar jasmon constitution. *Helv. Chim. Acta* 26: 673–679.

96 Stetter, H. and Kuhlmann, H. (1975). Addition of aldehydes to activated double-bonds. 7. New simple synthesis of *cis*-jasmon and dihydrojasmon. *Synthesis-Stuttgart* (6): 379–380.

97 Roth, B.D., Blankley, C.J., Chucholowski, A.W. et al. (1991). Inhibitors of cholesterol-biosynthesis. 3. Tetrahydro-4-hydroxy-6-[2-(1h-pyrrol-1-Yl)ethyl]-2h-pyran-2-one inhibitors of Hmg-Coa reductase. 2. Effects of introducing substituents at positions 3 and 4 of the pyrrole nucleus. *J. Med. Chem.* 34 (1): 357–366.

98 Baumann, K.L., Butler, D.E., Deering, C.F. et al. (1992). The convergent synthesis of Ci-981, an optically-active, highly potent, tissue selective inhibitor of Hmg-Coa reductase. *Tetrahedron Lett.* 33 (17): 2283–2284.

99 Hayakawa, Y., Kawakami, K., Seto, H., and Furihata, K. (1992). Structure of a new antibiotic Roseophilin. *Tetrahedron Lett.* 33 (19): 2701–2704.

100 Furstner, A., Reinecke, K., Prinz, H., and Waldmann, H. (2004). The core structures of roseophilin and the prodigiosin alkaloids define a new class of protein tyrosine phosphatase inhibitors. *ChemBioChem* 5 (11): 1575–1579.

101 Harrington, P.E. and Tius, M.A. (1999). A formal total synthesis of roseophilin: cyclopentannelation approach to the macrocyclic core. *Org. Lett.* 1 (4): 649–651.

102 Harrington, P.E. and Tius, M.A. (2001). Synthesis and absolute stereochemistry of roseophilin. *J. Am. Chem. Soc.* 123 (35): 8509–8514.

103 Daly, J.W., Green, F.C., and Eastman, R.H. (1958). Sabinene hydrate: a constituent of American peppermint oil. *J. Am. Chem. Soc.* 80 (23): 6330–6336.

104 Karasawa, D. and Shimizu, S. (1978). Studies on chemical constituents of genus Mentha. 2. Mentha-Candicans, a new chemical strain of section Spicatae, containing *trans*-Sabinene hydrate as principal component of essential oil. *Agric. Biol. Chem.* 42 (2): 433–437.
105 Novak, J., Bitsch, C., Langbehn, J. et al. (2000). Ratios of *cis*- and *trans*-sabinene hydrate in *Origanum majorana* L, and *Origanum microphyllum* (Bentham) Vogel. *Biochem. Syst. Ecol.* 28 (7): 697–704.
106 Galopin, C.C. (2001). A short and efficient synthesis of (±)-*trans*-sabinene hydrate. *Tetrahedron Lett.* 42 (33): 5589–5591.
107 Ritter, F.J., Rotgans, I.E.M., Talman, E. et al. (1973). 5-Methyl-3-butyl-octahydroindolizine – novel type of pheromone attractive to pharaohs ants (*Monomorium Pharaonis* (L)). *Experientia* 29 (5): 530–531.
108 Jones, T.H., Highet, R.J., Blum, M.S., and Fales, H.M. (1984). (5z,9z)-3-Alkyl-5-methylindolizidines from Solenopsis (diplorhoptrum) species. *J. Chem. Ecol.* 10 (8): 1233–1249.
109 Randl, S. and Blechert, S. (2003). Concise enantioselective synthesis of 3,5-dialkyl-substituted Indolizidine alkaloids via sequential cross-metathesis – double-reductive cyclization. *J. Org. Chem.* 68 (23): 8879–8882.
110 Gerle, B. (1964). Clinical observations of the side-effects of haloperidol. *Acta Psychiatr. Scand.* 40 (1): 65–76.
111 Itil, T.M. (1972). Clinical and laboratory side-effects of haloperidol, thiothixene, and fluphenazine in chronic schizophrenic patients. *Clin. Pharmacol. Ther.* 13 (1): 141.
112 Janzen, G. and Moleman, P. (1988). Parkinsonian side-effects of haloperidol. *Pharmacopsychiatry* 21 (1): 62.
113 Allan, E.R., Sison, C.E., Alpert, M. et al. (1998). The relationship between negative symptoms of schizophrenia and extrapyramidal side effects with haloperidol and olanzapine. *Psychopharmacol. Bull.* 34 (1): 71–74.
114 Jakovcevska-Kujundziska, M., Filovska, V., Lazarova, V., and Serafimova, T. (2002). Side effects and efficacy with combined therapy of clozapine and haloperidol in the treatment-resistant schizophrenic patients. *Eur. Neuropsychopharmacol.* 12: S330.
115 Lyles-Eggleston, M., Altundas, R., Xia, J. et al. (2004). Design, synthesis, and evaluation of metabolism-based analogues of haloperidol incapable of forming MPP+-like species. *J. Med. Chem.* 47 (3): 497–508.
116 Barbui, C., Accordini, S., Nose, M. et al. (2011). Aripiprazole versus haloperidol in combination with clozapine for treatment-resistant schizophrenia in routine clinical care a randomized, controlled trial. *J. Clin. Psychopharmacol.* 31 (3): 266–273.
117 Anjaiah, S., Chandrasekhar, S., and Gree, R. (2004). Stetter reaction in room temperature ionic liquids and application to the synthesis of haloperidol. *Adv. Synth. Catal.* 346 (11): 1329–1334.
118 Ratnayake, R., Covell, D., Ransom, T.T. et al. (2009). Englerin A, a selective inhibitor of renal cancer cell growth, from *Phyllanthus engleri*. *Org. Lett.* 11 (1): 57–60.

119 Xu, J., Caro-Diaz, E.J.E., and Theodorakis, E.A. (2010). Enantioselective formal synthesis of (−)-Englerin A via a Rh-catalyzed [4+3] cycloaddition reaction. *Org. Lett.* 12 (16): 3708–3711.
120 Zhou, Q.H., Chen, X.F., and Ma, D.W. (2010). Asymmetric, protecting-group-free total synthesis of (−)-Englerin A. *Angew. Chem. Int. Ed.* 49 (20): 3513–3516.
121 Parmar, V.S., Jain, S.C., Bisht, K.S. et al. (1997). Phytochemistry of the genus Piper. *Phytochemistry* 46 (4): 597–673.
122 Adams, D.J., Boskovic, Z.V., Theriault, J.R. et al. (2013). Discovery of small-molecule enhancers of reactive oxygen species that are nontoxic or cause genotype-selective cell death. *ACS Chem. Biol.* 8 (5): 923–929.
123 Boskovic, Z.V., Hussain, M.M., Adams, D.J. et al. (2013). Synthesis of piperlogs and analysis of their effects on cells. *Tetrahedron* 69 (36): 7559–7567.
124 Kubo, M., Ishii, R., Ishino, Y. et al. (2013). Evaluation of constituents of *Piper retrofractum* fruits on neurotrophic activity. *J. Nat. Prod.* 76 (4): 769–773.
125 Sommerwerk, S., Kern, S., Heller, L., and Csuk, R. (2014). First total synthesis of piperodione and analogs. *Tetrahedron Lett.* 55 (45): 6243–6244.
126 Nakamura, E. and Kuwajima, I. (1984). Copper-catalyzed acylation and conjugate addition of zinc homoenolate – synthesis of 4-oxo and 6-oxo esters. *J. Am. Chem. Soc.* 106 (11): 3368–3370.
127 Oshino, H., Nakamura, E., and Kuwajima, I. (1985). Catalytic homo-Reformatsky reaction. Ambident chemical reactivities of the zinc homoenolate of propionate. *J. Org. Chem.* 50 (15): 2802–2804.
128 Nakamura, E. and Kuwajima, I. (1986). Palladium catalyzed-reactions of propionate homoenolate – arylation, vinylation, and acylation. *Tetrahedron Lett.* 27 (1): 83–86.
129 Nakamura, E., Aoki, S., Sekiya, K. et al. (1987). Carbon–carbon bond-forming reactions of zinc homoenolate of esters – a novel 3-carbon nucleophile with general synthetic utility. *J. Am. Chem. Soc.* 109 (26): 8056–8066.
130 DiMauro, E. and Fry, A.J. (1999). Convenient synthesis of a reactive ester homoenolate. *Tetrahedron Lett.* 40 (45): 7945–7949.
131 Burstein, C. and Glorius, F. (2004). Organocatalyzed conjugate umpolung of α,β-unsaturated aldehydes for the synthesis of γ-butyrolactones. *Angew. Chem. Int. Ed.* 43 (45): 6205–6208.
132 Sohn, S.S., Rosen, E.L., and Bode, J.W. (2004). N-heterocyclic carbene-catalyzed generation of homoenolates: γ-butyrolactones by direct annulations of enals and aldehydes. *J. Am. Chem. Soc.* 126 (44): 14370–14371.
133 Chan, A. and Scheidt, K.A. (2005). Conversion of α,β-unsaturated aldehydes into saturated esters: an umpolung reaction catalyzed by nucleophilic carbenes. *Org. Lett.* 7 (5): 905–908.
134 Nair, V., Vellalath, S., Poonoth, M. et al. (2006). N-heterocyclic carbene catalyzed reaction of enals and 1,2-dicarbonyl compounds: stereoselective synthesis of spiro γ-butyrolactones. *Org. Lett.* 8 (3): 507–509.

135 Nair, V., Vellalath, S., Poonoth, M., and Suresh, E. (2006). N-heterocyclic carbene-catalyzed reaction of chalcones and enals via homoenolate: an efficient synthesis of 1,3,4-trisubstituted cyclopentenes. *J. Am. Chem. Soc.* 128 (27): 8736–8737.

136 Chan, A. and Scheidt, K.A. (2007). Highly stereoselective formal [3+3] cycloaddition of enals and azomethine imines catalyzed by N-heterocyclic carbenes. *J. Am. Chem. Soc.* 129 (17): 5334–5335.

137 He, M. and Bode, J.W. (2008). Enantioselective, NHC-catalyzed bicyclo-β-lactam formation via direct annulations of enals and unsaturated N-sulfonyl ketimines. *J. Am. Chem. Soc.* 130 (2): 418–419.

138 Phillips, E.M., Reynolds, T.E., and Scheidt, K.A. (2008). Highly diastereo- and enantioselective additions of homoenolates to nitrones catalyzed by N-heterocyclic carbenes. *J. Am. Chem. Soc.* 130 (8): 2416–2417.

139 Chan, A. and Scheidt, K.A. (2008). Direct amination of homoenolates catalyzed by N-heterocyclic carbenes. *J. Am. Chem. Soc.* 130 (9): 2740–2741.

140 Rommel, M., Fukuzumi, T., and Bode, J.W. (2008). Cyclic ketimines as superior electrophiles for NHC-catalyzed homoenolate additions with broad scope and low catalyst loadings. *J. Am. Chem. Soc.* 130 (51): 17266–17267.

141 Maki, B.E., Chan, A., and Scheidt, K.A. (2008). Protonation of homoenolate equivalents generated by N-heterocyclic carbenes. *Synthesis-Stuttgart* 2008 (8): 1306–1315.

142 Maki, B.E., Patterson, E.V., Cramer, C.J., and Scheidt, K.A. (2009). Impact of solvent polarity on N-heterocyclic carbene-catalyzed β-protonations of homoenolate equivalents. *Org. Lett.* 11 (17): 3942–3945.

143 Raup, D.E.A., Cardinal-David, B., Holte, D., and Scheidt, K.A. (2010). Cooperative catalysis by carbenes and Lewis acids in a highly stereoselective route to γ-lactams. *Nat. Chem.* 2 (9): 766–771.

144 Cohen, D.T., Cardinal-David, B., Roberts, J.M. et al. (2011). NHC-catalyzed/titanium(IV)-mediated highly diastereo- and enantioselective dimerization of enals. *Org. Lett.* 13 (5): 1068–1071.

145 Cohen, D.T., Cardinal-David, B., and Scheidt, K.A. (2011). Lewis acid activated synthesis of highly substituted cyclopentanes by the N-heterocyclic carbene catalyzed addition of homoenolate equivalents to unsaturated ketoesters. *Angew. Chem. Int. Ed.* 50 (7): 1678–1682.

146 Phillips, E.M., Chan, A., and Scheidt, K.A. (2009). Discovering new reactions with N-heterocyclic carbene catalysis. *Aldrichimica Acta.* 42 (3): 55–66.

147 Nair, V., Menon, R.S., Biju, A.T. et al. (2011). Employing homoenolates generated by NHC catalysis in carbon–carbon bond-forming reactions: state of the art. *Chem. Soc. Rev.* 40 (11): 5336–5346.

148 Feling, R.H., Buchanan, G.O., Mincer, T.J. et al. (2003). Salinosporamide A: a highly cytotoxic proteasome inhibitor from a novel microbial source, a marine bacterium of the new genus Salinospora. *Angew. Chem. Int. Ed.* 42 (3): 355–357.

149 Struble, J.R. and Bode, J.W. (2009). Formal synthesis of salinosporamide a via NHC-catalyzed intramolecular lactonization. *Tetrahedron* 65 (26): 4957–4967.

150 Margalef, I.V., Rupnicki, L., and Lam, H.W. (2008). Formal synthesis of salinosporamide a using a nickel-catalyzed reductive aldol cyclization-lactonization as a key step. *Tetrahedron* 64 (34): 7896–7901.
151 Abe, N., Onoda, R., Shirahat, K. et al. (1968). Structure of bakkenolide-A. *Tetrahedron Lett.* 9 (3): 369–373.
152 Wu, T.S., Kao, M.S., Wu, P.L. et al. (1999). The bakkenolides from the root of *Petasites formosanus* and their cytotoxicity. *Chem. Pharm. Bull.* 47 (3): 375–382.
153 Silva, L.F. (2001). Total synthesis of bakkanes. *Synthesis-Stuttgart* 2001 (5): 671–689.
154 Brocksom, T.J., Brocksom, U., and Constantino, M.G. (2008). Synthesis of the bakkenolide sesquiterpenes. *Quim Nova* 31 (4): 937–941.
155 Jamieson, G.R., Reid, E.H., Turner, B.P., and Jamieson, A.T. (1976). Bakkenolide-A – its distribution in Petasites species and cytotoxic properties. *Phytochemistry* 15 (11): 1713–1715.
156 Kano, K., Hayashi, K., and Mitsuhashi, H. (1982). Syntheses of steroids having a bakkenolide-type Spirolactone ring.1. Synthesis of 4′-methylenedihydrospiro-[5-α-cholestane-3,3′(2'H)-furan]-2'-one. *Chem. Pharm. Bull.* 30 (4): 1198–1203.
157 Phillips, E.M., Roberts, J.M., and Scheidt, K.A. (2010). Catalytic enantioselective total syntheses of bakkenolides I, J, and S: application of a carbene-catalyzed desymmetrization. *Org. Lett.* 12 (12): 2830–2833.
158 Greene, A.E., Coelho, F., Depres, J.P., and Brocksom, T.J. (1988). Enantiocontrolled total synthesis of (+)-bakkenolide-A. *Tetrahedron Lett.* 29 (44): 5661–5662.
159 Hamelin, O., Wang, Y., Depres, J.P., and Greene, A.E. (2000). The first entry to complex bakkanes: a highly effective retroaldol – aldol-based approach to (−)-bakkenolides III, B, C, and H. *Angew. Chem. Int. Ed.* 39 (23): 4314–4316.
160 Balkbindseil, W., Helmke, E., Weyland, H., and Laatsch, H. (1995). Marine bacteria. 8. Maremycin-a and Maremycin-B, new diketopiperazines from a marine Streptomyces Sp. *Liebigs Ann.* 2 (7): 1291–1294.
161 Tang, Y.Q., Sattler, I., Thiericke, R. et al. (2001). Maremycins C and D, new diketopiperazines, and maremycins E and F, novel polycyclic spiro-indole metabolites isolated from Streptomyces sp. *Eur. J. Org. Chem.* 2001 (2): 261–267.
162 Ueda, T., Inada, M., Okamoto, I. et al. (2008). Synthesis of maremycins a and D1 via cycloaddition of a nitrone with (*E*)-3-ethylidene-1-methylindolin-2-one. *Org. Lett.* 10 (10): 2043–2046.
163 Bergonzini, G. and Melchiorre, P. (2012). Dioxindole in asymmetric catalytic synthesis: routes to enantioenriched 3-substituted 3-hydroxyoxindoles and the preparation of maremycin A. *Angew. Chem. Int. Ed.* 51 (4): 971–974.
164 Liu, Y., Zhang, L., and Jia, Y. (2012). Total synthesis of maremycins A, B, C1/C2, D1, and D2. *Tetrahedron Lett.* 53 (6): 684–687.
165 Dugal-Tessier, J., O'Bryan, E.A., Schroeder, T.B.H. et al. (2012). *Angew. Chem. Int. Ed.* 51: 4963–4967.
166 Zhu, C. and Zhang, J. (2003). Enzyme kinetics of clausenamide enantiomers in rat liver microsomes. *Acta Pharm. Sin.* 9: 654–657.

167 Chen, Y.R., Yang, M.H., Huang, L., et al., inventors (1986). Chinese Academy of Medical Sciences, Peop. Rep. China; Bayer A.-G (assignee). New δ-butyrolactam, its pharmacologically active formulations, preparation and medicinal use patent DE3431257A1.
168 Zhu, X.Z., Li, X.-Y., and Liu, J. (2004). Recent pharmacological studies on natural products in China. *Eur. J. Pharmacol.* 500 (1–3): 221–230.
169 He, M., Rommel, M., and Bode, J.W. (2012). Formal synthesis of (±)-clausenamide by NHC-catalyzed γ-lactam formation. *Heterocycles* 86 (2): 1689–1696.
170 Hartwig, W. and Born, L. (1987). Diastereoselective and enantioselective total synthesis of the hepatoprotective agent clausenamide. *J. Org. Chem.* 52 (19): 4352–4358.
171 Bourin, M., Chue, P., and Guillon, Y. (2001). Paroxetine: a review. *CNS Drug Rev.* 7 (1): 25–47.
172 Gunasekara, N.S., Noble, S., and Benfield, P. (1998). Paroxetine: An update of its pharmacology and therapeutic use in depression and a review of its use in other disorders. *Drugs* 55 (1): 85–120.
173 White, N.A., Ozboya, K.E., Flanigan, D.M., and Rovis, T. (2014). Rapid construction of (−)-paroxetine and (−)-femoxetine via an N-heterocyclic carbene catalyzed homoenolate addition to nitroalkenes. *Asian J. Org. Chem.* 3 (4): 442–444.
174 Fu, Z., Xu, J., Zhu, T. et al. (2013). β-Carbon activation of saturated carboxylic esters through N-heterocyclic carbene organocatalysis. *Nat. Chem.* 5 (10): 835–839.
175 Olpe, H.R., Demiéville, H., Baltzer, V. et al. (1978). The biological activity of d-baclofen (Lipresal®). *Eur. J. Pharmacol.* 52 (1): 133–136.
176 Zhu, J., Mix, E., and Winblad, B. (2001). The antidepressant and antiinflammatory effects of rolipram in the central nervous system. *CNS Drug Rev.* 7 (4): 387–398.
177 Krause, W., Kühne, G., and Sauerbrey, N. (1990). Pharmacokinetics of (+)-rolipram and (−)-rolipram in healthy volunteers. *Eur. J. Clin. Pharmacol.* 38 (1): 71–75.
178 Oba, M., Saegusa, T., Nishiyama, N., and Nishiyama, K. (2009). Synthesis of non-proteinogenic amino acids using Michael addition to unsaturated orthopyroglutamate derivative. *Tetrahedron* 65 (1): 128–133.
179 Diaz, A., Siro, J.G., Garcia-Navio, J.L. et al. (1997). A stereoselective synthesis of (*R*)-(−)-rolipram from L-glutamic acid. *Synthesis* 1997 (5): 559–562.
180 Ohsaki, A., Ishiyama, H., Yoneda, K., and Ji, K. (2003). Secu'amamine A, a novel indolizidine alkaloid from *Securinega suffruticosa* var. amamiensis. *Tetrahedron Lett.* 44 (15): 3097–3099.
181 Rognan, D., Boulanger, T., Hoffmann, R. et al. (1992). Structure and molecular modeling of GABAA receptor antagonists. *J. Med. Chem.* 35 (11): 1969–1977.
182 Saito, S., Yoshikawa, H., Sato, Y. et al. (1966). Total synthesis of Securinine. *Chem. Pharm. Bull.* 14 (3): 313–314.
183 ElSohly, A.M., Wespe, D.A., Poore, T.J., and Snyder, S.A. (2013). An efficient approach to the securinega alkaloids empowered by cooperative

N-heterocyclic carbene/Lewis acid catalysis. *Angew. Chem. Int. Ed.* 52 (22): 5789–5794.

184 Lee, A. and Scheidt, K.A. (2014). A cooperative N-heterocyclic carbene/chiral phosphate catalysis system for allenolate annulations. *Angew. Chem. Int. Ed.* 53 (29): 7594–7598.

185 Miyashita, A., Matsuda, H., Iijima, C., and Higashino, T. (1990). Catalytic action of azolium salts. 1. Aroylation of 4-chloro-1h-pyrazolo[3,4-D]pyrimidines with aromatic-aldehydes catalyzed by 1,3-dimethylbenzimidazolium iodide. *Chem. Pharm. Bull.* 38 (5): 1147–1152.

186 Miyashita, A., Matsuda, H., Iijima, C., and Higashino, T. (1992). Catalytic action of azolium salts. 2. Aroylation of 4-chloroquinazolines with aromatic-aldehydes catalyzed by 1,3-dimethylbenzimidazolium iodide. *Chem. Pharm. Bull.* 40 (1): 43–48.

187 Miyashita, A., Matsuda, H., and Higashino, T. (1992). Catalytic action of azolium salts. 3. Aroylation of N-phenylbenzimidoyl chlorides with aromatic-aldehydes in the presence of 1,3-dimethylimidazolium iodide. *Chem. Pharm. Bull.* 40 (10): 2627–2631.

188 Miyashita, A., Matsuda, H., Suzuki, Y. et al. (1994). Catalytic action of azolium salts. 4. Preparations of 4-aroylquinazolines and 4-aroyl-1h-pyrazolo[3,4-D]pyrimidines by catalytic action of 1,3-dimethylimidazolium iodide. *Chem. Pharm. Bull.* 42 (10): 2017–2022.

189 Suzuki, Y., Toyota, T., Miyashita, A., and Sato, M. (2006). Synthesis of heterocyclic compounds *via* nucleophilic aroylation catalyzed by imidazolidenyl carbene. *Chem. Pharm. Bull.* 54: 1653–1658.

190 Suzuki, Y., Fukuta, Y., Ota, S. et al. (2011). Xanthone natural products via *N*-heterocyclic carbene catalysis: total synthesis of atroviridin. *J. Org. Chem.* 76 (10): 3960–3967.

191 Suzuki, Y., Ota, S., Fukuta, Y. et al. (2008). N-heterocyclic carbene-catalyzed nucleophilic aroylation of fluorobenzenes. *J. Org. Chem.* 73 (6): 2420–2423.

192 Kosin, J., Ruangrungsi, N., Ito, C., and Furukawa, H. (1998). A xanthone from *Garcinia atroviridis*. *Phytochemistry* 47 (6): 1167–1168.

193 Castells, J., Llitjos, H., and Morenomanas, M. (1977). Nitrobenzene aldehyde oxidations catalyzed by conjugate bases of thiazolium ions. *Tetrahedron Lett.* (2): 205–206.

194 Inoue, H. and Higashiura, K. (1980). Electron-transfer reductions by active aldehydes catalyzed by thiazolium salt in the presence of triethylamine. *J. Chem. Soc. Chem. Commun.* (12): 549–550.

195 Tam, S.W., Jimenez, L., and Diederich, F. (1992). One-pot synthesis of aromatic methyl-esters by electrochemical oxidation of aldehydes mediated by biscoenzyme catalysis. *J. Am. Chem. Soc.* 114 (4): 1503–1505.

196 Miyashita, A., Suzuki, Y., Nagasaki, I. et al. (1997). Catalytic action of azolium salts. 8. Oxidative aroylation with arenecarbaldehydes catalyzed by 1,3-dimethylbenzimidazolium iodide. *Chem. Pharm. Bull.* 45 (8): 1254–1258.

197 Chan, A. and Scheidt, K.A. (2006). Hydroacylation of activated ketones catalyzed by N-heterocyclic carbenes. *J. Am. Chem. Soc.* 128 (14): 4558–4559.

198 Maki, B.E., Chan, A., Phillips, E.M., and Scheidt, K.A. (2007). Tandem oxidation of allylic and benzylic alcohols to esters catalyzed by N-heterocyclic carbenes. *Org. Lett.* 9 (2): 371–374.

199 Maki, B.E. and Scheidt, K.A. (2008). N-heterocyclic carbene-catalyzed oxidation of unactivated aldehydes to esters. *Org. Lett.* 10 (19): 4331–4334.

200 De Sarkar, S., Grimme, S., and Studer, A. (2010). NHC catalyzed oxidations of aldehydes to esters: chemoselective acylation of alcohols in presence of amines. *J. Am. Chem. Soc.* 132 (4): 1190–1191.

201 Reynolds, N.T., de Alaniz, J.R., and Rovis, T. (2004). Conversion of α-haloaldehydes into acylating agents by an internal redox reaction catalyzed by nucleophilic carbenes. *J. Am. Chem. Soc.* 126 (31): 9518–9519.

202 Chow, K.Y.K. and Bode, J.W. (2004). Catalytic generation of activated carboxylates: direct, stereoselective synthesis of β-hydroxyesters from epoxyaldehydes. *J. Am. Chem. Soc.* 126 (26): 8126–8127.

203 Sohn, S. and Bode, J.W. (2006). *Angew. Chem. Int. Ed.* 45: 6021–6024.

204 Vora, H.U. and Rovis, T. (2007). Nucleophilic carbene and HOAt relay catalysis in an amide bond coupling: an orthogonal peptide bond forming reaction. *J. Am. Chem. Soc.* 129 (45): 13796–13797.

205 Bode, J.W. and Sohn, S.S. (2007). N-heterocyclic carbene-catalyzed redox amidations of α-functionalized aldehydes with amines. *J. Am. Chem. Soc.* 129 (45): 13798–13799.

206 Thai, K., Wang, L., Dudding, T. et al. (2010). NHC-catalyzed intramolecular redox amidation for the synthesis of functionalized lactams. *Org. Lett.* 12 (24): 5708–5711.

207 Kuwano, S., Harada, S., Oriez, R., and Yamada, K. (2012). Chemoselective conversion of α-unbranched aldehydes to amides, esters, and carboxylic acids by NHC-catalysis. *Chem. Commun.* 48 (1): 145–147.

208 Vora, H.U., Moncecchi, J.R., Epstein, O., and Rovis, T. (2008). Nucleophilic carbene catalyzed synthesis of 1,2 amino alcohols via azidation of epoxy aldehydes. *J. Org. Chem.* 73 (24): 9727–9731.

209 De Sarkar, S. and Studer, A. (2010). Oxidative amidation and azidation of aldehydes by NHC catalysis. *Org. Lett.* 12 (9): 1992–1995.

210 Ryan, S.J., Candish, L., and Lupton, D.W. (2009). N-heterocyclic carbene-catalyzed generation of α,β-unsaturated acyl imidazoliums: synthesis of dihydropyranones by their reaction with enolates. *J. Am. Chem. Soc.* 131 (40): 14176–14177.

211 De Sarkar, S. and Studer, A. (2010). NHC-catalyzed Michael addition to α,β-unsaturated aldehydes by redox activation. *Angew. Chem. Int. Ed.* 49 (48): 9266–9269.

212 Kaeobamrung, J., Mahatthananchai, J., Zheng, P.G., and Bode, J.W. (2010). An enantioselective Claisen rearrangement catalyzed by N-heterocyclic carbenes. *J. Am. Chem. Soc.* 132 (26): 8810–8812.

213 Sun, F.G., Sun, L.H., and Ye, S. (2011). N-heterocyclic carbene-catalyzed enantioselective annulation of bromoenal and 1,3-dicarbonyl compounds. *Adv. Synth. Catal.* 353 (17): 3134–3138.

214 Biswas, A., De Sarkar, S., Frohlich, R., and Studer, A. (2011). Highly stereoselective synthesis of 1,2,3-trisubstituted indanes via oxidative N-heterocyclic carbene-catalyzed cascades. *Org. Lett.* 13 (18): 4966–4969.

215 Wanner, B., Mahatthananchai, J., and Bode, J.W. (2011). Enantioselective synthesis of dihydropyridinones via NHC-catalyzed aza-Claisen reaction. *Org. Lett.* 13 (19): 5378–5381.

216 Wang, B. and Forsyth, C.J. (2009). Total synthesis of largazole – devolution of a novel synthetic strategy. *Synthesis-Stuttgart* (17): 2873–2880.

217 Wang, B., Huang, P.H., Chen, C.S., and Forsyth, C.J. (2011). Total syntheses of the histone deacetylase inhibitors largazole and 2-epi-Largazole: application of N-heterocyclic carbene mediated acylations in complex molecule synthesis. *J. Org. Chem.* 76 (4): 1140–1150.

218 Sipma, G. and van der Wal, B. (1968). The structure of davanone a new sesquiterpene from savana: (*Artemisia Pallens*, Wall.). *Recl. Trav. Chim. Pays-Bas.* 87 (7): 715–720.

219 Vajs, V., Trifunovic, S., Janackovic, P. et al. (2004). Antifungal activity of davanone-type sesquiterpenes from *Artemisia lobelii* var. *conescens*. *J. Serb. Chem. Soc.* 69 (11): 969–972.

220 Morrison, K.C., Litz, J.P., Scherpelz, K.P. et al. (2009). A concise, biomimetic total synthesis of (+)-davanone. *Org. Lett.* 11 (10): 2217–2218.

221 Zhao, C., Li, F., and Wang, J. (2016). N-heterocyclic carbene catalyzed dynamic kinetic resolution of pyranones. *Angew. Chem. Int. Ed.* 55 (5): 1820–1824.

222 Cohen, D.T., Eichman, C.C., Phillips, E.M. et al. (2012). Catalytic dynamic kinetic resolutions with N-heterocyclic carbenes: asymmetric synthesis of highly substituted β-lactones. *Angew. Chem. Int. Ed.* 51 (29): 7309–7313.

223 Goodman, C.G. and Johnson, J.S. (2014). Dynamic kinetic asymmetric cross-benzoin additions of β-stereogenic α-keto esters. *J. Am. Chem. Soc.* 136 (42): 14698–14701.

224 Goodman, C.G., Walker, M.M., and Johnson, J.S. (2015). Enantioconvergent synthesis of functionalized γ-butyrolactones via (3 + 2)-annulation. *J. Am. Chem. Soc.* 137 (1): 122–125.

225 Wu, Z., Li, F., and Wang, J. (2015). Intermolecular dynamic kinetic resolution cooperatively catalyzed by an *N*-heterocyclic carbene and a Lewis acid. *Angew. Chem. Int. Ed.* 54 (5): 1629–1633.

226 Chen, X., Fong, J.Z.M., Xu, J. et al. (2016). Carbene-catalyzed dynamic kinetic resolution of carboxylic esters. *J. Am. Chem. Soc.* 138 (23): 7212–7215.

227 Lin, L.-Z., Cordell, G.A., Ni, C.-Z., and Clardy, J. (1991). Oxindole alkaloids from *Gelsemium elegans*. *Phytochemistry* 30 (4): 1311–1315.

228 Kitajima, M., Kogure, N., Yamaguchi, K. et al. (2003). Structure reinvestigation of gelsemoxonine, a constituent of *Gelsemium elegans*, reveals a novel, azetidine-containing indole alkaloid. *Org. Lett.* 5 (12): 2075–2078.

229 Shimokawa, J., Harada, T., Yokoshima, S., and Fukuyama, T. (2011). Total synthesis of gelsemoxonine. *J. Am. Chem. Soc.* 133 (44): 17634–17637.

230 Singh, I.P., Milligan, K.E., and Gerwick, W.H. (1999). Tanikolide, a toxic and antifungal lactone from the marine cyanobacterium *Lyngbya majuscula*. *J. Nat. Prod.* 62 (9): 1333–1335.

231 Nambu, H., Jinnouchi, H., Fujiwara, T., and Yakura, T. (2016). Total synthesis of (+)-tanikolide by a traceless stereoinduction method using rhodium(II)-catalyzed oxonium ylide formation-[2,3]-sigmatropic rearrangement and NHC-catalyzed ring-expansion lactonization. *Synlett* 27 (7): 1106–1009.

232 Wang, L., Thai, K., and Gravel, M. (2009). NHC-catalyzed ring expansion of oxacycloalkane-2-carboxaldehydes: a versatile synthesis of lactones. *Org. Lett.* 11 (4): 891–893.

233 Szpilman, A.M., Cereghetti, D.M., Wurtz, N.R. et al. (2008). Synthesis of 35-deoxy amphotericin B methyl ester: a strategy for molecular editing. *Angew. Chem. Int. Ed.* 47 (23): 4335–4338.

234 Corey, E.J. and Schmidt, G. (1979). Useful procedures for the oxidation of alcohols involving pyridinium dichromate in aprotic media. *Tetrahedron Lett.* 20 (5): 399–402.

235 Kraus, G.A. and Roth, B. (1980). Synthetic studies toward verrucarol. 2. Synthesis of the AB ring system. *J. Org. Chem.* 45 (24): 4825–4830.

236 Corey, E.J., Gilman, N.W., and Ganem, B.E. (1968). New methods for the oxidation of aldehydes to carboxylic acids and esters. *J. Am. Chem. Soc.* 90 (20): 5616–5617.

237 Maki, B.E., Chan, A., Phillips, E.M., and Scheidt, K.A. (2009). N-heterocyclic carbene-catalyzed oxidations. *Tetrahedron* 65 (16): 3102–3109.

238 Finney, E.E., Ogawa, K.A., and Boydston, A.J. (2012). Organocatalyzed anodic oxidation of aldehydes. *J. Am. Chem. Soc.* 134 (30): 12374–12377.

239 Wu, X., Zhang, Y., Wang, Y. et al. (2017). Polyhalides as efficient and mild oxidants for oxidative carbene organocatalysis by radical processes. *Angew. Chem. Int. Ed.* 56 (11): 2942–2946.

240 Cutignano, A., Bruno, I., Bifulco, G. et al. (2001). Dactylolide, a new cytotoxic macrolide from the Vanuatu sponge *Dactylospongia* sp. *Eur. J. Org. Chem.* 4 (4): 775–778.

241 Lee, K., Kim, H., and Hong, J. (2012). N-heterocyclic carbene catalyzed oxidative macrolactonization: total synthesis of (+)-dactylolide. *Angew. Chem. Int. Ed.* 51: 5735–5738.

242 Kharasch, M.S. and Joshi, B.S. (1957). Reactions of hindered phenols. 2. Base-catalyzed oxidations of hindered phenols. *J. Org. Chem.* 22 (11): 1439–1443.

243 Pereira, A.R., McCue, C.F., and Gerwick, W.H. (2010). Cyanolide A, a glycosidic macrolide with potent Molluscicidal activity from the Papua New Guinea cyanobacterium *Lyngbya bouillonii*. *J. Nat. Prod.* 73 (2): 217–220.

244 Ross, A.G.P., Bartley, P.B., Sleigh, A.C. et al. (2002). Schistosomiasis. *N. Engl. J. Med.* 346 (16): 1212–1220.

245 Kim, H. and Hong, J. (2010). Total synthesis of Cyanolide A and confirmation of its absolute configuration. *Org. Lett.* 12 (12): 2880–2883.

246 Baker, J.B., Kim, H., and Hong, J. (2015). Total synthesis of clavosolide a via tandem allylic oxidation/oxa-conjugate addition reaction. *Tetrahedron Lett.* 56 (23): 3120–3122.

247 Rao, M.R. and Faulkner, D.J. (2002). Clavosolides A and B, dimeric macrolides from the Philippines sponge Myriastra clavosa. *J. Nat. Prod.* 65 (3): 386–388.

248 Duan, J.J.W. and Smith, A.B. (1993). Iodine monobromide (IBr) at low temperature: enhanced diastereoselectivity in electrophilic cyclizations of homoallylic carbonates. *J. Org. Chem.* 58 (14): 3703–3711.

249 Yadav, J.S., Rao, P.P., Reddy, M.S. et al. (2007). Stereoselective synthesis of (+)-cryptocarya diacetate by an iterative Prins cyclisation and reductive cleavage sequence. *Tetrahedron Lett.* 48 (8): 1469–1471.

250 Pettit, G.R., Herald, C.L., Doubek, D.L. et al. (1982). Isolation and structure of bryostatin 1. *J. Am. Chem. Soc.* 104 (24): 6846–6848.

251 Sudek, S., Lopanik, N.B., Waggoner, L.E. et al. (2007). Identification of the putative Bryostatin polyketide synthase gene cluster from "Candidatus Endobugula sertula", the uncultivated microbial symbiont of the marine bryozoan *Bugula neritina*. *J. Nat. Prod.* 70 (1): 67–74.

252 Hale, K.J., Hummersone, M.G., Manaviazar, S., and Frigerio, M. (2002). The chemistry and biology of the bryostatin antitumour macrolides. *Nat. Prod. Rep.* 19 (4): 413–453.

253 Sun, M.-K., Hongpaisan, J., Nelson, T.J., and Alkon, D.L. (2008). Poststroke neuronal rescue and synaptogenesis mediated in vivo by protein kinase C in adult brains. *Proc. Natl. Acad. Sci.* 105 (36): 13620–13625.

254 Etcheberrigaray, R., Tan, M., Dewachter, I. et al. (2004). Therapeutic effects of PKC activators in Alzheimer's disease transgenic mice. *Proc. Natl. Acad. Sci. U. S. A.* 101 (30): 11141–11146.

255 Lu, Y., Woo, S.K., and Krische, M.J. (2011). Total synthesis of Bryostatin 7 via C–C bond-forming hydrogenation. *J. Am. Chem. Soc.* 133 (35): 13876–13879.

256 Lu, Y., Kim, I.S., Hassan, A. et al. (2009). 1,*n*-glycols as dialdehyde equivalents in iridium-catalyzed enantioselective carbonyl allylation and iterative two-directional assembly of 1,3-polyols. *Angew. Chem. Int. Ed.* 48 (27): 5018–5021.

257 Han, S.B., Kim, I.S., Han, H., and Krische, M.J. (2009). Enantioselective carbonyl reverse prenylation from the alcohol or aldehyde oxidation level employing 1,1-dimethylallene as the prenyl donor. *J. Am. Chem. Soc.* 131 (20): 6916–6917.

258 Kim, I.S., Ngai, M.-Y., and Krische, M.J. (2008). Enantioselective iridium-catalyzed carbonyl Allylation from the alcohol or aldehyde oxidation level using allyl acetate as an allyl metal surrogate. *J. Am. Chem. Soc.* 130 (20): 6340–6341.

259 Kim, I.S., Ngai, M.-Y., and Krische, M.J. (2008). Enantioselective iridium-catalyzed carbonyl allylation from the alcohol or aldehyde oxidation level via transfer hydrogenative coupling of allyl acetate: departure from chirally modified allyl metal reagents in carbonyl addition. *J. Am. Chem. Soc.* 130 (44): 14891–14899.

260 Keck, G.E., Covel, J.A., Schiff, T., and Yu, T. (2002). Pyran annulation: asymmetric synthesis of 2,6-disubstituted-4-methylene tetrahydropyrans. *Org. Lett.* 4 (7): 1189–1192.
261 Yu, C.-M., Lee, J.-Y., So, B., and Hong, J. (2002). Sequential catalytic asymmetric allylic transfer reaction: enantioselective and diastereoselective construction of tetrahydropyran units. *Angew. Chem. Int. Ed.* 41 (1): 161–163.
262 Elnaggar, L.J. and Beal, J.L. (1980). Iridoids – a review. *J. Nat. Prod.* 43 (6): 649–707.
263 Boros, C.A. and Stermitz, F.R. (1991). Iridoids – an updated review. 2. *J. Nat. Prod.* 54 (5): 1173–1246.
264 Dinda, B., Debnath, S., and Harigaya, Y. (2007). Naturally occurring iridoids. A review, part 1. *Chem. Pharm. Bull.* 55 (2): 159–222.
265 Tietze, L.F. (1983). Secologanin, a biogenetic key compound – synthesis and biogenesis of the iridoid and secoiridoid glycosides. *Angew. Chem. Int. Ed. Engl.* 22 (11): 828–841.
266 Tundis, R., Loizzo, M.R., Menichini, F. et al. (2008). Biological and pharmacological activities of iridoids: recent developments. *Mini Rev. Med. Chem.* 8 (4): 399–420.
267 Dinda, B., Chowdhury, D.R., and Mohanta, B.C. (2009). Naturally occurring iridoids, secoiridoids and their bioactivity. An updated review, part 3. *Chem. Pharm. Bull.* 57 (8): 765–796.
268 Candish, L. and Lupton, D.W. (2010). The total synthesis of (−)-7-deoxyloganin via *N*-heterocyclic carbene catalyzed rearrangement of α,β-unsaturated enol esters. *Org. Lett.* 12 (21): 4836–4839.
269 Rong, Z.Q., Zhang, W., Yang, G.Q., and You, S.L. (2011). N-heterocyclic carbene-catalyzed redox reactions of α-functionalized aldehydes. *Curr. Org. Chem.* 15 (17): 3077–3090.
270 Candish, L. and Lupton, D.W. (2011). Concise formal synthesis of (−)-7-deoxyloganin via N-heterocyclic carbene catalysed rearrangement of α,β-unsaturated enol esters. *Org. Biomol. Chem.* 9 (23): 8182–8189.
271 Grasa, G.A., Kissling, R.M., and Nolan, S.P. (2002). N-heterocyclic carbenes as versatile nucleophilic catalysts for transesterification/acylation reactions. *Org. Lett.* 4 (21): 3583–3586.
272 Nyce, G.W., Lamboy, J.A., Connor, E.F. et al. (2002). Expanding the catalytic activity of nucleophilic N-heterocyclic carbenes for transesterification reactions. *Org. Lett.* 4 (21): 3587–3590.
273 Movassaghi, M. and Schmidt, M.A. (2005). N-heterocyclic carbene-catalyzed amidation of unactivated esters with amino alcohols. *Org. Lett.* 7 (12): 2453–2456.
274 Connor, E.F., Nyce, G.W., Myers, M. et al. (2002). First example of N-heterocyclic carbenes as catalysts for living polymerization: organocatalytic ring-opening polymerization of cyclic esters. *J. Am. Chem. Soc.* 124 (6): 914–915.
275 Kano, T., Sasaki, K., and Maruoka, K. (2005). Enantioselective acylation of secondary alcohols catalyzed by chiral N-heterocyclic carbenes. *Org. Lett.* 7 (7): 1347–1349.

276 Phillips, E.M., Riedrich, M., and Scheidt, K.A. (2010). N-heterocyclic carbene-catalyzed conjugate additions of alcohols. *J. Am. Chem. Soc.* 132 (38): 13179–13181.

277 Chen, J. and Huang, Y. (2014). Asymmetric catalysis with N-heterocyclic carbenes as non-covalent chiral templates. *Nat. Commun.* 5: 3437.

278 Sheu, J.-H., Hung, K.-C., Wang, G.-H., and Duh, C.-Y. (2000). New cytotoxic sesquiterpenes from the gorgonian *Isis hippuris*. *J. Nat. Prod.* 63 (12): 1603–1607.

279 Ren, Y., Presset, M., Godemert, J. et al. (2016). A switchable dual organocatalytic system and the enantioselective total synthesis of the quadrane sesquiterpene suberosanone. *Chem. Commun.* 52 (39): 6565–6568.

Index

N-Heterocyclic Carbenes in Organocatalysis, First Edition. Edited by Akkattu T. Biju.
© 2019 Wiley-VCH Verlag GmbH & Co. KGaA. Published 2019 by Wiley-VCH Verlag GmbH & Co. KGaA.

b

e

t